AF779633

Human Factors Reference Guide for Process Plants

The McGraw-Hill Engineering Reference Guide Series

This series makes available to professionals and students a wide variety of engineering information and data available in McGraw-Hill's library of highly acclaimed books and publications. The books in the Series are drawn directly from this vast resource of titles. Each one is either a condensation of a single title or a collection of sections culled from several titles. The Project Editors responsible for the books in the Series are highly respected professionals in the engineering areas covered. Each Editor selected only the most relevant and current information available in the McGraw-Hill library, adding further details and commentary where necessary.

Hicks • PLUMBING DESIGN AND INSTALLATION REFERENCE GUIDE

Covers the fundamentals of plumbing design and installation for a wide variety of industrial buildings and structures. Culled by Tyler G. Hicks from several McGraw-Hill books.

Hicks • POWER PLANT EVALUATION AND DESIGN REFERENCE GUIDE

Provides concise evaluation and design information for power plants serving many different needs—utility, industrial, and commercial. Culled by Tyler G. Hicks from several McGraw-Hill books and magazine articles.

Johnson & Jasik • ANTENNA APPLICATIONS REFERENCE GUIDE

Includes practical information and guidelines to antenna applications in all areas of communication. Comprised of one full section of Johnson and Jasik's Antenna Engineering Handbook, *Second Edition. Prepared by Richard C. Johnson.*

Markus and Weston • CLASSIC CIRCUITS REFERENCE GUIDE

Collects in one source hundreds of electronic circuits immediately useful in a wide variety of applications. Culled by Charles D. Weston from Markus's Sourcebook of Electronic Circuits, Electronics Circuits Manual, *and* Guidebook of Electronics Circuits.

Merritt • CIVIL ENGINEERING REFERENCE GUIDE

Offers quick reference to major civil engineering fields: structural design, surveying; geotechnical, environmental, and water engineering. A condensation by Max Kurtz of Merritt's Standard Handbook for Civil Engineers, *Third Edition.*

Woodson • HUMAN FACTORS REFERENCE GUIDE FOR ELECTRONICS AND COMPUTER PROFESSIONALS

Presents all essential data on human factors (ergonomics) relevant to the electronics and computer fields. Compiled by Wesley E. Woodson from his Human Factors Design Handbook.

Woodson • HUMAN FACTORS REFERENCE GUIDE FOR PROCESS PLANTS

Makes available to engineers and specialists all essential data on human factors (ergonomics) relevant to the process industries. Compiled by Nicholas P. Chopey from Woodson's Human Factors Design Handbook.

06120

TA
166
.W57
1987
c.2

Human Factors Reference Guide for Process Plants

Wesley E. Woodson
President, Man Factors, Inc.

Nicholas P. Chopey *Project Editor*

McGraw-Hill Book Company
New York St. Louis San Francisco Auckland Bogotá Hamburg
Johannesburg London Madrid Mexico Milan Montreal New Delhi
Panama Paris São Paulo Singapore Sydney Tokyo Toronto

Shell Oil Company
Products Technical Information Ctr.
Room. T-3231
P. O. Box 1380
Houston, TX 77251-1380

Library of Congress Cataloging-in-Publication Data

Woodson, Wesley E.
Human factors reference guide for process plants.

(The McGraw-Hill engineering reference guide series)
Rev. ed. of: Human factors design handbook. c1981.
Includes index.
1. Human engineering—Handbooks, manuals, etc.
I. Woodson, Wesley E. Human factors design handbook.
II. Title. III. Series.
TA166.W57 1986 620.8′2 86-7329
ISBN 0-07-071767-2

HUMAN FACTORS REFERENCE GUIDE FOR PROCESS PLANTS
Copyright © 1987 by McGraw-Hill, Inc. All rights reserved.
Printed in the United States of America. Except as permitted under the United States Copyright Act of 1976, no part of this publication may be reproduced or distributed in any form or by any means, or stored in a data base or retrieval system, without the prior written permission of the publisher.

1234567890 SEM/BKP 8932109876

ISBN 0-07-071767-2

The material in this volume has been published previously in *Human Factors Design Handbook,* by Wesley E. Woodson. Copyright © 1981 by McGraw-Hill, Inc. All rights reserved.

Printed by Semline, Inc. and bound by The Book Press.

Contents

Preface

The purpose of this handbook is to help optimize the interface between process-industries plants and the people who work in them. Thus, it applies the principles of human-factors engineering, or ergonomics, to process plants. It is intended for mechanical and civil engineers, designers, architects, and others who design such plants—or the equipment and systems for them.

This book does *not* deal with the design of processes themselves. That topic is already covered widely in the engineering (particularly chemical engineering) literature—and rightly so, because process design is at the heart of process-plant design. The present book instead focuses on an important corollary: that taking human factors into account is vital for the successful operation of the process plant, as well as for the job satisfaction of the process-plant worker (indeed, successful operation and job satisfaction are mutually reinforcing).

The material in this book is taken—verbatim, for the most part—from a larger and broader work: *Human Factors Design Handbook,* by Wesley E. Woodson, published by McGraw-Hill in 1981. That book applied human-factors engineering to the design of a wide range of industrial, commercial, recreational, and even military projects. The aim of the present volume is to capture and reproduce those principles and data that relate especially closely to process plants. However, in this regard the boundary-line between "process" and "nonprocess" is not sharp, and we have tried to err on the side of presenting possibly extraneous information rather than omitting useful information.

By definition, the human body is a focal point for human-factors engineering. Thus, the first part of this book presents extensive data on the body, and its capabilities and limitations. The second part focuses on various aspects of the workplace, i.e., the process plant.

A most important consideration in process plants is *process* safety, which derives from principles of chemistry, chemical engineering, and toxicology. That very complex topic receives virtually no attention in the present volume. If the reader's work should bring him or her into contact with the topic, a handbook specializing in it should be consulted—and, most likely, a chemical engineer should also be consulted.

Human Factors Reference Guide for Process Plants

1

The Person

GENERAL BODY CHARACTERISTICS

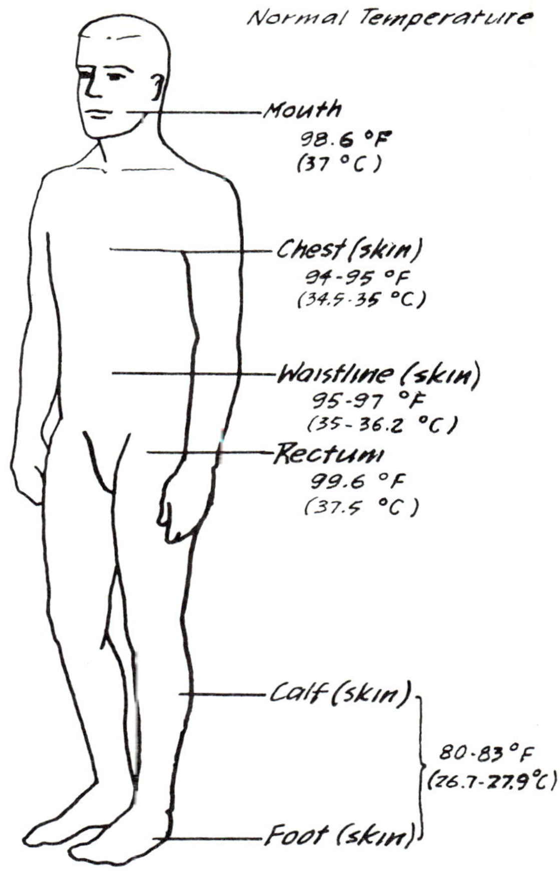

1. Temperature
As indicated in the accompanying sketch, normal body temperatures range from a low skin temperature of about 80°F (26.6°C) to a high internal temperature of about 99.6°F (37.5°C). At about 79.5°F (26.4°C), the heart fails.

2. Weight
Adult males normally weigh from about 110 to 215 lb (50 to 97 kg); adult females weigh from about 92 to 175 lb (42 to 79 kg).

The adult brain weight varies from about 38.8 to 60 oz (1100 to 1700 g) for males and from about 37 to 55 oz (198 to 1560 g) for females.

The adult male heart weighs about 10 oz (280 g); the adult female heart weighs about 8 oz (226 g).

Water makes up about 54 to 60 percent of total body weight; muscles, 36 to 42 percent; fat, 18 to 20 percent; and bone, about 18 percent.

Other elementary composition percentages are as follows:

Carbon	50 percent
Oxygen	20 percent
Hydrogen	10 percent
Nitrogen	8.5 percent
Calcium	4 percent
Phosphorus	2.5 percent
Potassium	1 percent
Sulfur	0.8 percent
Sodium	0.4 percent
Chlorine	0.4 percent
Magnesium	0.1 percent
Iron	0.01 percent
Manganese	0.001 percent
Iodine	0.00005 percent
Trace elements—	

3. Quantity of Blood
Males—1.5 gal (567 cm³); females—0.875 gal (330 cm³)

4. Surface Area of the Skin
Males—2.2 m²; females—1.9 m²

5. Lung Capacity
Males—4.5 to 9.5 qt; females—3.3 to 5.7 qt

6. Air Intake per Breath
Resting: males—0.79 qt; females—0.36 qt
Light work: males—1.77 qt; females—0.91 qt
Heavy work: males—2.15 qt; females—0.93 qt

7. Number of Breaths per Minute at Rest
Males—14 to 18; females—20 to 22

The deepest intake (vital capacity) for males is about 5.18 qt; for females it is 3.17 qt.

8. Heartbeats per Minute
Sleeping—55 to 60 beats per minute
Awake—70 beats per minute

9. Blood Pressure
120/80 mm at age 20

BODY DEVELOPMENT

Certain general body proportion relationships are observable between children and adults. Similar relationships are observable between males and females during the maturing or growth cycle. The accompanying sketches illustrate some of these relationships.

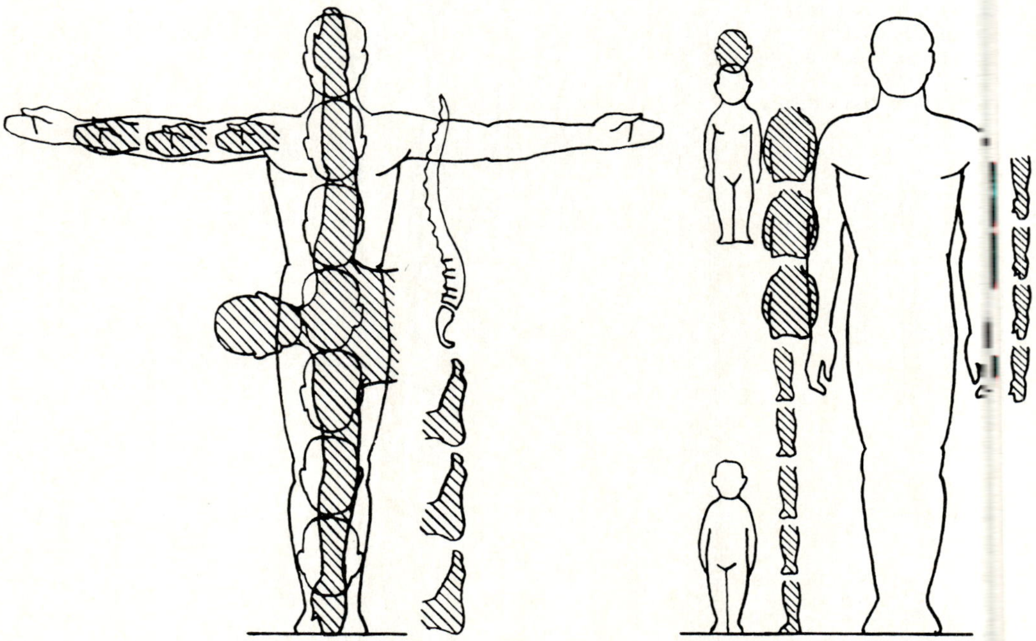

Proportional relationships between body components and between child and adult size.

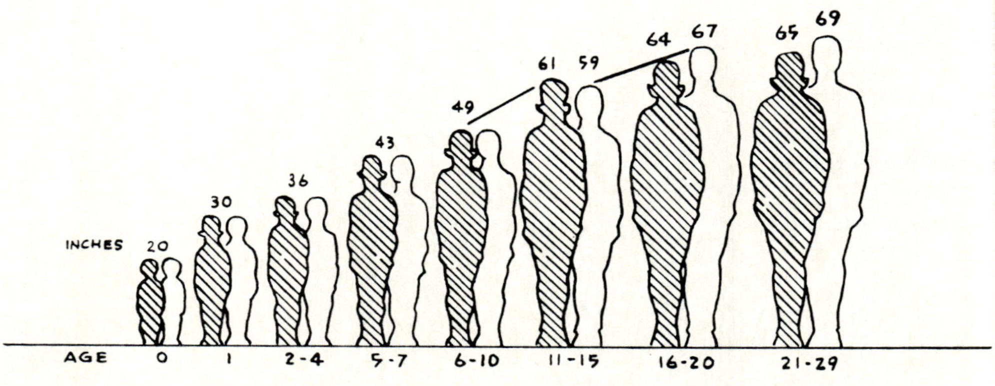

Male-female growth pattern. (Note the reversal that occurs during the years 11 through 20.)

BODY BUILD

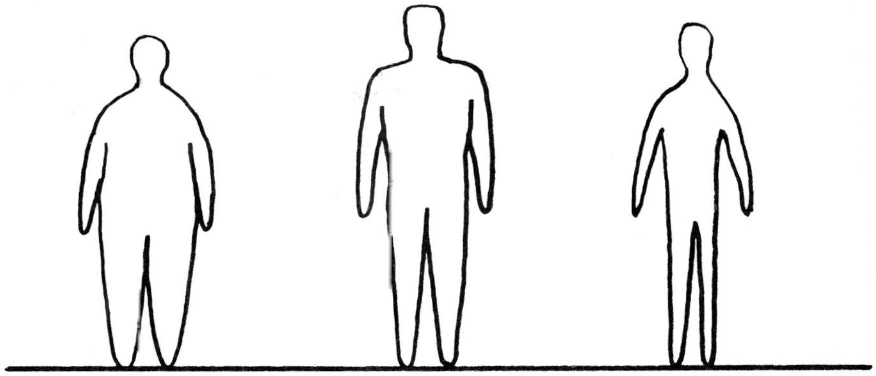

ENDOMORPH MESOMORPH ECTOMORPH

Sometimes referred to as "somatotype," one's dominant body build can be described as either "endomorphy," "mesomorphy," or "ectomorphy." The accompanying sketches illustrate the general differences between these body types.

Endomorphs tend to have a soft, round form with loose and flabby tissue, small bones, and a spherical head. The body has a relatively low density and is physically weak. Mean stature is around 66 in (168 cm), and mean weight is 179 lb (81 kg).

Mesomorphs tend to have a more massive, solid form with a squarish head, large bones, and heavy muscles. They are often muscle-bound and somewhat physically awkward. Mean stature is about 68 in (173 cm), and mean weight is 141 lb (64 kg).

Ectomorphs tend to have a slender body and limbs, a small head, and a small face with delicate features. They generally are more agile. Mean stature is around 69 in (175 cm), and mean weight is 141 lb (64 kg).

The principal value in recognizing differences in body build is that accommodating the human body requires careful consideration of more than just size. That is, in addition to differences in need for clearance, one must sometimes account for variations in mobility and agility.

BODY SURFACE AREA

Body Surface Area in Relation to Stature and Weight[1]

Area, m²

Body Weight, kg	Height, cm																								
	20	30	40	50	60	70	80	90	100	110	120	130	140	150	160	170	180	190	200	210	220	230	240	250	260
5	.18	.20	.23	.26	.29	.33	.37	.42	.48	.55	.62														
10		.35	.36	.38	.41	.44	.48	.52	.57	.64	.69	.76													
15					.54	.57	.60	.63	.67	.72	.77	.83	.89												
20							.68	.72	.76	.80	.85	.91	.97	1.03											
25								.80	.84	.88	.93	.98	1.03	1.09	1.15										
30									.92	.96	1.01	1.05	1.10	1.16	1.22	1.28									
35										1.04	1.08	1.12	1.17	1.23	1.29	1.35	1.42								
40										1.11	1.15	1.20	1.25	1.30	1.36	1.42	1.48	1.55							
45											1.23	1.27	1.32	1.37	1.43	1.48	1.54	1.61							
50											1.30	1.34	1.39	1.44	1.49	1.54	1.60	1.67	1.74						
55											1.37	1.42	1.46	1.50	1.55	1.61	1.67	1.73	1.80						
60											1.44	1.48	1.52	1.57	1.62	1.67	1.73	1.79	1.85	1.92					
65												1.54	1.58	1.63	1.68	1.73	1.79	1.85	1.91	1.97					
70												1.61	1.65	1.70	1.75	1.80	1.85	1.91	1.96	2.02	2.08				
75												1.68	1.72	1.76	1.81	1.86	1.91	1.96	2.02	2.07	2.13				
80												1.74	1.78	1.82	1.86	1.91	1.96	2.02	2.07	2.13	2.18	2.25			
85												1.81	1.84	1.88	1.92	1.97	2.02	2.07	2.13	2.18	2.24	2.31			
90												1.87	1.90	1.94	1.98	2.03	2.08	2.13	2.18	2.24	2.30	2.36			
95													1.97	2.01	2.05	2.09	2.14	2.18	2.24	2.30	2.36	2.42	2.48		
100													2.03	2.07	2.12	2.16	2.20	2.24	2.30	2.35	2.41	2.47	2.54		
105													2.10	2.14	2.18	2.22	2.26	2.31	2.35	2.41	2.47	2.53	2.60		
110													2.17	2.21	2.24	2.28	2.32	2.36	2.41	2.47	2.53	2.58	2.65	2.73	
115													2.23	2.27	2.30	2.33	2.38	2.42	2.47	2.53	2.58	2.64	2.71	2.78	
120														2.33	2.36	2.39	2.43	2.48	2.53	2.58	2.63	2.70	2.77	2.84	2.93
125														2.39	2.42	2.45	2.49	2.53	2.58	2.63	2.69	2.76	2.83	2.90	2.97
130														2.44	2.47	2.51	2.54	2.59	2.63	2.68	2.75	2.82	2.88	2.95	3.02
135														2.50	2.53	2.56	2.60	2.64	2.69	2.74	2.81	2.87	2.93	3.00	3.08
140														2.55	2.58	2.62	2.66	2.70	2.74	2.80	2.87	2.93	2.98	3.06	
145														2.61	2.63	2.67	2.71	2.75	2.80	2.86	2.92	2.98	3.04		
150														2.66	2.69	2.73	2.77	2.81	2.86	2.92	2.97	3.03	3.09		
155														2.72	2.74	2.78	2.83	2.87	2.92	2.97	3.03	3.08			
160														2.77	2.80	2.83	2.88	2.92	2.97	3.02	3.08				
165															2.86	2.89	2.93	2.97	3.02	3.07					
170															2.91	2.94	2.98	3.03	3.07						
175															2.96	2.99	3.03	3.08							
180															3.01	3.04	3.08								
185															3.06	3.09									

[1]Albert Damon et al., *The Human Body in Equipment Design*, Harvard University Press, Cambridge, Mass., 1966, p. 155. Area in square meters.

SKELETAL LINKAGE SYSTEM

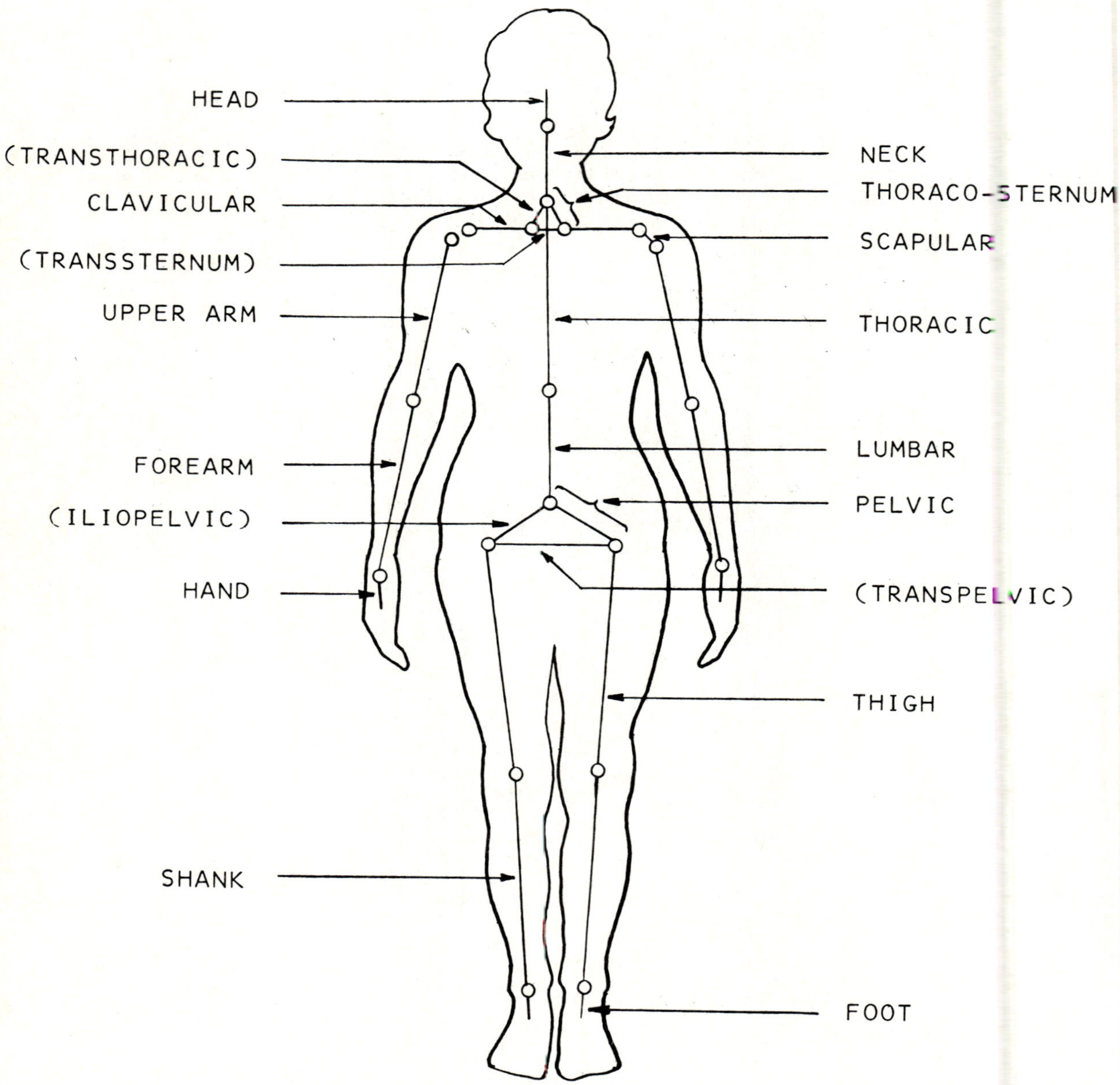

SKELETAL NOMENCLATURE AND CHARACTERISTICS

The skeleton is the basic support structure for the body, and thus it is an important consideration when designing equipment that interfaces with the body.

The dry, fat-free adult male skeleton weighs approximately 8 lb (3.6 kg); the adult female skeleton weighs approximately 6 lb (2.7 kg).

Selected skeletal nomenclature is shown in the accompanying sketches. These are the terms that have been found to be most useful when working with special scientific references dealing with typical body functions and equipment design.

In addition, certain terms are commonly used in connection with particular design problems; e.g., the lumbar vertebra is a key reference when working with seating design.

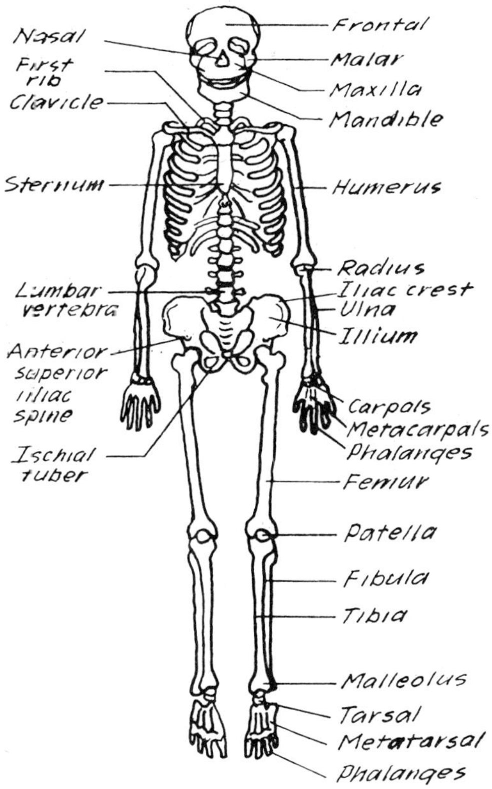

Of the total weight of the skeleton, the skull makes up about 20 percent; the ribs, 18 percent; the arms, 18 percent; and the legs, about 45 percent.

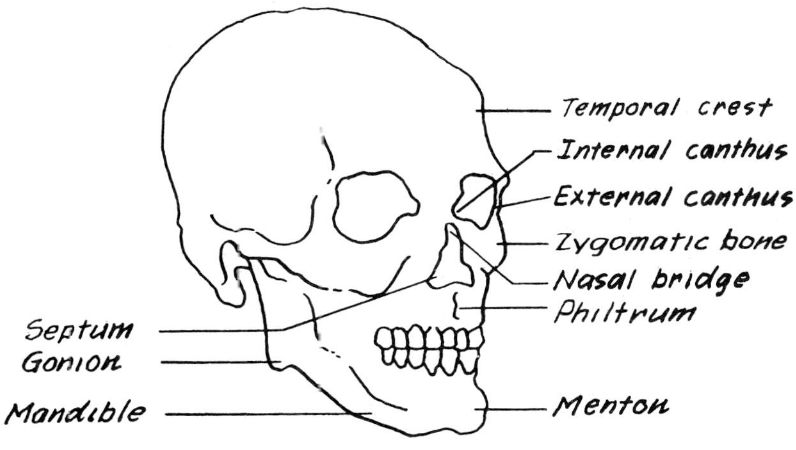

BODY MOBILITY: HINGE POINTS AND CENTERS OF GRAVITY

Mobility and posture control depend on the skeletal hinge-point locations and the weight and centers of gravity of various body components. The accompanying table and illustration present key hinge-point and c/g locations and the approximate weight of key body segments.

	Weight, lb (kg)	Percent of Total Weig
Head	10.7 (4.86)	6.9
Trunk and neck	70.7 (32.1)	46.1
Upper arms	10.1 (4.59)	6.6
Lower arms	6.4 (2.9)	4.2
Hands	2.6 (1.18)	1.7
Upper legs	33.0 (14.98)	21.5
Lower legs	14.7 (6.67)	9.6
Feet	5.2 (2.36)	3.4
Note: Example total—153.4		100.0

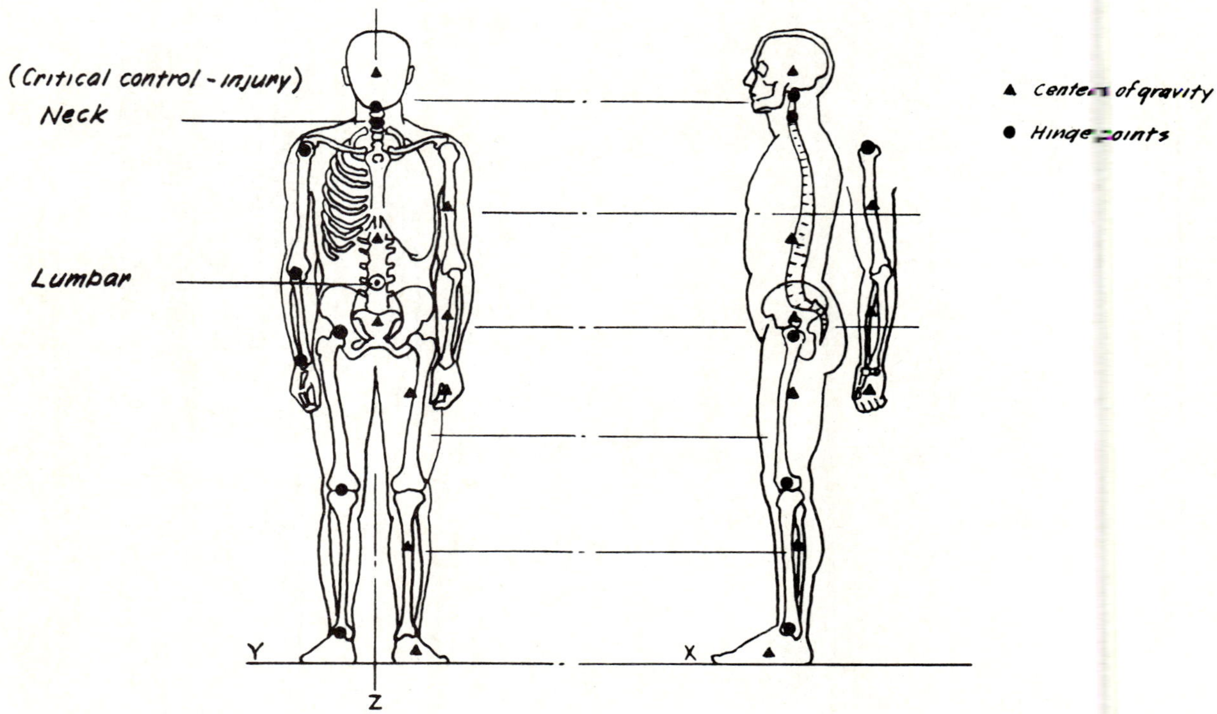

TYPICAL LIMITS OF BODY MOVEMENTS[2]

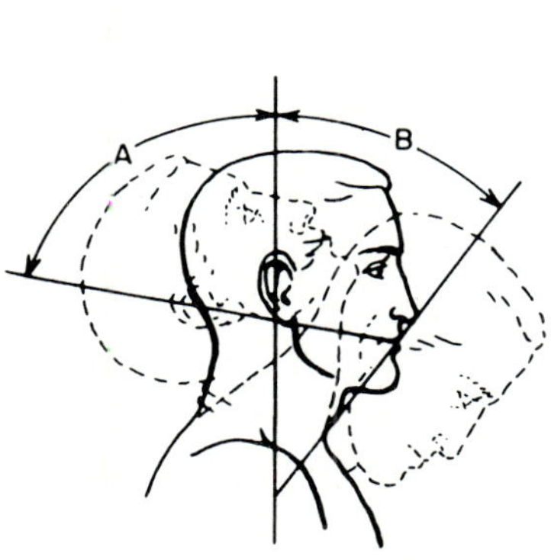

NECK FLEXION, DORSAL (A), VENTRAL (B)

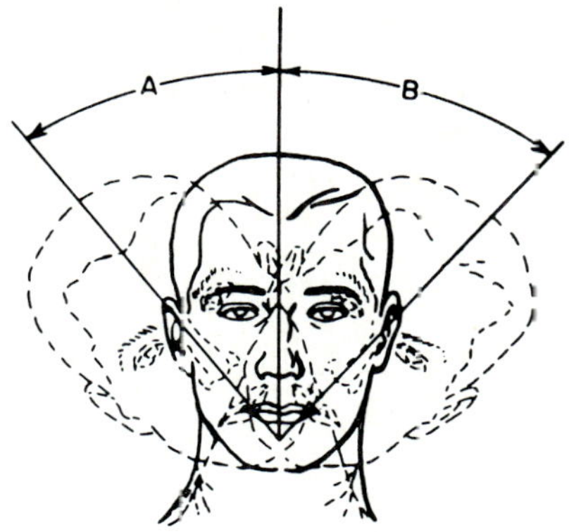

NECK FLEXION, RIGHT (A), LEFT (B)

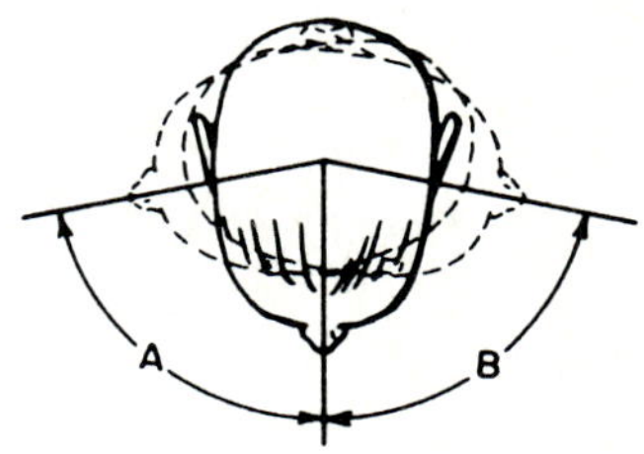

NECK ROTATION, RIGHT (A), LEFT (B)

Other body movement characteristics are illustrated by the accompanying sketches and tables. Although joint mobility decreases only slightly with age for healthy individuals between the ages of 20 and 60, the incidence of arthritis beyond age 45 suggests that designers give serious consideration to this limiting factor.

Range of Movement at the Neck Joint*

	Average	SD
Ventral flexion	60°	12
Dorsal flexion	61°	27
Right-left flexion	41°	7
Right-left rotation	79°	14

*Males.

ANTHROPOMETRY

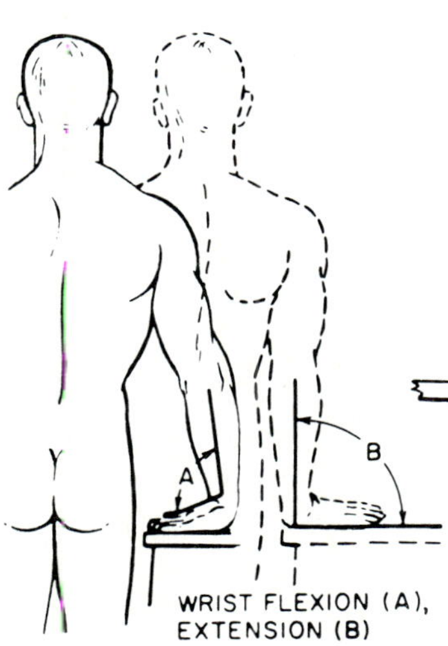

WRIST FLEXION (A), EXTENSION (B)

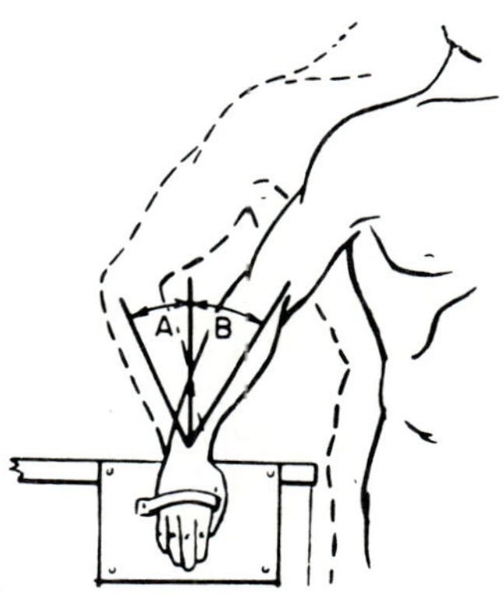

WRIST ABDUCTION (A), ADDUCTION (B)

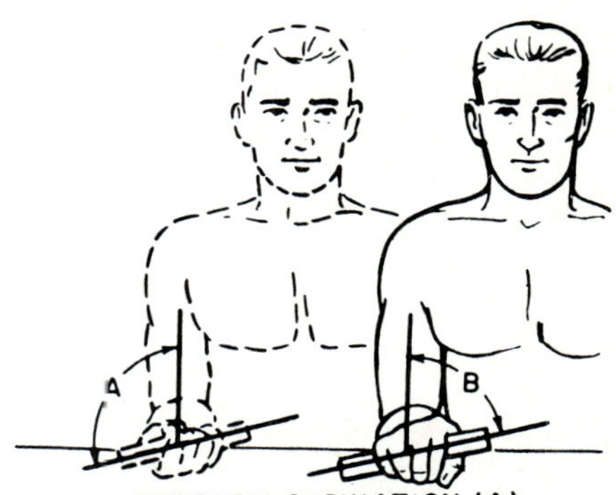

FOREARM SUPINATION (A), PRONATION (B)

[2]C. T. Morgan et al. (eds.), *Human Engineering Guide to Equipment Design,* McGraw-Hill Book Company, New York, 1963, pp. 553–555.

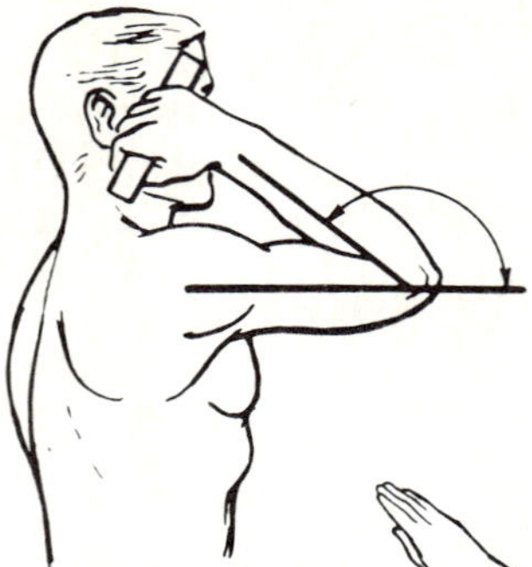
ELBOW FLEXION

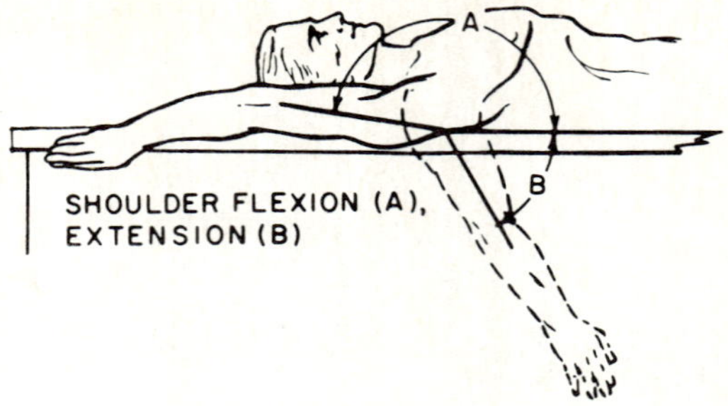
SHOULDER FLEXION (A), EXTENSION (B)

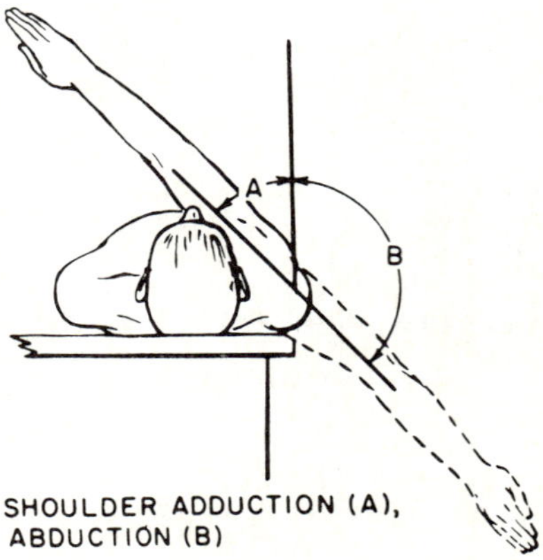
SHOULDER ADDUCTION (A), ABDUCTION (B)

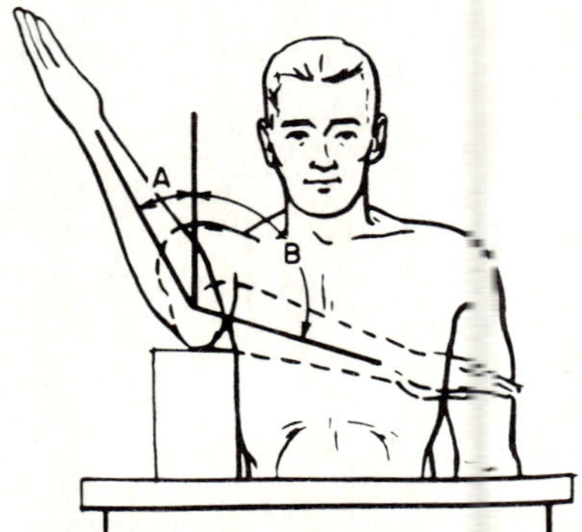
SHOULDER ROTATION, LATERAL (A), MEDIAL (B)

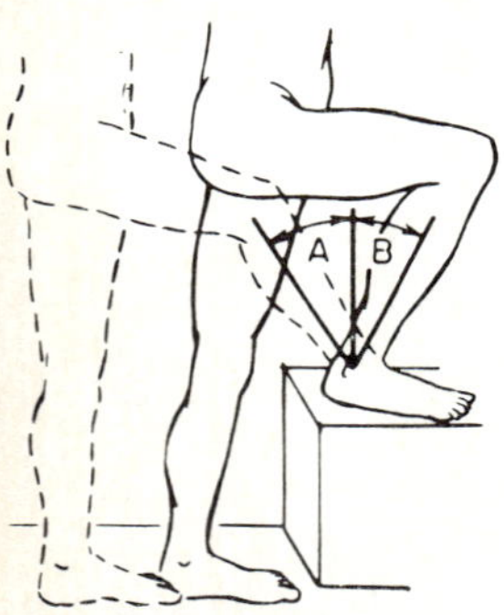
ANKLE EXTENSION (A), FLEXION (B)

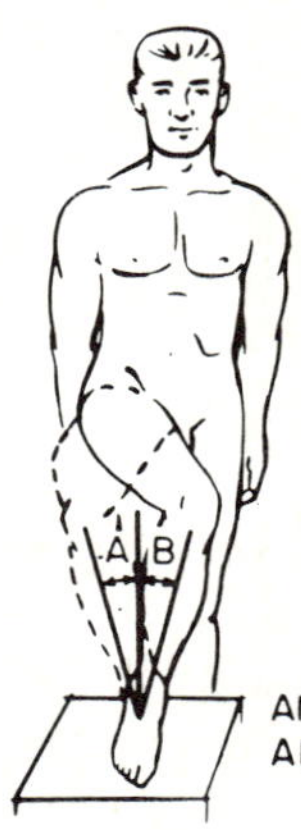
ANKLE ABDUCTION (A), ADDUCTION (B)

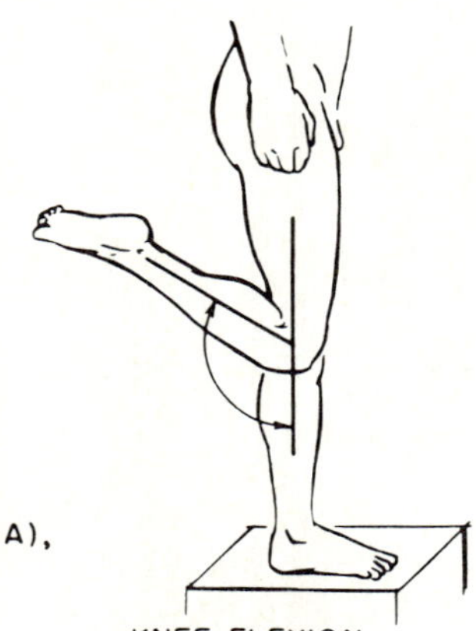
KNEE FLEXION, STANDING

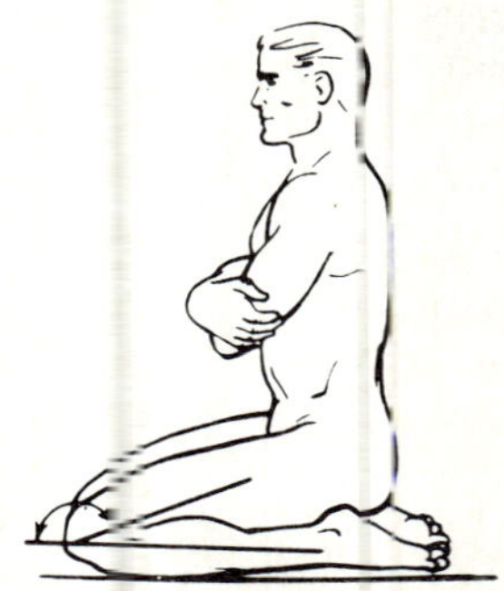
KNEE FLEXION, KNEELING

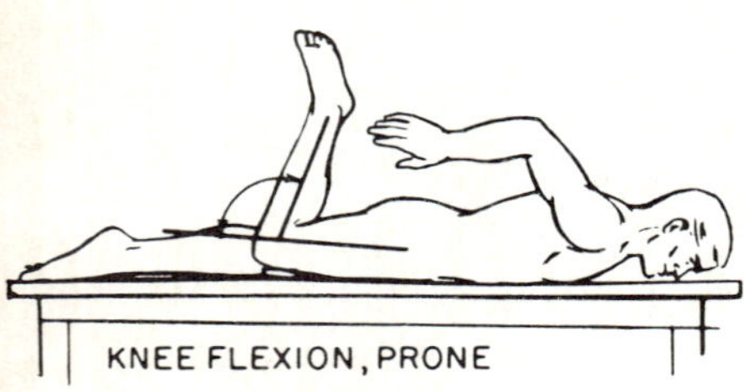
KNEE FLEXION, PRONE

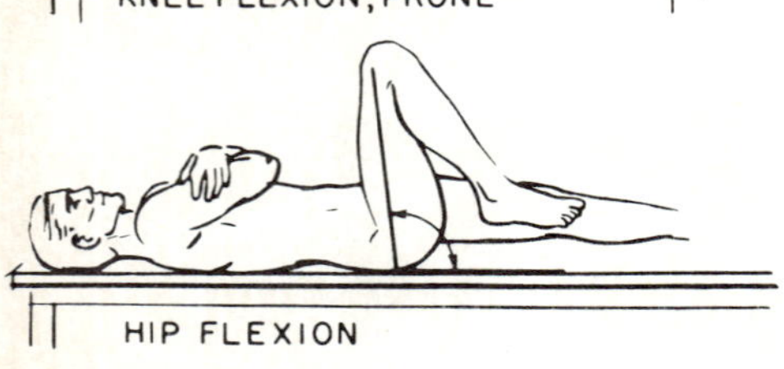
HIP FLEXION

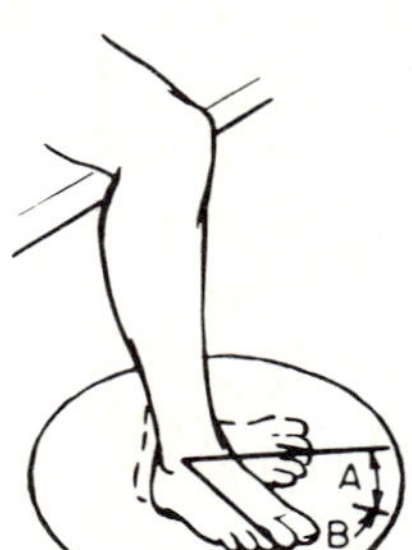
KNEE ROTATION, MEDIAL (A), LATERAL (B)

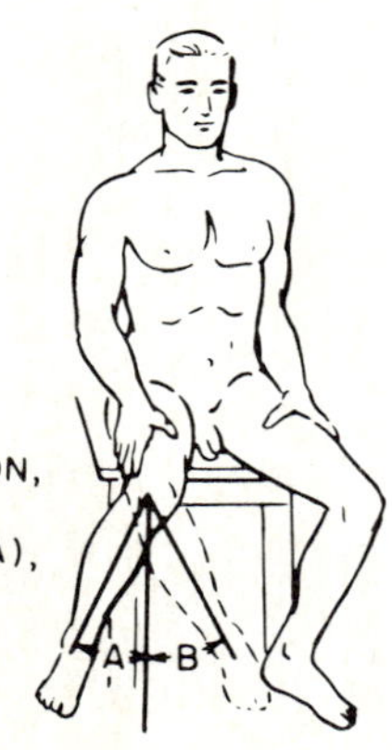
HIP ROTATION, SITTING, LATERAL (A), MEDIAL (B)

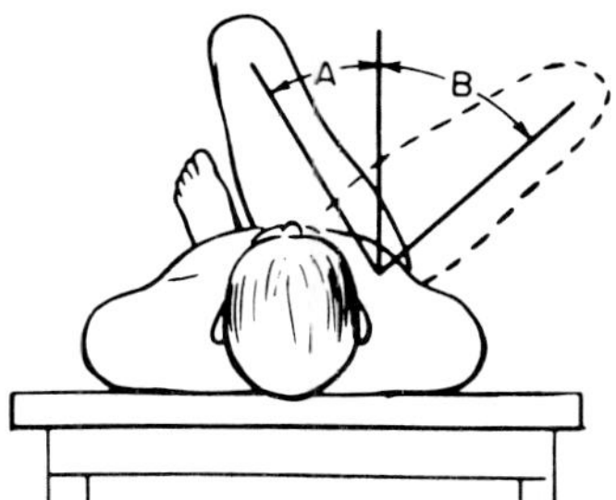

HIP ADDUCTION (A), ABDUCTION (B)

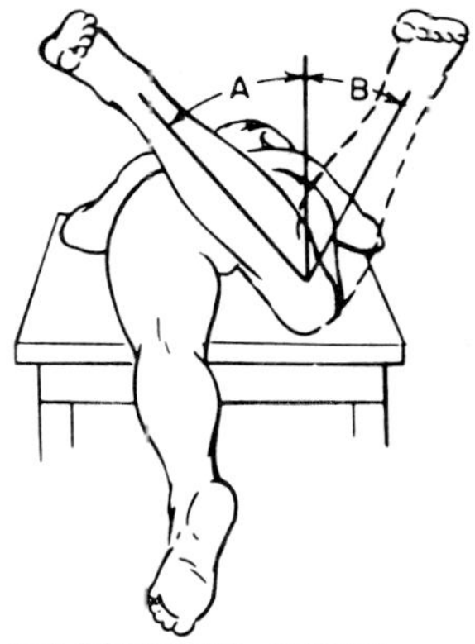

HIP ROTATION, PRONE, MEDIAL (A), LATERAL (B)

Range of Movement at the Joints of the Hand and Arm of Male Air Force Personnel

Movement[2]	Range (deg) Avg.	S.D.
Wrist flexion	90	12
Wrist extension	99	13
Wrist adduction	27	9
Wrist abduction	47	7
Forearm supination	113	22
Forearm pronation	77	24
Elbow flexion	142	10
Shoulder flexion	188	12
Shoulder extension	61	14
Shoulder adduction	48	9
Shoulder abduction	134	17

Note: Females generally exceed males with respect to the range of joint movements, except in the case of the knee joint (see the accompanying table). Ordinarily, fat men and women have less movement capability than slender ones. People with muscular body builds have less than those with average builds; however, because of the effect of special exercise or training, one should not assume that these generalizations always apply.

Range of Movement at the Joints of the Foot and Leg of Male Air Force Personnel

Movement	Range (deg) Avg.	S.D.
Ankle flexion	35	7
Ankle extension	38	12
Ankle adduction	24	9
Ankle abduction	23	7
Knee flexion		
Standing	113	13
Kneeling	159	9
Prone	125	10
Knee rotation		
Medial	35	12
Lateral	43	12
Hip flexion	113	13
Hip adduction	31	12
Hip abduction	53	12
Hip rotation (sitting)		
Medial	31	9
Lateral	30	9
Hip rotation (prone)		
Medial	39	10
Lateral	34	10

Average Increase in Range of Joint Movement of Women over Men

Movement	Difference (deg)
Wrist flexion and extension	14
Wrist adduction and abduction	11
Elbow flexion and extension	8
Shoulder abduction (rearward)	2
Ankle flexion and extension	4
Knee flexion and extension	0
Hip flexion	3

WHOLE-BODY WEIGHT (NUDE)

Whole-body weights for various populations are shown in the accompanying table. Values for only the 5th, 50th and 95th percentiles are provided, since these are the most significant for the majority of design requirements. For additional population percentiles, refer to the references provided in the section on body dimensions.

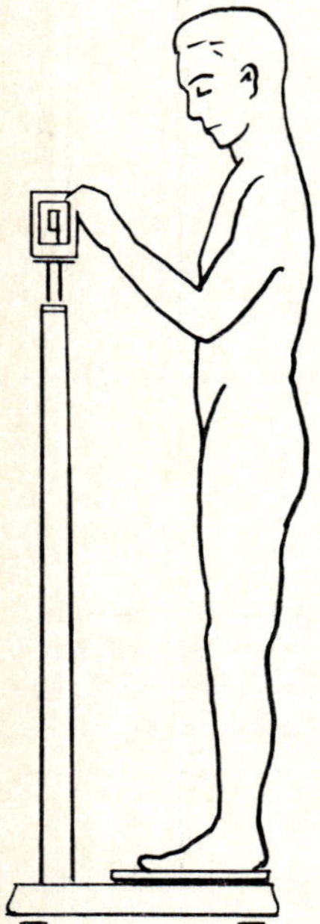

	Percentile		
	5th	50th	95th
Adults:			
Males	124 lb	168 lb	224 lb
Females	104 lb	139 lb	208 lb
Boys:			
Age 17	120 lb	149 lb	195 lb
Age 14	84 lb	112 lb	153 lb
Age 12	67 lb	84 lb	121 lb
Age 6	37 lb	44 lb	59 lb
Age 2	23 lb	27 lb	31 lb
Girls:			
Age 17	92 lb	120 lb	162 lb
Age 14	81 lb	108 lb	146 lb
Age 12	62 lb	88 lb	120 lb
Age 6	35 lb	44 lb	54 lb
Age 2	22 lb	24 lb	31 lb
Adults age 70 and over:			
Males	105 lb	144 lb	196 lb
Females	93 lb	135 lb	191 lb
Truck and bus drivers:			
Male	138 lb	175 lb	227 lb
Females	100 lb	128 lb	181 lb
Airline pilots (Male)	134 lb	168 lb	201 lb
Flight attendants (Female)	102 lb	117 lb	133 lb
Law enforcement officers:			
Males	146 lb	181 lb	230 lb
Females	—	—	—

Average Male Adult Weights for Other Populations

Nationality	Weight, lb
Chinese	120
English	143
Filipino	118
French	149
Greek	143
Hawaiian	172
Indian	123
Japanese	115
Korean	124
American Indians:	
Hopi	135
Navajo	139
Zuni	125

Note: 1.0 lb = 0.45 kg

PHYSICAL DIMENSIONS

The science of measuring the human body is *anthropometry.* Anthropometrists have been measuring a wide variety of people in many ways for many years. More than 350 different measurements have been taken at one time or another. Although measuring techniques have become more or less standardized, there is no assurance that each anthropometrist measured his or her particular sample of subjects, or each particular dimension, in the same manner. Thus one has to assume a certain caution in comparing data from various populations.

A great share of anthropometric data has been generated for the purpose of scientific comparison of populations, and not necessarily for the purpose of providing design-related information. This is evident from the fact that subjects are measured in stiff and unnatural positions, and in the nude. Also, since anthropometric surveys are extremely expensive, they are not repeated periodically to update information as people and populations change. Because of these and other unique conditions, it is important that the user of anthropometric data recognize the limitations of such data and treat the information merely as a point of departure. That is, anthropometric data are most useful as an aid in selecting a sample of "live" test subjects, who then should be used to test any proposed design.

Unfortunately, not all surveys include all the dimensions one might need for a given design problem. The most complete surveys have almost always been conducted on military personnel, and most of these have dealt with males.

In spite of these shortcomings, designers should make a point of referring to anthropometric data whenever a design involves a "fit" problem. Also, care should be taken to refer to data that are representative of the pertinent user population, since a particular survey may not have examined the dimension in question and thus may provide no data on it. To assist the designer, the following information on body dimensions has been organized according to the population groups in which product designers are most commonly interested. It will be noted that there are some obvious gaps in the tables where data for a given dimension and population are unavailable. Also, it should be pointed out that, in some cases, a substitute value has been entered when (in the author's opinion) a similar population seemed to justify using a particular dimensional value for the population in question. Although such liberty is often frowned upon by conservative anthropometrists, it is excused here on the basis that *none of these data should be used directly for design purposes, but rather as starting points to establish requirements for selecting a dynamic (live) test subject sample.* There is no substitute for a good, full-scale mockup evaluation to validate "fit" of the human body.

BASIC U.S. ADULT CONSUMER ANTHROPOMETRIC REFERENCE[3]

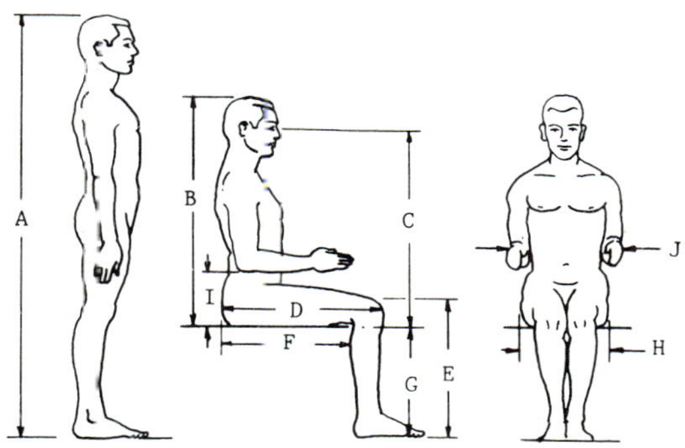

A - Standing Height
B - Sitting Height
C - Seated Eye Height
D - Upper Leg Length
E - Knee Height
F - Seat Length
G - Seat Height
H - Seat Width
I - Elbow Rest Height
J - Elbow Room

Basic Percentile Dimensions of U. S. Adult Population (18 to 79 years)

%	A	B	C	D	E	F	G	H	I	J	Weight
5	63.6 59.0	33.2 30.9	28.7 27.4	21.3 20.4	19.3 17.9	17.3 17.0	15.5 14.0	12.2 12.3	7.4 7.1	13.7 12.3	126 104
10	64.5 59.8	33.8 31.4	29.3 27.8	21.8 20.9	20.0 18.2	17.9 17.3	16.0 14.2	12.5 12.7	8.0 7.6	14.3 12.9	134 111
20	66.0 61.1	34.4 32.2	30.0 28.4	22.3 21.3	20.4 18.6	18.4 17.9	16.4 14.7	13.1 13.3	8.5 8.2	15.0 13.5	144 118
30	66.8 61.8	34.9 32.6	30.5 28.7	22.7 21.7	20.7 19.1	18.8 18.2	16.7 15.1	13.4 13.6	8.9 8.5	15.5 14.1	152 125
40	67.6 62.4	35.3 33.1	30.9 29.0	23.0 22.1	21.1 19.3	19.2 18.6	17.0 15.4	13.7 14.0	9.2 8.9	16.0 14.6	159 131
50	68.3 62.9	35.7 33.4	31.3 29.3	23.3 22.4	21.4 19.6	19.5 18.9	17.3 15.7	14.0 14.3	9.5 9.2	16.5 15.1	166 137
60	68.8 63.7	36.0 33.8	31.7 29.6	23.6 22.6	21.7 19.8	19.8 19.2	17.6 16.0	14.3 14.7	9.8 9.5	17.0 15.6	173 144
70	69.7 64.4	36.5 34.2	32.0 29.8	23.9 22.9	22.0 20.1	20.1 19.5	17.8 16.3	14.6 15.1	10.2 9.7	17.5 16.3	181 152
80	70.6 65.1	36.9 34.6	32.5 30.2	24.4 23.4	22.4 20.5	20.5 19.9	18.2 16.6	14.9 15.6	10.6 10.1	18.1 17.1	190 164
90	71.8 66.4	37.6 35.2	33.0 30.7	24.8 24.0	22.9 21.0	21.0 20.6	18.8 17.0	15.5 16.4	11.0 10.7	19.0 18.3	205 182
95	72.8 67.1	38.0 35.7	33.5 31.0	25.2 24.6	23.4 21.5	21.6 21.0	19.3 17.5	15.9 17.1	11.6 11.0	19.9 19.3	217 199

[3] H. W. Stoudt et al., *Weight, Height and Selected Body Dimensions of Adults, United States 1960–1962,* Public Health Service Publication no. 1000, ser. 11, no. 8.

STANDING HEIGHT (STATURE)[4]

Standing height is the primary indicator by which one selects general test subjects to evaluate various designs when there is no specific aspect of the design pertaining to an individual body component. The dimension is pertinent for adjusting head clearances.

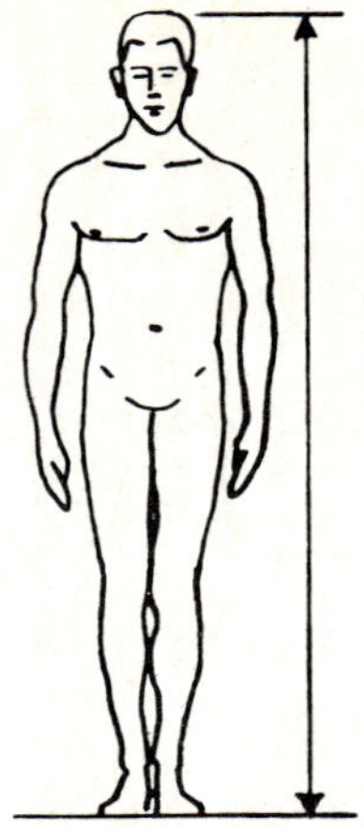

	Percentile		
	5th	50th	95th
Adults:			
Males	63.6 in	68.3 in	72.8 in
Females	59.0 in	62.9 in	67.1 in
Boys:			
Age 17	65.1 in	69.4 in	72.6 in
Age 14	58.0 in	63.2 in	69.1 in
Age 12	54.4 in	58.0 in	63.5 in
Age 6	41.8 in	45.1 in	49.1 in
Age 2	31.7 in	33.2 in	35.8 in
Girls:			
Age 17	60.0 in	64.1 in	67.6 in
Age 14	57.8 in	62.3 in	66.5 in
Age 12	54.1 in	58.8 in	63.0 in
Age 6	41.3 in	45.0 in	47.9 in
Age 2	30.4 in	33.0 in	35.4 in
Adults age 70 and over:			
Males	61.3 in	66.2 in	69.5 in
Females	55.3 in	61.8 in	64.9 in
Truck and bus drivers:			
Males	65.1 in	69.8 in	74.3 in
Females	58.9 in	63.5 in	68.0 in
Airline pilots (Male)	66.0 in	70.0 in	73.9 in
Flight attendants (Female)	62.5 in	65.4 in	68.8 in
Law enforcement officers:			
Males	66.6 in	70.0 in	74.0 in
Females	—	—	—

EYE HEIGHT (STANDING)

This dimension is pertinent to the location of visual displays and/or the sizing of visual obstructions, where a small person may have to see over the obstruction or over someone who is taller. As a general "rule of thumb," the eye height of males is about 5.2 in (13 cm) less than their standing height; for females it is about 4.8 in (12.2 cm) less. Similarly, when people stand normally (i.e., with some slump), their eye height lowers by about 1.6 in (4.1 cm) for males and by about 1.2 in (3.0 cm) for females.[5]

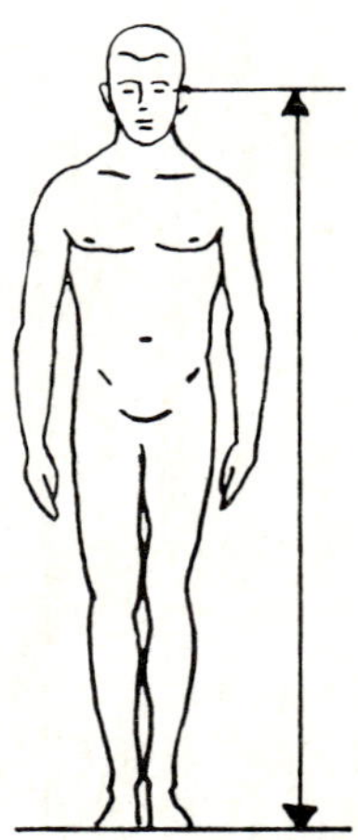

	Percentile		
	5th	50th	95th
Adults:			
Males	60.8 in	64.7 in	68.6 in
Females	57.3 in	60.3 in	65.3 in
Boys:			
Age 17	—	—	—
Age 14	—	—	—
Age 12	—	—	—
Age 6	—	—	—
Age 2	—	—	—
Girls:			
Age 17	—	—	—
Age 14	—	—	—
Age 12	—	—	—
Age 6	—	—	—
Age 2	—	—	—
Adults age 70 and over:			
Males	58.3 in	61.1 in	64.9 in
Females	50.8 in	55.8 in	60.8 in
Truck and bus drivers:			
Males	59.6 in	63.4 in	67.5 in
Females	—	—	—
Airline pilots (Male)	61.0 in	65.0 in	68.9 in
Flight attendants (Female)	58.0 in	60.9 in	64.3 in
Law enforcement officers:			
Males	61.6 in	65.0 in	69.0 in
Females	—	—	—

[4]The data in this table and the following tables have been compiled from the several references cited at the end of this section.

[5]Rules of thumb apply only to adults.

OVERHEAD REACH (STANDING)

This dimension is pertinent to locating controls that are overhead. It should be used in conjunction with stature because, although a short person must be able to reach a control, it should not be so low that it becomes an obstruction for the taller person. Use this dimension to select subjects for evaluating the accessibility of objects on high shelves.

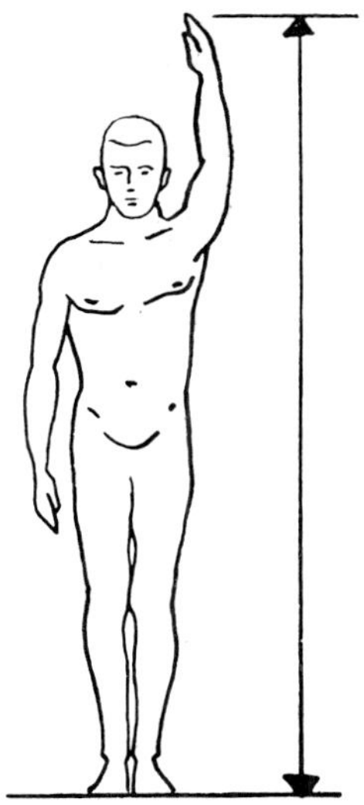

	Percentile		
	5th	50th	95th
Adults:			
Males*	82.0 in	88.0 in	94.0 in
Females*	73.0 in	79.0 in	86.0 in
Boys:			
Age 17	77.7 in	84.1 in	88.6 in
Age 14	70.5 in	77.3 in	83.1 in
Age 12	64.7 in	70.7 in	78.9 in
Age 6	50.3 in	53.3 in	58.2 in
Age 2	38.2 in	41.7 in	48.1 in
Girls:			
Age 17	72.6 in	77.7 in	81.9 in
Age 14	66.1 in	75.2 in	83.1 in
Age 12	65.0 in	72.1 in	77.4 in
Age 6	48.1 in	53.0 in	56.8 in
Age 2	38.7 in	41.6 in	45.6 in
Adults age 70 and over:			
Males	—	—	—
Females	69.5 in	75.5 in	82.5 in
Truck and bus drivers:			
Males*	81.0 in	88.1 in	93.7 in
Females	—	—	—
Airline pilots* (Male)	85.0 in	91.5 in	95.1 in
Flight attendants* (Female)	65.8 in	68.9 in	72.3 in
Law enforcement officers:			
Males*	85.0 in	89.7 in	95.2 in
Females	—	—	—

*Estimated.

FORWARD REACH (STANDING)

This dimension is pertinent to the selection of test subjects who will be evaluating reach conditions within the design situation. This dimension should be used in conjunction with leg-reach dimensions when the design situation calls for a seated operator to operate both hand and foot controls.

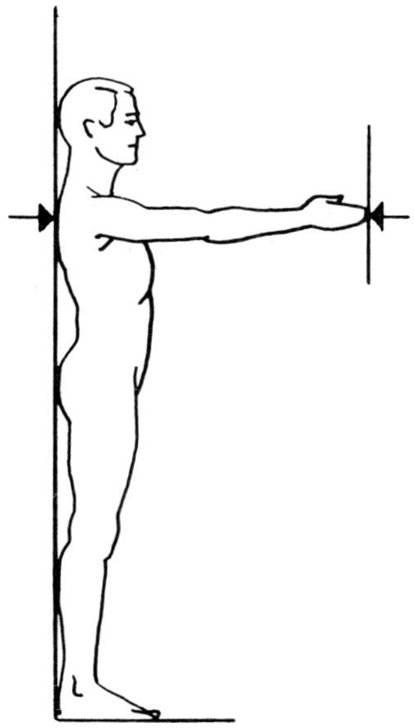

	Percentile		
	5th	50th	95th
Adults:			
Males	31.9 in	34.6 in	37.3 in
Females	29.7 in	31.8 in	34.1 in
Boys:			
Age 17	25.7 in	28.7 in	32.0 in
Age 14	23.8 in	26.9 in	29.3 in
Age 12	22.2 in	24.2 in	26.8 in
Age 6	17.6 in	19.1 in	20.9 in
Age 2	14.0 in	15.8 in	18.1 in
Girls:			
Age 17	23.9 in	26.7 in	28.8 in
Age 14	22.8 in	25.5 in	28.4 in
Age 12	21.8 in	24.3 in	26.4 in
Age 6	16.6 in	18.7 in	20.6 in
Age 2	13.7 in	15.1 in	16.8 in
Adults age 70 and over:			
Males	—	33.5 in	—
Females	—	—	—
Truck and bus drivers:			
Males	33.0 in	35.8 in	38.4 in
Females	—	—	—
Airline pilots (Male)	32.9 in	34.3 in	37.0 in
Flight attendants (Female)	29.0 in	31.0 in	33.3 in
Law enforcement officers:			
Males	32.3 in	34.8 in	37.7 in
Females	—	—	—

MAXIMUM BODY WIDTH

As indicated by the lack of data below, this measurement is seldom taken during typical anthropometric surveys. However, it is pertinent to lateral clearance requirements in design work. It would be useful, for instance, in selecting test subjects to evaluate lateral corridor clearance.

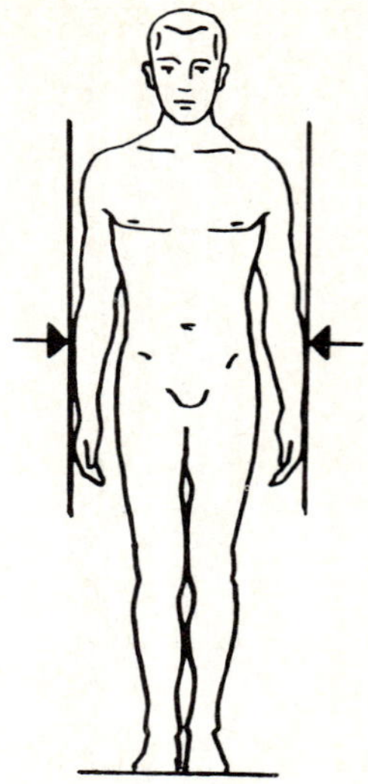

	Percentile		
	5th	50th	95th
Adults:			
Males*	18.8 in	20.9 in	22.8 in
Females			
Boys:			
Age 17			
Age 14			
Age 12			
Age 6			
Age 2			
Girls:			
Age 17			
Age 14			
Age 12			
Age 6			
Age 2			
Adults age 70 and over:			
Males			
Females			
Truck and bus drivers:			
Males			
Females			
Airline pilots			
Flight attendants			
Law enforcement officers:			
Males			
Females			

*Combined Air Force personnel and college students.

CROTCH HEIGHT

This dimension is a secondary indicator of leg length and is pertinent to design situations in which a person is required to step over an obstacle.

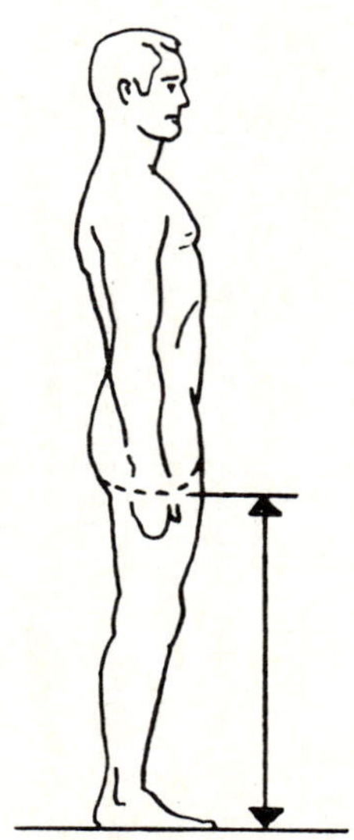

	Percentile		
	5th	50th	95th
Adults:			
Males	30.0 in	33.0 in	36.1 in
Females	26.8 in	29.3 in	32.0 in
Boys:			
Age 17	—	—	—
Age 14	—	—	—
Age 12	25.1 in	27.8 in	30.7 in
Age 6	17.9 in	19.9 in	21.7 in
Age 2	11.5 in	13.4 in	17.3 in
Girls:			
Age 17	—	—	—
Age 14	—	—	—
Age 12	24.9 in	28.0 in	30.6 in
Age 6	17.8 in	20.2 in	21.8 in
Age 2	11.5 in	13.4 in	17.3 in
Adults age 70 and over:			
Males			
Females			
Truck and bus drivers:			
Males			
Females			
Airline pilots			
Flight attendants			
Law enforcement officers:			
Males			
Females			

SITTING HEIGHT

This dimension is pertinent to the establishment of proper overhead clearances for seated persons. It is particularly important in the design of driver work stations.

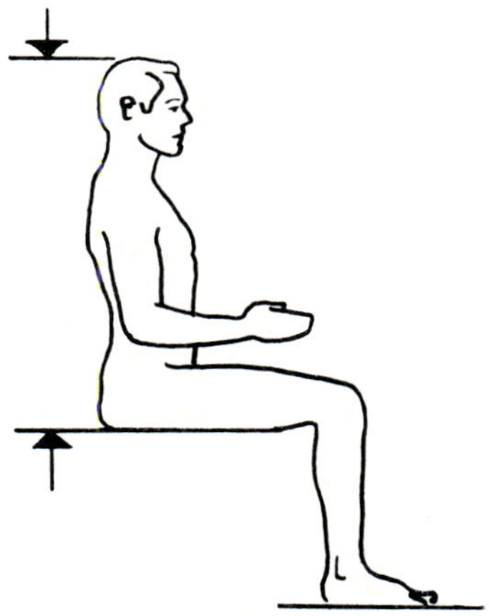

	Percentile		
	5th	50th	95th
Adults:			
Males	33.2 in	35.7 in	38.0 in
Females	30.9 in	33.4 in	35.7 in
Boys:			
Age 17	33.7 in	35.9 in	37.8 in
Age 14	29.4 in	32.3 in	35.6 in
Age 12	28.2 in	29.9 in	32.5 in
Age 6	23.0 in	25.0 in	26.8 in
Age 2	19.0 in	20.6 in	21.9 in
Girls:			
Age 17	31.8 in	33.7 in	35.9 in
Age 14	29.9 in	32.4 in	34.8 in
Age 12	28.0 in	30.4 in	32.9 in
Age 6	22.8 in	24.9 in	26.5 in
Age 2	18.9 in	19.8 in	21.5 in
Adults age 70 and over:			
Males	31.8 in	34.3 in	36.7 in
Females	28.1 in	32.1 in	34.8 in
Truck and bus drivers:			
Males	34.3 in	36.3 in	38.2 in
Females	—	—	—
Airline pilots (Male)	—	—	—
Flight attendants (Female)	32.4 in	34.3 in	36.1 in
Law enforcement officers:			
Males	34.1 in	36.3 in	38.5 in
Females	—	—	—

EYE HEIGHT (SITTING)

This dimension is pertinent to the design of work stations in which visual displays and/or outside viewing requires accommodation of a range of operator sizes. As a rule of thumb the eye height of males is about 5.2 inch (13 cm) less than their sitting height; that of females is about 4.8 in (12.2 cm) less. Similarly, when people sit normally (with some slump), their eye height lowers by about 1.2 in (3.0 cm) for males and by about 1.2 in (3.0 cm) for females.

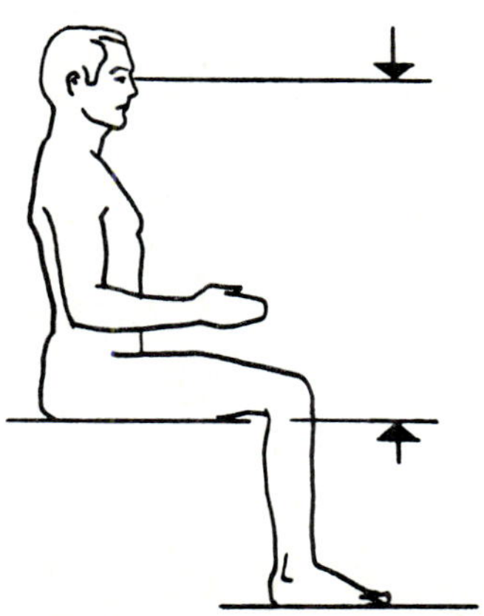

	Percentile		
	5th	50th	95th
Adults:			
Males	28.7 in	31.3 in	33.5 in
Females	27.4 in	29.3 in	31.0 in
Boys:			
Age 17	29.7 in	31.6 in	33.2 in
Age 14	25.2 in	28.2 in	30.7 in
Age 12	23.6 in	25.4 in	27.0 in
Age 6	18.3 in	20.5 in	22.2 in
Age 2	15.9 in	17.6 in	18.8 in
Girls:			
Age 17	26.9 in	29.0 in	31.1 in
Age 14	25.1 in	28.0 in	30.4 in
Age 12	23.6 in	25.6 in	27.6 in
Age 6	17.8 in	20.6 in	22.4 in
Age 2	15.2 in	16.7 in	18.1 in
Adults age 70 and over:			
Males	—	26.8 in	—
Females	—	—	—
Truck and bus drivers:			
Males	27.7 in	29.6 in	31.6 in
Females	—	—	—
Airline pilots (Male)	29.4 in	31.5 in	33.5 in
Flight attendants (Female)	28.1 in	29.9 in	31.7 in
Law enforcement officers:			
Males	29.1 in	31.3 in	33.5 in
Females	—	—	—

BUTTOCK-TO-POPLITEAL LENGTH

This dimension is pertinent to seat length. Although it is desirable to provide adequate support for the larger person, it is the shorter person who will have the most problems if this dimension is ignored in the test subject sample.

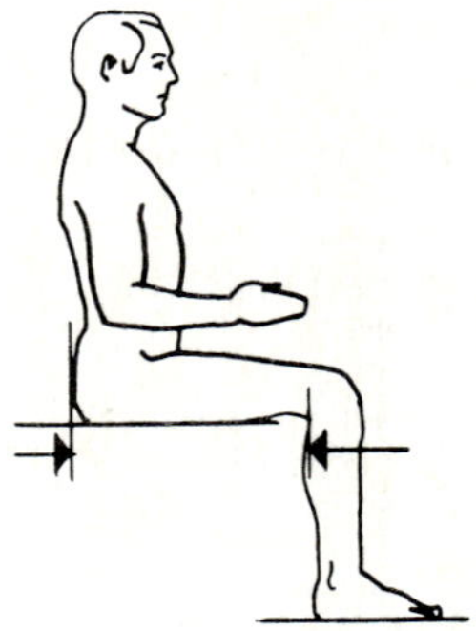

	Percentile		
	5th	50th	95th
Adults:			
Males	17.3 in	19.5 in	21.6 in
Females	17.0 in	18.9 in	21.0 in
Boys:			
Age 17			
Age 14			
Age 12			
Age 6			
Age 2			
Girls:			
Age 17			
Age 14			
Age 12			
Age 6			
Age 2			
Adults age 70 and over:			
Males	17.0 in	18.9 in	21.2 in
Females	17.0 in	18.7 in	20.0 in
Truck and bus drivers:			
Males	20.2 in	22.2 in	24.0 in
Females	18.4 in	20.5 in	22.3 in
Airline pilots			
Flight attendants (Female)	17.4 in	19.0 in	20.6 in
Law enforcement officers:			
Males	—	—	—
Females	—	—	—

MIDSHOULDER HEIGHT (SITTING)

This dimension is pertinent to the location of controls (i.e., most controls should not be located above the shoulder) and to the location of the anchor point for seat belts (i.e., the belt should not depart aft of the shoulder horizontally or at a negative angle; otherwise, it will create discomfort and tend to push the occupant downward during a frontal-impact crash).

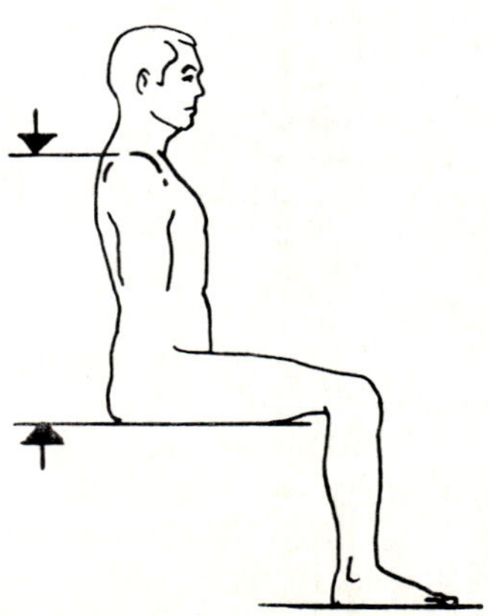

	Percentile		
	5th	50th	95th
Adults:			
Males	21.0 in	—	25.0 in
Females	18.0 in	—	25.0 in
Boys:			
Age 17	—	—	—
Age 14	18.1 in	20.1 in	21.2 in
Age 12	17.8 in	19.4 in	21.5 in
Age 6	13.7 in	15.4 in	17.0 in
Age 2	11.1 in	12.3 in	13.3 in
Girls:			
Age 17	—	—	—
Age 14	19.1 in	21.2 in	22.6 in
Age 12	17.6 in	19.5 in	21.0 in
Age 6	13.6 in	15.1 in	16.5 in
Age 2	11.1 in	12.3 in	14.0 in
Adults age 70 and over:			
Males	20.9 in	24.0 in	27.2 in
Females	—	—	—
Truck and bus drivers:			
Males	21.2 in	24.3 in	27.5 in
Females	—	—	—
Airline pilots	—	—	—
Flight attendants (Female)	20.1 in	22.9 in	25.7 in
Law enforcement officers:			
Males	—	—	—
Females	—	—	—

BUTTOCK-TO-KNEE LENGTH

This dimension is pertinent to establishing knee clearance for the seated operator. It should be used in conjunction with knee-height and thigh-clearance dimensions.

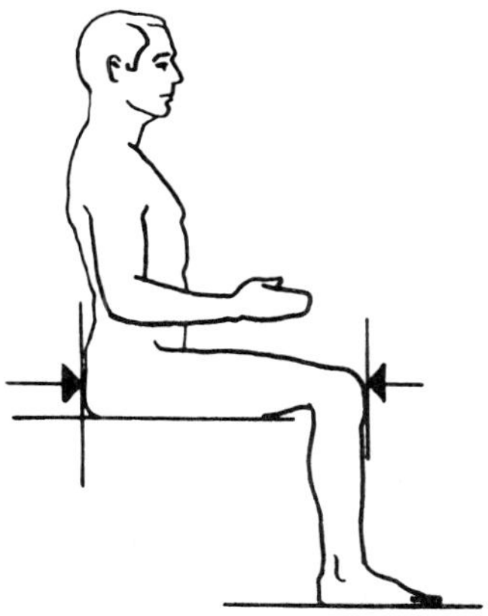

	Percentile		
	5th	50th	95th
Adults:			
Males	21.3 in	23.3 in	25.2 in
Females	20.4 in	22.4 in	24.6 in
Boys:			
Age 17	21.2 in	23.1 in	25.0 in
Age 14	19.2 in	21.4 in	23.5 in
Age 12	17.7 in	19.5 in	21.5 in
Age 6	12.9 in	14.1 in	15.6 in
Age 2	8.5 in	9.3 in	10.5 in
Girls:			
Age 17	20.0 in	21.7 in	23.5 in
Age 14	19.4 in	21.2 in	23.2 in
Age 12	17.9 in	20.1 in	22.0 in
Age 6	12.5 in	14.2 in	15.4 in
Age 2	7.4 in	9.4 in	10.6 in
Adults age 70 and over:			
Males	21.0 in	22.6 in	24.4 in
Females	19.9 in	22.2 in	23.9 in
Truck and bus drivers:			
Males	22.7 in	24.6 in	26.8 in
Females	20.6 in	22.9 in	24.9 in
Airline pilots (Male)	22.0 in	23.6 in	25.6 in
Flight attendants (Female)	21.2 in	22.6 in	24.2 in
Law enforcement officers:			
Males	22.6 in	24.2 in	26.1 in
Females	—	—	—

POPLITEAL HEIGHT (SITTING)

This dimension is pertinent to the establishment of appropriate seat heights. It is also pertinent to the selection of test subjects who will be used to evaluate the relationships between a vehicle seat and foot controls.

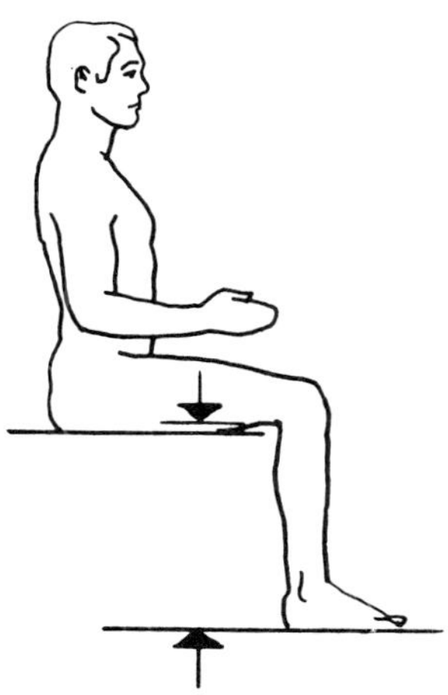

	Percentile		
	5th	50th	95th
Adults:			
Males	15.5 in	17.3 in	19.3 in
Females	14.0 in	15.7 in	17.5 in
Boys:			
Age 17	—	—	—
Age 14	—	—	—
Age 12	13.2 in	14.6 in	16.1 in
Age 6	10.4 in	11.5 in	12.6 in
Age 2	—	—	—
Girls:			
Age 17	—	—	—
Age 14	—	—	—
Age 12	13.0 in	14.7 in	16.3 in
Age 6	10.2 in	11.3 in	12.5 in
Age 2	—	—	—
Adults age 70 and over:			
Males	15.2 in	16.6 in	17.9 in
Females	13.5 in	15.6 in	17.2 in
Truck and bus drivers:			
Males	15.7 in	17.5 in	19.7 in
Females	—	—	—
Airline pilots (Male)	15.7 in	17.0 in	18.2 in
Flight attendants (Female)	15.9 in	17.1 in	18.5 in
Law enforcement officers:			
Males	—	—	—
Females	—	—	—

KNEE HEIGHT (SITTING)

This dimension is pertinent to the establishment of knee clearance. It should be used in conjunction with buttock-to-knee length and thigh-clearance dimensions.

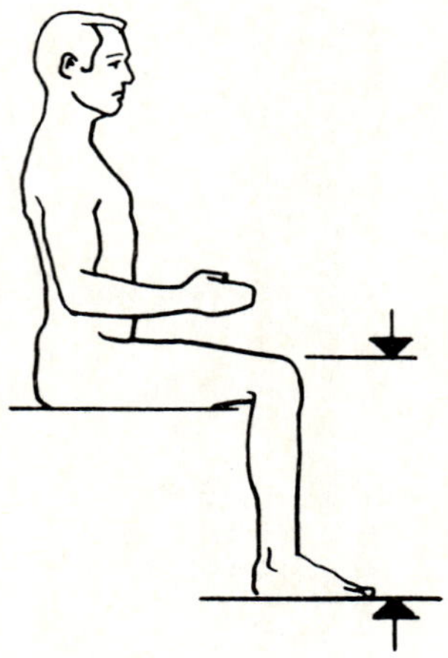

	Percentile		
	5th	50th	95th
Adults:			
Males	19.3 in	21.4 in	23.4 in
Females	17.9 in	19.6 in	21.5 in
Boys:			
Age 17	19.4 in	21.8 in	23.4 in
Age 14	18.5 in	20.3 in	22.6 in
Age 12	17.2 in	18.6 in	20.7 in
Age 6	12.4 in	13.9 in	15.3 in
Age 2	8.1 in	9.4 in	10.2 in
Girls:			
Age 17	17.8 in	19.5 in	21.2 in
Age 14	17.7 in	19.5 in	21.1 in
Age 12	17.0 in	18.6 in	20.2 in
Age 6	12.4 in	13.5 in	14.8 in
Age 2	8.0 in	9.4 in	10.0 in
Adults age 70 and over:			
Males	19.0 in	20.7 in	22.7 in
Females	17.3 in	19.4 in	20.9 in
Truck and bus drivers:			
Males	20.1 in	21.7 in	23.5 in
Females	—	—	—
Airline pilots (Male)	20.1 in	21.7 in	23.3 in
Flight attendants (Female)	19.1 in	20.4 in	21.9 in
Law enforcement officers:			
Males	20.4 in	22.0 in	23.7 in
Females	—	—	—

THIGH CLEARANCE (SITTING)

This dimension is pertinent to the establishment of the proper clearance between a seat and the lower edge of a desk and to the establishment of the space between the driver's seat and a vehicle steering wheel.

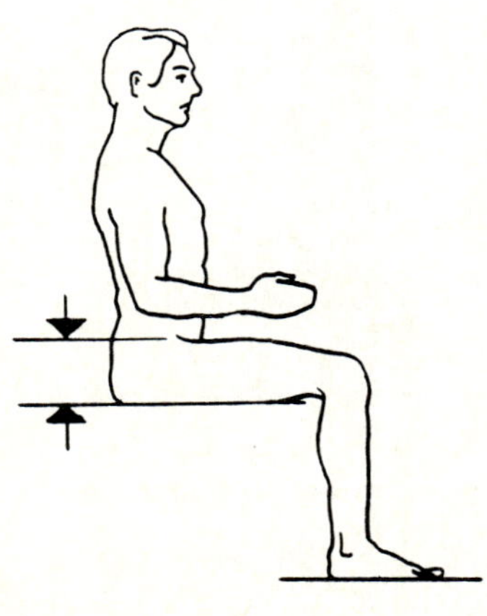

	Percentile		
	5th	50th	95th
Adults:			
Males	4.3 in	5.7 in	6.9 in
Females	4.1 in	5.4 in	6.9 in
Boys:			
Age 17	4.8 in	6.1 in	6.9 in
Age 14	4.3 in	5.5 in	6.4 in
Age 12	4.2 in	5.0 in	5.8 in
Age 6	3.1 in	3.6 in	4.3 in
Age 2	2.5 in	3.3 in	3.9 in
Girls:			
Age 17	4.6 in	5.4 in	6.8 in
Age 14	4.6 in	5.4 in	6.1 in
Age 12	4.2 in	4.9 in	6.1 in
Age 6	3.0 in	3.7 in	4.5 in
Age 2	2.1 in	3.1 in	3.7 in
Adults age 70 and over:			
Males	4.1 in	5.2 in	6.6 in
Females	4.0 in	5.2 in	6.5 in
Truck and bus drivers:			
Males	5.0 in	5.9 in	6.9 in
Females	—	—	—
Airline pilots (Male)	4.8 in	5.6 in	6.5 in
Flight attendants (Female)	—	—	—
Law enforcement officers:			
Males	—	—	—
Females	—	—	—

ELBOW-TO-FINGERTIP LENGTH

This dimension is pertinent to special problems such as evaluating whether a control may have been placed too close to the operator, with the result that (because of the seat) the operator cannot get his or her arm back far enough to use the control properly.

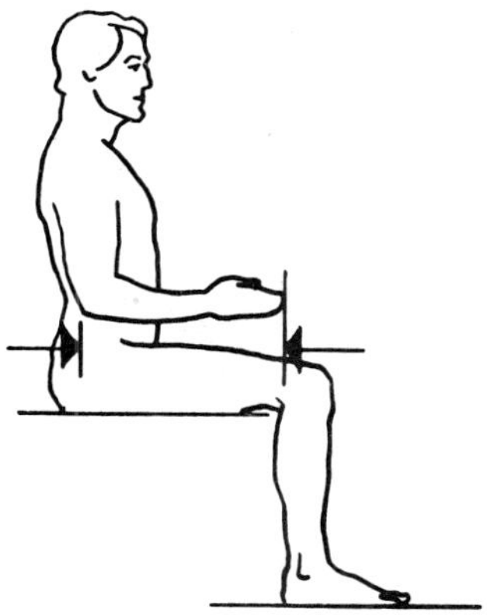

	Percentile 5th	50th	95th
Adults:			
Males	—	—	—
Females	—	—	—
Boys:			
Age 17	17.5 in	18.7 in	20.0 in
Age 14	15.6 in	17.3 in	19.1 in
Age 12	14.4 in	15.6 in	17.3 in
Age 6	10.9 in	12.0 in	13.2 in
Age 2	7.6 in	8.9 in	9.9 in
Girls:			
Age 17	15.5 in	16.8 in	18.0 in
Age 14	15.2 in	16.5 in	17.9 in
Age 12	14.0 in	15.7 in	17.3 in
Age 6	10.7 in	11.7 in	12.8 in
Age 2	7.7 in	8.8 in	9.4 in
Adults age 70 and over:			
Males	17.2 in	18.3 in	19.5 in
Females	—	—	—
Truck and bus drivers:			
Males	17.4 in	18.8 in	20.4 in
Females	15.1 in	16.7 in	18.2 in
Airline pilots (Male)	17.6 in	18.9 in	20.2 in
Flight attendants (Female)	16.0 in	17.2 in	18.3 in
Law enforcement officers:			
Males	—	—	—
Females	—	—	—

WAIST DEPTH

This dimension is pertinent to the establishment of clearance requirements between the backrest of a chair or seat and the leading edge of a worktable, desk, or steering wheel.

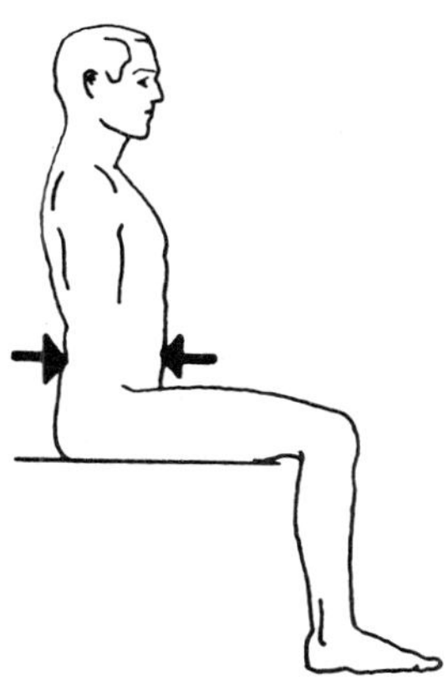

	Percentile 5th	50th	95th
Adults:			
Males	7.1 in	9.7 in	12.3 in
Females	5.8 in	6.6 in	7.9 in*
Boys:			
Age 17			
Age 14			
Age 12			
Age 6			
Age 2			
Girls:			
Age 17			
Age 14			
Age 12			
Age 6			
Age 2			
Adults age 70 and over:			
Males			
Females			
Truck and bus drivers:			
Males			
Females			
Airline pilots (Male)	7.6 in	9.9 in	10.4 in
Flight attendants (Female)	5.2 in	5.8 in	6.5 in
Law enforcement officers:			
Males			
Females			

*In certain working situations, one should consider the additional requirement for clearance when the pregnant woman must be accommodated. Unfortunately, to our knowledge, there are no data on this additional waist-clearance factor.

ELBOW REST HEIGHT

Not only is this dimension pertinent to the establishment of armrest heights, but it also provides a basis for establishing the level of a writing surface and/or the approximate position of the middle row of a keyboard, the location of a joy-stick handle or control wheel, etc.

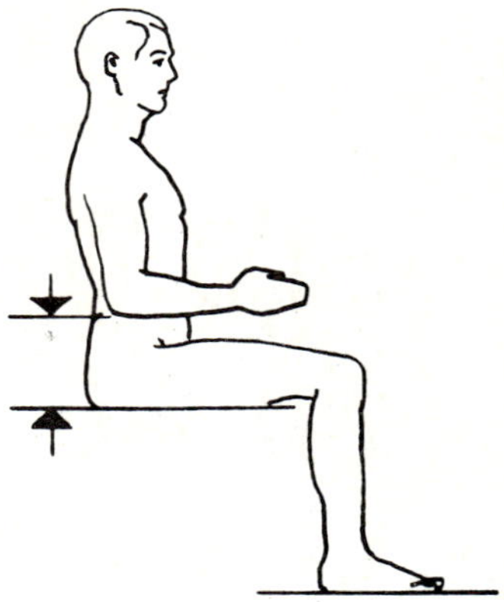

	Percentile		
	5th	50th	95th
Adults:			
Males	7.4 in	9.5 in	11.6 in
Females	7.1 in	9.2 in	11.0 in
Boys:			
Age 17			
Age 14			
Age 12			
Age 6			
Age 2			
Girls:			
Age 17			
Age 14			
Age 12			
Age 6			
Age 2			
Adults age 70 and over:			
Males	6.5 in	8.6 in	10.6 in
Females	6.4 in	8.4 in	10.0 in
Truck and bus drivers			
Males	—	—	—
Females	—	—	—
Airline pilots (Male)	7.4 in	9.1 in	10.8 in
Flight attendants (Female)	7.7 in	9.6 in	11.0 in
Law enforcement officers:			
Males	—	—	—
Females	—	—	—

BUTTOCK-TO-HEEL LENGTH

This dimension is pertinent to the establishment of the distance of foot-operated controls from the seat reference point.

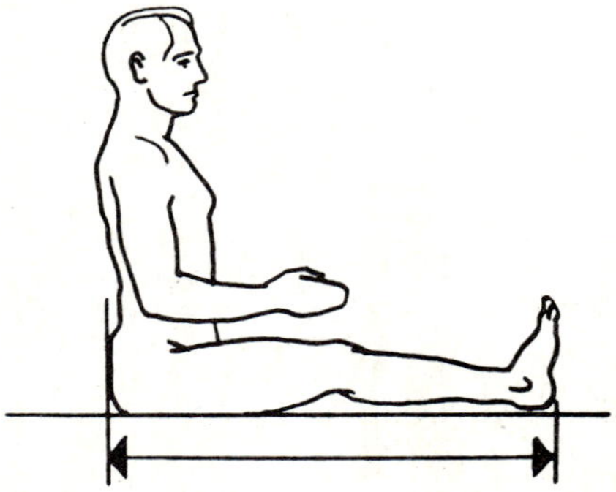

	Percentile		
	5th	50th	95th
Adults:			
Males	39.0 in	42.0 in	46.0 in
Females	34.0 in	37.8 in	41.2 in
Boys:			
Age 17	—	—	—
Age 14	31.7 in	34.1 in	39.1 in
Age 12	30.4 in	33.7 in	37.3 in
Age 6	22.5 in	24.7 in	26.4 in
Age 2	15.1 in	16.5 in	18.3 in
Girls:			
Age 17	—	—	—
Age 14	33.1 in	36.8 in	39.0 in
Age 12	30.2 in	33.7 in	37.1 in
Age 6	22.4 in	24.6 in	27.2 in
Age 2	14.1 in	17.0 in	18.0 in
Adults age 70 and over:			
Males	—	—	—
Females	—	—	—
Truck and bus drivers			
Males	—	—	—
Females	—	—	—
Airline pilots (Male)	38.3 in	42.3 in	46.3 in
Flight attendants (Female)	—	—	—
Law enforcement officers:			
Males	—	—	—
Females	—	—	—

SHOULDER BREADTH

This dimension is pertinent to the establishment of lateral clearance between persons who may be required to sit side by side and to the establishment of the lateral clearance requirement for a worker who may have to squeeze into a tight space to work on an item of equipment.

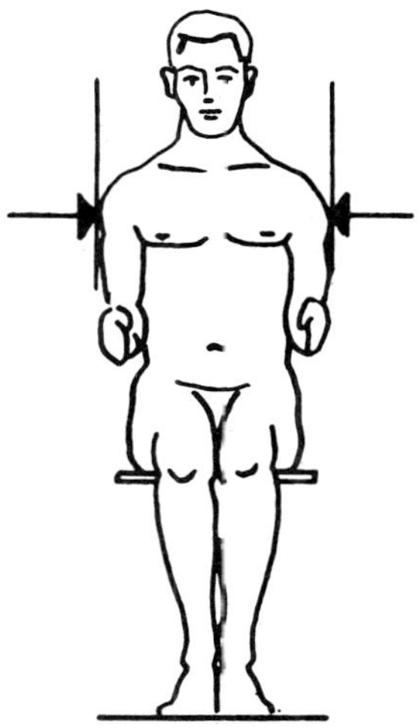

	Percentile		
	5th	50th	95th
Adults:			
Males	16.4 in	17.9 in	19.6 in
Females	14.4 in	15.7 in	17.6 in
Boys:			
Age 17	15.5 in	17.4 in	19.2 in
Age 14	13.7 in	15.3 in	17.4 in
Age 12	12.3 in	13.8 in	15.6 in
Age 6	9.3 in	11.0 in	12.5 in
Age 2	8.0 in	8.9 in	9.8 in
Girls:			
Age 17	14.2 in	15.6 in	17.3 in
Age 14	13.4 in	14.8 in	16.7 in
Age 12	12.2 in	13.8 in	15.8 in
Age 6	10.0 in	10.9 in	11.9 in
Age 2	7.6 in	9.0 in	9.7 in
Adults Age 70 and over:			
Males	15.6 in	17.0 in	18.5 in
Females	—	—	—
Truck and bus drivers:			
Males	16.9 in	18.3 in	19.9 in
Females	—	—	—
Airline pilots (Male)	16.5 in	17.9 in	19.4 in
Flight attendants (Female)	14.9 in	16.0 in	17.0 in
Law enforcement officers:			
Males	17.6 in	19.4 in	21.4 in
Females	—	—	—

HIP BREADTH (SITTING)

This dimension is pertinent to the establishment of seat widths, particularly the lateral clearance between armrests.

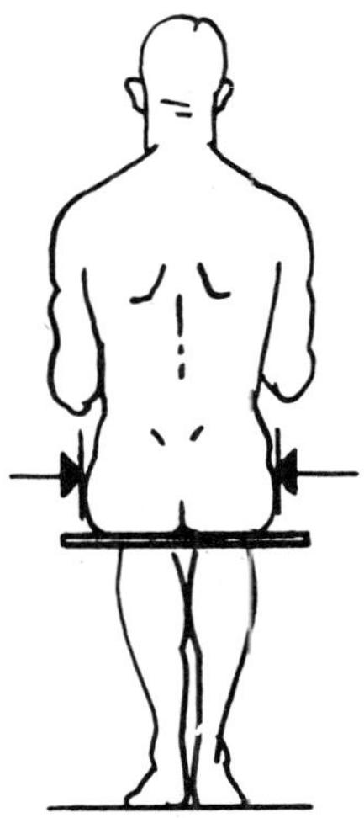

	Percentile		
	5th	50th	95th
Adults:			
Males	12.2 in	14.0 in	15.9 in
Females	12.3 in	14.3 in	17.1 in
Boys:			
Age 17	11.9 in	13.1 in	14.8 in
Age 14	10.4 in	11.8 in	13.6 in
Age 12	9.2 in	10.7 in	12.8 in
Age 6	7.5 in	8.3 in	9.4 in
Age 2	6.9 in	7.5 in	8.4 in
Girls:			
Age 17	12.0 in	13.5 in	15.7 in
Age 14	11.2 in	12.7 in	14.4 in
Age 12	9.6 in	11.3 in	13.5 in
Age 6	7.4 in	8.4 in	9.3 in
Age 2	6.6 in	7.4 in	8.4 in
Adults age 70 and over:			
Males	12.1 in	13.6 in	15.5 in
Females	11.7 in	14.0 in	16.8 in
Truck and bus drivers:			
Males	13.2 in	14.5 in	16.3 in
Females	—	—	—
Airline pilots (Male)	12.1 in	13.2 in	14.4 in
Flight attendants (Female)	13.3 in	14.5 in	15.6 in
Law enforcement officers:			
Males	—	—	—
Females	—	—	—

FOREARM-TO-FOREARM BREADTH

This dimension is pertinent to the establishment of seat and armrest separation. It also should be considered along with shoulder breadth to establish lateral body clearance for side-by-side seating.

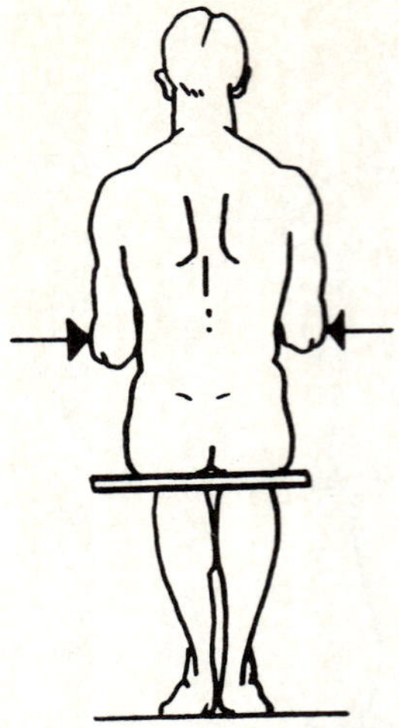

	Percentile		
	5th	50th	95th
Adults:			
Males	13.7 in	16.5 in	19.9 in
Females	12.3 in	15.1 in	19.3 in
Boys:			
Age 17			
Age 14			
Age 12			
Age 6			
Age 2			
Girls:			
Age 17			
Age 14			
Age 12			
Age 6			
Age 2			
Adults age 70 and over:			
Males	14.0 in	16.4 in	18.7 in
Females	13.1 in	15.7 in	19.1 in
Truck and bus drivers:			
Males	16.6 in	19.4 in	23.1 in
Females	14.1 in	16.5 in	20.2 in
Airline pilots (Male)	15.2 in	17.2 in	19.8 in
Flight attendants (Female)	11.6 in	13.0 in	14.6 in
Law enforcement officers:			
Males	—	—	—
Females	—	—	—

INTERPUPILLARY BREADTH

This dimension is pertinent to the design of eyeglasses, binoculars, and other optical aids.

	Percentile		
	5th	50th	95th
Adults:			
Males	2.27 in	—	2.74 in
Females	1.41 in	—	2.94 in
Boys:			
Age 17			
Age 14			
Age 12			
Age 6			
Age 2			
Girls:			
Age 17			
Age 14			
Age 12			
Age 6			
Age 2			
Adults age 70 and over:			
Males			
Females			
Truck and bus drivers:			
Males	2.28 in	2.44 in	2.80 in
Females	—	—	—
Airline pilots (Male)	2.27 in	2.49 in	2.74 in
Flight attendants (Female)	—	—	—
Law enforcement officers:			
Males	—	—	—
Females	—	—	—

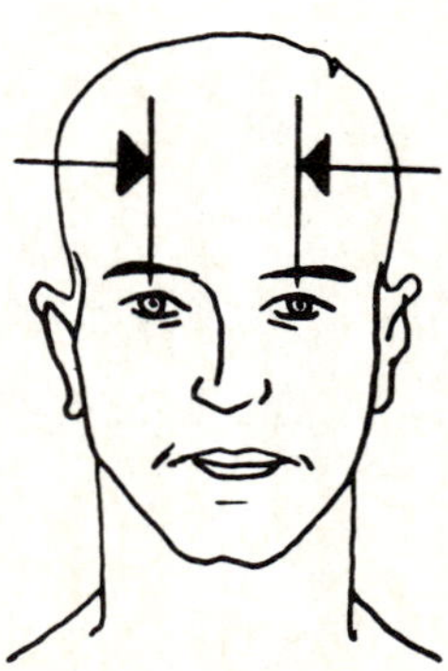

HEAD BREADTH

This dimension is pertinent to the design of safety helmets and other head-mounted gear.

	Percentile		
	5th	50th	95th
Adults:			
Males	5.6 in	6.0 in	6.4 in
Females	5.4 in	5.7 in	6.1 in
Boys:			
Age 17	5.7 in	6.0 in	6.4 in
Age 14	5.5 in	5.9 in	6.2 in
Age 12	5.4 in	5.7 in	6.1 in
Age 6	5.2 in	5.5 in	5.9 in
Age 2	4.9 in	5.2 in	5.5 in
Girls:			
Age 17	5.5 in	5.8 in	6.1 in
Age 14	5.5 in	5.7 in	6.1 in
Age 12	5.4 in	5.7 in	5.9 in
Age 6	5.1 in	5.4 in	5.7 in
Age 2	4.7 in	5.0 in	5.4 in
Adults age 70 and over:			
Males	5.8 in	6.1 in	6.4 in
Females	—	—	—
Truck and bus drivers:			
Males	5.7 in	6.0 in	6.3 in
Females	—	—	—
Airline pilots (Male)	5.7 in	6.1 in	6.4 in
Flight attendants (Female)	5.4 in	5.7 in	6.1 in
Law enforcement officers:			
Males	5.7 in	6.1 in	6.5 in
Females	—	—	—

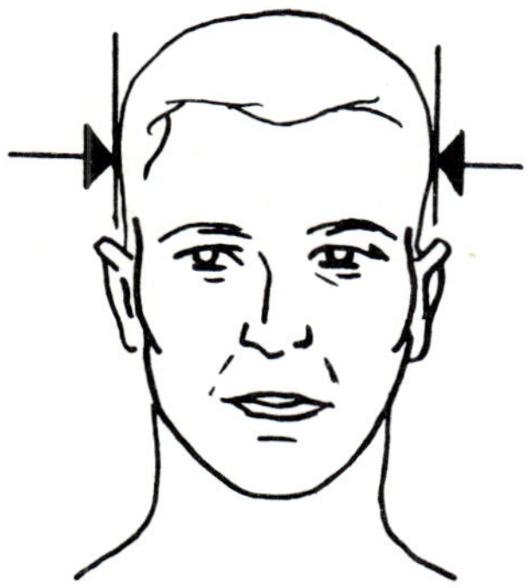

HEAD LENGTH

This dimension is pertinent to the design of safety helmets and other head-mounted gear. It should be used in conjunction with head breadth.

	Percentile		
	5th	50th	95th
Adults:			
Males	7.3 in	7.7 in	8.2 in
Females	7.0 in	7.4 in	7.7 in
Boys:			
Age 17	7.2 in	7.7 in	8.0 in
Age 14	7.2 in	7.6 in	8.1 in
Age 12	6.9 in	7.5 in	7.7 in
Age 6	6.6 in	7.2 in	7.6 in
Age 2	6.5 in	6.9 in	7.3 in
Girls:			
Age 17	6.9 in	7.4 in	7.8 in
Age 14	7.0 in	7.4 in	7.9 in
Age 12	6.8 in	7.3 in	7.8 in
Age 6	6.6 in	7.1 in	7.4 in
Age 2	6.1 in	6.7 in	6.9 in
Adults age 70 and over:			
Males	7.3 in	7.7 in	8.1 in
Females	—	—	—
Truck and bus drivers:			
Males	7.2 in	7.6 in	8.1 in
Females	—	—	—
Airline pilots (Male)	7.2 in	7.8 in	8.6 in
Flight attendants (Female)	6.9 in	7.3 in	7.7 in
Law enforcement officers:			
Males	7.3 in	7.8 in	8.2 in
Females	—	—	—

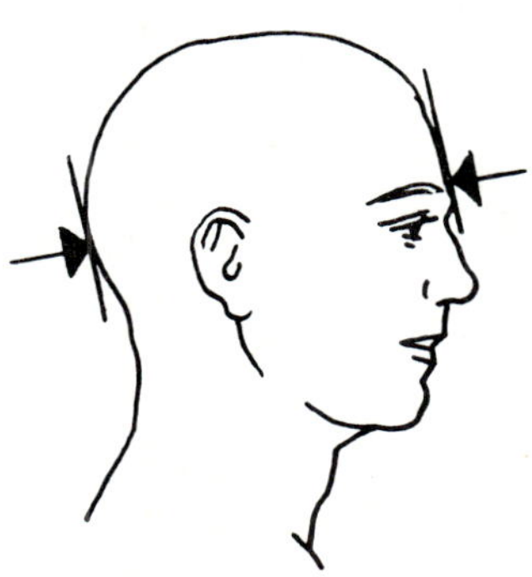

HEAD LENGTH (MAXIMUM)

This dimension is useful in the design of special headgear that requires a full face mask. These values include an arbitrary 1-in (2.5-cm) additional clearance.

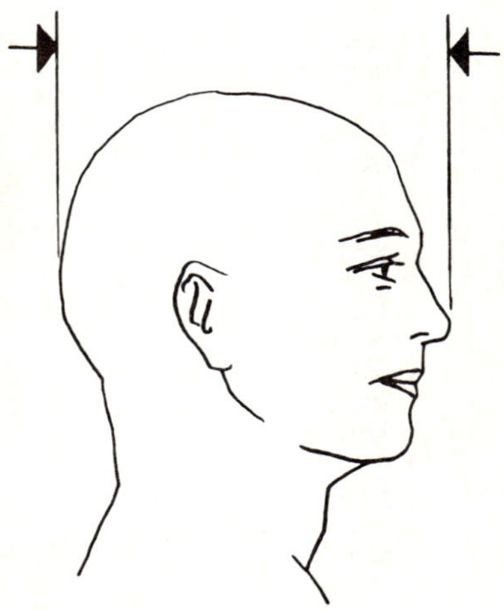

	Percentile		
	5th	50th	95th
Adults:			
Males			
Females			
Boys:			
Age 17			
Age 14			
Age 12			
Age 6			
Age 2			
Girls:			
Age 17			
Age 14			
Age 12			
Age 6			
Age 2			
Adults age 70 and over:			
Males			
Females			
Truck and bus drivers:			
Males	8.1 in	8.7 in	9.2 in
Females			
Airline pilots			
Flight attendants			
Law enforcement officers:			
Males			
Females			

HEAD HEIGHT (MAXIMUM)

This dimension is pertinent to the design of fully enclosed headgear or helmets. As suggested by the lack of data for various populations, the values indicated below were derived from a very small sample of male and female subjects selected at random and provide some idea of the larger head height characteristics.

	Percentile		
	5th	50th	95th
Adults:			
Males	—	—	10.2 in
Females	—	—	9.0 in
Boys:			
Age 17	7.8 in	8.6 in	9.1 in
Age 14	7.5 in	8.1 in	8.9 in
Age 12	7.5 in	7.8 in	8.5 in
Age 6	6.6 in	7.3 in	8.0 in
Age 2	6.3 in	6.9 in	7.5 in
Girls:			
Age 17	7.5 in	8.1 in	8.7 in
Age 14	7.3 in	8.0 in	8.5 in
Age 12	7.0 in	7.8 in	8.5 in
Age 6	6.3 in	7.2 in	7.6 in
Age 2	6.1 in	6.6 in	7.2 in
Adults age 70 and over:			
Males			
Females			
Truck and bus drivers:			
Males			
Females			
Airline pilots			
Flight attendants			
Law enforcement officers:			
Males			
Females			

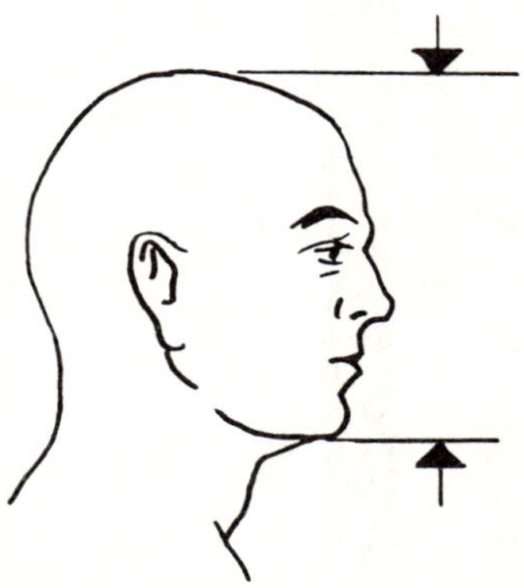

CHIN-TO-EYE HEIGHT

This dimension is pertinent to the design of headgear in which eye protective elements are integrated into the helmet. Although these values represent a very small subject sample, they provide some idea of the larger chin-to-eye heights. They should be considered in conjunction with head height.

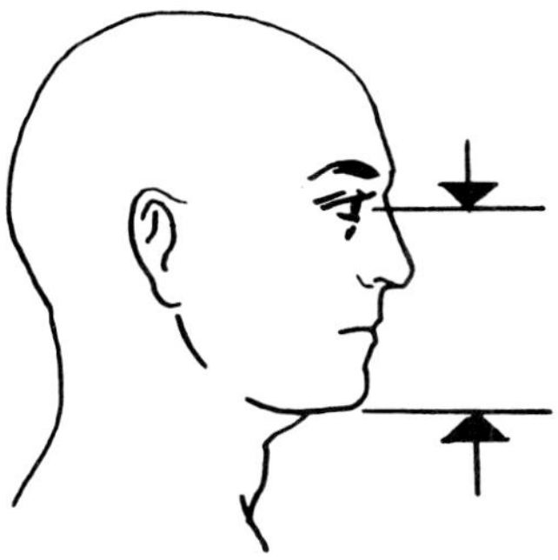

	Percentile		
	5th	50th	95th
Adults:			
Males	4.3 in	—	5.0 in
Females	3.8 in	—	4.25 in
Boys:			
Age 17	4.1 in	4.4 in	4.8 in
Age 14	3.8 in	4.3 in	4.7 in
Age 12	3.7 in	4.0 in	4.3 in
Age 6	3.2 in	3.5 in	3.8 in
Age 2	2.9 in	3.2 in	3.6 in
Girls:			
Age 17	3.7 in	4.2 in	4.5 in
Age 14	3.9 in	4.1 in	4.5 in
Age 12	3.6 in	4.0 in	4.4 in
Age 6	3.1 in	3.5 in	3.8 in
Age 2	2.8 in	3.1 in	3.4 in
Adults age 70 and over:			
Males			
Females			
Truck and bus drivers:			
Males			
Females			
Airline pilots (Male)	4.3 in	—	5.1 in
Flight attendants (Female)	3.8 in	—	4.6 in
Law enforcement officers:			
Males			
Females			

Note: By subtracting the above dimensions from maximum head height it can be observed that the top of a person's head is approximately 5 in (12.7 cm) above his or her eye height; this gives some idea of how much clearance should be provided above an eye reference for a seated operator. A good rule of thumb is to provide a minimum of 7 in (18 cm) of head clearance for bareheaded operators.

HAND LENGTH

This dimension is pertinent to the design of gloves mittens, and other devices for protecting the hands. This dimension is also useful in selecting test subjects for evaluating hand-held or hand-manipulated devices such as handrails and joy-stick grips.

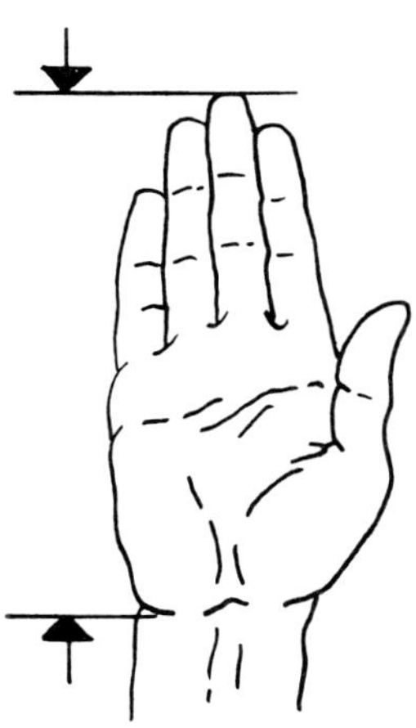

	Percentile		
	5th	50th	95th
Adults:			
Males	7.0 in	7.6 in	8.2 in
Females	6.4 in	6.9 in	7.4 in
Boys:			
Age 17	6.8 in	7.4 in	7.9 in
Age 14	6.3 in	7.0 in	7.6 in
Age 12	5.7 in	6.3 in	7.0 in
Age 6	4.6 in	5.0 in	5.7 in
Age 2	3.6 in	3.9 in	4.2 in
Girls:			
Age 17	6.1 in	6.7 in	7.2 in
Age 14	6.1 in	6.7 in	7.2 in
Age 12	5.6 in	6.4 in	7.0 in
Age 6	4.5 in	4.9 in	5.4 in
Age 2	3.3 in	3.8 in	4.3 in
Adults age 70 and over:			
Males	7.0 in	7.4 in	8.0 in
Females	—	—	—
Truck and bus drivers:			
Males	7.1 in	7.6 in	8.0 in
Females	—	—	—
Airline pilots (Male)	6.9 in	7.5 in	8.0 in
Flight attendants (Female)	6.3 in	6.8 in	7.3 in
Law enforcement officers:			
Males	7.0 in	7.6 in	8.2 in
Females	—	—	—

HAND BREADTH AT THUMB

This dimension is pertinent to the design of hand protective gear and is also useful in defining lateral hand clearance.

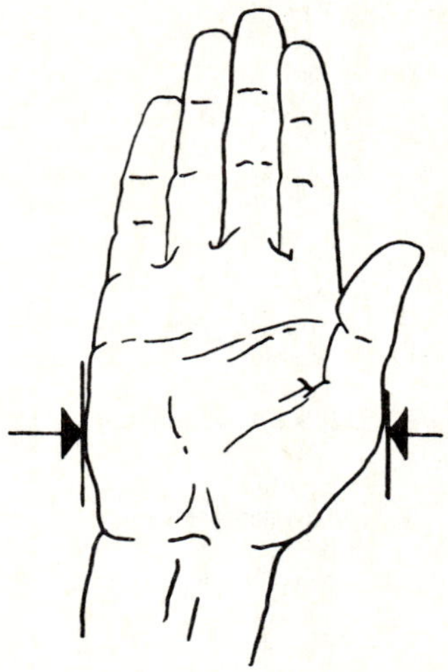

	Percentile		
	5th	50th	95th
Adults:			
Males			
Females			
Boys:			
Age 17			
Age 14			
Age 12			
Age 6			
Age 2			
Girls:			
Age 17			
Age 14			
Age 12			
Age 6			
Age 2			
Adults age 70 and over:			
Males			
Females			
Truck and bus drivers			
Males			
Females			
Airline pilots (Male)	3.7 in	4.1 in	4.4 in
Flight attendants (Female)	3.2 in	3.6 in	4.0 in
Law enforcement officers:			
Males			
Females			

HAND BREADTH AT METACARPAL

This dimension is pertinent to the design of hand protective gear.

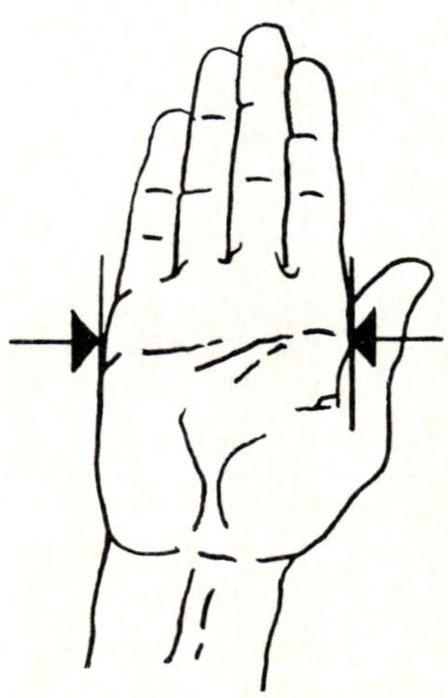

	Percentile		
	5th	50th	95th
Adults:			
Males	3.1 in	3.4 in	3.8 in
Females	2.7 in	3.0 in	3.4 in
Boys:			
Age 17	3.2 in	3.5 in	3.7 in
Age 14	2.8 in	3.2 in	3.6 in
Age 12	2.6 in	2.9 in	3.3 in
Age 6	2.1 in	2.4 in	2.6 in
Age 2	1.7 in	1.9 in	2.2 in
Girls:			
Age 17	2.7 in	2.9 in	3.3 in
Age 14	2.7 in	2.9 in	3.2 in
Age 12	2.5 in	2.8 in	3.1 in
Age 6	2.0 in	2.3 in	2.5 in
Age 2	1.6 in	1.8 in	2.1 in
Adults age 70 and over:			
Males	3.1 in	3.3 in	3.6 in
Females	—	—	—
Truck and bus drivers:			
Males	3.2 in	3.5 in	3.8 in
Females	—	—	—
Airline pilots (Male)	3.2 in	3.5 in	3.8 in
Flight attendants (Female)	2.7 in	2.9 in	3.1 in
Law enforcement officers:			
Males	3.3 in	3.5 in	3.8 in
Females	—	—	—

HAND THICKNESS AT METACARPAL

This dimension is pertinent to the design of hand protective gear and also to the establishment of hand clearance.

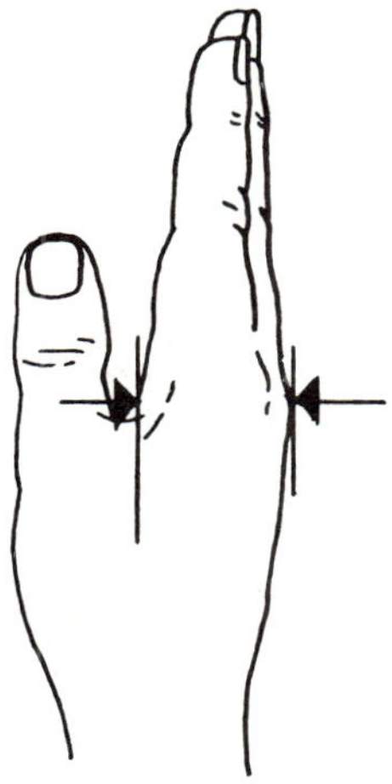

	Percentile		
	5th	50th	95th
Adults:			
Males	1.1 in	1.2 in	1.3 in
Females	0.8 in	1.0 in	1.1 in
Boys:			
Age 17			
Age 14			
Age 12			
Age 6			
Age 2			
Girls:			
Age 17			
Age 14			
Age 12			
Age 6			
Age 2			
Adults age 70 and over:			
Males			
Females			
Truck and bus drivers:			
Males			
Females			
Airline pilots			
Flight attendants			
Law enforcement officers:			
Males			
Females			

FOOT LENGTH[6]

This dimension is pertinent to the design of footgear and also to the establishment of clearances for the foot (see the discussion of clothing effects).

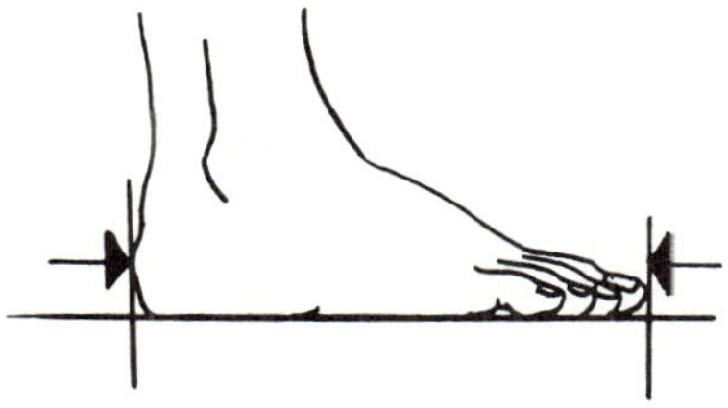

	Percentile		
	5th	50th	95th
Adults:			
Males	9.7 in	10.4 in	11.2 in
Females	8.7 in	9.4 in	10.2 in
Boys:			
Age 17	9.5 in	10.6 in	11.1 in
Age 14	9.0 in	10.0 in	11.0 in
Age 12	8.4 in	9.2 in	10.1 in
Age 6	6.5 in	7.0 in	8.0 in
Age 2	4.5 in	5.1 in	5.7 in
Girls:			
Age 17	8.6 in	9.2 in	10.1 in
Age 14	8.5 in	9.2 in	10.2 in
Age 12	8.3 in	9.1 in	9.8 in
Age 6	6.3 in	7.0 in	7.6 in
Age 2	4.2 in	5.0 in	5.5 in
Adults age 70 and over:			
Males	9.7 in	10.2 in	10.9 in
Females	—	—	—
Truck and bus drivers:			
Males	9.6 in	10.4 in	11.3 in
Females	—	—	—
Airline pilots (Male)	9.8 in	10.5 in	11.3 in
Flight attendants (Female)	8.7 in	9.4 in	10.1 in
Law enforcement officers:			
Males	—	—	—
Females	—	—	—

[6]Measured with subjects standing.

BALL OF FOOT WIDTH[7]

This dimension is pertinent to the design of footgear and also to the establishment of foot clearance.

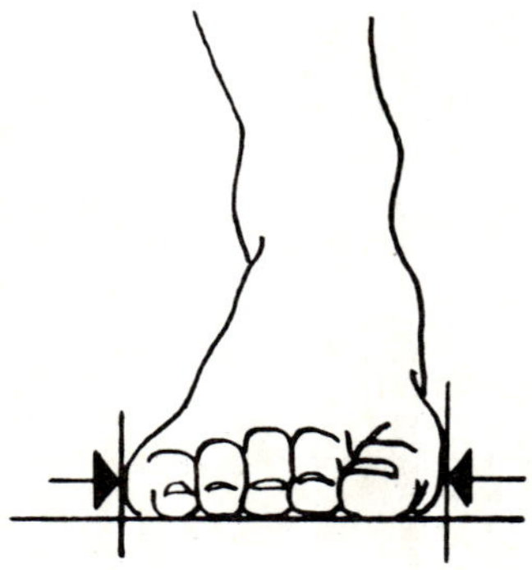

	Percentile		
	5th	50th	95th
Adults:			
Males	3.5 in	3.9 in	4.3 in
Females	3.2 in	3.6 in	4.0 in
Boys:			
Age 17	3.7 in	4.1 in	4.4 in
Age 14	3.4 in	3.9 in	4.3 in
Age 12	3.1 in	3.5 in	4.0 in
Age 6	2.5 in	2.8 in	3.1 in
Age 2	1.8 in	2.1 in	2.4 in
Girls:			
Age 17	3.2 in	3.5 in	3.9 in
Age 14	3.1 in	3.5 in	3.9 in
Age 12	3.0 in	3.4 in	3.8 in
Age 6	2.4 in	2.7 in	3.0 in
Age 2	1.6 in	1.9 in	2.2 in
Adults age 70 and over:			
Males	3.6 in	3.9 in	4.3 in
Females	—	—	—
Truck and bus drivers:			
Males	3.7 in	4.0 in	4.3 in
Females	—	—	—
Airline pilots (Male)	3.5 in	3.8 in	4.1 in
Flight attendants (Female)	3.2 in	3.4 in	3.8 in
Law enforcement officers:			
Males			
Females			

SHOULDER CIRCUMFERENCE

This dimension is pertinent to the design of garments. It also is useful in establishing the general description of a particular test subject who may be used for dynamic testing of various physical mockups.

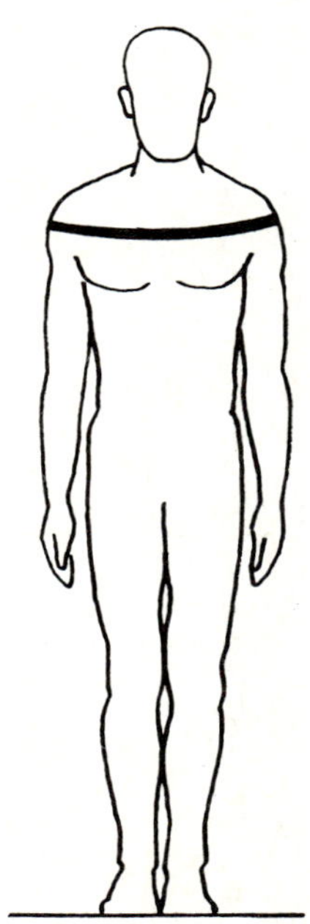

	Percentile		
	5th	50th	95th
Adults:			
Males			
Females			
Boys:			
Age 17			
Age 14			
Age 12			
Age 6			
Age 2			
Girls:			
Age 17			
Age 14			
Age 12			
Age 6			
Age 2			
Adults age 70 and over:			
Males			
Females			
Truck and bus drivers:			
Males			
Females			
Airline pilots (Male)	41.9 in	45.3 in	47.7 in
Flight attendants (Female)	35.6 in	37.6 in	39.7 in
Law enforcement officers:			
Males			
Females			

[7]Measured with subjects standing.

CHEST CIRCUMFERENCE

This dimension can be used in conjunction with shoulder circumference for garment design and the definition of the body characteristics of test subjects.

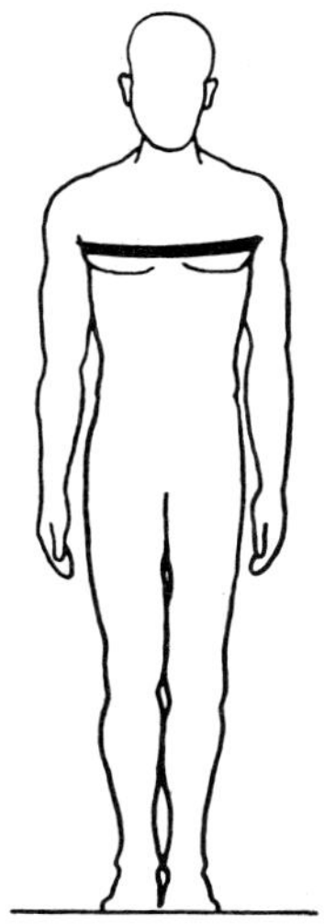

	Percentile		
	5th	50th	95th
Adults:			
Males	34.3 in	39.0 in	44.9 in
Females	30.4 in	34.3 in	40.5 in
Boys:			
Age 17	32.5 in	36.2 in	40.8 in
Age 14	27.5 in	31.3 in	36.1 in
Age 12	25.6 in	28.1 in	32.1 in
Age 6	20.7 in	22.6 in	25.3 in
Age 2	18.0 in	19.0 in	20.6 in
Girls:			
Age 17	29.7 in	32.3 in	36.3 in
Age 14	27.6 in	30.8 in	34.6 in
Age 12	25.2 in	28.7 in	33.4 in
Age 6	20.3 in	22.3 in	24.3 in
Age 2	17.3 in	18.4 in	20.5 in
Adults age 70 and over:			
Males	33.0 in	38.1 in	44.1 in
Females	30.1 in	35.0 in	39.9 in
Truck and bus drivers:			
Males	—	—	—
Females	—	—	—
Airline pilots (Male)	36.3 in	38.8 in	41.3 in
Flight attendants (Female)	31.2 in	33.7 in	36.4 in
Law enforcement officers:			
Males	35.4 in	40.1 in	45.6 in
Females	—	—	—

WAIST CIRCUMFERENCE

This dimension can be used in conjunction with shoulder and chest circumference for garment design and the definition of the body characteristics of test subjects.

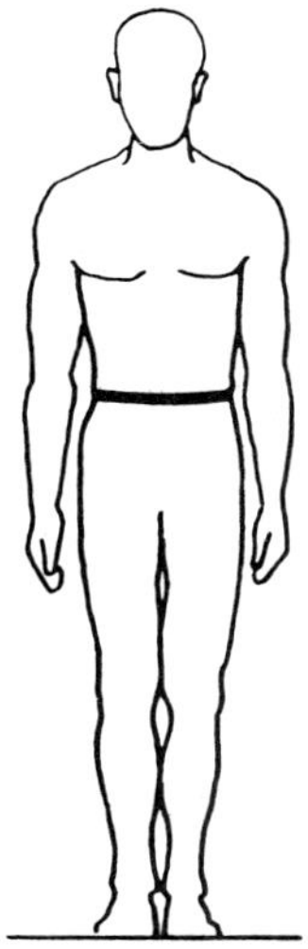

	Percentile		
	5th	50th	95th
Adults:			
Males	28.4 in	34.8 in	42.9 in
Females	24.1 in	29.2 in	39.1 in
Boys:			
Age 17	26.5 in	29.6 in	35.9 in
Age 14	23.7 in	26.8 in	32.7 in
Age 12	22.1 in	24.8 in	29.7 in
Age 6	18.0 in	20.3 in	23.0 in
Age 2	15.5 in	17.3 in	19.3 in
Girls:			
Age 17	24.9 in	27.8 in	34.4 in
Age 14	23.1 in	26.8 in	31.9 in
Age 12	21.8 in	25.1 in	32.0 in
Age 6	18.1 in	20.2 in	23.1 in
Age 2	14.8 in	17.4 in	19.4 in
Adults age 70 and over:			
Males	28.3 in	35.8 in	41.3 in
Females	25.9 in	33.2 in	39.9 in
Truck and bus drivers:			
Males	—	—	—
Females	—	—	—
Airline pilots (Male)	29.0 in	32.0 in	35.0 in
Flight attendants (Female)	22.8 in	24.5 in	26.4 in
Law enforcement officers:			
Males	30.1 in	35.3 in	42.3 in
Females	—	—	—

HIP (BUTTOCK) CIRCUMFERENCE (STANDING)

This dimension can be used in conjunction with hip circumference considerations for garment design and the definition of the body characteristics of test subjects.

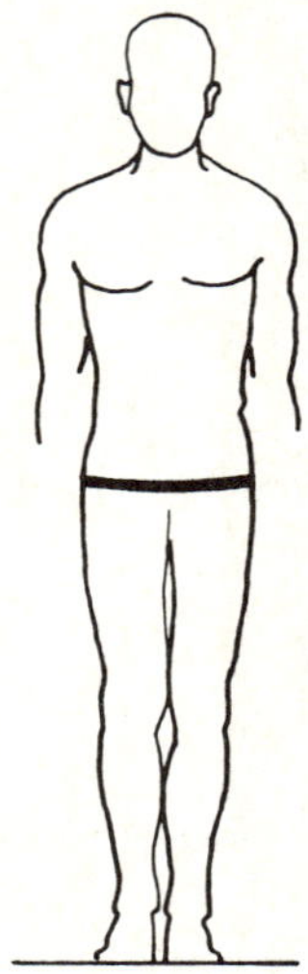

	Percentile		
	5th	50th	95th
Adults:			
Males	—	—	—
Females	—	—	—
Boys:			
Age 17	32.8 in	35.7 in	40.7 in
Age 14	28.3 in	32.4 in	37.4 in
Age 12	26.0 in	29.2 in	34.5 in
Age 6	20.5 in	22.6 in	26.0 in
Age 2	18.6 in	20.5 in	22.7 in
Girls:			
Age 17	32.1 in	36.1 in	41.6 in
Age 14	29.9 in	34.1 in	38.8 in
Age 12	26.1 in	30.7 in	35.9 in
Age 6	20.5 in	23.1 in	25.5 in
Age 2	18.1 in	20.2 in	22.1 in
Adults age 70 and over:			
Males	—	—	—
Females	—	—	—
Truck and bus drivers:			
Males	—	—	—
Females	—	—	—
Airline pilots (Male)	35.5 in	37.8 in	40.1 in
Flight attendants (Female)	33.4 in	35.4 in	37.8 in
Law enforcement officers:			
Males	—	—	—
Females	—	—	—

HIP (BUTTOCK) CIRCUMFERENCE (SITTING)

This dimension is pertinent to the design of seats and seatbelts. When it is used in conjunction with hip breadth, subjects can be selected to represent a larger user population for evaluating seat widths and lap belt lengths.

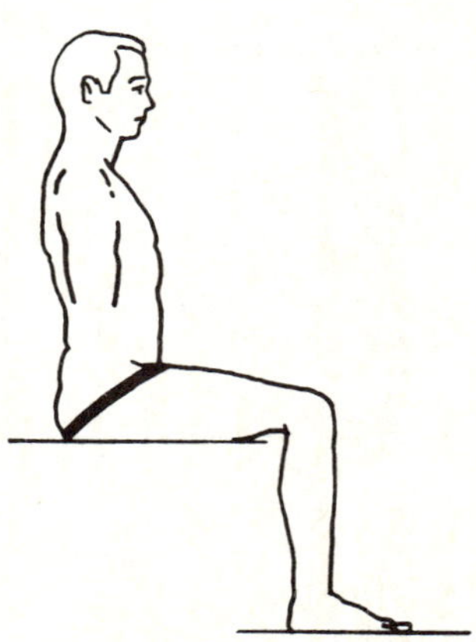

	Percentile		
	5th	50th	95th
Adults:			
Male			
Female			
Boys:			
Age 17			
Age 14			
Age 12			
Age 6			
Age 2			
Girls:			
Age 17			
Age 14			
Age 12			
Age 6			
Age 2			
Adults age 70 and over:			
Males			
Females			
Truck and bus drivers:			
Males			
Females			
Airline pilots (Male)	38.9 in	41.7 in	44.5 in
Flight attendants (Female)	37.0 in	39.2 in	41.4 in
Law enforcement officers:			
Males			
Females			

HEAD CIRCUMFERENCE

This dimension is pertinent to the design of headgear.

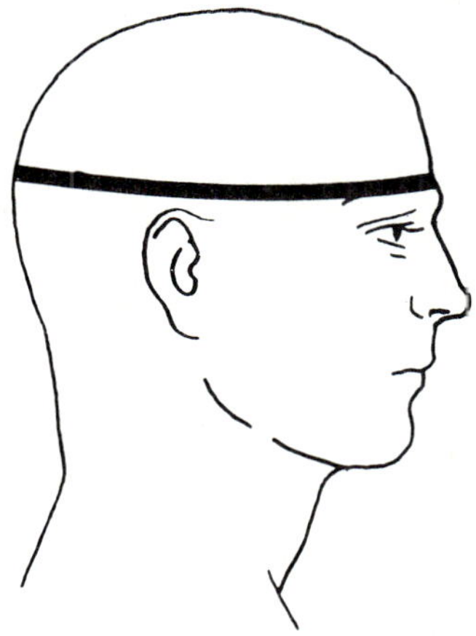

	Percentile		
	5th	50th	95th
Adults:			
Males	—	—	—
Females	—	—	—
Boys:			
Age 17	21.1 in	22.3 in	23.3 in
Age 14	20.4 in	21.5 in	22.8 in
Age 12	20.1 in	21.1 in	22.3 in
Age 6	19.2 in	20.4 in	21.2 in
Age 2	18.5 in	19.3 in	20.4 in
Girls:			
Age 17	20.4 in	21.5 in	22.6 in
Age 14	20.3 in	21.2 in	22.3 in
Age 12	19.8 in	20.9 in	21.9 in
Age 6	19.1 in	19.8 in	20.8 in
Age 2	18.1 in	18.8 in	19.7 in
Adults age 70 and over:			
Males	—	—	—
Females	—	—	—
Truck and bus drivers:			
Males	—	—	—
Females	—	—	—
Airline pilots (Male)	21.9 in	22.5 in	23.1 in
Flight attendants (Female)	20.8 in	21.5 in	22.2 in
Law enforcement officers:			
Males	21.6 in	22.6 in	23.6 in
Females	—	—	—

NECK CIRCUMFERENCE

This dimension may be pertinent to the design of special garments and/or devices worn about the user's neck.

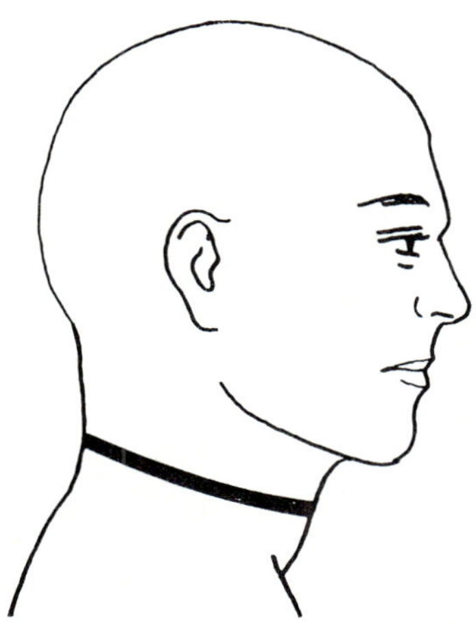

	Percentile		
	5th	50th	95th
Adults:			
Males	—	—	—
Females	—	—	—
Boys:			
Age 17	12.4 in	14.2 in	15.0 in
Age 14	11.5 in	12.7 in	13.8 in
Age 12	10.7 in	11.6 in	12.6 in
Age 6	9.2 in	10.2 in	11.0 in
Age 2	8.6 in	9.6 in	10.1 in
Girls:			
Age 17	11.3 in	12.2 in	13.3 in
Age 14	10.9 in	11.7 in	13.1 in
Age 12	10.4 in	11.4 in	12.8 in
Age 6	9.1 in	10.0 in	10.9 in
Age 2	8.2 in	9.2 in	9.7 in
Adults age 70 and over:			
Males	—	—	—
Females	—	—	—
Truck and bus drivers:			
Males	—	—	—
Females	—	—	—
Airline pilots (Male)	14.3 in	15.0 in	15.7 in
Flight attendants (Female)	11.1 in	11.8 in	12.6 in
Law enforcement officers:			
Males	—	—	—
Females	—	—	—

STATURE STATISTICS FOR OTHER POPULATIONS

With the exception of anthropometric surveys conducted by military organizations, most other population surveys include only a few dimensional statistics. The most common (and one which provides a point of departure for comparing populations) is stature. The following compilation of statures for different populations presented in the accompanying table is useful when a particular product design is intended for one or more of the specific populations noted.

Group or Nationality	Percentile, cm 5th	50th	95th
Australian Army males	164.0	173.0	184.0
Bulgarian civilian males	163.6	170.2	176.8
Bulgarian women	151.6	157.8	164.0
Canadian civilian males	153.9	159.8	173.5
Chinese civilian males	—	166.4	—
English, RAF male pilots	166.1	177.0	185.9
English, male Navy recruits	164.7	174.5	184.2
Filipino civilian males	—	163.1	—
French civilian males	—	172.5	—
French civil pilots	168.1	173.1	178.0
German civilian males	167.1	176.8	187.5
Greek Army males	160.9	170.5	180.3
Hawaiian civilian males	—	171.2	—
Indian civilian males	—	167.9	—
Iranian Army males	157.5	166.7	176.6
Israeli civilian males	—	168.7	—
Italian Army males	160.2	170.7	180.7
Japanese Air Force, male pilots	159.4	166.7	175.0
Korean Army males	154.5	164.0	172.7
Korean Air Force, male pilots	159.6	165.6	173.7
Latin American Army males	157.0	165.9	177.0
Norwegian (unknown males)	171.5	177.5	183.5
Swedish civilian males	—	174.0	—
Swedish civilian females	164.1	—	—
Thai Army males	155.0	163.5	172.0
Turkish Army males	160.6	169.0	179.2
Vietnamese Army males	151.6	160.4	169.6
U.S. military:			
Army ground forces, male	163.8	174.4	185.6
Army ground forces, female	152.6	162.8	174.1
Army aviators, male	164.2	174.6	185.0
Air Force, male pilots	167.2	177.3	187.7
Air Force, ground, female	156.8	162.7	168.6
Marine Corps, male	168.2	174.5	180.8
Navy, enlisted, male	171.7	177.6	183.5
Navy, pilots, male	168.2	177.5	187.7
Navy, divers, male	170.0	176.2	182.2
U.S. FAA civil pilots, male	170.4	176.7	183.2

RACIAL AND ETHNIC ORIGINS OF THE U.S. POPULATION

Racial and Ethnic Origins of U.S. Population (from Census Bureau Data, April 1970)

Group	Number, in Thousands	Percent
White	177,784	87.5
Spanish-speaking	10,115	4.9
Black	22,580	11.1
Other	2,883	1.4
Indian	793	0.4
Japanese	591	0.3
Chinese	435	0.2
Filipino	343	0.2
Other	720	0.4

ANTHROPOMETRIC REFERENCES

The following references were used to compile the previous data on weight and body dimensions. Note that some liberties have been taken; i.e., when a certain dimension was not available for a population, a substitution was made from the most likely (similar) population. This was done, however, only when it was considered likely that the borrowed data would be quite similar to that for the population designated.

Anthropometric Source Book, NASA Reference Publication 1024, vols. I, II, and III, 1978 (prepared for the National Aeronautics and Space Administration, Scientific and Technical Information Office, by Webb Associates, Yellow Springs, Ohio).

DAMON, ALBERT, et al.: *The Human Body in Equipment Design,* Harvard University Press, Cambridge, Mass., 1966.

GARRETT, JOHN W., and KENNETH W. KENNEDY: *A Collation of Anthropometry,* AMRL-TR-68-1, vols. I and II, Aerospace Medical Research Laboratory, Aerospace Medical Division, Air Force Systems Command, Wright-Patterson Air Force Base, Ohio, 1971.

HAMILL, P. V. V., et al.: *Body Weight, Stature, and Sitting Height: White and Negro Youths 12–17 Years,* U.S. Department of Health, Education, and Welfare, Public Health Service ser. 11, no. 126, 1973.

MALINA, R. M., et al.: *Body Dimensions and Proportions: White and Negro Children 6–11 Years,* U.S. Department of Health, Education, and Welfare, Public Health Service ser. 11, no. 143, 1973.

MALINA, R. M., et al.: *Selected Body Measurements of Children 6–11 Years,* U.S. Department of Health, Education, and Welfare, Public Health Service ser. 11, no. 123, 1973.

MARTIN, J. I., et al.: *Anthropometry of Law Enforcement Officers,* Technical Document 442, U.S. Naval Electronics Laboratory Center, San Diego, 1975.

SNYDER, RICHARD G, et al.: *Physical Characteristics of Children as Related to Death and Injury for Consumer Product Safety Design,* UM-HSRI-BI-75-5, Highway Safety Research Institute, Ann Arbor, Mich., 1975.

WHITE, ROBERT M.: "Anthropometric Measurements on Selected Populations of the World," in Alphonse Chapanis (ed.), *Ethnic Variables in Human Factors Engineering,* The Johns Hopkins Press, Baltimore, 1975.

The reader is referred to the following for selected military population anthropometrics:

CHURCHILL, EDMUND, et al.: *Anthropometry of U.S. Army Aviators—1970,* U.S. Army Natick Laboratories Technical Report 72-52-CE, Natick, Mass., 1971.

CHURCHILL, EDMUND, et al.: *Anthropometry of Women of the U.S. Army—1977,* Report no. 2, Natick-TR-77-024, U.S. Army Natick Research and Development Command, Natick, Mass.

CLAUSER, CHARLES E., et al.: *Anthropometry of Air Force Women,* AMRL-TR-70-5, Aerospace Medical Research Laboratory, Aerospace Medical Division, Air Force Systems Command, Wright-Patterson Air Force Base, Ohio, 1972.

GIFFORD, E. C.: *Compilation of Anthropometric Measures on U.S. Navy Pilots,* NAMC-AGEL-437, Naval Air Material Command, Air Crew Equipment Laboratory, Philadelphia, 1960.

WHITE, ROBERT M., and EDMUND CHURCHILL: *The Body Size of Soldiers: U.S. Army Anthropometry—1966,* U.S. Army Natick Laboratories, Natick, Mass., 1971.

Special acknowledgment is made of the invaluable assistance of Dr. Robert M. White in reviewing and contributing to the foregoing special anthropometric tables. Mr. White, formerly with the U.S. Army Laboratory in Natick, Massachusetts, has specialized for many years in the collection of anthropometric data for various populations and is recognized as an authority in this area.

ANTHROPOMETRIC DATA RELATIVE TO TYPICAL WORKING POSITIONS[8]

The accompanying illustrations and tables provide useful information about work-space dimensioning. However, keep in mind that these data were derived from measurements of male Air Force personnel, and although they represent a somewhat functional dimensional guide, each work space must be given specific attention in terms of its own unique requirements.

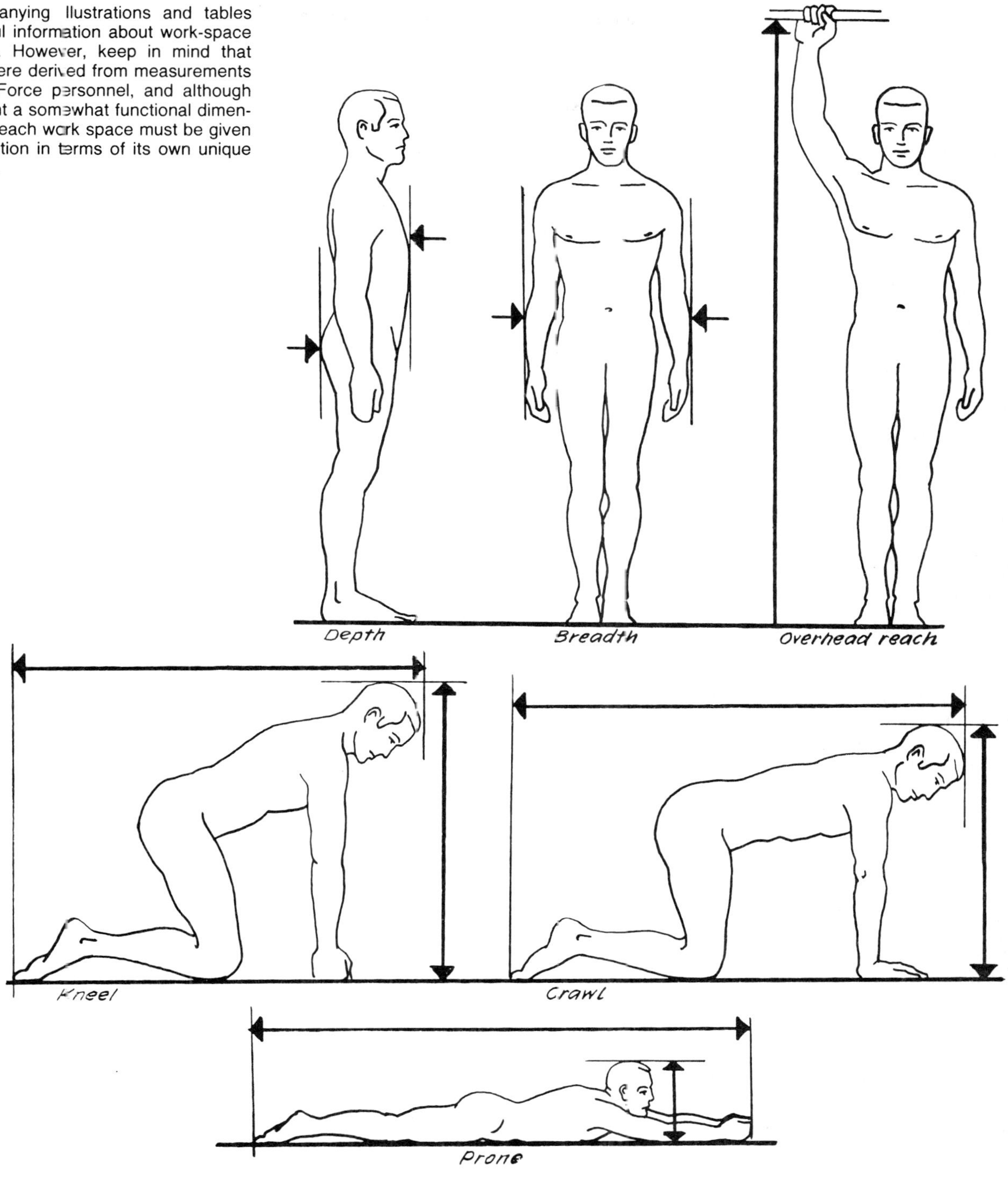

[8]H. T. E. Hertzberg et al., *The Anthropometry of Working Positions,* WADC Technical Report 54-520, Wright-Patterson Air Force Base, Ohio; and K. W. Kennedy and B. E. Filler, *Aperture Sizes and Depths of Reach for One- and Two-Handed Tasks,* AMRL-TR-66-27, Aerospace Medical Research Laboratory, Wright-Patterson Air Force Base, Ohio, 1966.

Measurement	Number of Subjects	5th Percentile	Mean	95th Percentile	SD
Maximum body depth	118	10.1	11.5	13.0	0.88
Maximum body breadth	40	18.8	20.9	22.8	1.19
Overhead reach	40	76.8	82.5	88.5	3.33
Kneeling height	40	29.7	32.0	34.5	1.57
Kneeling length	40	37.6	43.0	48.1	3.26
Crawling height	40	26.2	28.4	30.5	1.30
Crawling length	40	49.3	53.2	58.2	2.61
Prone height	40	12.3	14.5	16.4	1.28
Prone length	40	84.7	90.1	95.8	3.41

Standing, Forward Reach (Both Arms)

	Percentiles				
	5th	25th	50th	75th	95th
A. Depth of reach Range: 17.50 to 25.25 SD: 1.50	19.25 in	21.00 in	22.25 in	22.75 in	24.50 in
B. Breadth of aperture Range: 15.00 to 20.25 Mean: 17.69 SD: 1.19	15.50 in	17.00 in	17.75 in	18.50 in	19.50 in
C. Floor to top of aperture Range: 58.75 to 70.50 SD: 2.34	61.00 in	63.50 in	65.25 in	66.50 in	69.00 in
D. Floor to bottom of aperture Range: 51.25 to 61.75 Mean: 56.09 SD: 2.05	52.25 in	54.75 in	56.00 in	57.25 in	59.00 in
E. Vertical dimension of aperture	(1)	(1)	(1)	(1)	(1)

(1) 16.75 in.

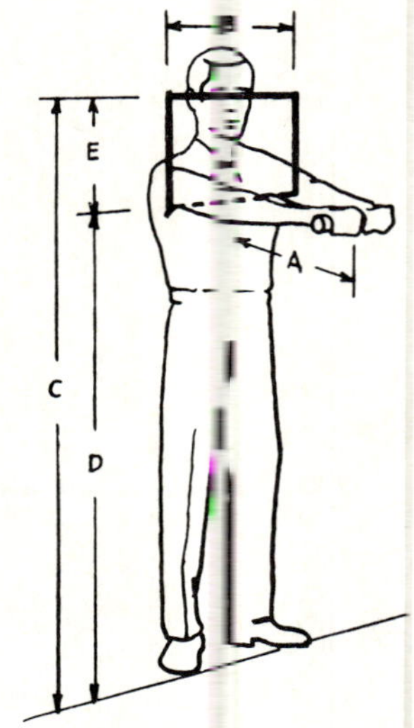

Standing, Forward Reach (Preferred Arm)

	Percentiles				
	5th	25th	50th	75th	95th
A. Depth of reach Range: 19.50 to 27.50 Mean: 23.61 SD: 1.82	20.25 in	22.25 in	23.75 in	25.00 in	26.75 in
B. Breadth of aperture: 12.00					
C. Floor to top of aperture Range: 58.25 to 70.50 Mean: 64.88 SD: 2.36	61.00 in	63.25 in	65.00 in	66.25 in	69.00 in
D. Floor to bottom of aperture Range: 51.25 Mean: 56.09 SD: 2.05	52.25 in	54.75 in	56.00 in	57.25 in	59.00 in
E. Vertical dimension of aperture	(1)	(1)	(1)	(1)	(1)

(1) 16.75 in.

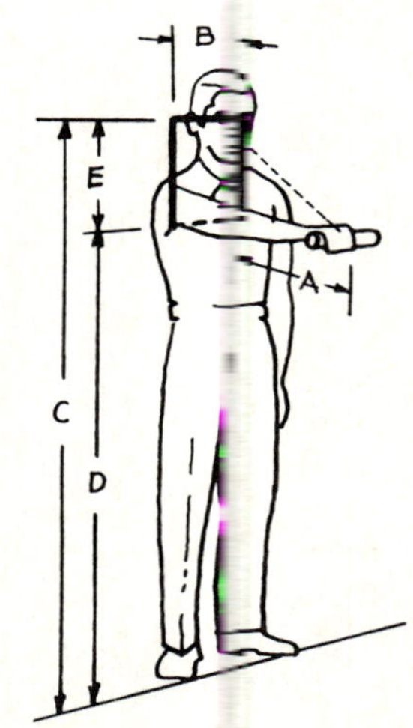

Standing, Lateral Reach (Preferred Arm)

	Percentiles				
	5th	25th	50th	75th	95th
A. Depth of reach Range: 21.75 to 28.53 Mean: 24.65 SD: 1.51	22.00 in	23.50 in	24.75 in	25.75 in	26.75 in
B. Breadth of aperture 10.00					
C. Floor to top of aperture Range: 58.25 to 70.00 Mean: 64.70 SD: 2.32	60.75 in	63.25 in	64.25 in	66.00 in	68.75 in
D. Floor to bottom of aperture Range: 51.25 to 61.75 Mean: 56.09 SD: 2.05	52.25 in	54.75 in	56.00 in	57.25 in	59.00 in
E. Vertical dimension of aperture	(1)	(1)	(1)	(1)	(1)

(1) 16.5 in.

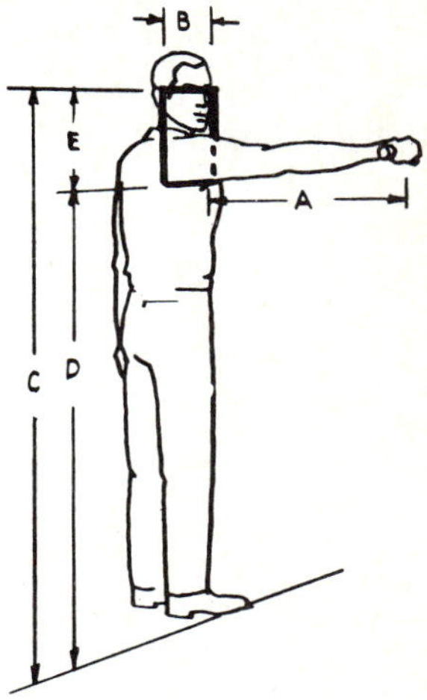

Seated, Forward Reach (Both Arms)

	Percentiles				
	5th	25th	50th	75th	95th
A. Depth of reach Range 14.00 to 23.50 Mean: 18.26 SD: 2.15	15.00 in	16.50 in	17.75 in	19.50 in	22.25 in
B. Breadth of aperture Range 13.50 to 18.75 Mean: 16.12 SD: 1.25	13.75 in	15.25 in	16.00 in	17.00 in	18.25 in
C. Floor to top of aperture Range 39.25 to 51.00 Mean: 43.25 SD: 2.05	19.75 in	41.75 in	43.00 in	44.25 in	46.50 in
D. Floor to bottom of aperture† Range: 32.50 to 41.75 Mean: 36.59 SD: 1.59	34.25 in	35.50 in	36.50 in	37.50 in	39.00 in
E. Vertical dimension of aperture	(1)	(1)	(1)	(1)	(1)

(1) 12.25 in.

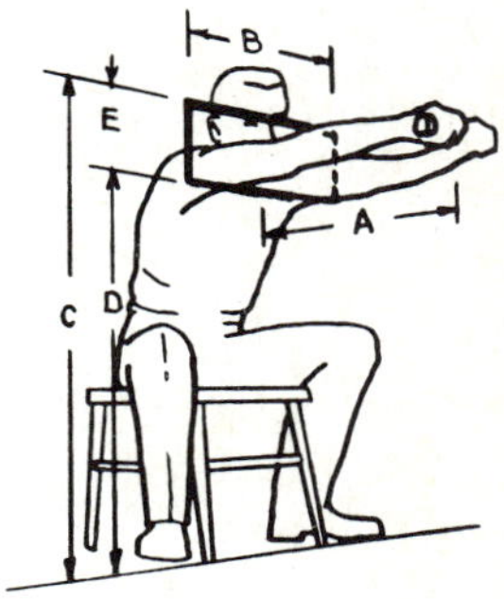

Cross-Legged Seated, Forward Reach (Both Arms)

	Percentiles				
	5th	25th	50th	75th	95th
A. Depth of reach Range: 13.50 to 22.25 Mean: 17.08 SD: 1.91	13.75 in	15.75 in	16.75 in	18.25 in	20.00 in
B. Breadth of aperture Range: 13.50 to 18.50 Mean: 15.89 SD: 1.54	13.75 in	14.75 in	16.00 in	16.75 in	17.75 in
C. Floor to top of aperture Range: 22.25 to 30.50 Mean: 25.30 SD: 1.54	22.75 in	24.25 in	25.25 in	26.25 in	28.00 in
D. Floor to bottom of aperture Range: 17.00 to 23.25 Mean: 19.23 SD: 1.19	17.00 in	18.50 in	19.25 in	20.00 in	21.25 in
E. Vertical dimension of aperture	(1)	(1)	(1)	(1)	(1)

(1) 11.0 in.

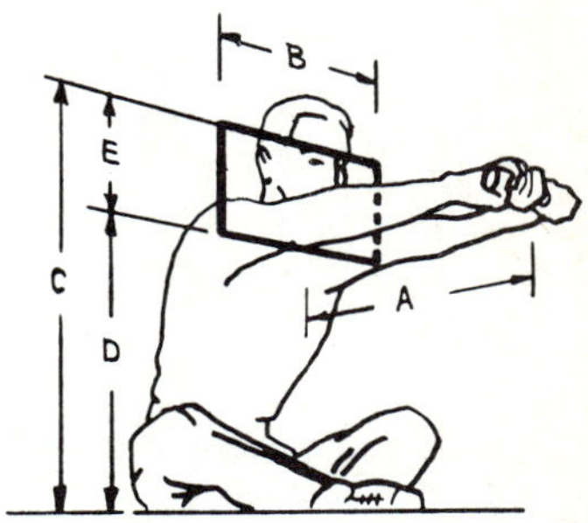

Selected Working-Position Data for U.S. Army Men and Women (Wearing Typical Army Clothing)[9]

Overhead reach height—standing with heels 23 cm apart and toes 15 cm from wall; arms extended overhead with fists touching and against wall; 1st phalanges horizontal. Measured from floor to highest point on 1st phalanges.

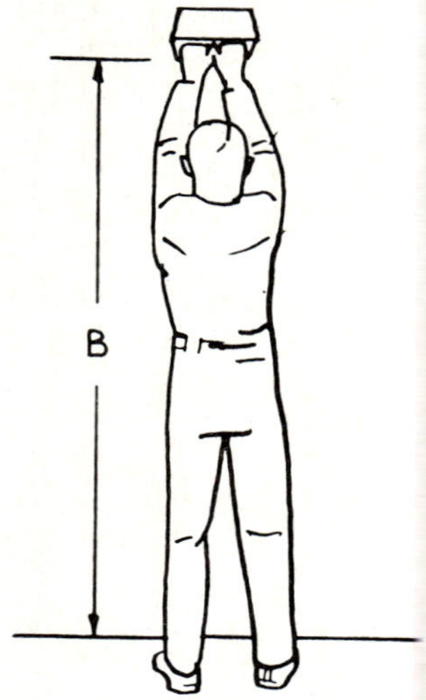

Overhead reach breadth—standing with heels 23 cm apart and toes 15 cm from wall; arms extended overhead with fists touching and against wall; 1st phalanges horizontal. Measured horizontally across arms or shoulders, whichever is wider.

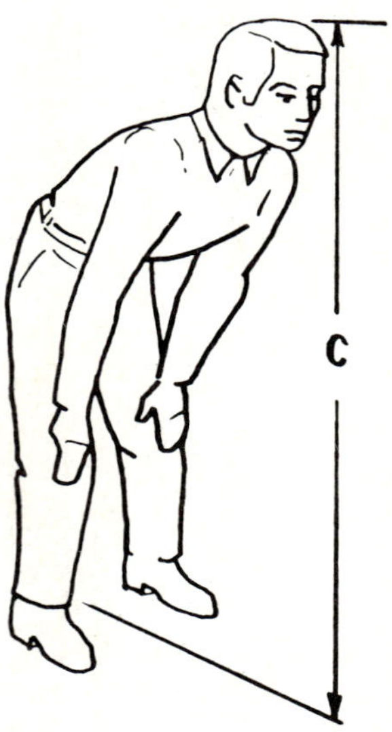

Bent torso height—standing with feet 30 cm apart; bending over and placing palms of the hands on kneecaps; elbows and knees locked; looking forward; head tilted as far back as possible. Measured from floor to top of head.

Bent torso breadth—standing with feet 30 cm apart; bending over and placing the palms of the hands on kneecaps; elbows and knees locked; looking forward; head tilted as far back as possible. Measured as maximum horizontal distance across shoulders.

[9]The illustrations on this page and on pp. 39 and 40 refer to position specifications and values in the table on p. 40.

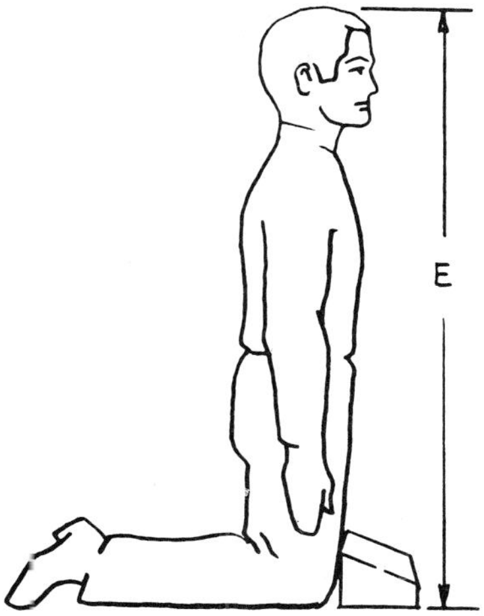

Kneeling height—kneeling with toes extended and lightly touching rear wall; torso erect with arms hanging loosely at sides. Measured from floor to top of head.

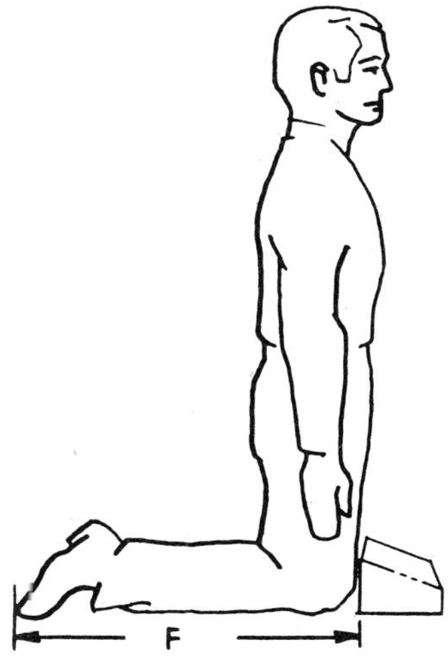

Kneeling leg length—kneeling with toes extended and lightly touching rear wall; torso erect with arms hanging loosely at sides. Measured from wall to anterior portion of both knees.

Overhead reach, sitting—sitting erect; right side against wall; right arm extended upward with palm flat against wall and fingers extended. Measured from sitting surface to tip of middle finger.

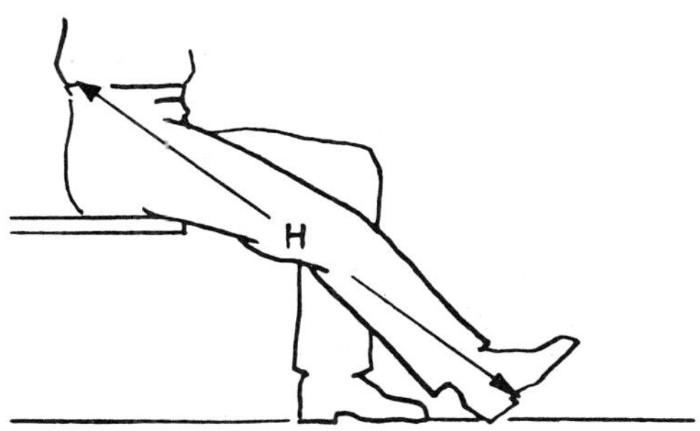

Functional leg length—sitting erect on edge of chair; right leg extended forward with knee straightened. Measured from heel along axis of leg to posterior waist.

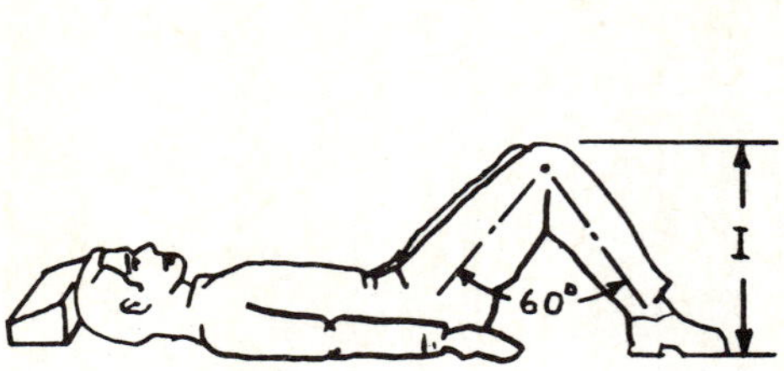

Bent knee height, supine—lying supine; knees raised until angle between upper and lower legs approximates 60°; toes lightly touching wall. Measured from floor to highest point on knees.

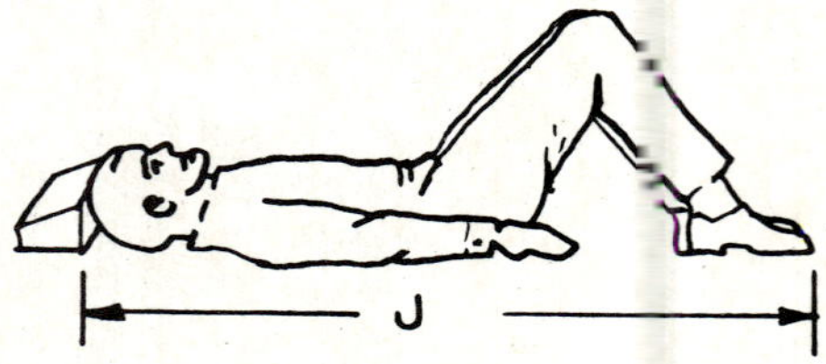

Horizontal length, knees bent—lying supine; knees raised until angle between upper and lower legs approximates 60°; toes lightly touching wall. Measured from wall to top of head.

Functional reach, extended—standing erect; looking straight ahead; right shoulder extended as far forward as possible while back of left shoulder firmly against wall; arm horizontal. Measured from wall to tip of index finger.

Anthropometric Data for U.S. Male and Female Personnel: Common Working Positions, Percentile Values

	5th Percentile		95th Percentile	
	Men	Women	Men	Women
A. Overhead reach	78.9 in	73.0 in	90.8 in	84.7 in
B. Overhead reach, breadth	13.9 in	12.4 in	16.5 in	14.9 in
C. Bent torso height	49.4 in	44.4 in	59.0 in	54.6 in
D. Bent torso breadth	16.1 in	14.5 in	19.0 in	17.1 in
E. Kneeling height	48.0 in	45.1 in	53.9 in	51.3 in
F. Kneeling leg length	25.2 in	23.3 in	29.7 in	27.8 in
G. Overhead reach, sitting	50.3 in	46.2 in	57.9 in	54.9 in
H. Functional leg length	43.5 in	39.2 in	50.3 in	46.7 in
I. Bent knee height, supine	17.6 in	16.3 in	21.1 in	19.5 in
J. Horizontal length, knee bent	59.4 in	55.2 in	68.1 in	64.5 in
K. Functional reach	33.2 in	28.9 in	39.8 in	36.5 in

Source: MIL-STD-1472B, *Notice—2,* May 10, 1978.

VISUAL SYSTEM CHARACTERISTICS

The primary organ through which visual responses are made is, of course, the eye. The accompanying sketches show the key elements of the visual receptor and liken them to the camera and the photographic process.

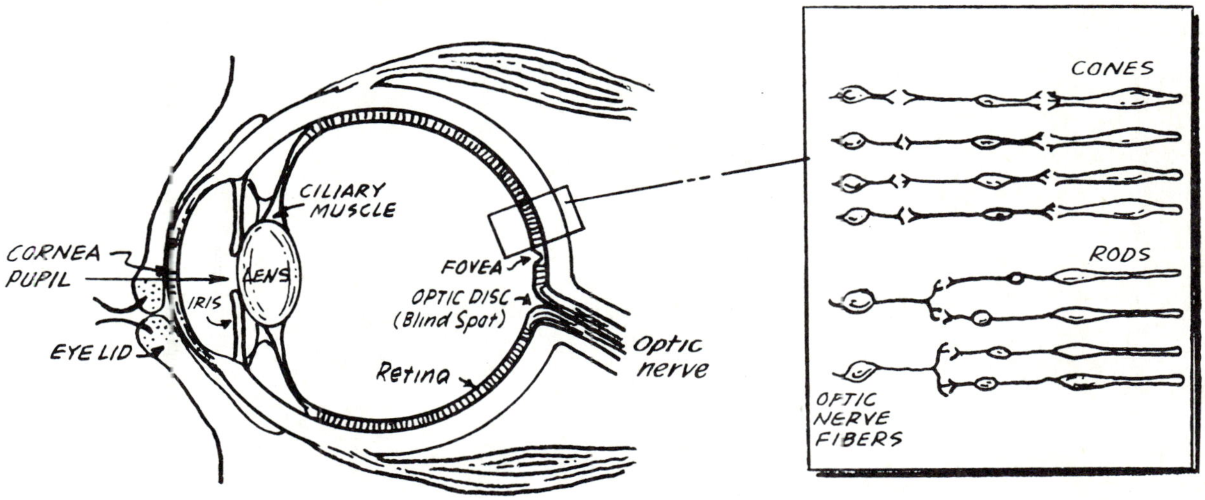

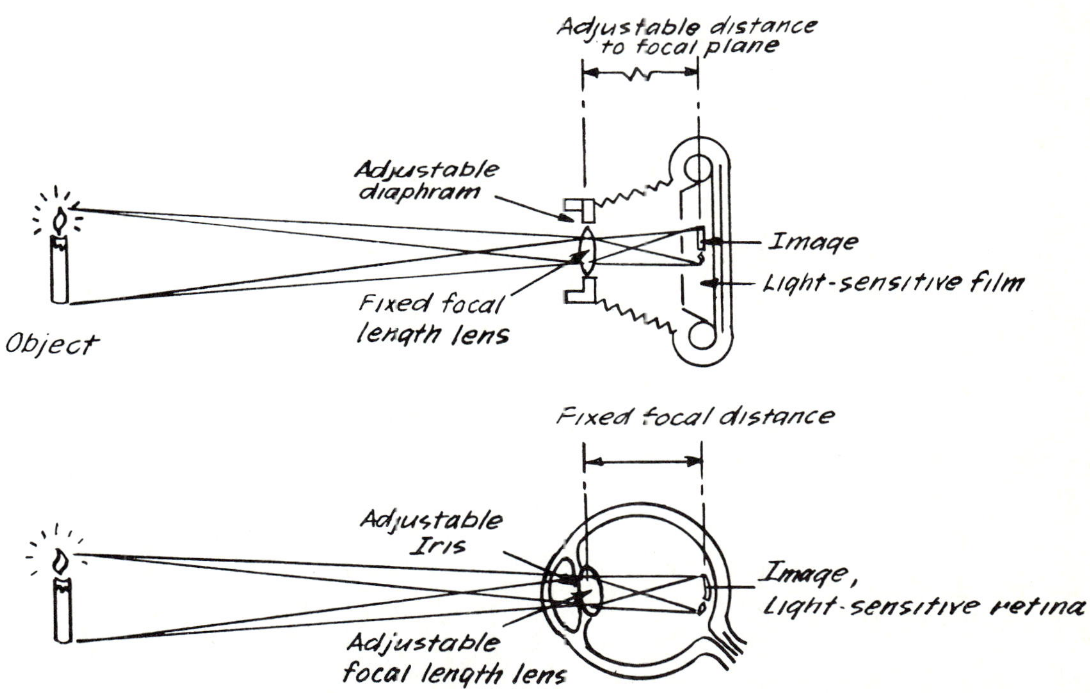

Image and Retinal System

The image of a moving object moves across the retina when the eyes are held still, providing information of the movement through sequential receptor stimulation along the path.

Eye and Head Movement System

When the eye follows a moving object, the image remains more or less stationary upon the retina, although the person still sees movement.

The image and retinal system and the eye and head movement system sometimes disagree, leading to visual illusions that are often confusing and misleading.

Size Constancy

The image of an object on the retina halves in size with each doubling of viewing distance of the object. However, it does not appear to shrink so much because the brain compensates for the shrinkage of the image with distance.

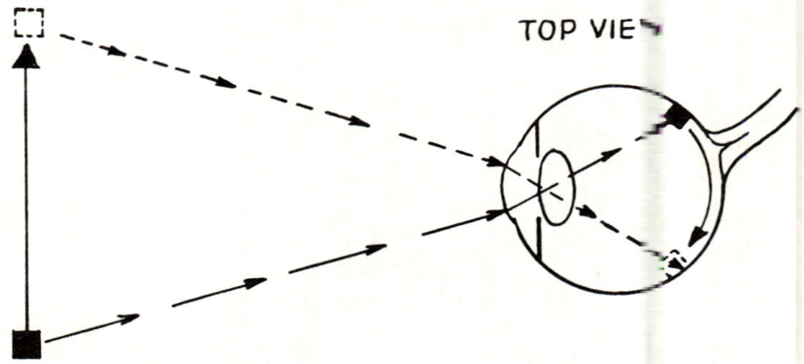

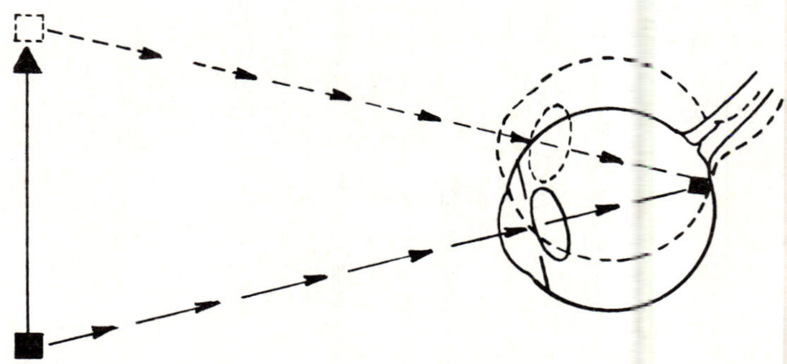

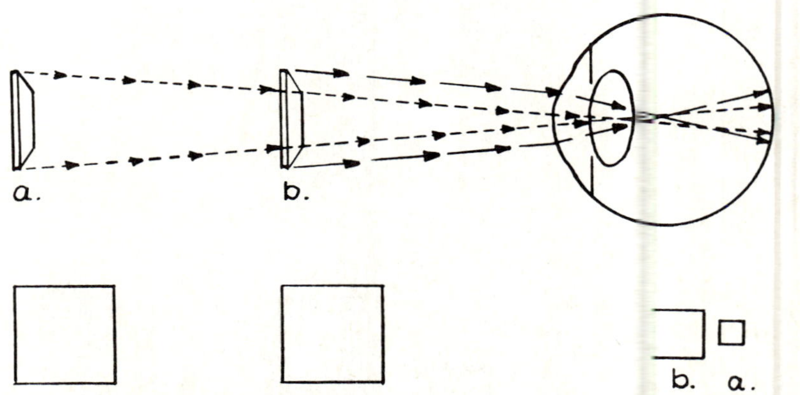

Reality and perception

Retinal images

ACCOMMODATION AND CONVERGENCE

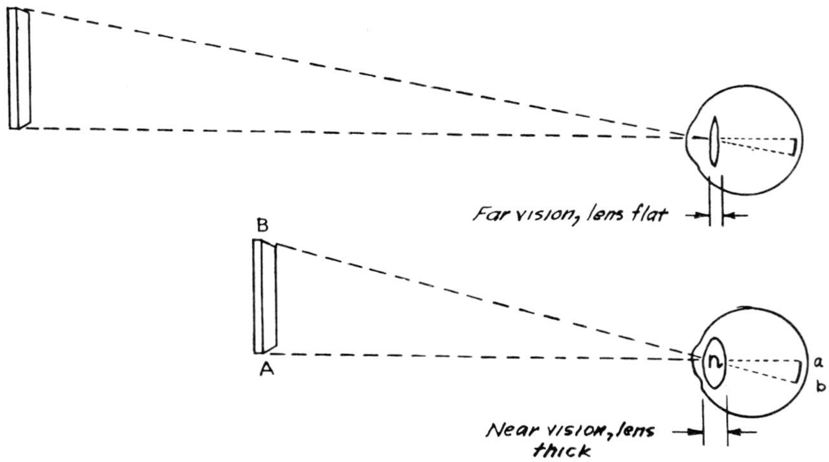

The size of a retinal image is calculated by using the equation $AB/ab = An/an$, when the size of the object and its viewing distance are known.

Amplitude of Accommodation

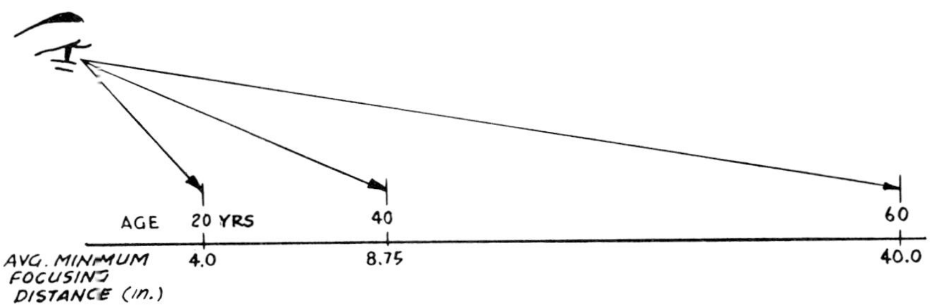

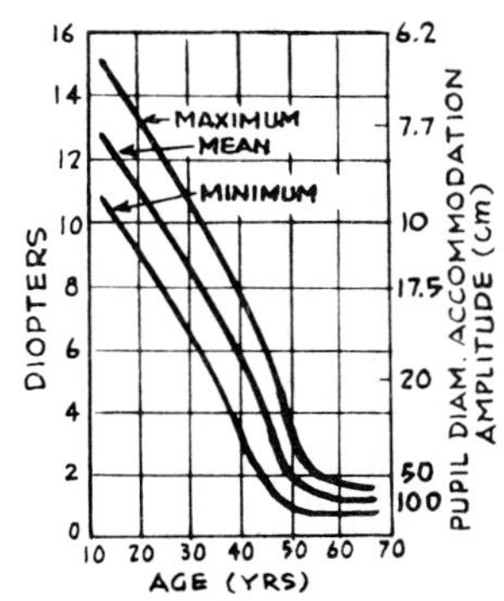

Accommodation versus age.

Although a child can focus on an object that is as close as 2.4 in (6.1 cm), an adult can focus no closer than 6 in (15.2 cm). Hardening of the lens with age reduces the focusing capacity of older persons' eyes.

Convergence

Convergence acts like a range finder, providing depth perception.

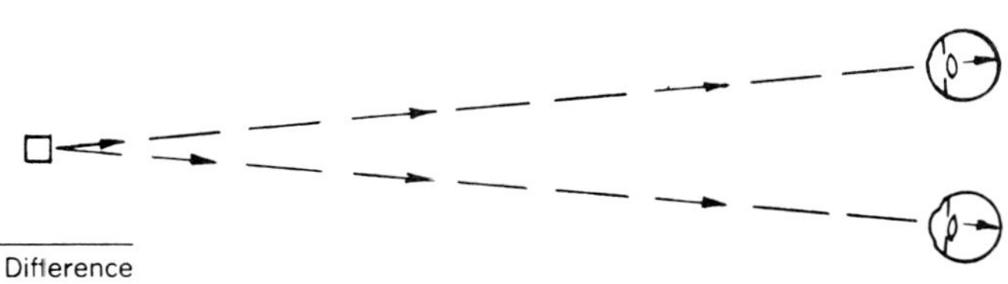

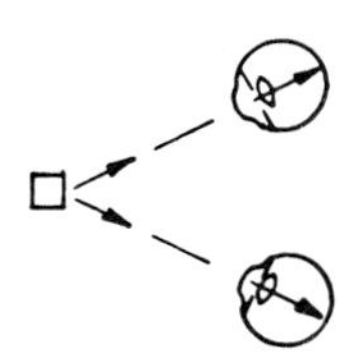

Diameter of the Pupil in Millimeters

Age	In Daylight	At Night	Difference
20	4.7	8.0	3.3
30	4.3	7.0	2.7
40	3.9	6.0	2.1
50	3.5	5.0	1.5
60	3.1	4.1	1.0
70	2.7	3.2	0.5
80	2.3	2.5	0.2

Accommodation in Relation to Age

The nearest points at which an individual can see objects distinctly when using maximum accommodation are shown in the accompanying table; i.e., the object must be as close as, or closer than, that corresponding to the specified accommodative power (in diopters).

Age	Minimum for Age	Normal for Age
17	8.8	11.8
20	8.1	11.1
25	6.9	9.9
30	5.7	8.7
40	2.8	5.8
45	0.6	3.6

Source: Air Force Manual 160-1.

EXTERNAL AND INTERNAL CUES TO DEPTH AND DISTANCE

External Cues	
Linear perspective	Apparent convergence of parallel lines and related effects
Apparent size	A strong cue to distance of objects of known size and texture
Motion parallax	Relative angular motion as either head or objects move
Interposition	Nearer objects eclipse more distant ones
Aerial perspective	Contrast and color loss due to aerosols; useless in free space
Shading	A cue to three-dimensional form of objects (not to distance)
Apparent intensity	A cue only to distance of effective "point sources"
Internal Cues	
Accommodation	Relatively unimportant
Convergence	Useful limit is about 20 m
Binocular disparity	Most important intrinsic cue to depth and distance

Note: All cues except for the last two can be utilized by a single eye, and by extension, in uniocular optical devices.

RADIANT ENERGY AND THE EYE

Light is visually evaluated radiant energy. As shown in the accompanying illustration, the visible portion of the electromagnetic radiant energy spectrum is extremely narrow—about 380 to 770 nm (1 nm = 1 mu = 10^{-9} m).

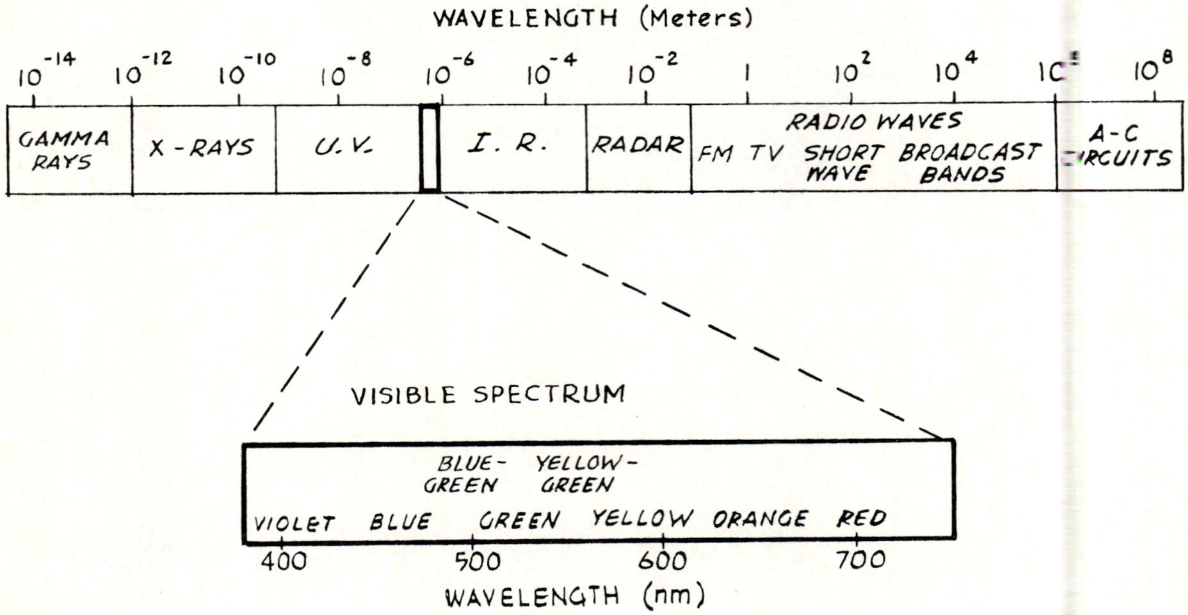

CHARACTERISTICS OF VISION

Parameter	Vision
Sufficient stimulus	Light-radiated electromagnetic energy in the visible spectrum
Spectral range	Wavelengths from 400 to 700/μ (violet to red)
Spectral resolution	120 to 160 steps in wavelength (hue) varying from 1 to 20/μ
Dynamic range	~90 dB (useful range) for rods = 0.00001 to 0.004 mL; cones = 0.004 mL to 10,000 mL
Amplitude resolution $\Delta I/I$	Contrast = $\Delta I/I$ = 0.015
Response rate for successive stimuli	~0.1 s
Reaction time for simple muscular movement	~0.22 s
Best operating range	500 to 600/μ (green-yellow) 10 to 200 fc
Indications for use	1. Spatial orientation required. 2. Spatial scanning or search required. 3. Simultaneous comparisons. 4. Multidimensional material presented. 5. High ambient noise levels.

Intensity Relationships

It takes only one-billionth of a lambert to excite the eye (absolute threshold).

The minimum ratio at which light can be seen is called the "contrast threshold." A light of very low intensity can be seen against a dark background. However, the light must be much brighter to be seen against a bright background, as shown in the accompanying table.

Log of Illuminance Just Visible at the Eye, fc	Log of Background Brightness, mL
Photopic vision:	
− 5	4
− 6.25	2
− 7.60	0
Mesopic vision:	
− 8.3	− 2
− 8.8	− 4
Scotopic vision:	
− 9.6	− 6
− 9.8	Total darkness

Visibility also depends on light target area (visual angle), as shown in the accompanying table.

Just visible brightness, log fL	− 2.2	− 2.7	−3.25	−4.5	−4.75	−5.6	−5.8
Diameter of target, min/arc	0.3	0.6	1.0	6.0	10.0	60.0	100.0

The threshold visibility of a flashing light depends on the total energy in the flash, which is the intensity of the flash multiplied by the duration. (For flashes longer than 0.1 s, visibility depends on intensity alone.)

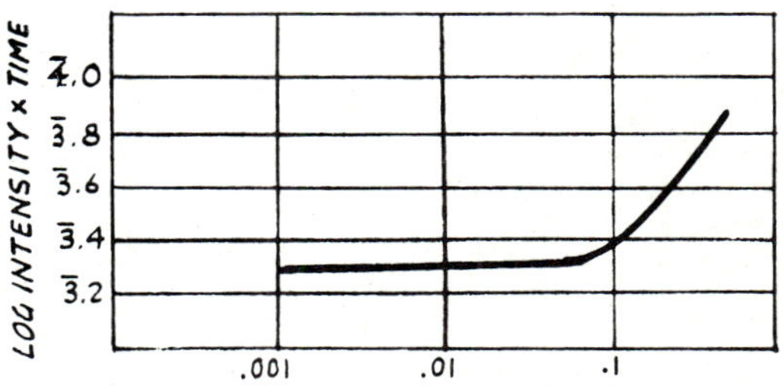

PHOTOPIC AND SCOTOPIC RESPONSE FUNCTIONS

Photopic (foveal) vision occurs under high illumination conditions, where the luminosity function is mediated by the cones. Scotopic (peripheral) vision occurs under low illumination conditions, where the luminosity function is mediated by the rods. A third type of vision (mesopic) occurs when both cones and rods are activated, i.e., under low or intermediate illumination conditions, such as at dusk or dawn.

In the fovea, where visual acuity is at a maximum, vision is mediated entirely by cones. Although cones are distributed throughout the peripheral retina, cone density is considerably less than in the foveal region, and therefore acuity is reduced in the periphery. At the limits of peripheral vision, cone vision may be completely absent, and only the rods mediate. As a result, the eye detects only light. Therefore, a faint light will disappear if one looks directly at it.

It should be noted that while each of the curves is shown at the same luminosity scale in the accompanying graph, this does not mean that the eye has equal peak luminous responses in the photopic and scotopic modes.

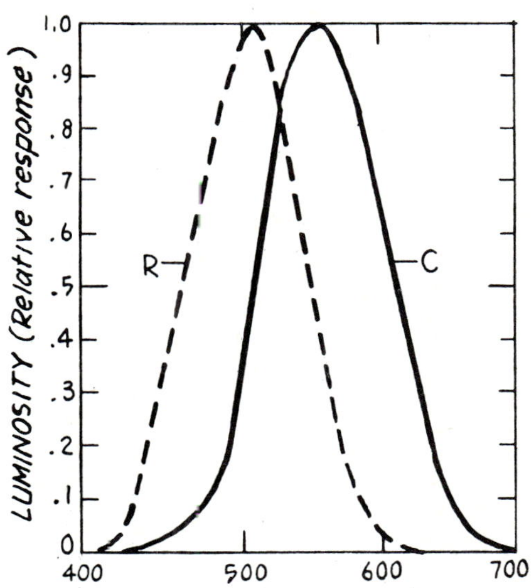

Standard luminosity functions (R—Rod, or photopic vision; C—Cone, or scotopic vision).

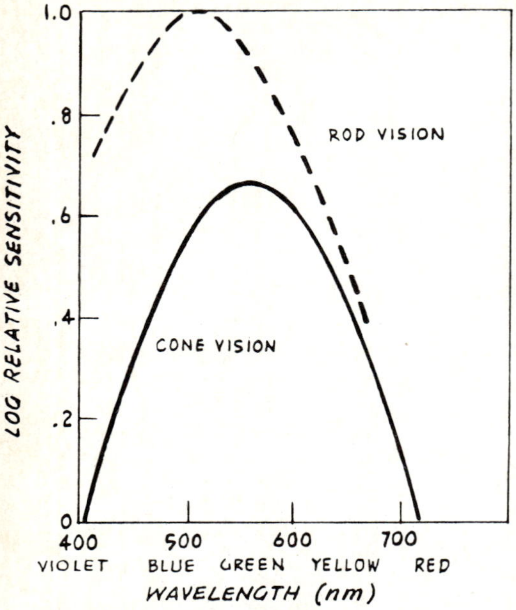

Spectral response of retinal rods and cones, with relative amount of radiant energy for absolute threshold vision as a function of the wavelength of light.

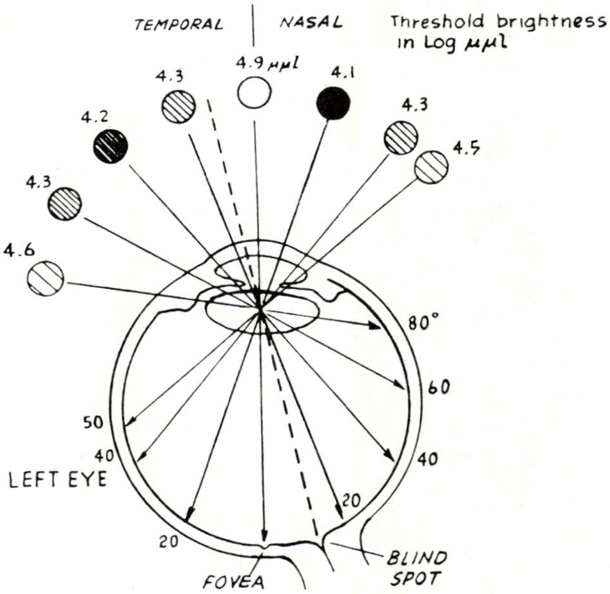

Sensitivity of different parts of the retina at night.

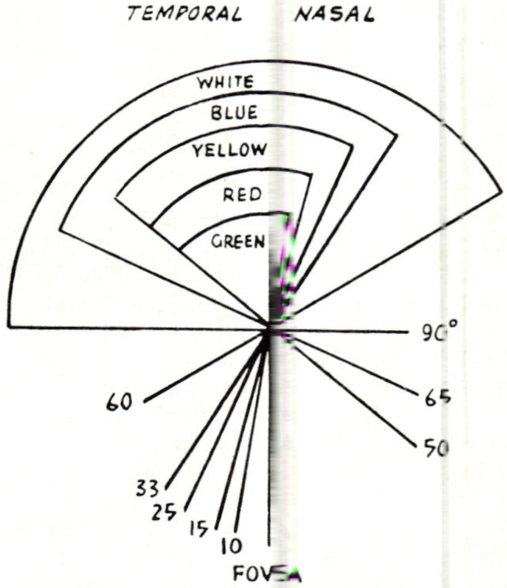

Sensitivity of different parts of the retina in daylight.

VISUAL ACUITY[1]

The term "visual acuity" refers to the resolving power of the eye or its ability to detect fine detail.

There is a marked relationship between visual acuity and the distribution of retinal rods and cones, as illustrated in the first graph.

Since there are more cones in the central area, acuity (photopic vision) is better in this area.

Photopic vision, in which rods are most used, is better farther from the foveal area of the retina.

The relationship between field brightness and minimum perceptible brightness difference (another measure of visual acuity) is shown in the second graph.

Although there is no appreciable change in the sensitivity of the cones (B) used for daylight conditions, there is a considerable change in the sensitivity of the rods (A), which are used for nighttime conditions.

At the lowest intensity of light, the eye can see a line whose thickness subtends a visual angle of 10 min. At very high intensities, the eye can see a line subtending less than 1 s of visual angle. These values assume maximum contrast between the lines and the background.

The relationship between contrast and image size is illustrated by a standard "parallel bar" target under a brightness level of 30 mL. As shown in the accompanying table, as contrast is increased, minimum size and spacing between parallel bars can be decreased without rendering the separation invisible. However, with decreasing contrast, target size and separation must be increased in order to maintain threshold acuity.

Contrast, %	45	8	5	3	2.8
Visual angle min	1	2	4	10	16

Visual acuity also varies under different spectral illuminants. When the background brightness level is 0.075 fc, the relationships shown in the accompanying table hold true.

Visual acuity, %	52	70	75	68	63
Wavelength (Mμ)	485	520	590	625	665

When higher illumination levels are available, however, the relationship between illuminant color and acuity for black-and-white targets is negligible. Luminance contrast, color contrast, illumination level, and exposure time are more important to acuity than color is.

Moving targets need to be larger as the rate of movement increases in order to provide the necessary acuity to distinguish target detail (see the accompanying table).

Visual acuity, %	100	56	30	19	10
Velocity, °/s	0	50	100	150	200

Acuity deteriorates with age, as shown in the accompanying graph. The designer must remember that older persons' eyes require more light and larger detail.

[1] J. W. Wulfeck et al., *Vision in Military Aviation,* WADC Technical Report 58-399, 1958.

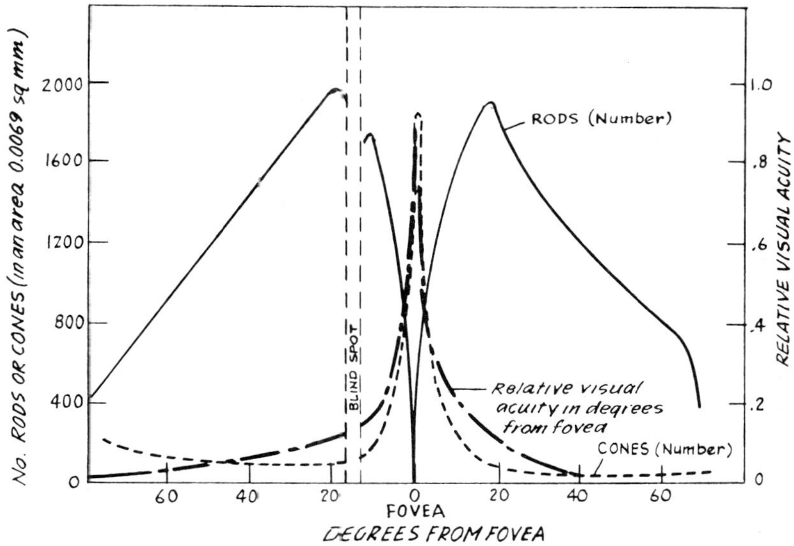

The relationship of visual acuity to the distribution of rods and cones.

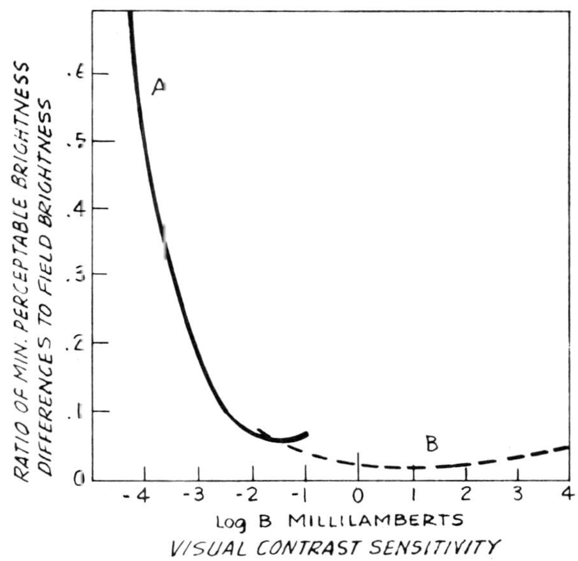

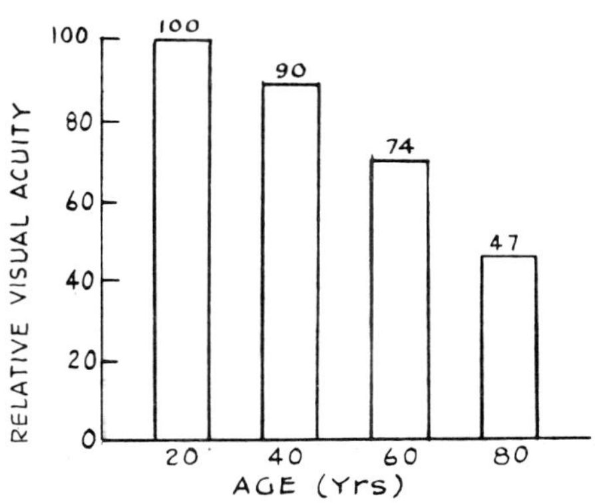

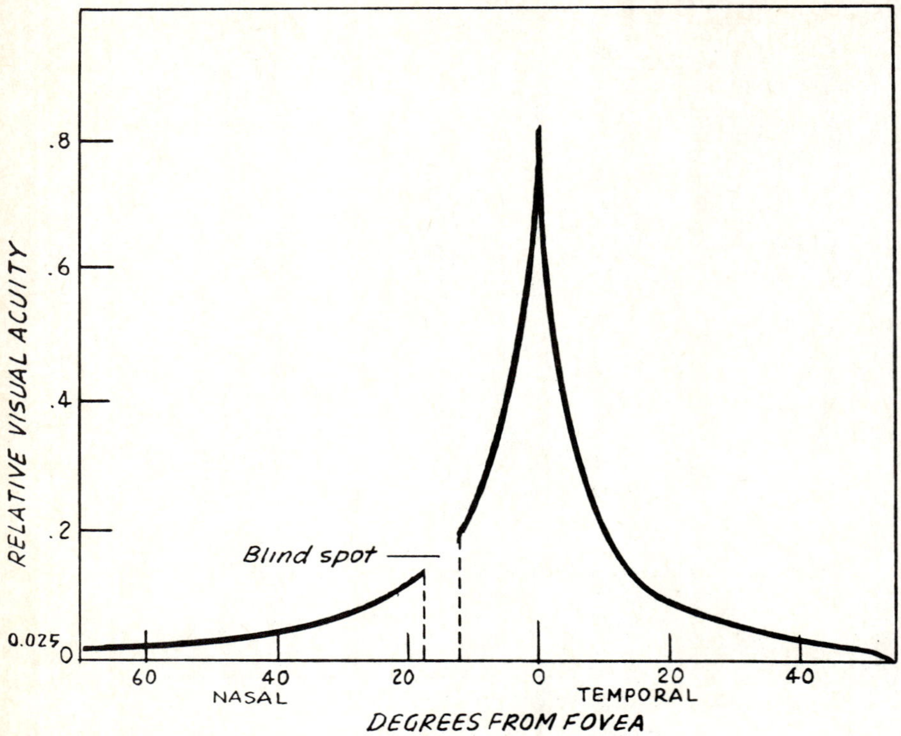

Daylight visual acuity for different parts of the eye.

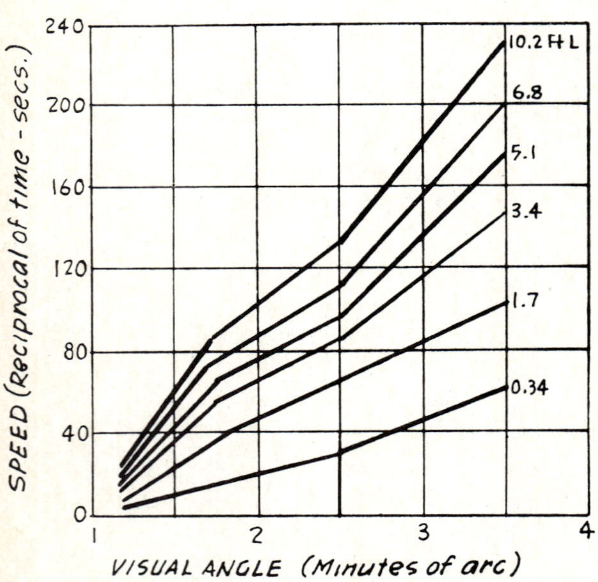

Acuity versus time.

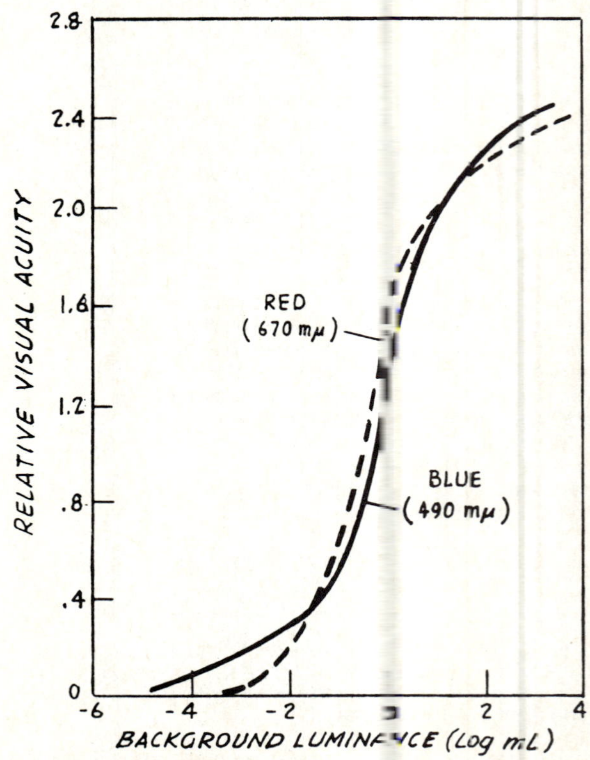

Acuity versus color of illuminant.

RELATIONSHIP BETWEEN CONTRAST, VISUAL ANGLE, AND BRIGHTNESS IN DETERMINING WHETHER WE SEE OR NOT

The accompanying three-dimensional chart illustrates the relationship between three critical variables that determine whether a person sees or does not see an object. When the combination of variables intersects behind the curve, the eye cannot discriminate, whereas any intersecting point in front of the curve provides good seeing conditions.

Good seeing is also affected by the brightness relationships between all elements within the visual field. As the accompanying table shows, it is desirable to maintain different brightness relationships between the primary visual task and immediate and distant visual phenomena.

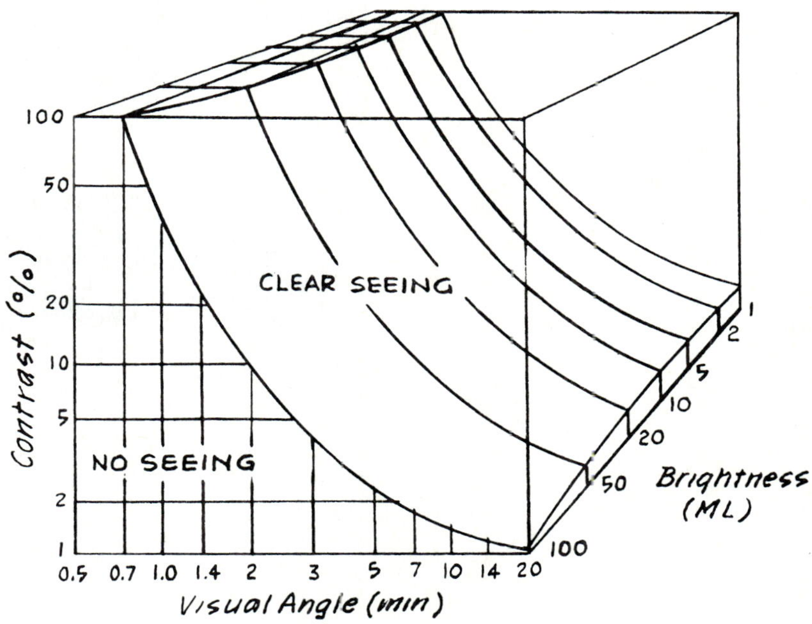

Recommended Brightness Ratios within the Visual Field

Areas	Recommended Maximum Luminance Ratio: Office	Recommended Maximum Luminance Ratio: Industrial
Task and adjacent surroundings	3:1	
Task and adjacent darker surroundings		3:1
Task and adjacent lighter surroundings		1:3
Task and more remote darker surfaces	5:1	10:1
Task and more remote lighter surfaces	1:5	1:10
Luminaires (or windows, etc.) and surfaces adjacent to them		20:1
Anywhere within normal field of view		40:1

VISUAL FIELD

Monocular Field

The general field that can be seen by the single eye is illustrated in the accompanying chart, which shows how the brow and the nose limit the field above and toward the nose. This chart also shows the extent to which some colors are more perceptible than others as a result of the distribution of retinal sensors; i.e., one can perceive blue farther out on the periphery than red. The outer limit denotes that only white light is recognizable.

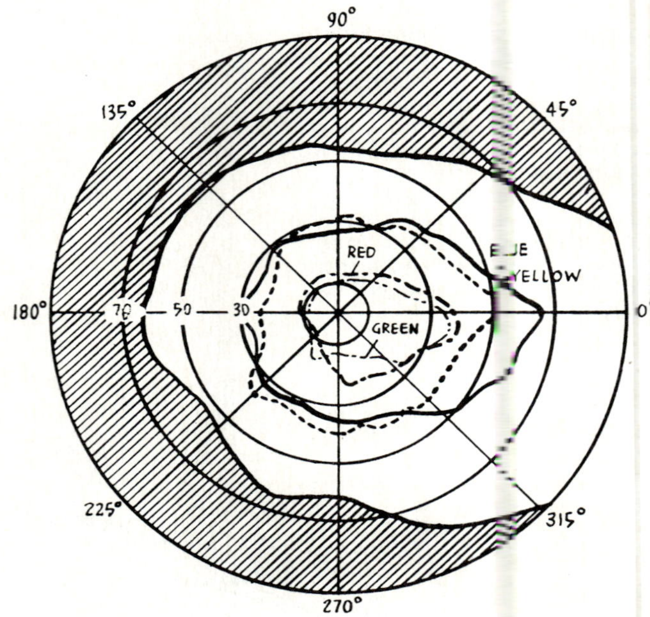

Binocular Field

Overlapping of the two binocular fields is illustrated in the accompanying chart and sketch. Note that the 15° cone about the major visual axis is always recommended as the limit for locating warning lights that must be seen while watching a specific display.

The accompanying sketch illustrates the general limiting variable for peripheral vision.

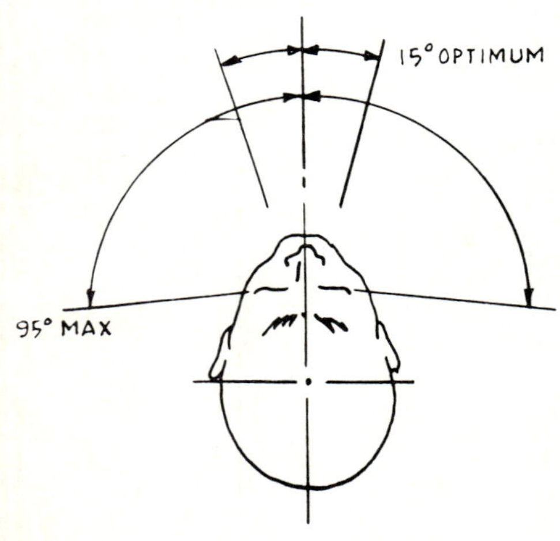

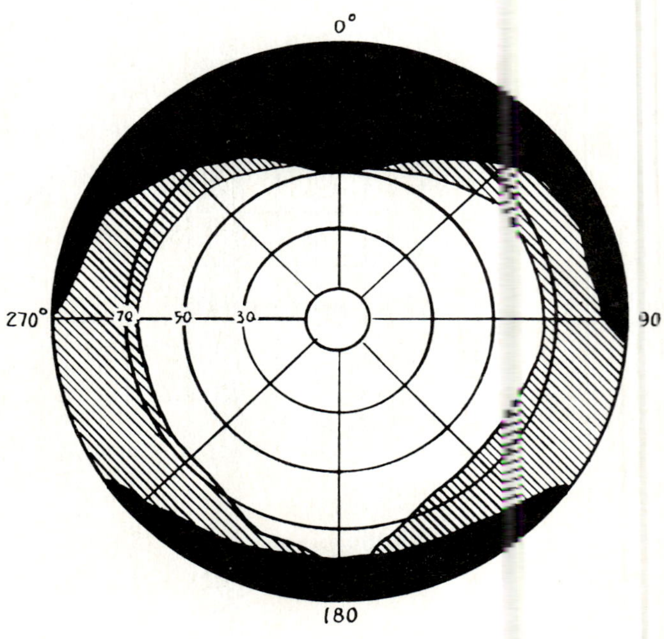

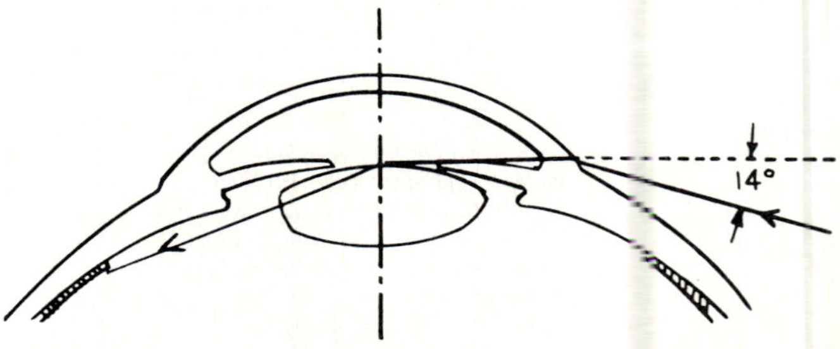

The Limits of the Visual Field under Various Kinds of Restraint

Movement Permitted	Type of Field and Factors Limiting Field	Horizontal Limits: Temporal Ambinocular Field (Each Side)	Horizontal Limits: Nasal Binocular Field (Each Side)	Vertical Limits: Field Angle Up	Vertical Limits: Field Angle Down
Moderate movements of head and eyes assumed as:	Range of fixation	60°		45°	
Eyes: 15° right or left; 15° up or down	Eye deviation (assumed)	15°	15°	15°	15°
	Peripheral field from point of fixation	95°	(45°)	46°	67°
Head 45° right or left; 30° up or down	Net peripheral field from central fixation	110°	60°	61°	82°
	Head rotation (assumed)	45°	45°	30°	30°
	Total peripheral field (from central body line)	155°	105°	91°	112°
Head fixed; Eyes fixed (central position with respect to head)	Field of peripheral vision (central fixation)	95°	60°	46°	67°
Head fixed; Eye maximum deviation	Limits of eye deviation (= range of fixation)	74°	55°	48°	66°
	Peripheral field (from point of fixation)	91°	Approx(5°)	18°	16°
	Total peripheral field (from central head line)	165°	60°	66°	82°
Head maximum movement; Eyes fixed (central with respect to head)	Limits of head motion	60°		45°	
	(= range of fixation)	72°	72°	80°	90°
	Peripheral field (from point of fixation)	95°	60°	46°	67°
	Total peripheral field (from central body line)	167°	132°	126°	157°
Maximum movement of head and eyes	Limits of head motion	72°	72°	80°	90°
	Maximum eye deviation	74°	55°	48°	66°
	Range of fixation (from central body line)	146°	127°	128°	156°
	Peripheral field (from point of fixation)	91°	Approx (5°)	18°	16°
	Total peripheral field (from central body line)	237°	132°	146°	172°

J. W. Wulfeck, et al., *Vision in Military Aviation,* WADC Technical Report 58-399, 1958.

MONOCULAR DAYLIGHT ACUITY RELATIVE TO CENTRAL ACUITY[2]

The relative acuity for one-eye daytime vision in various parts of the visual field is shown in the accompanying chart. Note that the best vision is within the small ring at the center of the field. Under the most favorable conditions, ½ min of arc may be discriminated. The figures on the larger rings illustrate how many times larger an object must be to be seen clearly.

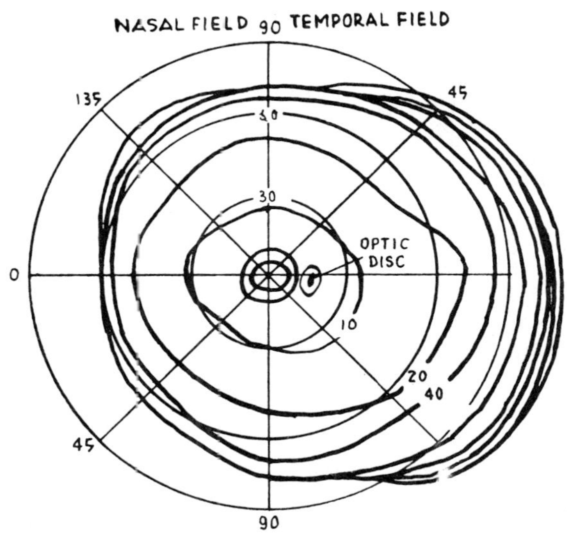

[2]After W. S. Duke-Elder, *Test Book of Ophthalmology,* The C. V. Mosby Company, St. Louis, 1934.

Loss of Peripheral Vision with Increases in Observer Motion[3]

As an observer (e.g., vehicle operator) moves forward within a visual field, his or her peripheral field sensitivity is reduced. It has been estimated that an individual can lose up to 40 percent of side vision when concentrating on a central visual task, as one typically does when driving. Objects at the side of the road tend to become blurred as speed increases and as the operator fixates farther down the road, as illustrated in the accompanying graph.

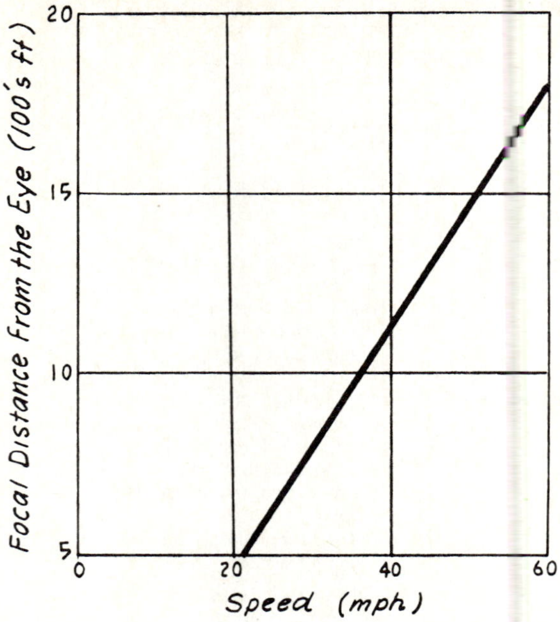

It has been determined that 75°/s is the angular speed at which visual acuity begins to deteriorate. This is called "dynamic acuity." The accompanying table gives angular speeds of objects along the path of travel for various speeds at various distances from the line of travel. Objects as close as 10 ft (3.1 m) are easily seen at walking and running speeds of about 8 mi/h (12.9 km/h), but at speeds of 18 mi/h (29 km/h), an individual begins to lose clear vision of nearby objects, e.g., objects 10 ft (3.1 m) away or closer.

Angular Speeds of Objects along Path of Travel for Various Speeds and Distances

Miles per Hour	Feet per Second	Degrees and Seconds									
100	146	164° 24′	149° 22′	135° 20′	122° 34′	111° 24′	101° 10′	92° 24′	84° 4[illegible]	78° 6′	72° 16′
80	118	160° 46′	142° 34′	126° 6′	111° 4′	99° 26′	89° 2′	80° 16′	72° 5[illegible]	66° 30′	61° 6′
70	102	157° 10′	136° 24′	119° 6′	103° 8′	91° 8′	80° 44′	72° 10′	65° [illegible]	59° 6′	54° 4′
60	88	154° 24′	131° 8′	111° 26′	95° 28′	82° 42′	72° 32′	64° 18′	57° 3[illegible]	52° 8′	47° 30′
50	74	149° 6′	123° 14′	101° 16′	85° 34′	72° 62′	63° 20′	55° 44′	49° 4[illegible]	44° 42′	40° 38′
40	58	141° 18′	110° 50′	88° 4′	71° 14′	60° 14′	51° 36′	45° 2′	39° 5[illegible]	35° 44′	32° 22′
30	44	131° 8′	95° 28′	72° 32′	57° 38′	47° 30′	40° 18′	34° 54′	30° 4[illegible]	27° 30′	24° 50′
20	29	110° 50′	71° 44′	51° 36′	39° 12′	32° 22′	27° 12′	23° 24′	20° 3[illegible]	18° 18′	16° 32′
15	22	95° 28′	57° 38′	40° 18′	30° 46′	24° 50′	20° 48′	17° 52′	15° 4[illegible]	14°	12° 38′
10	15	73° 6′	41° 8′	28° 6′	21° 16′	17° 4′	14° 16′	12° 34′	10° 4[illegible]	9° 32′	8° 36′
8	12	61° 56′	33° 24′	22° 38′	17° 4′	13° 42′	11° 26′	9° 48′	8° 3[illegible]	7° 38′	6° 54′
5	7	38° 36′	19° 12′	13° 20′	10°	8° 2′	6° 42′	5° 50′	5° [illegible]	4° 28′	4° 2′
2	3	17° 4′	8° 36′	5° 4′	4° 18′	3° 26′	2° 52′	2° 28′	2° 1[illegible]	1° 56′	1° 44′
		10	20	30	40	50	60	70	80	90	100

The accompanying table[4] shows acuity ranges for various angular speeds of objects at the side of the road and illustrates variations among groups of observers. It can be seen that visual acuity falls off even at relatively low angular velocities. It should be noted, however, that increased illumination considerably improves dynamic visual acuity. And when vehicle windows are larger, acuity is increased because the eye has a better opportunity to pick up and track the object.

Visual Acuity Ranges for Various Angular Speeds of Roadside Objects

Distance to Object, ft	60 mi/h: Angular Velocity, °/s	60 mi/h: Visual Acuity Gr 1	60 mi/h: Visual Acuity Gr 2	60 mi/h: Visual Acuity Gr 3	30 mi/h: Angular Velocity, °/s	30 mi/h: Visual Acuity Gr 1	30 mi/h: Visual Acuity Gr 2	30 mi/h: Visual Acuity Gr 3
1000	5.0	20/51	20/48	20/38	2.5	20/51	20/48	20/38
500	10.0	20/51	20/48	20/39	5.0	20/51	20/48	20/38
100	47.5	20/64	20/54	20/42	24.75	20/53	20/48	20/29
80	57.5	20/73	20/50	20/45	30.75	20/54	20/49	20/39
60	72.5	20/96	20/70	20/52	40.25	20/59	20/51	20/41
40	95.5	20/154	20/99	20/70	57.5	20/73	20/59	20/45
20	131.0	20/317	20/317	20/121	95.5	20/154	20/99	20/70

[3]P. L. Connolly, *Visual Considerations of Man, the Vehicle and the Highway,* part II, SAE Paper SP-279, 1966.
[4]P. L. Connolly, *Visual Considerations of Man, the Vehicle, and the Highway,* part II, SAE Paper SP-279, 1966.

DARK ADAPTATION

The transition from high- to low-level light requires adaptation time. Adaptation from low to higher light levels is approximately instantaneous, although it can be annoying or even slightly painful.

Although the time required to become dark-adapted varies both among subjects and within an individual eye, the general form of the dark adaptation curve remains somewhat characteristic. Note the transition point at which adaptation is fairly rapid, after which it levels off and then proceeds farther, but at a slower pace.

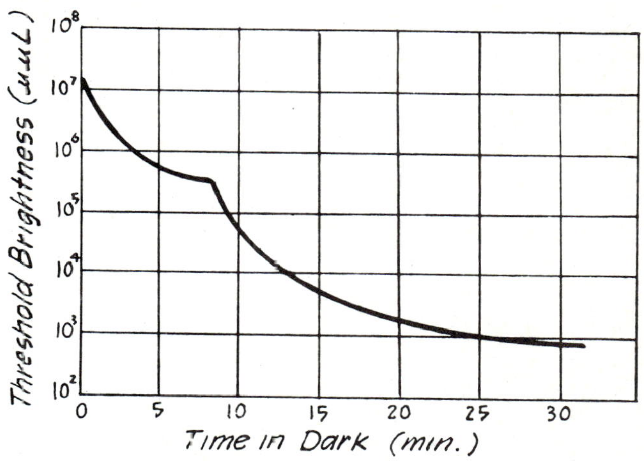

Course of dark adaptation showing intensity just barely discernible as function of time spent in dark.

Dark adaptation patterns vary depending on the color of the light, as illustrated in the accompanying graph.

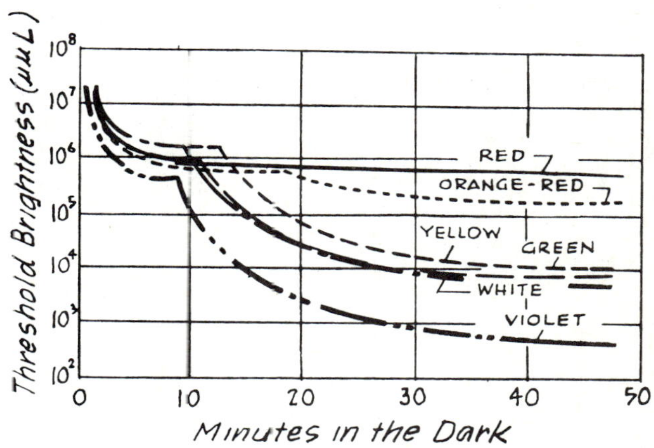

Dark adaptation measured with light of different wavelengths. Although lights were all equated in brightness initially, they are no longer equally bright even at cone threshold. The differences are further exagerated during rod dark adaptation.

The accompanying graph shows dark adaptation as a function of the region of the retina being stimulated.

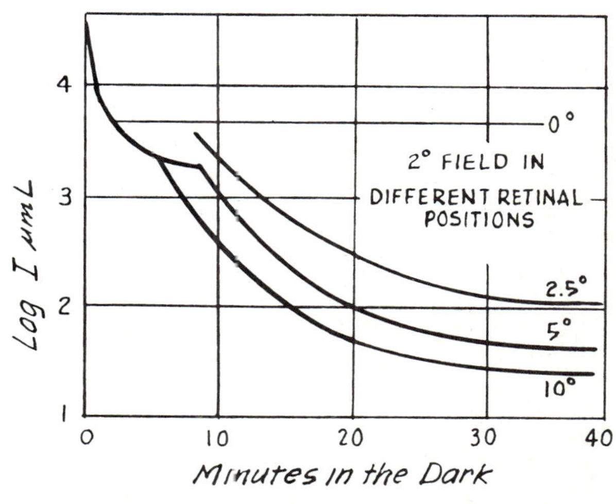

CRITICAL FUSION FREQUENCY (CFF)

CFF is the point at which successive light flashes blend into a continuous light. A flickering light that is on 50 percent of the time and off 50 percent of the time flashes at the rate of 10 Hz; it appears brighter than a steady light.

The relation between CFF and retinal illumination *(I)* with white light for three retinal locations is shown in the accompanying chart. Crossing of the curves is due to the shift from rod to cone vision.

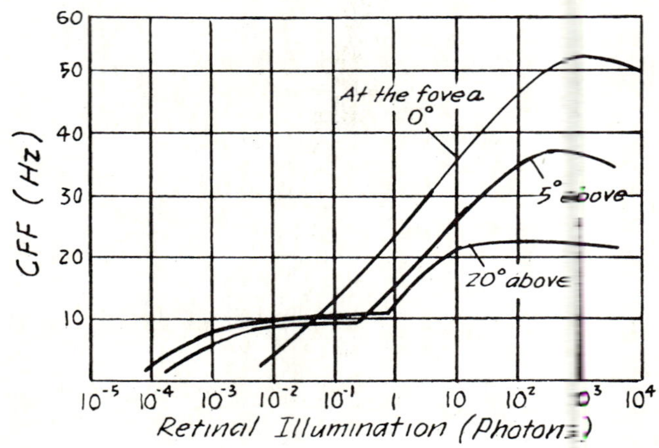

The relation between CFF and various target-area sizes is shown in the accompanying graph. The small areas (2 to 3°) stimulate only the cones, whereas for larger areas (6 and 19°) a double function yields a higher CFF.

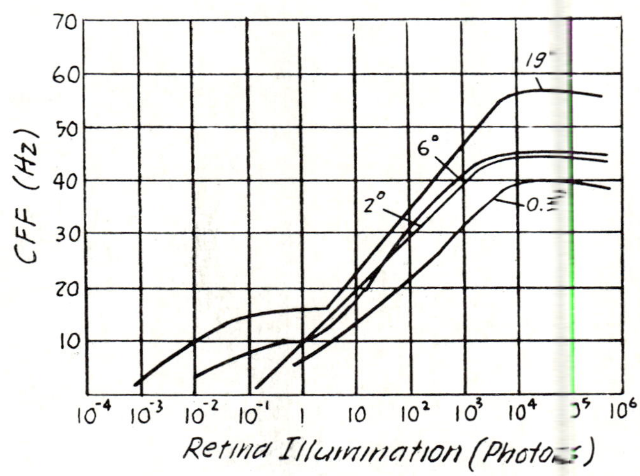

The accompanying graph shows the relation between CFF and retinal illumination for different spectral regions.

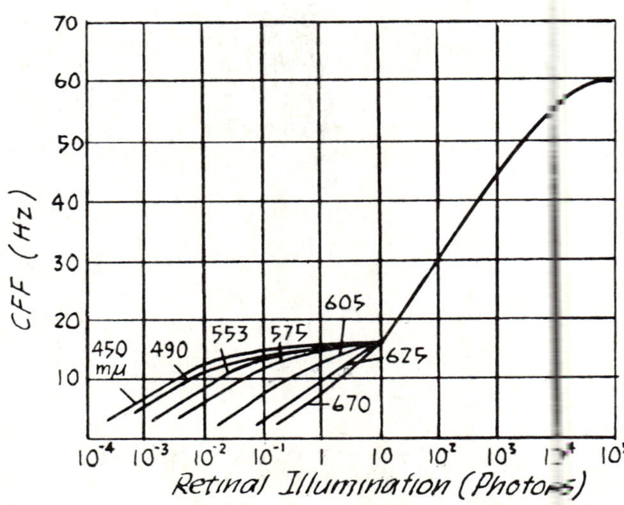

GLARE

Effect of Glare on Performance

Function	Effects
Visibility	Glare of 5 fc located 5° above the line of vision reduces visibility about as much as replacing a 100-fc lighting system with a 1-fc system. An object that can be seen at 20 ft with 5-fc glare at 5° could be seen at 45 ft with no glare. Glare becomes less significant as the brightness of an object is raised, and is greater for a large object seen in low contrast to its background than for a small, high contrast object, such as a printed letter.
Size threshold	Glare greatly increases size threshold, particularly for objects of low contrast. The effect of glare diminishes as contrast and illumination of surrounds are increased, and thus is more easily controlled at high levels of illumination. Threshold size for an object of 5% contrast under 1 fL with a 5-fc glare 5° above line of vision = 30.0 min visual angle; for an object at 100 fL with no glare = 0.63 min.
Contrast threshold	Glare greatly increases the contrast threshold. The degree of increase depends in complex fashion upon the brightness of task background, the glare illumination produced by each glare source at the eye on a plane perpendicular to the line of sight to the task, and the angle of separation between the glare source and the line of sight.
Muscular tension	Glare significantly increases muscular tension—in some cases up to 30%. Although visual tasks may be carried out under severe handicaps with no loss in immediate accuracy, a decrease in efficiency is apparent with a considerable increase in nervous muscular tension. Subjects not compensating for glare effects may show a considerable drop in performance but no increase in muscular tension.
Vision through eyeglasses	The number of persons wearing glasses is sufficiently large and the glare effect sufficiently severe so that reflections from eyeglasses should be considered in placing light sources. When the light source is behind the head, no glare from eyeglasses will exist if the light source is 30° or more above the line of vision, 40° or more below the line of vision, or at an angle of 15° or more with the axis of symmetry of the head.

COLOR PERCEPTION

We can only generalize with respect to what each individual perceives in the case of a particular color. The accompanying chromaticity diagram has been created to identify particular wavelength characteristics that normally lead to an individual's identification of colors. The points A, B, C, and E represent standardized light sources under which colors are designated. The dominant wavelength and purity of a color are determined by reference to a particular illuminant. This diagram is based on the concept that the eye actually has three separate mechanisms of color response and that any wavelength stimulates all these mechanisms, not just one, in varying degrees.

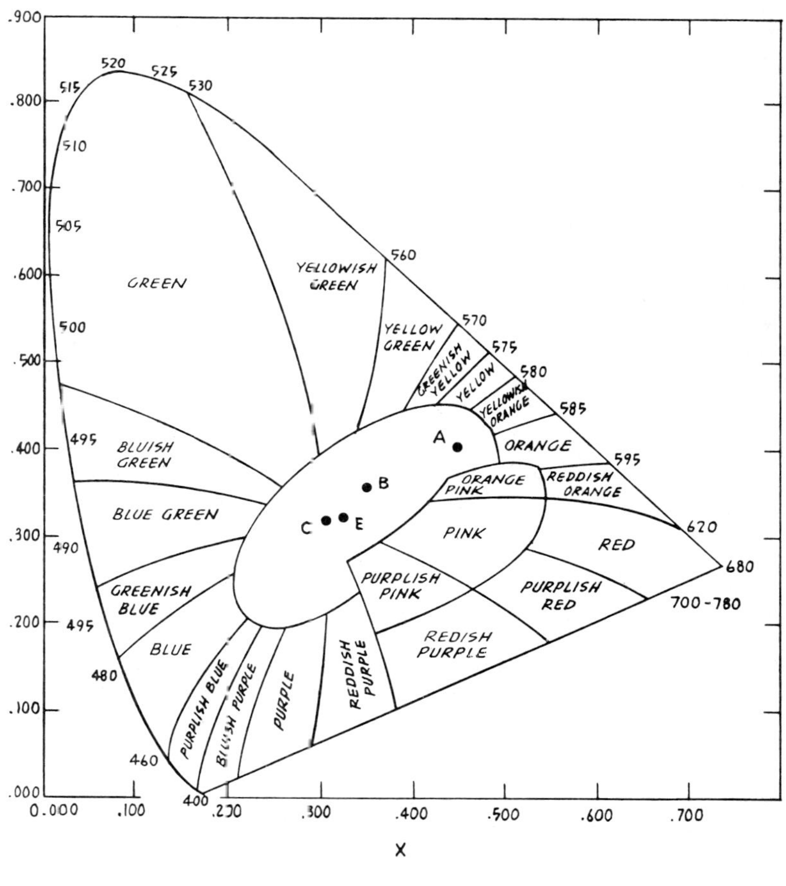

Color Identification Capability

People use the term "color" to identify only one aspect of color, i.e., hue. As the accompanying color model illustrates, color consists of three attributes—hue, brightness, and saturation. A fully saturated *red* hue can be modified considerably and still be perceived as red by color-normal observers; i.e., when one moves toward the center of the model, the so-called red color becomes less saturated (less intense), and when one moves toward the upper end of the model, the color becomes pale. In addition, as one moves around the circumference of the model, a given color tends to blend with its neighboring hue; e.g., red moving toward yellow first appears orange-red, then orange, then reddish yellow, and finally yellow.

Although some observers can discriminate 150 or more hues under proper viewing conditions, the average person can accurately and reliably verbalize only about eight or nine hues. The accompanying table and nomogram[5] indicate probable color discrimination limits.

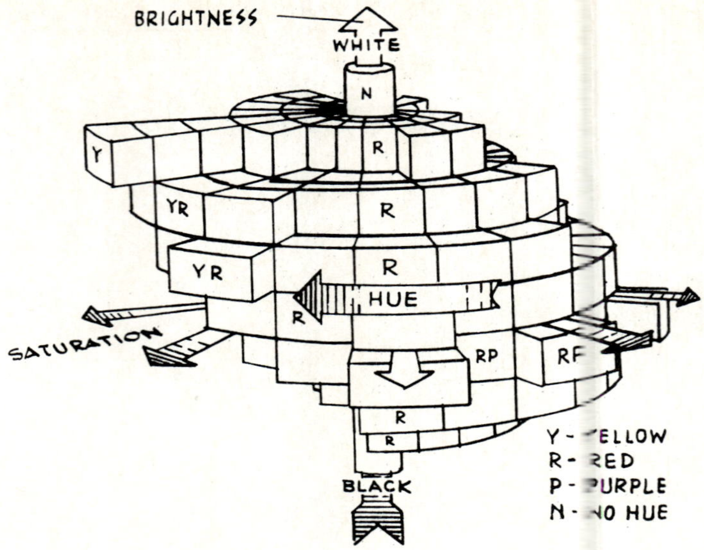

Nine Equally Discriminable Surface Colors

Hues	Code Number	Munsell Book Number	Excitation Purity	Dominant Wavelength
1	1.5	3R	37.2	629
2	3	9R	65.8	596
3	5.5	9YR	81.8	582
4	8.5	1GY	76.0	571
5	11.5	3G	27.5	538
6	15	7BG	35.0	491
7	18	9B	56.5	481
8	20.5	9PB	52.7	460
9	24	3RP	36.5	510

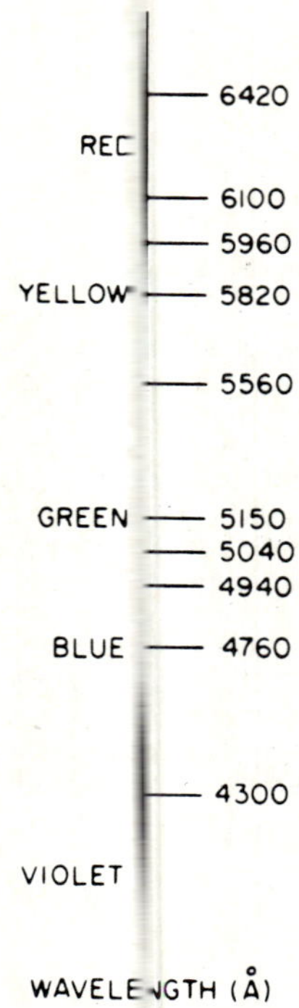

Since many people have some level of color deficiency, it is desirable to select colors that both color-normal and color-deficient individuals can safely identify. The accompanying table defines such colors.

Surface Colors for Color-Normal and Color-blind People

Color	Spec. No.*	Color	Spec. No.*
Red	1110	Gray	1625
Orange	1210	Buff	1745
Yellow	1310	White	1755
Blue	10B 7/6	Black	1770
Purple	2715		

*From Fed. Spec. II-C-595 except for blue, which is from Munsell (1942).

[5] C. T. Morgan et al. (eds.), *Human Engineering Guide to Equipment Design*, McGraw-Hill Book Company, New York, 1963, p. 83.

What People Do Perceive versus What They Can Perceive

Surface color recognition depends on three distinct factors: the color of the light source, the color of the reflecting surface or surfaces, and the state of the observer's visual system. Even though the object—say, a red box—may be a single basic color, this color can be modified both by the color of the light falling directly on it and/or by the color of the light reflected by nearby surfaces. And, of course, the color of the top surface (which receives the full impact of the light source) will not appear the same as the color of the side of the box that is in shadow or the side that is receiving bounce light from another surface (which may be of a different color). Fortunately, there is usually enough color constancy so that these variations are still accounted for, and thus the observer compensates and calls the box "red."

The paler a color is, the more easily it is influenced by the color of the light source and by the color of the light being reflected by nearby surfaces. In addition, colors are greatly influenced by the level of illumination and by the inherent reflectivity characteristics of the surface being viewed. For example, in a dim light, colors tend to take on a grayish and less vital appearance. A highly polished surface may reflect high-intensity illumination so effectively that the basic hue is indistinguishable.

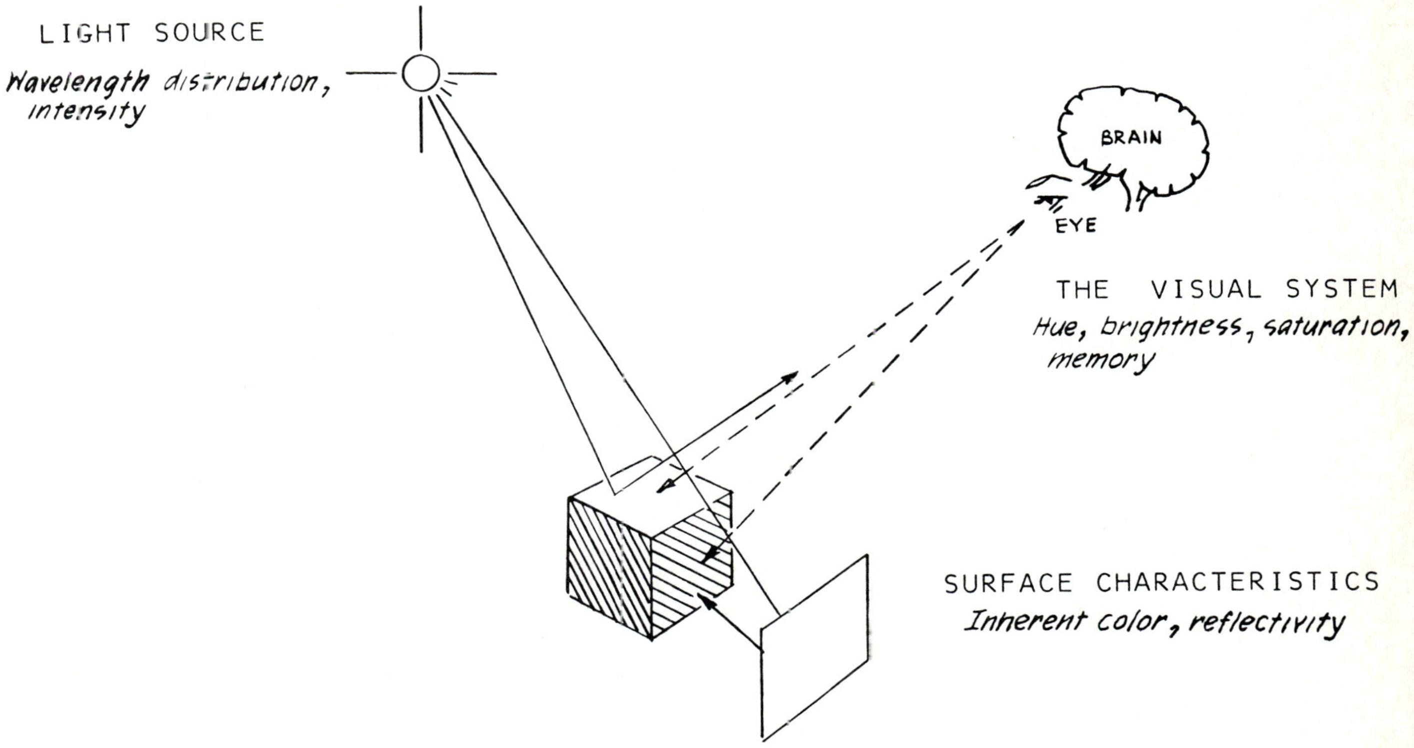

Effect of Monochromatic Light on Colored Objects

If the wavelength distribution of the light source is modified so that certain wavelengths are missing or considerably reduced (red light, yellow light, green light, etc.) the inherent color characteristics of a surface reflect a different color impression. Thus, as indicated in the accompanying table, a yellow object will appear to be red-orange when viewed under a red light.

Effect of Standard Luminaires[6]

The accompanying tables give a general description of the effects that common fluorescent and mercury or filament luminaires have on colored surfaces. Both the lighting level and the color of the light affect the appearance of colored surfaces. Filament lamps and warm fluorescent lamps, which are deficient in blue, emphasize the redness of a surface color and thus accent warm hues.

Effect of Colored Light on Colored Objects

Object Color	Red Light	Blue Light	Green Light	Yellow Light
White	Light pink	Very light blue	Very light green	Very light yellow
Black	Reddish black	Blue black	Greenish black	Orange black
Red	Brilliant red	Dark bluish red	Yellowish red	Bright red
Light blue	Reddish blue	Bright blue	Greenish blue	Light reddish blue
Dark blue	Dark reddish purple	Brilliant blue	Dark greenish blue	Light reddish purple
Green	Olive green	Green blue	Brilliant green	Yellow green
Yellow	Red orange	Light reddish brown	Light greenish yellow	Brilliant light orange
Brown	Brown red	Bluish brown	Dark olive brown	Brownish orange

Color Effects of White Fluorescent Lamps

	Cool White	Deluxe Cool White	Warm White	Deluxe Warm White	Daylight	White	Soft White—Natural
Lamp appearance; effect on neutral surfaces	White	White	Yellowish white	Yellowish white	Bluish white	Pale yellowish white	Pinkish white
Effect on "atmosphere"	Neutral to moderately cool	Neutral to moderately cool	Warm	Warm	Very cool	Moderately warm	Warm, pinkish
Colors strengthened	Orange, yellow, blue	All nearly equal	Orange, yellow	Red, orange, yellow, green	Green, blue	Orange, yellow	Red, orange
Colors grayed	Red	None appreciably	Red, green, blue	Blue	Red, orange	Red, green, blue	Green, blue
Remarks	Blends with natural daylight	Best overall color rendition; simulates natural daylight	Blends with incandescent light	Excellent color rendition; simulates incandescent light	Usually replaceable with CW	Usually replaceable with CW or WW	Usually replaceable with CWX or WWX

[6] A. O. Halse, *The Use of Color in Interiors*, McGraw-Hill Book Company, New York, 1968, pp. 41, 42.

Color Effects of Mercury and Filament Lamps

	Mercury	White Mercury	Color-Improved Mercury	Deluxe White Mercury	Filament
Lamp appearance; effect on neutral surfaces	Greenish blue white	Greenish white	Yellowish white	White	Yellowish white
Effect on "atmosphere"	Very cool, greenish	Moderately cool, greenish	Warm, yellowish	Moderately cool	Warm
Colors strengthened	Yellow, green, blue	Yellow, green, blue	Yellow, green	Orange, yellow, blue	Red, orange, yellow
Colors grayed	Red, orange	Red, orange	Blue	Green	Blue
Remarks	Poor overall color rendering		Color rendering often acceptable, but not equal to any white fluorescent	Color rendering good; compares favorably with CWX fluorescent	Excellent color rendering

Classification of Color Vision

Type of Color Vision	Incidence (Men),* %	Number of Color Receptors†	Spectral Colors Confused	Colors Probably Seen
Trichromatic:				
Normal		3		
Anomalous:				
Deuteranomalous	4.69	3	Some reduction in wavelength discrimination	Reds look darker than normal
Protanomalous	0.89	3	Some reduction in wavelength discrimination	Reds look darker than normal
Tritanomalous	Not known			
Dichromatic:				
Deuteranope	1.42	2	Red, yellow, and green	Yellow, white, and blue
Protanope	1.21	2	Red, and yellow, and green	Green (?), white, and blue
Tritanope	0.008	2	Blue and green	Reds look darker than normal
Monochromatic	0.001	1	All	White and black

* Average of several surveys.
† Assuming trichromatic color vision theory.

Average Reaction Times for Each Class of Color Vision

Subjects	Reaction Time, s
Normals	0.497
Deuteranomals	0.788
Protanomals	0.708
Deuteranopes	0.855
Protanopes	0.984
All abnormals	0.945

Psychological Response to Color

Although color researchers have not always been able to quantify the precise effects of various colors or light levels on humans, their research and experience seem to indicate that certain colors and light conditions often elicit typical repeatable reactions. Illumination level and quality do contribute to impressions of compartment size, to an atmosphere of warmth or coolness, or generate moods or feelings of feeling repose, pleasantness, excitement, boredom, or depression. Color can have a considerable bearing on whether food looks appetizing and on the appearance of one's skin.

The following four basic psychological response characteristics are perhaps most important to the development of appropriate color and lighting schemes:

1. Certain colors make a space appear larger than it actually is, while others cause a space to "close in" on the observer.
2. Certain colors cause a space to seem warm, while others make it seem cool.
3. Some colors appear to have a definite effect on the mood of the observer; i.e., some colors may be stimulating, while others are quieting.
4. Some colors seem to clash with each other and therefore produce a feeling of irritation to observers who are especially sensitive to color incompatibilities.

Color in Relation to Size of Space

Average observers perceive dark and/or saturated colors as advancing toward them, and light and/or desaturated colors as receding away from them. If we wish to make a room appear less crowded, we should use light and unsaturated colors. Any structure or piece of furniture that is dark in color will make a space appear crowded. The level of illumination contributes to this appearance of being crowded in that a dimly lit space generally appears smaller than one that has a higher, more evenly distributed level of light. Therefore, with few exceptions, good illumination and lighter colors are desirable.

Color in Relation to Temperature

So-called cool colors (e.g., blues and greens) make a space seem cooler, while so-called warm colors (such as reds, browns, and oranges) make a space seem warmer. Although little quantitative evidence of any direct correlation between cool or warm colors and actual physiological temperature variation exists, cool colors usually are preferred for facilities located in hot climates, and warmer colors are preferred for facilities located in cold climates. However, since many colors considered cool or warm also have other mood-generating properties, it is important not to overemphasize these temperature-related characteristics. Thus a warm color may be more important in terms of its ability to generate a feeling of comfort or relaxation, especially if a compartment is well air-conditioned and therefore does not need "cooling off" by means of color.

Color in Relation to Mood

Most individuals tend to associate various "artificial" colors with their personal experiences with nature's own color schemes. The accompanying table provides a fairly complete list of known color-mood associations as well as other psychological responses in terms of food, odors, etc. Perhaps the most important consideration in selecting a color scheme is avoidance of traditional institutional gray and white. A variety of colors helps enliven the environment and provide relief from the boredom of neutral colors.

Typical Colors and Their Effect on Humans

Description	Response
Pale pink	Slightly warm feeling, pleasing when associated with odors.
Pink	Warm, soft, pleasing when associated with odors and taste.
Reddish-pink	Warm feeling, slightly stimulating, pleasing association with taste.
Scarlet	Very warm feeling, stimulating and exciting.
RED	Warm feeling, stimulating, generally exciting, but calm to the extent of indicating a protective quality, pleasing to the appetite, associated with danger.
Reddish-orange	Warm, stimulating, exciting, cheerful, pleasing association with taste.
ORANGE	Warm, stimulating, exciting, cheerful, pleasing association with taste.

Any of the above reflect light that is flattering to skin and complexion, thus making people appear healthy and normal (as opposed to cool colors which create an appearance of being pale). Scarlet through ORANGE colors advance slightly toward the observer.

Description	Response
Yellowish-orange	Warm, somewhat exciting, cheerful, express a feeling of comfort, associated with pleasing taste.

(continued)

Typical Colors and Their Effect on Humans ***(continued)***

Description	Response
YELLOW	Warm, somewhat exciting, cheerful, comfortable, associated with pleasing taste.
Pale yellow	Warm, cheerful, associated with pleasing odor and taste, softness and comfort.
Greenish-yellow	May be associated with feeling of slight warmth or coolness depending on other colors used; not associated with good taste.
Yellowish-green	May be associated with feeling of slight warmth or coolness depending on other colors used; calming, somewhat neutral, may or may not be associated with good taste.
GREEN	Generally cool, slightly cheerful, comfortable, calming, associated with pleasing refreshing odor.
Bluish-green	Cool, calming, associated with good taste
BLUE	Cool, comfort, protective, calming, although may be slightly depressing if other colors are dark, associated with bad taste.
GREEN through BLUE will advance toward observer if dark shades are used.	
Pale blue	Cool, soft, calming, tends to neutralize if other colors are pale, reflected light makes skin appear pale.
Lavender	Slightly cool, calming, soft, associated with pleasing odor, but with bad taste.
Violet	Slightly warm, calming, associated with bad taste.
Royal blue	Rich, substantial, may be slightly depressing if used with other dark colors, associated with bad taste.
Purple	Rich, protective, calming, may be depressing, associated with pleasing odor, but bad taste.
Violet through Purple will advance toward observer, creates a feeling of heaviness.	
Hot pink (a yellowish cast)	Very warm, stimulating, exciting, cheerful
Rose (a bluish cast in pink range)	Neutral relative to warmth, comfortable, calm, associated with pleasing odor.
Fluorescent orange	Warm, stimulating, exciting, cheerful, extremely conspicuous.
Fluorescent red	Warm, stimulating, exciting, cheerful, extremely conspicuous.
Fluorescent yellow	Warm, stimulating, slightly irritating but cheerful.
Chartreuse	Slightly warm, cheerful, associated with bad taste.
Olive	Warm, comfortable, slightly depressing, associated with bad taste.
Cream	Slightly warm, comfortable, calming, clean, reflected light enhances skin tone.
Buff	Warm, comfortable, calm, soft; good blend with other colors.
Tan	Warm, very comfortable; good blend with other colors.
Reddish-brown	Warm, comfortable, cheerful, slightly stimulating.
Brown	Warm, comfortable, rich, substantial, protective, may be slightly depressing.
Fluorescent Orange and Red, Olive, Reddish-brown and Browns advance toward observer.	
WHITE	Neutral, sterile, clean, fresh, stark, crisp; may appear hard or soft depending on lighting color; may appear harsh, glaring.
Off-white	Neutral, clean, fresh.
Light gray	Neutral, clean, fresh, calming, soft, comfortable; critical that the tint is compatible with other colors (e.g., bluish, pinkish, yellowish, etc.)
GRAY	Neutral, comfortable, calming, slightly hard; critical that the tint is compatible with other colors used (e.g., bluish, pinkish, yellowish, etc.)
Dark gray	Neutral, comfortable, may be depressing; critical that the tint is compatible with other colors used (e.g., bluish, pinkish, yellowish, etc.); substantial, heavy, advances toward the observer.
Flat black	Solid, heavy, comfortable, generally neutral, advances toward the observer, gives the impression of being dirty, vagueness, recedes.
Deep black	Protective, depressing, heavy, substantial advances toward observer in small amounts, but may recede in large amounts (e.g., painting a ceiling black).
Gold/brass	Rich, comfortable, warm (tint important for compatibility with other colors); slightly advancing.
Silver/aluminum	Neutral, cold, hard, clean, stiff and uncomfortable, sterile, lifeless, recedes.
Light wood grain	Warm, comfortable, quiet.
Dark wood grain	Warm, comfortable, quiet, protective, slightly depressing.
Light leather grain	Warm, comfortable, cheerful, soft.
Dark leather grain	Warm, comfortable, protective, soft, slightly depressing, advancing.

Source: U.S. Navy Shipboard Color Coordination Guidance Manual, NAVSEA 0929-002-7010 (prepared by Man Factors, Inc., 1975).

PSYCHOLOGICAL RESPONSE TO AMOUNT AND COLOR OF LIGHT

A certain amount of confusion has been generated as a result of conflicting specifications for illumination for good seeing versus desired psychological response. The designer of interiors, in particular, must accommodate consumers in terms of both aspects; i.e., good seeing is essential to ensure adequate visual task performance, but it is also necessary to create a lighting environment that produces a desired psychological response. For example, although a high level of illumination may not be necessary in order for the occupants of a space to perform a visual task, the illumination level may have to be high so that they will "feel" that there is adequate light.

An important aspect of the lighting of interior spaces is the use of interior surfaces that provide adequate reflectance, as indicated in the accompanying table.

Suggested Reflectance Ranges for Interiors

Interior Surface or Element	Percentage Reflectance Desired
Ceilings	60–90%. Surface should be white, or a very pale tint that is compatible with the hue of the interior wall. A minimum of [illegible]0% is required for effective visual performance of indirect lighting methods.
Walls	50–85%. Higher reflectance is required above standing waist height when the wall consists of two different reflectance values (i.e., a lower "dado" is used to reduce the expected soiling of lower wall surfaces).
Windows or glass walls	Fabric or other drapery provisions should maintain an average reflectance of 15–45% (or more).
Furniture	30–40%. It is important to use higher reflectance on the tops of desks or tables to avoid extreme brightness contrast between the top and reading materials. Matte finishes are required to prevent glare.
Floors	15–35%. Middle to high values are preferred in order to ensure visual detection of possible obstructions on the floor that might cause tripping.

Although little quantitative evidence exists regarding the psychological effect on general alertness of light levels, the accompanying table indicates suggested requirements.

Alertness-Level Requirement	Average Light Level, fL
Maximum mental alertness required for highly complex mental task performance	50 and above
Medium mental alertness required for routine manual tasks, leisure reading, and/or stimulating social activity	40
Minimum mental alertness required for nondemanding social intercourse and/or perceptual-motor performance (dining, dressing, personal hygiene, etc.)	30
Rest, mental alertness for minimal interaction with other people (e.g., as in a bar or private dining environment)	15
Sleep	Below 3

The color of the light source also affects an individual's interpretation of brightness or intensity (see the accompanying graph). If the level is too high (A), surface colors seem to be faded and unnatural; if the level is too low, the colors appear dim and cold (C). Illuminants falling within the area (B) appear normal. The graph suggests that warmer light is more acceptable to most people when the brightness level is low.

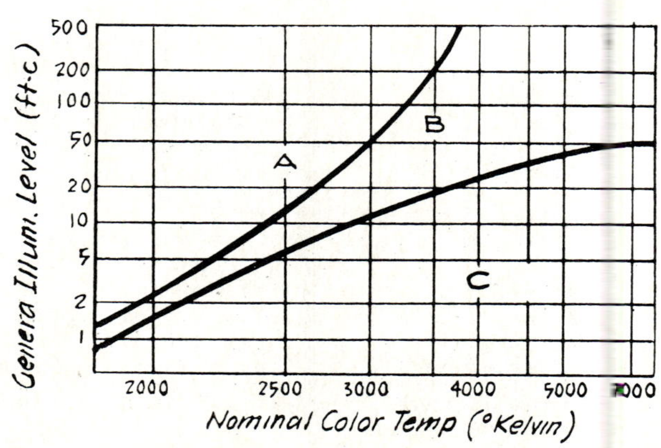

AUDITORY SYSTEM CHARACTERISTICS[1]

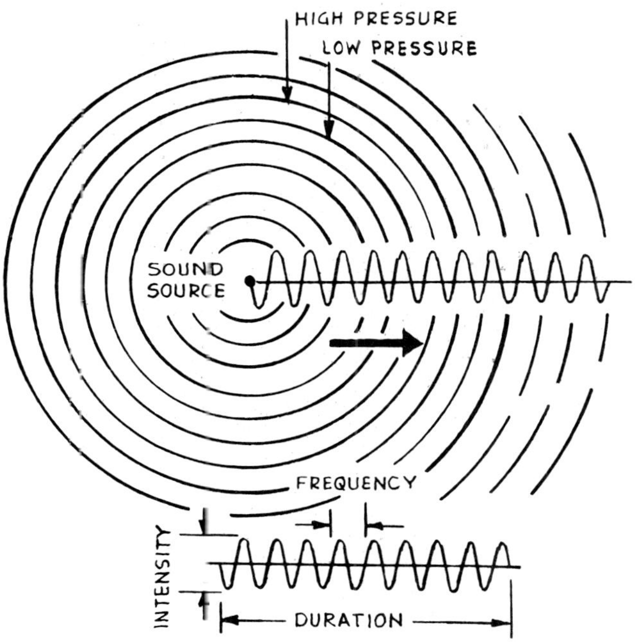

PURE TONE

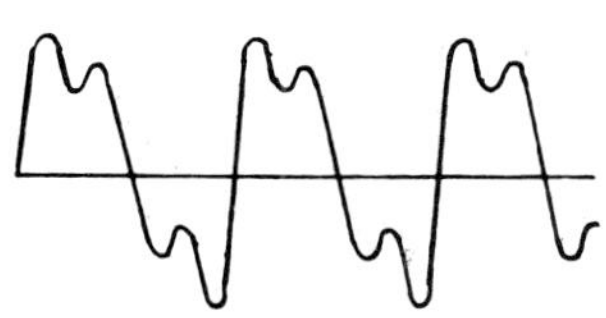

PERIODIC SOUND

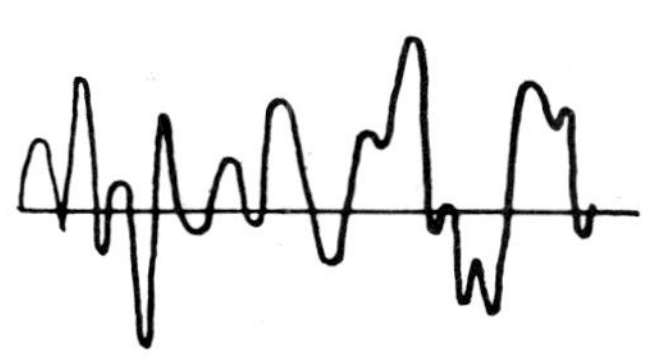

NON-PERIODIC SOUND

WHITE NOISE

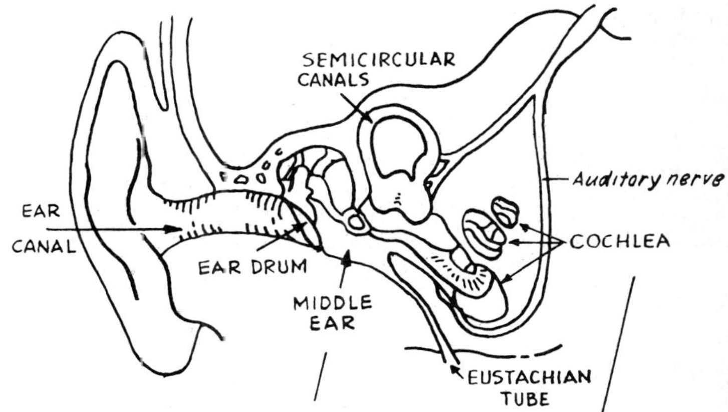

CROSS SECTION OF EAR, WITH MAJOR STRUCTURES

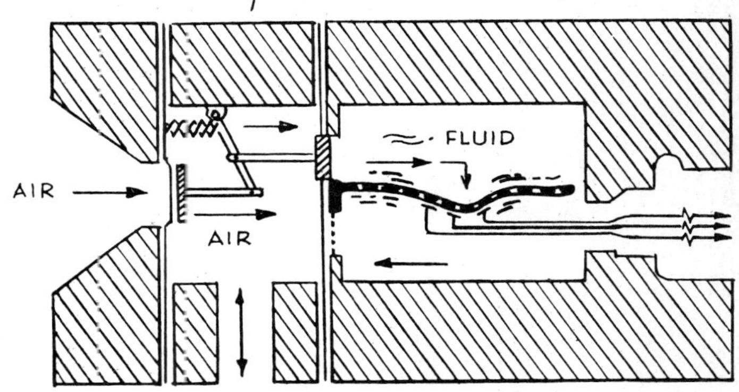

Functional diagram of ear. Sound waves impinge on outer ear and cause ear drum to vibrate. Vibrations are conducted via lever action of middle-ear bones to oval window, which actuates fluid-filled inner ear. Portions of nerve endings in cochlea are selectively excited, and their outputs terminate in brain.

[1]After W. E. Woodson and D. W. Conover, *Human Engineering Guide for Equipment Designers,* University of California Press Berkeley, 1964.

AUDITION PARAMETERS

The accompanying graph shows thresholds for hearing (B) and feeling (A). Generally, hearing is considered a subjective phenomenon, while sound is considered an objective phenomenon. The ear typically responds in a relatively predictable manner to physical sounds, as shown in the accompanying table.

Sound	Hearing
Amplitude Pressure Intensity	Loudness
Frequency	Pitch
Energy distribution	Quality

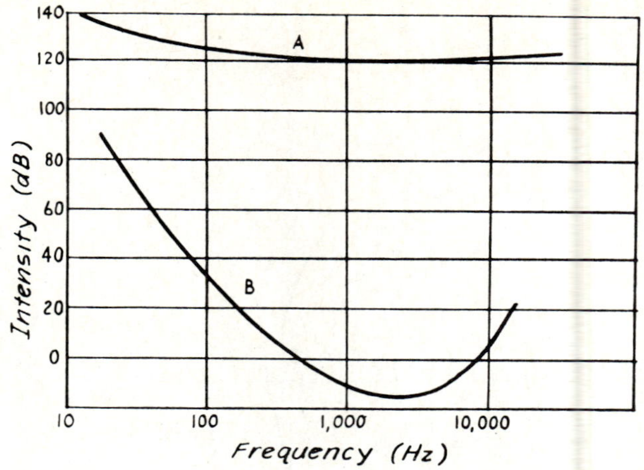

Human audibility curve.

Loudness is the physical response to sound pressure and intensity, and it varies directly with the sound pressure and intensity, as indicated in the accompanying graph.

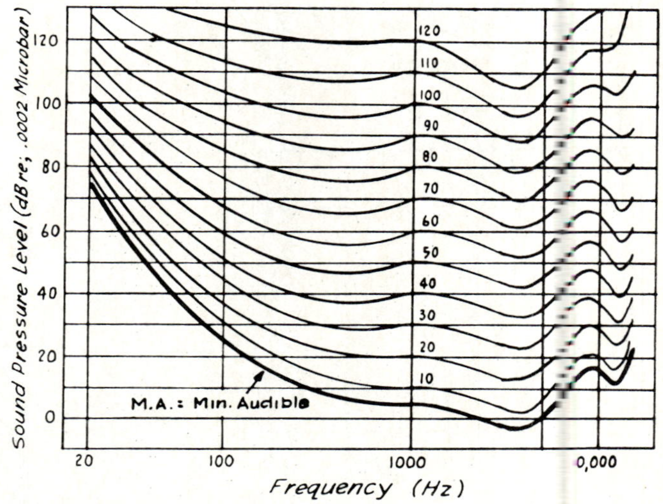

Equal loudness contours.

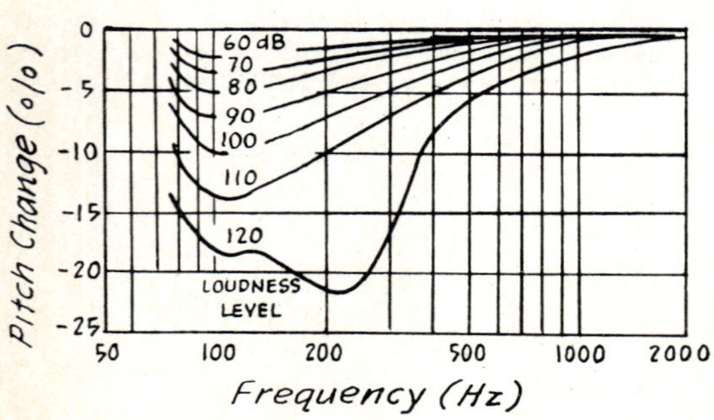

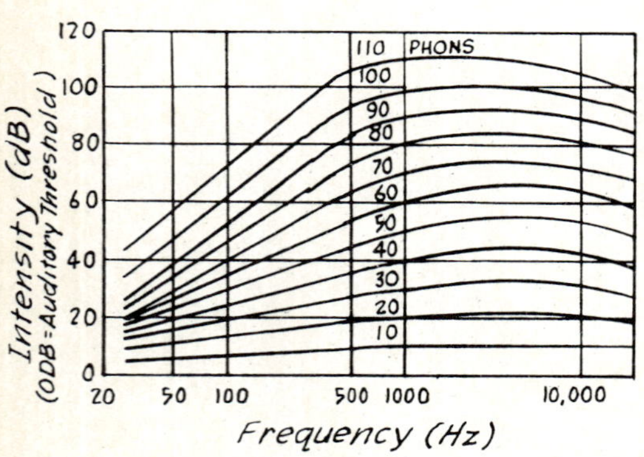

Contours of constant loudness level are shown in the accompanying graph. The curves illustrate the amount by which the pitch of a pure tone is shifted when the loudness level is changed from 40 dB to the level of contour. For example, a 100-Hz tone will seem 10 percent lower in pitch when the loudness level shifts from 40 to 100 dB, but one of 500 Hz will shift only 2 percent lower.

Loudness-level contours showing all pure tones on each curve sound equally loud. Observations with tones that differ considerably from the reference tone, however, are more difficult to make, and the error for the combined result is on the order of 1 to 2 dB.

Loudness as a function of intensity and frequency is illustrated in the upper diagram. Subjective loudness in "sones" is represented vertically above the intensity-frequency plane. The heavier curves progressing from front to rear are equal-loudness contours for pure tones.

Pitch as a function of frequency is illustrated in the lower diagram. The upper curve shows that subjective pitch (in "mels") increases less and less rapidly as the stimulus frequency increases in a linear fashion. The lower curve shows that subjective pitch increases as stimulus frequency is increased logarithmically. Our tempered musical scale is a logarithmic scale. The pitch of a 1000-Hz tone, 40 dB above threshold, is defined as 1000 mels.

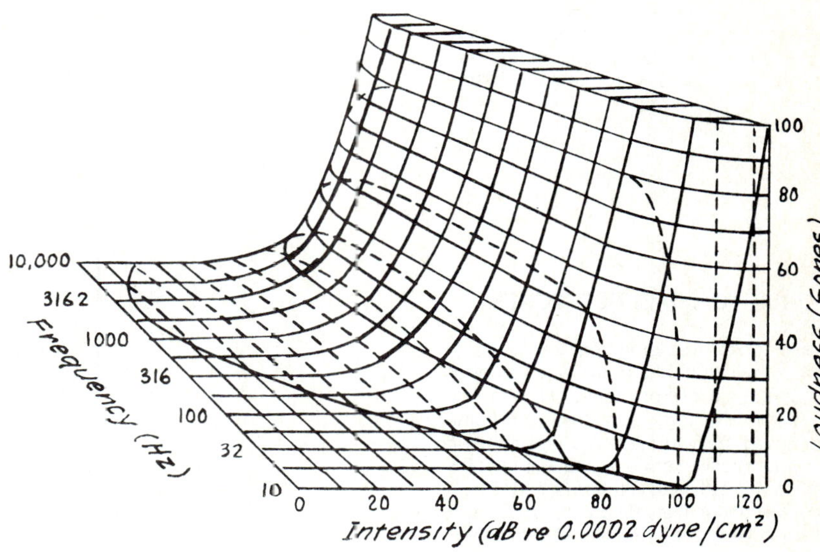

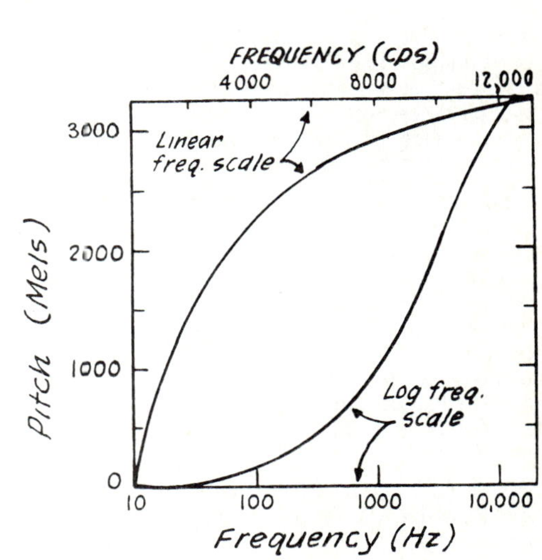

EFFECT OF FREQUENCY ON DIFFERENTIAL THRESHOLD Δ*f*/*f* as a function of frequency for five different sensation levels is shown in the accompanying graph. Between 500 and 8000 Hz, Δ*f*/*f* remains relatively constant. Between 62 and 500 Hz, Δ*f* is approximately constant. At 45 Hz, Δ*f*/*f* versus *f* shows an inflection point and the tendency to bend downward, with the tendency being more marked at higher sensation levels (because of subjective harmonics).

EFFECT OF INTENSITY OF TEST TONES Δ*f*/*f* as a function of sensation level for seven different frequencies is shown in the accompanying graph. Sensitivity increases as sensation increases, up to 40 dB (0= threshold of hearing at that frequency). Note that the effect is more pronounced for lower tones.

When the loudness level of a 1024-Hz tone (as well as its duration) is varied, accuracy of discrimination is reduced by the decrease in loudness; the effect is most pronounced for short durations.

SENSITIVITY TO DIFFERENCES IN LOUDNESS: EFFECT OF FREQUENCY Sensitivity to differences in loudness of pure tones as a function of frequency for seven different intensity levels is shown in the accompanying graph (energy of the wave is the intensity measure). On the basis of these data, it is apparent that the greatest number of loudnesses that are distinguishable at any frequency is 370 at 1300 Hz.

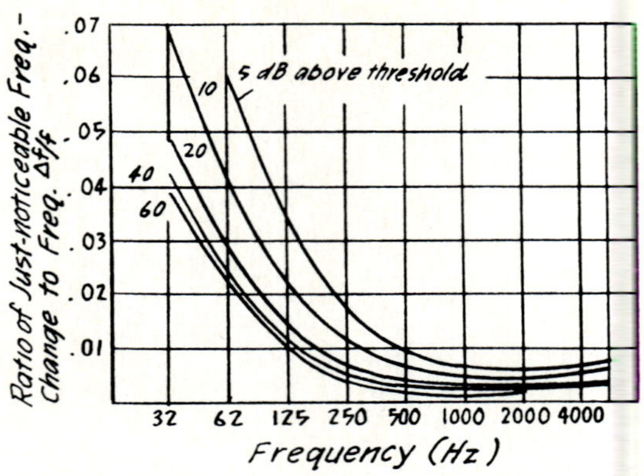

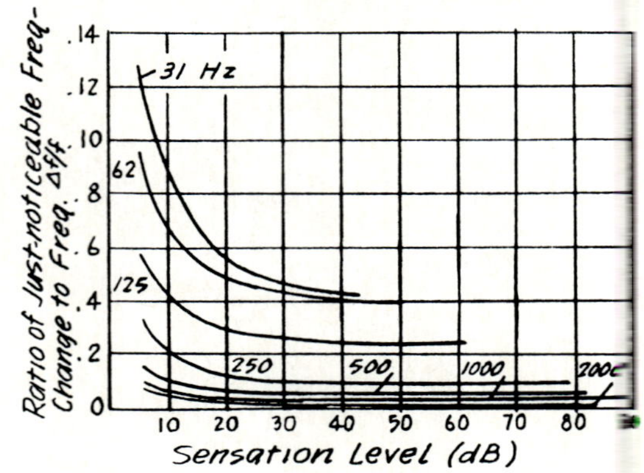

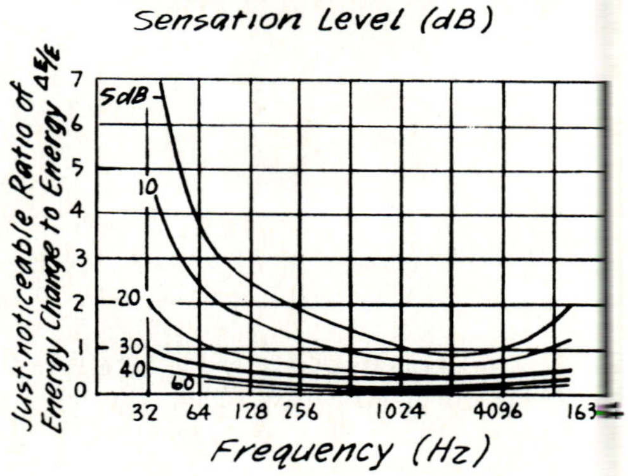

RESPONSE TO COMBINED TONES

When two tones are combined, even though they are of the same frequency, their loudness is the summation of both tones. However, if the phase of one tone is altered, the same pitch is heard, but with less loudness. If two tones are identical in terms of phase, they cancel each other out, and neither is heard.

Beats will occur (i.e., loudness builds and fades) when two tones that are slightly different in frequency are presented at the same time. Beats occur at a frequency equal to the difference between the two tones. The pitches of the two tones merge and will be heard as a single tone, the pitch of which is halfway between the pitches of the separate tones.

When the separation of tones is varied, beats will occur in three distinct stages:

At close frequencies (6 beats per second) beats will be very distinct and will appear as smooth changes in loudness.

As the frequency interval between tones increases (8 beats per second), beats become faster, and thus the resultant tone appears to throb or pulsate.

The two tones become more prominent as the frequency differential becomes greater (20 beats per second), resulting in a "buzz" rather than a throb.

The accompanying diagram illustrates the several phenomena that occur when tones are combined.

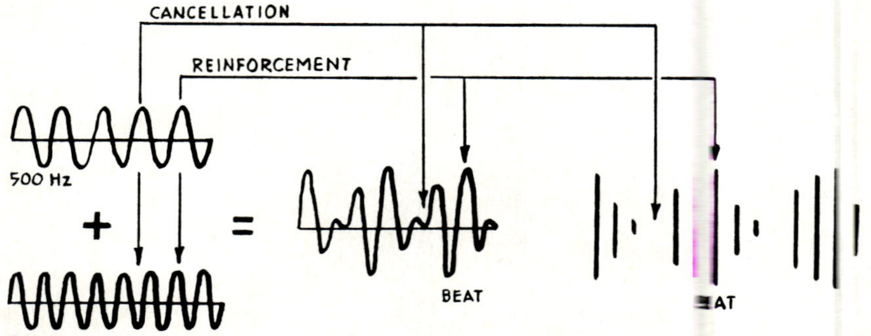

The effect of different intensity interactions upon the experiencing of beats is shown in the first graph. At the upper and lower extremes, only one tone is heard (a strong tone masking out a weaker one). Between 20 and 60 dB, the absolute intensity is unimportant. Beats are most apparent when the two generators are approximately equal in intensity.

THRESHOLDS FOR SHORT TONES WITH DIFFERENT RATES OF REPETITION

Threshold signal-to-noise ratio for masked thresholds as a function of repetition rate of the sound is shown in the second graph. Differences between curves for different durations are almost exactly equivalent to the differences in total energy produced (difference is independent of repetition rate). Note that the shape of curves is similar regardless of frequency and that threshold decreases continuously as repetition rate increases. Discontinuity between the 2- and 5-per-second rates is more marked, indicating that the ear does little integrating of energy beyond 200-ms duration.

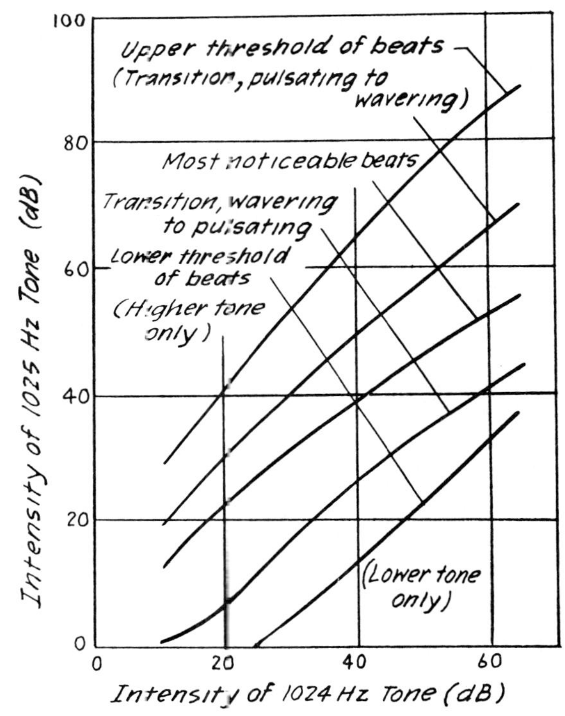

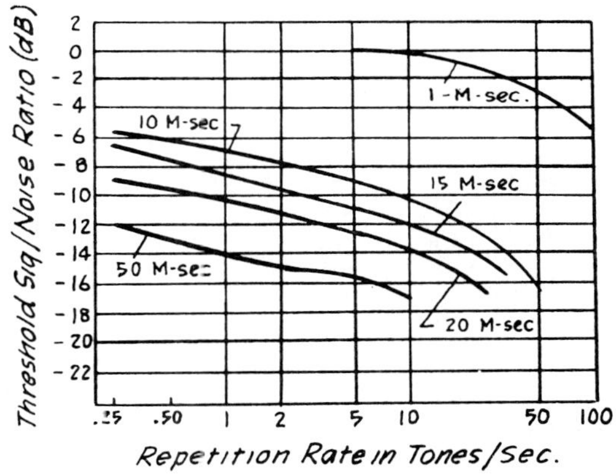

LOCALIZATION OF SOUND

Listeners can determine the direction from which a sound emanates accurately when the sound source is to their right or left or directly above or below them but less accurately when the sound source is in front of them, to the rear, or at intermediate levels above or below their heads.

It is often disconcerting to listeners if they can see the lips of an orator or the motions of an orchestra but the sound emanates (via loudspeakers) from another direction.

DEPENDENCE OF LOCALIZATION UPON FREQUENCY The average error of localization in degrees at various frequencies is illustrated in the graph at the bottom of this page. Accuracy is worst between 2000 and 4000 Hz; otherwise, localization is relatively constant. Note that the average error is smallest nearest the median plane and increases toward the side of the listener's head.

For stimuli of 200 and 2000 Hz, the least amount of intensity difference at the two ears required to produce a shift in localization is 0.9 dB at 200 Hz and 0.2 dB at 2000 Hz.

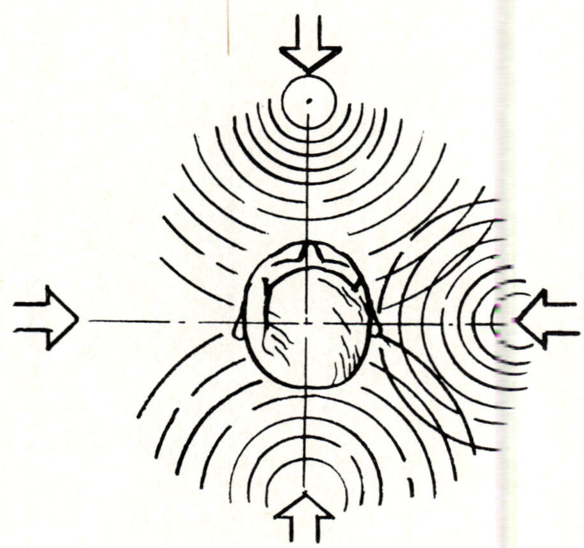

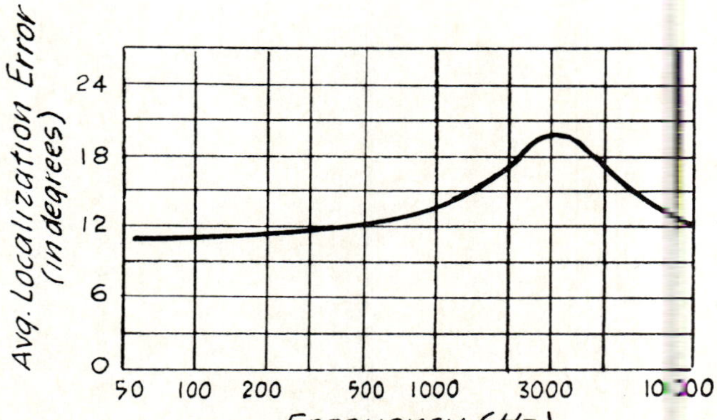

MASKING

When one sound masks another, intensity and frequency relationships determine the extent of the masking.

The accompanying graph shows the relation between masking produced by a white noise and the effective level of the noise; i.e., the effective level is the amount of noise power in a narrow frequency band (the critical band) centered about the frequency of the masked threshold. The graph indicates that the function is essentially independent of the frequency of the masked sinusoid.

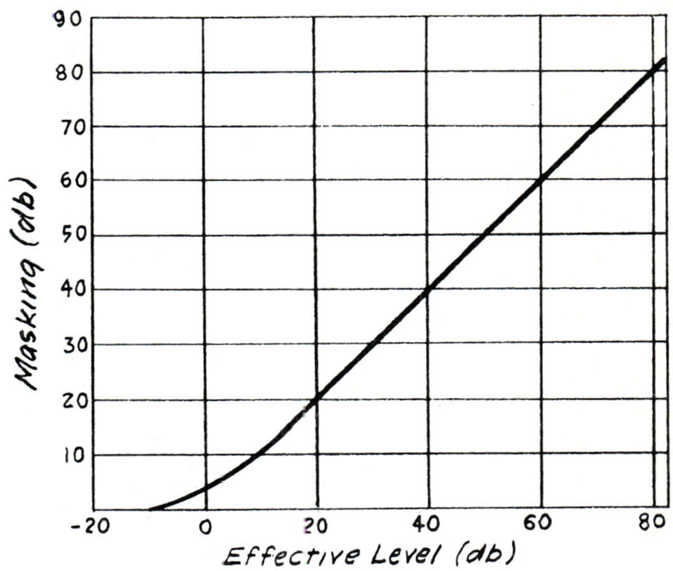

A pure tone more effectively masks a tone higher than itself than one that is lower. Masking is greatest when the masking frequency is close to the "wanted" signal frequency.

Masking increases with increases in the intensity of the masking tone. The accompanying graph illustrates masking at frequencies from 200 to 1200 Hz produced by a 1200-Hz tone.

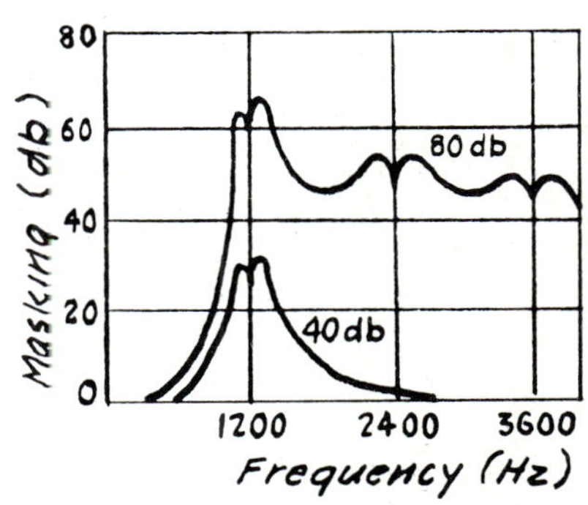

LOUDNESS THRESHOLD OF A MASKED TONE Contours for masked threshold (i.e., the sound pressure necessary to make a pure tone just audible in the presence of an interfering sound) at various noise levels are shown in the accompanying graph. Note that thresholds generally parallel one another, with approximately 20-dB intervals between masking noise levels.

Pitch discrimination becomes worse as the signal-to-noise ratio changes from infinity (no noise) to zero. Pitch discrimination is, however, better at some frequencies than others; e.g., 500 Hz is superior to 800 Hz.

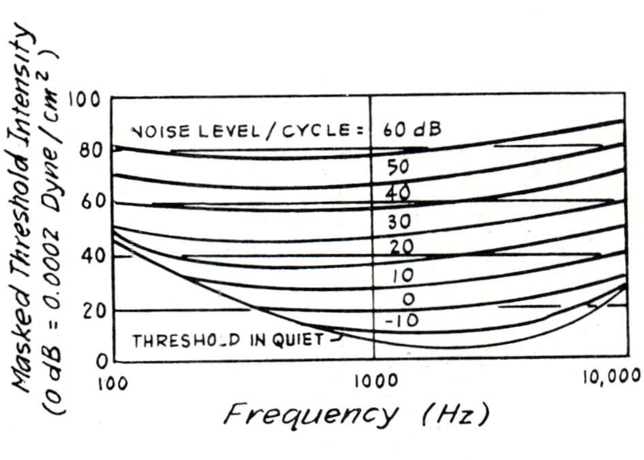

EFFECTS OF NOISE ON HUMAN PERFORMANCE

The term "noise" refers to "unwanted" sounds. One should consider not only the annoyance aspects of unwanted sounds but also the deleterious effects that noise may have on individual task performance. The accompanying table presents a general analysis of how various noise levels affect humans and their ability to perform typical tasks.

Noise Level, dB	Effects
100	Serious reduction in alertness. Attention lapses occur, although attention duration is usually not affected. Temporary hearing loss occurs if no protection is provided in the region 600–1200 Hz. Most people will consider this level unacceptable, and 8 hr is the maximum duration they will accept.
95	Considered to be the upper acceptance level for occupied areas where people expect the environment to be noisy. Temporary hearing loss often occurs in the range of 300–1200 Hz. Speech will be extremely difficult, and people will be required to shout, even though they may be talking directly into a listener's ear.
90	At least half of the people in any given group will judge the environment as being too noisy, even though they expected a noisy environment. Some temporary hearing loss in the range of 300–1200 Hz occurs. Skill errors and mental decrements will be frequent. The annoyance factor is high, and certain physiological changes often occur (e.g., the pupils dilate, the blood pressure increases, and the stroke volume of the heart may decrease). Listening to a radio is impossible without good earphones. The maximum duration that most people will accept is 8 hr.
85	The upper acceptance level (noise expected) in the range of 150–1200 Hz. Some hearing loss occurs in the range of 300–1200 Hz. This is considered the upper comfort level, although some cognitive performance decrement can be expected, especially where decision making is necessary.
80	Conversation is difficult (i.e., people have to converse in a loud voice less than 1 ft apart). It is difficult to think clearly after about 1 hr. There may be some stomach contraction and an increase in metabolic rate. Strong complaints can be expected from those exposed to this level in confined spaces, and [illegible] hr is the maximum duration acceptable within the frequency range 1200–4[illegible]0 Hz.
75	Too noisy for adequate telephone conversation. A raised voice is required for conversants 2 ft apart. Most people will still judge the environment as being too noisy.
70	The upper level for normal conversation, even when conversants are close together (at a distance of 6 ft people will have to shout). Although persons such as industrial workers and shipboard personnel who are used to working in a noisy environment will accept this noise level, unprotected telephone conversation will be difficult (upper phone level is 68 dB).
65	The acceptance level when people expect a generally noisy environment. Intermittent personal conversation is acceptable. About half of the people in a given population will experience difficulty sleeping.
60	The upper limit for spaces used for dining, social conversation, and sedentary recreational activities. Most people will rate the environment as "good" for general daytime living conditions.
55	The upper acceptance level for spaces where quiet is expected (150–2400 Hz). People will have to raise their voices slightly to converse over distances greater than 8 ft. This level of noise will awaken about half of a given population about half of the time. It is still annoying to people who are especially sensitive to noise.
50	Acceptable to most people where quiet is expected. About [illegible]5 percent will be awakened or delayed in falling asleep. Normal conversation is possible at distances up to 8 ft.
40	Very acceptable to all. The recommended upper level for quiet living spaces, although a few people may still have sleep problems.
30	Necessary for specialized listening tasks (e.g., threshold signal detection).
Below 30	Introduces additional problems; i.e., low-level intermittent sounds become disturbing. Some people have difficulty getting used to the extreme quiet, and a few may become psychologically disturbed.

Note: The above represents an amalgamation of many studies and contains a general interpretation of a wide variety of subject samples and testing conditions.

When noise levels exceed 100 dB, potentially serious consequences occur as shown in the accompanying table.

Noise Level, dB	Spectrum	Duration	Effects
105	Jet engine	2 min	Reduced visual acuity, stereoscopic acuity, and near-point accommodation and permanent hearing loss when exposure continues over a long period (months)
110	Machinery noise	8 hr	Chronic fatigue and digestive disorders
120	Broadband	1 hr	Loss of equilibrium
150	1–100 Hz	2 min	Reduced visual acuity, chest-wall vibration, changes in respiratory rhythm, and a "gagging" sensation*

*Subjects were wearing so-called protective aids to prevent hearing loss.

GENERAL NOISE EXPOSURE LIMITS

Deafness Avoidance

Octave Band, Frequency in Hz	Maximum Permissible Sound Pressure Level, dB: Occasional (1 hr or Less)	Continuous (Period of Months)
38–75	125	115
75–150	120	110
150–300	120	110
300–600	120	105
600–1200	115	100
1200–2400	110	95
2400–4800	105	90
4800–9600	110	95

Exposure Time	With Ears Unprotected	With Earplugs	With Earplugs and Earmuffs
8 hr	100	112	120
1 hr	108	120	128
5 min	120	132	140
30 s	130	142	150

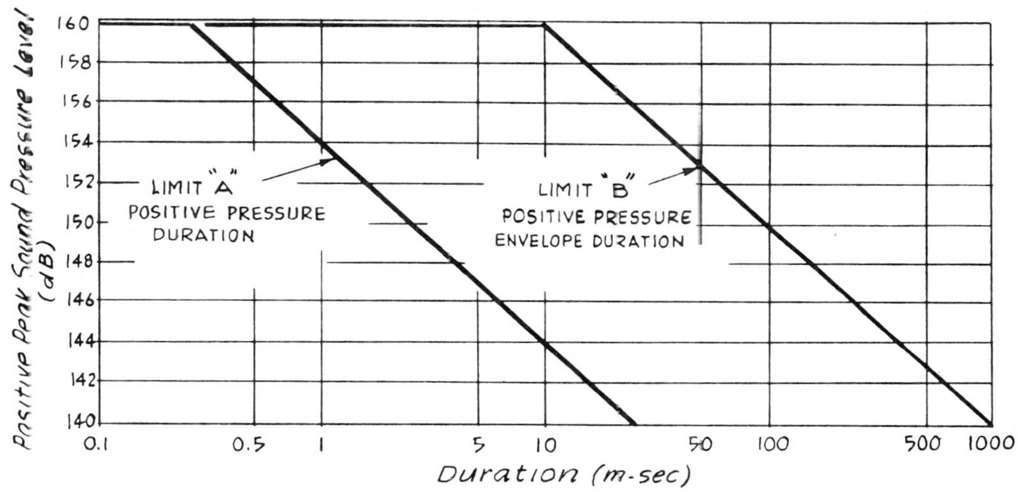

Duration (milliseconds) Impulse noise limits. Positive peak sound pressure level versus positive pressure duration should be below limit A; positive peak sound pressure level versus positive pressure envelope duration should be below level B (HEL Std. 1-63B).

Office of Safety and Health Administration (OSHA) Noise Standards

Occupational Safety and Health Act Permissible Daily Noise Exposure

Duration, hr	Sound Level, dBA
8	90
6	92
4	95
3	97
2	100
1.5	102
1	105
0.5	110
0.25	115

Note: When exposure is intermittent at different levels, the fraction $C_1/T_1 + C_2/T_2 \ldots C_n/T_n$ should not exceed unity to meet the exposure limit. C_n = total exposure time at the specified level; T_n = total exposure time permitted at the specified level.

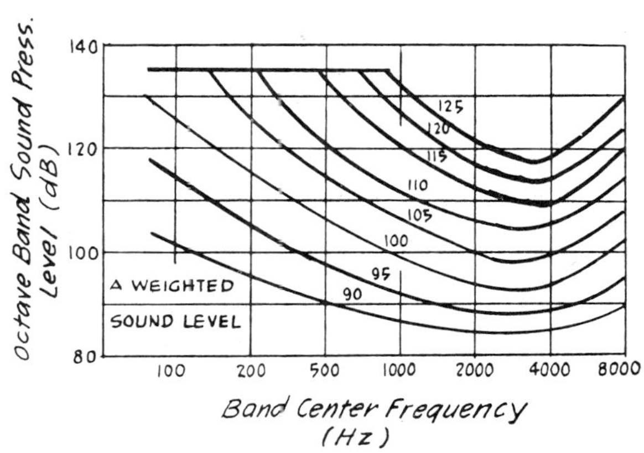

Contours for determining equivalent A-weighted sound level. This graph is used in interpreting octave-band sound levels according to OSHA provisions.

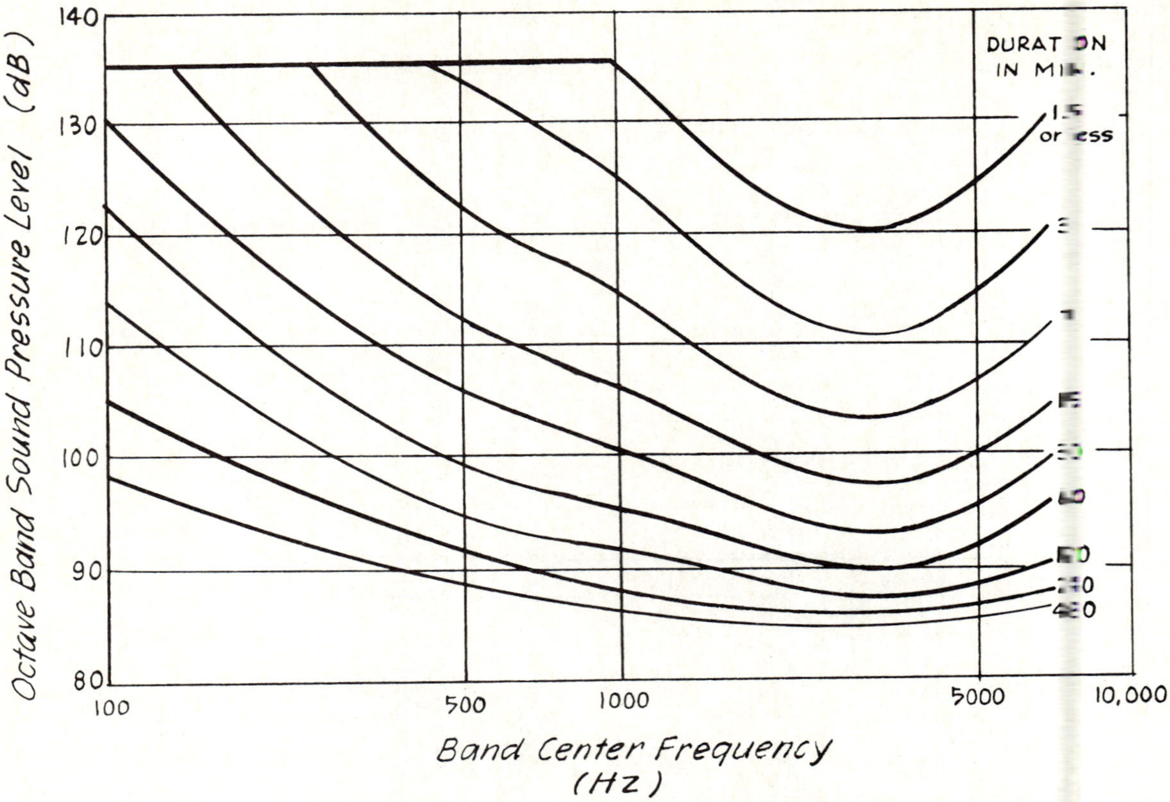

Damage risk contours for one exposure per day (left ordinate) and/or one-third octave or narrower (right ordinate) bands of noise. (From Committee on Hearing, Bioacoustics and Biomechanics, National Academy of Science, 1964.)

Speech Interference Limits

Distance, ft	Acoustic Absorption of Area, Sabines*	Speech Level, dB			
		Shout	Very Loud	Raised Voice	Normal
0.5	Any value	90	84	78	72
1	Any value	84	78	72	66
2	Any value	78	72	66	60
4	Below 100	78	72	66	60
	Above 100	72	66	60	54
8 or more	Below 100	78	72	66	60
	100–400	72	66	60	54
	400–1600	66	60	54	48

*The sabine is a unit of absorption equal to the absorption of 1 ft² of surface that is totally sound absorbent. As more absorption is present, the less interference can be tolerated because speech energy is partially absorbed. The above table is for the male voice. For the female voice, decrease values 5 db.

Comfort Limits

Octave Band, Frequency in Hz	Comfort Criteria, dB	
	Normally Noisier Environment	Normally Quiet Environment
38–75	100	80
75–150	95	70
150–300	90	60
300–600	85	55
600–1200	75	50
1200–2400	65	50
2400–4800	60	50
4800–9600	55	45

Speech Intensity Characteristics Relative to Establishment of Speech Interference Noise Limits

Description	Typical Intensity of Source, dB	Distance	Message Characteristic
Soft whisper		3–6 in	Secret communication
Audible whisper	44–69	8–20 in	Confidential or personal
Normal voice	50–75	20–60 in	Personal or small group
Loud voice	56–81	5–8 ft	Nonpersonal, intermediate group
Very loud voice	62–87	To 20 ft	Group address
Shouting	68–93	Upper limit	Hailing or emergency communication

Subjective Effects of Changes in Sound Level

Sound-Level Change, dB	Change in Apparent Loudness
3	Barely perceptible
5	Clearly noticeable
10	Either twice as loud or twice as soft
20	Much louder or softer

Steady-State Noise-Level Limits for Person-to-Person Communications (U.S. Army Standards)

Octave Band Limits, Hz		Center Frequency		Noise Level, dB	
A	B	A	B	A	B
37.5– 75	44– 87	53	63	79	77
75 – 150	87– 175	106	125	73	72
150 – 300	175– 350	212	250	68	67
300 – 600	350– 700	425	500	64	63
600 –1200	700– 1400	850	1000	62	61
1200 –2400	1400– 2800	1700	2000	60	59
2400 –4800	2800– 5600	3400	4000	58	58
4800 –9600	5600–11200	6800	8000	57	57

Note: A—non-electrically aided (commercial frequencies ASA Z24.10-1953); B—non-electrically aided (preferred frequencies ASA S1.6-1960).

Source: Human Engineering Laboratory HEL Standard S-1-63B.

Commercial Work-Space Noise Criteria[2]

Noise criteria curves (NC) are widely used as work-space design criteria (see the accompanying graph). The accompanying table identifies typical work spaces with the appropriate NC curves that should be applied.

NC curves are referenced to preferred octave bands (lower abscissa) and to commercial bands (upper abscissa). When using the commercial frequency, the NC number is also the speech interference level (SIL) for that particular spectrum.

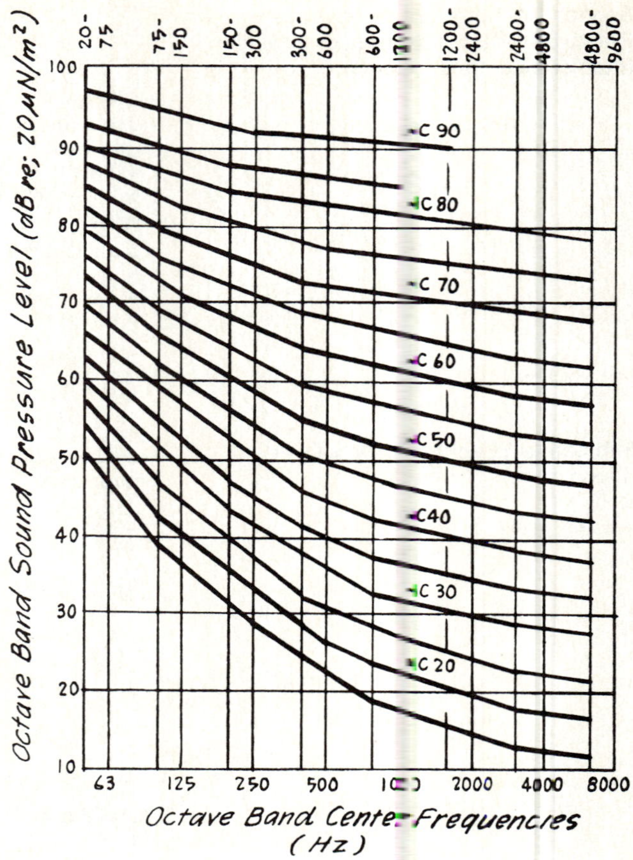

Recommended NC Curves for Various Work Spaces

NC Curve	Type of Work Space	Communication Equivalent	Office Application
90		Noise-attenuating headset required	Not recommended
80		Communication very difficult; telephone use unsatisfactory	Not recommended
70–80		Raised voice range 1–2 ft; shouting range 3–6 ft; telephone use very difficult	Not recommended
60–70		Raised voice range 1–2 ft; telephone use difficult	Not recommended
55–60		Very noisy, not suited for office; telephone use difficult	Not recommended

[2]D. C. Hodge, and G. R. Garinther, in J. F. Parker and V. R. West (eds.), *Bioastronautics Data Book,* 2d ed., National Aeronautics and Space Administration, 1973, pp. 693–750.

REVERBERATION

Reverberating sounds tend to mix with direct sounds, thus affecting the clarity with which the sounds are perceived by the listener.

Reverberation time in relation to speech intelligibility is illustrated in the accompanying graph.

Reverberation time in relation to room size for both speech and music sounds is illustrated in the accompanying graph.

Suggested reverberation objectives for specific listening requirements are noted in the accompanying table.

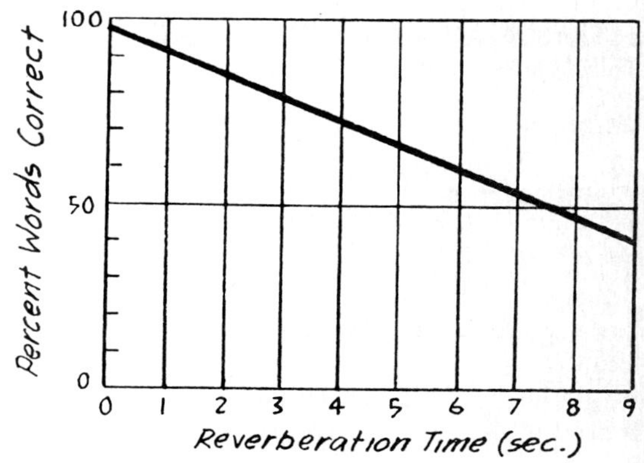

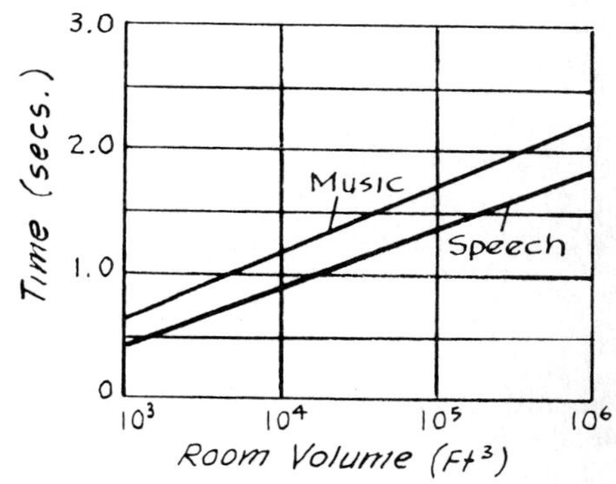

Listening Requirement	Reverberation Times, s
Optimum for speech (too dead for music)	Below 1.0
Good for speech, fair for music	1.0–1.5
Fair for speech, good for music	1.5–2.0
Poor for speech, good for liturgical music	2.0 +

ULTRASONIC, LOW-FREQUENCY, AND INFRASONIC SOUND

Ultrasonic Noise Limits

Sound pressure level (SPL) should not exceed 75 dB in one-third-octave bands centered at 8 to 16 kHz or at 20 to 31.5 kHz in order to preclude unpleasant subjective reactions and prevent TTS.

Low-Frequency and Infrasonic Noise Limits

To prevent injury in the low-frequency and infrasonic noise range (1 to 100 Hz), the limits noted in the accompanying table should not be exceeded.

Frequency, Hz	SPL, dB	Duration, min/day	
1–7	150	4	Should use earplugs to reduce unpleasant sensations
8–11	145	4	Should use earplugs to reduce unpleasant sensations
12–20	140	4	Should use earplugs to reduce unpleasant sensations
21–100	135	20	Without earplugs
21–100	150	20	With earplugs

Note: Refers to pure tones or octave bands with center frequencies as indicated. "Duration" refers to one exposure per day with a 24-hr gap between successive exposures.

Selected References: Noise Control Standards

Acoustics: Assessment of Occupational Noise Exposure for Hearing Conservation Purposes, Recommendation R 1999, International Standards Organization, Geneva, Switzerland, 1971.

Criteria for a Recommended Standard—Occupational Exposure to Noise, Report HSM 73-11—1, National Institute for Occupational Safety and Health, Cincinnati, 1972.

Hazardous Noise Exposure, Air Force Regulation 161-35, Department of the Air Force, 1973.

Hearing Conservation Program, BUMEDINST 6260.6B, U.S. Navy, Bureau of Medicine and Surgery, 1970.

Information on Levels of Environmental Noise Requisite to Protect Public Health and Welfare with Adequate Margin of Safety, Report 550/9-74-004, Environmental Protection Agency, Office of Noise Abatement and Control, 1974.

Noise and Conservation of Hearing, TB MED 251, The Surgeon General's Office, 1972.

Noise Limits for Army Material, MIL-STD-1474A(MI), U.S. Department of Defense, 1975.

HUMAN INFORMATION-PROCESSING CAPABILITIES

Humans can transmit only about 5 to 10 bits of information per second (b/s). They can transmit about 2 b/s when the stimuli they receive are fairly well structured, although this can often be doubled by adding appropriate coding or anchoring of the input. The relationship between bits per item and the bits-per-second limitation depends on what is referred to as "attention switch time" (e.g., 0.1 to 0.2 s). Thus an operator can accept no more than two or three items of data per second.

As the speed of information-processing demand increases, the number of errors typically increases. Thus the overall information transfer rate tends to remain constant at about 10^{-2} b/s. Through careful input display design, however, this rate can be increased to 10^{-3} or even 10^{-6} under good conditions. The following error classification scheme, which was suggested by Kidd,[1] helps characterize types of information-processing errors and why they occur:

1. Failure to detect a signal: input overload or underload and/or actual interference
2. Misidentification: insufficient cues
3. Improper weighting of informational factors and/or selection of input factors: poor or inadequate conceptualizations or evaluation of action choices
4. Action failure: A wrong action at the right time or a right action at the wrong time

INFORMATION STORAGE CAPACITY

Information storage is of two distinct types: long-term and short-term. Short-term memory storage capacity is generally limited to about 30 lb or eight individual items. The human generally organizes stored information in terms of sensory modality (visual, auditory, etc.). The most significant storage problem occurs because of the potential interference between old ("held") information and new items that present themselves during the holding period. This accounts for the frequent "reversal errors" in information processing. As a general principle, human memory is more effectively utilized as a means of orienting and sequencing information than as a depository for isolated data or symbolic items.

INPUT CAPACITY

The total sensory input capacity of the human system is about 10^9 b/s (as compared with output capacity of 10 b/s). The five basic input categories for the visual channel are:

Relative position
Shapes
Brightness
Color
Movement

[1] J. S. Kidd, "Some Sources of Load and Constraints on Operator Performance in a Simulated Radar Air Traffic Control Task," USAF WADD TR-60-612, 1961.

Skill Breakdown

Effect	Reaction
Failure of selective attention fails	Narrowing of attention
Perceptual disorganization	Reduction in the size of the data sample and actions upon it and reduction in filtering efficiency
Cumulative disruption	Temporary halts (complete stop-pages) with start-overs

For the auditory channel they are:

Pitch
Loudness
Rhythm
Timbre (the quality that allows one to distinguish different voices, instruments, or special auditory displays)

DECISION MAKING

Three basic kinds of information must be available for an operator to make good decisions:

Information regarding policies and objectives
Information regarding possible alternatives and consequences
Information about the current state of the system

It is important to recognize that, in making decisions, people pursue two different courses of action:

1. Evaluating likely outcomes
2. Keeping options open as long as possible

STRESS SUSCEPTIBILITY

Human information processing is subject to a variety of stresses that may affect the efficiency with which information is received, processed, and acted upon. Two basic factors are important to consider:

1. The state of arousal of the human system (alertness)
2. Potential skill deterioration due to disorganization, cumulative disruption, and/or fatigue

The effects of, and reactions to, skill breakdown have been categorized as shown in the table at the top of the page.

INFORMATION DISPLAY IN RELATION TO DECISION MAKING[2]

Although research has shown that, on the basis of information acquisition only, added display complexity generally degrades operator performance, decision-making performance is not always similarly affected:

The relationship between "fact density" and decision adequacy may be curvilinear; i.e., where there is a decreasing slope under low-density conditions and/or an increasing slope under high-density conditions, actual decision-making performance may improve. For example, although complexity of information acquisition degrades overall performance when fact density is low, decision-making performance may actually increase with increases in fact density.

When information is compressed, an inverted "U-shaped" relationship occurs, with moderate levels of compression producing the best decision-making performance; i.e., the critical variable is not merely symbol count, etc., but also the subjective weighting of the particular facts being compressed.

At levels of low fact density, high compression, and high display clutter, simple coding (e.g., color) may be superior to more elaborate encoding combinations. However, at other levels of fact compression and clutter, a double coding is usually more effective (e.g., color plus size).

Subject motivation (incentive) works only at low informational levels. At high levels of motivation, degraded performance is often mediated by scattered attention.

Increases in perceptual clutter sometimes increase performance rather than degrading it, as one might expect, because we are used to working in a particularly cluttered type of information-transfer environment. The more random an irrelevancy, the more performance is facilitated because of our need to have irrelevancies for a "figure-ground" decision-making background.

It is important, however, to recognize the importance of input validity to decision making, since humans quickly recognize and react to the futility of trying to decide on information that destroys or degrades the quality of their decisions.

[2] After W. T. Singleton, "The Ergonomics of Information Presentation," *Applied Ergonomics,* vol. 2, no. 4, pp. 213–220, 1971.

AMOUNT OF INFORMATION IN ABSOLUTE JUDGMENT FOR VARIOUS STIMULUS DIMENSIONS

Stimulus Dimension	Number of Levels Which Can Be Discriminated	Bits of Information Transmitted*
Audition (single dimension)		
Pure tones	5	2.3
Loudness	5	2.3
Audition (combination of six variables including frequency and intensity, on-time fraction, duration, spatial location, and rate/interruption	150	7.2
Vision (single dimension), e.g., pointer and linear scale	9	3.1
Visual size	7	2.8
Hue	9	3.1
Brightness	5	2.3
Vision (combinations):		
Size, brightness, and hue	17	4.1
Hue and saturation	15	3.9
Position of dot in square	24	4.6
Odor (single dimension)	4	2.0
Odor (combination, e.g., kind, intensity, and number)	16	4.0
Taste:		
Saltiness	4	1.9
Sweetness	3	1.7

*Lack of correspondence of bit values is due to rounding to the nearest whole number.

"Although we cannot neatly separate different sequential phases of information processing (i.e., sensation, perception, decision, response, etc.), it should be noted that the bottleneck probably is the brain rather than the sensory mechanisms."[3] It has been estimated that information reduction from reception through intermediate processes to permanent storage is as shown in the accompanying table.

Process	Maximum Flow of Information, b/s
Sensory reception	1,000,000,000
Nerve connections	3,000,000
Consciousness	16
Permanent storage	0.7

[3]E.J. McCormick, *Human Factors Engineering*, 3d ed., McGraw-Hill Book Company, New York, 1970, p. 92.

VISUAL AND AUDITORY CODING METHODS[4]

Summary of Certain Visual and Auditory Coding Methods

(Numbers refer to number of levels which can be discriminated on an absolute basis under optimum conditions.)

Alphanumeric	Single numerals, 10; single letters, 26; combinations, unlimited. Good; especially useful for identification; uses little space if there is good contrast. Certain items easily confused with each other.
Color	Hues, 9; hue, saturation, and brightness combinations, 15–24. Particularly good for searching and counting tasks; poorer for identification tasks; trained observers can use many codes (up to 24). Affected by some lights; problem with color-defective individuals.
Geometric shapes	15 or more. Generally useful coding system, particularly in symbolic representation; good for CRTs. Shapes used together need to be discriminable; some sets of shapes more difficult to discriminate than others.
Visual angle	24. Generally satisfactory for special purposes such as indicating direction, angle, or position on round instruments like clocks, CRTs, etc.
Size of forms (such as squares)	5. Takes considerable space. Use only when specifically appropriate; preferably use less than 5.
Visual number	6. Use only when specifically appropriate, such as to represent numbers of items. Takes considerable space; may be confused with other symbols.
Brightness of lights	3–4. Use only when specifically appropriate. Preferably limit to two levels; weaker signals may be masked.
Flash rate of lights	4. Limited applicability. Preferably limit to two levels; combination of individual flashes and controlled time intervals may have special application, such as lighthouse signals and naval communications.
Sound frequency	5. For untrained listeners, use less than 5 levels; space widely apart, but avoid multiples and low and high frequencies; intensity should be 30 dB above threshold. Frequency changes easier to detect than single frequencies; combinations usually require training except for clearly distinguishable sounds such as bells, buzzers, and sirens.
Sound intensity	4. Preferably use less than 4. Intensity changes easier to detect than single intensities; for pure tones restrict to 1000–4000 Hz, but preferably use wide band.
Sound duration	Use clear-cut differences, preferably 2 or 3.
Sound direction	Difference in intensity to two ears should be distinct; particularly useful for directional information (i.e., right versus left).

KEEPING TRACK OF SEQUENTIAL EVENTS[5]

At slow event-presentation rates (one per second or less), individual events falling into two categories can be discriminated one at a time and mentally tallied with considerable accuracy.

As either the number of categories or the trial length increases, the number of errors per trial increases, generally very rapidly. For a given trial length, one should present the operator with fewer categories with multiple occurrences rather than large numbers of categories with fewer occurrences.

At fast presentation rates, the ON/OFF ratio has little effect on performance as long as the ON time is sufficient for accurate stimulus perception. Intermediate-rate–short-stimulus ON time tends to optimize performance. Slow-rate–short-stimulus ON time degrades performance.

It is generally desirable to "pace" operator performance by means of cues as to the expected occurrence of a next event.

In general, the slower the rate of presentation, the better the performance, a truism that nevertheless is important to remember.

Irrelevant information along with relevant information does not necessarily degrade performance, provided that it is introduced in a manner that allows operators to pace themselves.

Irregular rates, on the other hand, disrupt performance because the operator cannot anticipate the occurrence of the next stimulus.

Stimuli with a "natural" or built-in order result in better performance. Whenever possible, sort out information categories prior to displaying information to the operator and align visual information elements spatially so that each category appears in its own unique location each time it is presented.

Subtractions are more difficult than additions in setting up informational content, and they should be kept to a minimum in any given sequence.

Information of greater value generally is "remembered" best, although typically at the expense of the remaining information.

The number of multiple categories displayed simultaneously can be increased to three with no significant degradation when the information rate is held constant.

SUMMARY GUIDELINES RELATIVE TO INFORMATION-PROCESSING OPTIMIZATION

Multidimensional coding typically results in more information transmission than single-dimension coding.

Information-processing capacity is adversely affected by both load and speed; i.e., the number of sources should be limited.

When time sharing of sensory inputs occurs, signals should be separated temporally (preferably by 0.5 s or more).

When operators can control the input rate (are self-pacing) and have some method of identifying the more important input (if they have a choice), auditory signals are generally more durable than visual signals.

Simultaneous presentation of the same information via two sensory modalities usually increases the probability of reception.

Signals presented on one channel (such as audition) can serve as cues to facilitate the use of another channel.

Time sharing of visual tasks adversely affects primarily those tasks with greater uncertainty and those which depend on short-term memory; thus speed stress does not affect all tasks equally.

[4] E. J. McCormick, *Human Factors Engineering,* 3d ed., McGraw-Hill Book Company, New York, 1970, p. 113.

[5] After R. A. Monty, "Keeping Track of Sequential Events: Implications for the Design of Displays," *Ergonomics,* vol. 16, no. 4, 1973.

INFORMATION TRANSFER FACILITATION

Certain information transfer facilitators are at the disposal of the designer and should be utilized whenever possible to maximize the human's processing response. The accompanying table reviews some of the key facilitators that can be used.

Facilitator	Remarks
Familiar patterns	Present information to operators in a form that is already familiar to them.
Visualization	Whenever practical, utilize the natural tendency of most individuals to try to visualize, even if the presentation mode is via channels other than the visual one.
Context	Provide as complete a picture as possible in a logically organized manner.
Normal relationships	Present informational components and component-observer relationships that are natural (alphanumeric characters right side up, high-low values oriented according to specific use format, etc.).
Minimal extrapolation	Avoid requiring the observer to extrapolate; i.e., present quantities, values, and patterns in their intended form
Reference bases	Provide a continuous reference (a scale value, a map, a comparison background, etc.).
Timing	Time inputs to avoid overload and delay; present them in a logical sequence and provide immediate feedback to the operator's responses or queries.
Noise	Control input competition, interference, and distractions.
Conspicuousness	Emphasize informational cues via adequate intensity, contrast, and special encoding to ensure maximum arousal and attention
Expectation	Anticipate the operator's mental set and prepare the operator to receive the information.
Meaningfulness	Clarify the value, necessity, and urgency of the information transfer.

Note: For additional enlightenment with regard to human information processing, see D. E. Rumelhart, *Introduction to Human Information Processing,* John Wiley & Sons, Inc., New York, 1977.

SIMPLE VERSUS COMPLEX REACTION TIME

In practical terms, one should think more in terms of total response time than in terms of simple reaction time. The term "reaction time" generally is considered at two levels. The first is called "simple reaction time," which is the shortest time between the moment a sensory receptor is stimulated and the time some body element reacts, as when a speck of dust strikes an eyelash, and the eyelid immediately closes to protect the eye. Practically speaking, however, scientists measuring simple reaction time typically have a test subject respond to a stimulus by pressing, say, a key—thus allowing the scientist to measure and compare the reaction time of various sensory channels (vision, audition, etc.).

The second level is called "complex reaction time," in which the scientist attempts to include human information-processing time; e.g., the subject may be asked to recognize one stimulus from among several and to respond by selecting one of several response modes.

Since complex reactions are more akin to real-world situations, the designer is more concerned with complex reaction time or, more realistically, total response time. In a sense, response time is a function of several factors, including the following:

1. The sensory channel through which the stimulus is initiated
2. The signal or stimulus characteristics
3. The complexity of the signal
4. The signal rate
5. Whether anticipatory provisions are present
6. The response mode, e.g., the body member used

SIMPLE REACTION-TIME COMPARISONS OF SENSORY INPUT CHANNELS[1]

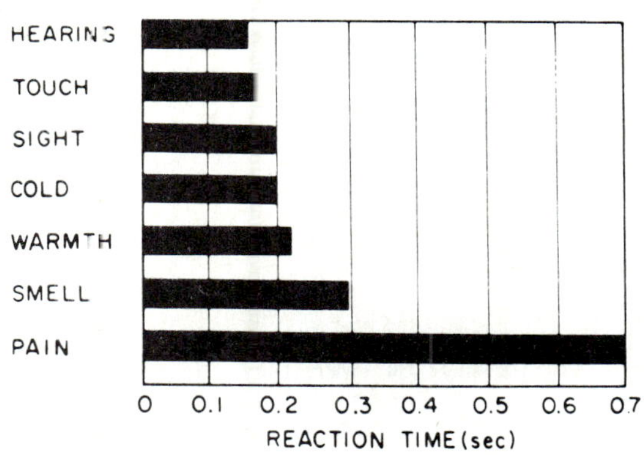

Using the typical key response technique, scientists have determined the reaction times for various sensory input channels shown in the accompanying graph.

[1]C. T. Morgan et al. (eds.), *Human Engineering Guide to Equipment Design,* McGraw Hill Book Company, New York, 1963, p. 229.

EFFECTS OF SIGNAL CHARACTERISTICS ON REACTION TIME

Signal intensity: The greater the intensity of the signal, the faster will be the reaction time.

Signal anticipation: When the signal is anticipated, reaction time is typically shorter.

Practice: Reaction time tends to be reduced with practice.

Pacing: If operators can set their own pace, they can often react faster to known signals.

Signal quality: Operators generally can react faster to a high-pitched sound, a brighter light, a larger visual target, a longer-duration signal, and a signal emanating from a particular location (i.e., a visual signal that strikes the center rather than the periphery of the eye). There is no difference in simple reaction time between a steady and a flashing light signal. However, when one intermittent light signal must be distinguished on the basis of flash rate, reaction time is related to the length of the flash.

Likelihood of signal appearance: The least likely signals will have the longest reaction times.

Signal format: When signals are arranged sequentially or are meaningfully grouped, reaction time is typically shortened; i.e., reaction time is a function of the number of signals that occur in the next sequential step or is proportional to the number of signal groups. When two successive signals occur within 0.1 s, they are generally treated as a single signal. When the interval between signals is 0.5 s or greater, the operator is capable of responding to each signal separately.

Overload: Although an operator can adjust to excessive signal rates by relying on memory for short bursts, total response failure may occur when rates are too high for too long.

Signals should not occur at rates faster than about two per second unless some means are provided for anticipating the signal. Avoid alerting periods shorter than 0.1 s.

COMPLEX REACTION TIME

The following estimates have been made regarding the several components of a basic, complex reaction time, i.e., speed of perception to action:

Stimulus detection and neural transit time 0.1 s
Brain recognition time 0.4 s
Decision-making time up to 4.0 s
Motor response time (minimum) 6.0 s

AGE AND SEX DIFFERENCES

The reaction time of adult males is generally faster than that of adult females, and, as one might expect, older people have a longer reaction time. Typical comparisons are shown in the accompanying table.

	Males		Female	
Age	Light, s	Sound, s	Light, s	Sound, s
20	0.24	0.23	0.32	0.31
30	0.22	0.19	0.26	0.20
40	0.26	0.24	0.34	0.30
50	0.27	0.25	0.36	0.30
60	0.38	0.37	0.44	0.42

SOME COMMON RESPONSE TIMES

Simple Movement Response	Time, s
Moving eyes (focusing on new spot)	0.48
Walking (one pace)	0.60
Sitting down in a chair	1.32
Standing up from a sitting position	1.62

Other generalizations of interest include the following:

It generally takes about 20 percent longer to respond with the feet than with the hands.
The preferred hand is usually about 3 percent faster than the nonpreferred one.
Everyone has a refractory period of about 2 to 3 s, regardless of a stimulus demand rate, which means that a second stimulus arriving within 0.5 s will be treated together with the first (e.g., two successive signals arriving 0.1 s apart will be treated as a single signal).
Simple reaction times usually can be reduced by as much as 40 percent by providing an alerting signal (approximately 2 to 8 s in advance for isolated response tasks and 2 to 3 s in advance for sequential tasks).

EFFECT OF NUMBER OF RESPONSE CHOICES

As one might expect, when the number of response choices increases, the reaction time is lengthened. The accompanying table illustrates this point.

Number of Choices	Mean Reaction Time, s
1	0.20
2	0.35
3	0.40
4	0.45
5	0.50
6	0.55
7	0.60
8	0.60
9	0.65
10	0.65

SELECTED MANUAL RESPONSE-TIME EXAMPLES

Although many studies have been carried out for the purpose of establishing response times for a variety of manual tasks, space permits citing only a few examples to demonstrate the variability that must naturally accompany the specifics of a given task.

Reach Distance and Terminal Movement Condition

	Terminal Conditions, s	
Reach Distance, in	Always the Same	Varied
10	0.03	0.041
20	0.045	0.066
30	0.062	0.093

EXPECTED VERSUS UNEXPECTED STIMULUS RESPONSE TIMES

Simple reaction-time reference 0.20 s
Forewarned 0.51 s
Surprised 0.61 s

MOVEMENT RESPONSE END-POINT CONTROL

When the operator has to control the final movement end point, his or her total response time is greater than when the end point is automatically controlled (e.g., the operator moves a stylus or control to a mechanical "stop").

Distance, in	Mechanical Stop, s	Operator-Controlled
6	0.28	0.47
12	0.36	0.57
20	0.46	0.67
30	0.58	0.79

END-POINT PRECISION

The more precise or accurate the operator has to be in making a control movement, the longer his or her response time will be.

Precision, in	Time, s
0.50	0.11
0.25	0.27
0.125	0.48
0.0625	0.67

EFFECT OF ADDED MANUAL FORCE ON RESPONSE TIME

Force Required, lb	Time, s
2	0.02
4	0.04
6	0.06
8	0.08
10	0.10
50	0.33

REACTION TIME TO A WARNING LIGHT RELATIVE TO ITS ANGULAR VARIATION FROM THE OPERATOR'S NOMINAL LINE OF SIGHT (WHILE TRACKING)

Angular Variation,	Time, s
20	0.40
40	0.40
80	0.80
100	1.30

EFFECTS OF DISTRACTION

Type of Distraction	Without Distraction	With Distraction
Continuous	0.218	0.220
Continuous with periodic interruptions	0.202	0.206
Random/intermittent	0.204	0.220

EFFECT OF REACH DISTANCE ON TOGGLE-SWITCH OPERATION

The response times shown in the accompanying table are those of an operator who, in addition to making the simple movement to the switch location in response to a warning light, had to decide which hand to take off a joy stick and then proceed to reach for and manipulate a toggle switch.

Distance, in	Time, s
6	0.76
9	0.66
15	0.84
18	0.77

Note the randomness of the responses, which is due, of course, to the decision-making step in the total response scenario.

EFFECT OF TOGGLE-SWITCH SEPARATION OR SPACING

Distance Apart, in	Time, s
1.0	0.28
2.0	0.24
4.0	0.35
8.0	0.40

Note that when switches are too close, response time is slightly longer; when they are too far apart, response time is significantly longer.

COMPARISON OF RESPONSE TIMES FOR VARIOUS DIGIT-INPUT DEVICES[2]

10-key pad (calculator type) 12.0 s
10×10 matrix keyboard 13.0 s
10 separate levers 17.0 s
10 separate rotary knobs 18.0 s

VISUAL SCANNING RESPONSE TIME

Two examples of visual scanning response time are shown below. The first involves the common task of pilots when they are required to transfer their line of sight from looking at instruments inside the cockpit to scanning outside the cockpit to look for other aircraft:

1. Vision shifts from outside to inside 1.5 s
2. Vision shifts from inside to outside 2.39 s

The difference is explained by the greater problem of accommodating from the near-vision instrument-watching task to the far-vision blank-sky target search task.

A further example of flying response deals with the difference between making a recovery flight maneuver using instruments and making the same recovery using a VFR (visual flight reference)

1. Instrument recovery 11.0 s
2. VFR recovery 9.2 s

The difference here is explained by the naturalness of a VFR, making it unnecessary for the pilot to perform extra data processing from abstract instrument graphics.

A second scanning example relates to the positioning of panel displays in the typical automobile. When instruments are clustered high, near the normal roadway line of sight, response to their indications is faster than when the instruments are farther down and are spread out on the instrument panel:

Close to line of sight 1.9 s
Normal panel position 2.2 s
Bottom of panel, spread out 2.9 s

Other Factors

Various other factors, of course, can obviously affect operator response time. These include physical and psychological stress, constraints within the workplace, and constrictions caused by clothing. Often, too, response times are shorter when the task is paced; i.e., operators can usually operate faster when they are paced than when they pace themselves. Response times vary less under paced conditions.

Finally, one must always consider the response of the mechanical system, since in many cases the operator's response is conditioned by feedback received from the machine.

SYSTEM RESPONSE-TIME CONSIDERATIONS

Whereas the discussion so far has dealt with the general question of how human reaction time affects a system, the following comments deal with how system response time affects the human—or how machine delays are perceived by the human operator.

Gallaway[3] notes that "silences longer than about 4 s are perceived as unacceptable breaks" in both human (conversation) and man-machine operations (especially in interactive human-computer systems). He defines the following general response-time considerations:

As a delay (response time) is extended beyond the user's concept of the amount of work being done by a system, frustration increases and affects the quantity and quality of the user's performance.

Response-time delays are more than a mere inconvenience to the user; i.e., rather than a straight-line decrease in efficiency as delay increases, there are sudden drops in mental efficiency.

Performing most tasks requires an operator to "hold" a body of information in mind, and the longer this information has to be held, the greater is the chance of forgetting (after 2.0 s, most individuals become aware of waiting, and short-term memory failure begins to occur).

Response times that exceed 2.0 s should follow only a task that an operator perceives as being completed or one that is structured to appear as a completion point.

Longer response times are more acceptable to an operator following completion of an important or long task than following routine, short tasks.

Although a response signal can communicate more than one message concurrently, the response time that should be sought is that demanded by the task in the group which demands the fastest response time.

When a response to an operator input exceeds about 5.0 s, it is desirable to indicate to the operator that some process is being performed by the machine, e.g., "transaction being processed."

Systems with a high degree of variation in response times for a given task are more frustrating to operators. In principle, the range of acceptable variation in delay is that range within which the human cannot detect differences, i.e.,:

1. Response time of 0 to 2 s: variance should be about +5 percent.
2. Response time of 2 to 4 s: variance should be about +10 percent.
3. Response time of 4 to 30 s: variance should be about +15 percent.

In general, delays greater than 15 s rule out conversational interactions. When delays exceed this amount, the operator should be freed from physical or mental captivity (waiting on the machine) and allowed to turn to another activity until some display requests that he or she return to the operation.

Response times greater than 4 to 15 s generally are too long for conversation requiring short-term memory and probably inhibit problem-solving activity and/or create frustration for a data-entry task.

Response times greater than 2 to 4 s may be inhibiting for tasks requiring a high level of concentration. A 2-s wait at a terminal when the operator is absorbed and emotionally committed to task completion will seem an interminable wait for most operators.

Response times less than 2 s are desirable when the operator has to remember information through several responses.

Response to a signal (e.g., feedback to a pressing key) should be set at less than 100 ms.

[2]The task was to enter 10 different digits.

[3]G. R. Gallaway, *Response Times to User Activities in Interactive Man-Machine Computer Systems,* Report no. 277-0014601, NCR Corp., Dayton, Ohio, 1978.

Other References for Human Reaction Time

BAILEY, G. B., and R. PRESGRAVE: *Basic Motion Timestudy,* McGraw-Hill Book Company, New York, 1958.

BROADBENT, D. E.: *Successive Responses to Simultaneous Stimuli,* Report FPRC-934, Flying Personnel Research Committee, RAF Institute of Aviation Medicine, Farnborough, England, 1955.

BROWN, J. S., et al.: *Discrete Movements toward and away from the Body in the Horizontal Plane,* Report no. 6, Project no. 2, U.S. Naval Training Devices Center, Port Washington, N.Y., 1948.

BROWN, J. W., and A. T. SLATER-HAMMEL: "Discrete Movements in the Horizontal Plane as a Function of Their Length and Direction," *Journal of Experimental Psychology,* vol. 39, no. 84, 1949.

DAMON, ALBERT, et al.: *The Human Body in Equipment Design,* Harvard University Press, Cambridge, Mass., 1966.

DEUPREE, R. H., and J. R. SIMON: "Reaction Time and Movement Time as a Function of Age, Stimulus Duration and Task Difficulty," *Ergonomics,* vol. 6, no. 403, 1963.

FORBES, G.: "The Effect of Certain Variables on Visual and Auditory Reaction Times," *Journal of Experimental Psychology,* vol. 35, no. 153, 1945.

GIBBS, C. B., et al.: *Reaction Times to the Flashing Light Signals of Cars: The Effects of Varying the Frequency and Duration of the Flash,* Report APU-245, Applied Psychology Research Unit, Medical Research Council, Cambridge, England, 1955.

HYMAN, R.: "Stimulus Information as a Determinant of Reaction Time," *Journal of Experimental Psychology,* vol. 45, no. 188, 1953.

MACKWORTH, J. F., and N. H. MACKWORTH: "The Overlapping of Signals for Decisions," *American Journal of Psychology,* vol. 69, no. 26, 1956.

MILTON, J. L., et al.: *Pilot Reaction Time: The Time Required to Comprehend and React to Contact and Instrument Recovery Problems,* Report TSEAA-694-13a, Aerospace Medical Research Laboratory, Wright Air Development Center, Wright-Patterson Air Force Base, Ohio, 1947.

MORGAN, C. T., et al. (eds.): *Human Engineering Guide to Equipment Design,* McGraw-Hill Book Company, New York, 1963.

PROVINS, K. A.: *A Study of Some Factors Affecting Speed of Cranking,* Report no. 53/755, Royal Naval Personnel Research Committee, Medical Research Council, London, 1953.

REID, L. S., and J. G. HOLLAND: *The Influence of Stimulus Similarity and Stimulus Rate: Third of a Series of Reports on Experimental Analysis of Complex Task Performance,* WADC-TR-54-146, Aerospace Medical Research Laboratory, Wright-Patterson Air Force Base, Ohio, 1954.

VOLUNTARY CONTROLS

Voluntary control of human body components, as distinguished from involuntary movement, is achieved by means of the human brain (cortex). Voluntary movements are generally classified as "tension movement" or "ballistic movement." In the first case, there is a slow, intense movement created by contraction of antagonistic muscles operating against one another with unequal tension. In the second case, movements are free and generally more rapid, since the simultaneous operation of antagonistic muscles is at a minimum.

Voluntary movements are part of a perceptual-motor process involving coordination between one or more sensors (visual, auditory, tactile, and/or proprioceptive), the cortex, and the musculoskeletal (motor response) system. The perceptual-motor system is a complex feedback system involving at least three major loops (see the accompanying diagram).[1] As this diagram illustrates, the human control system involves such concepts as oscillation, inertia, damping, reverberation, and lag, and its characteristics at any given time depend on many factors, including mental set, practice, instructions, motivation, the physical characteristics of the workplace, and the environment (temperature, light, sound, vibration, etc.).

In general, the human functions as a nonlinear system but becomes more linear as the task becomes more predictable.

The human system often introduces its own "noise," which may affect the ultimate response. Motor coordination cannot be generalized from task to task, although steadiness in any given individual remains constant.

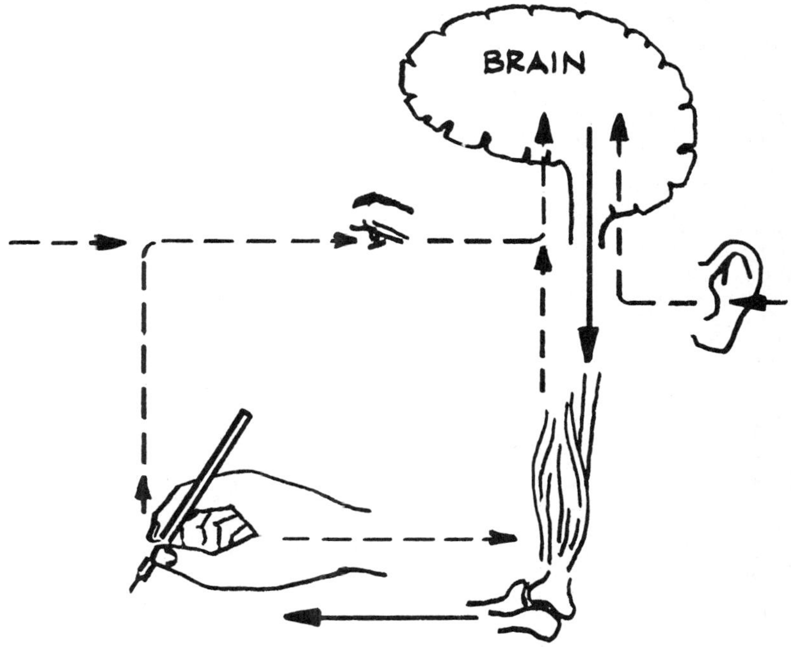

Although the perceptual-motor response may be driven primarily by specific sensory inputs (visual or auditory), continuous feedback is received by other sources (tactual, kinesthetic, proprioceptive).

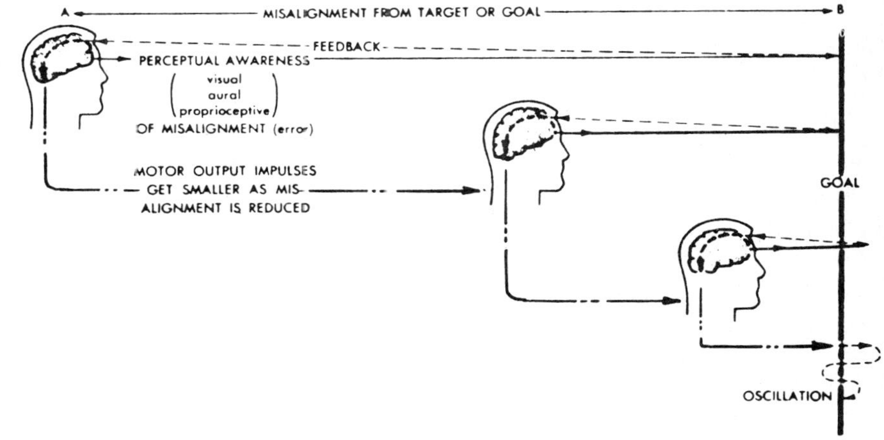

[1]W. E. Woodson and D. W. Conover, *Human Engineering Guide for Equipment Designers,* University of California Press, Berkeley, 1964.

GENERALIZATIONS

Eye-hand coordination is generally more accurate and reliable than ear-hand coordination.

Sensorimotor control between any of the sensory channels and the hands is generally more accurate and reliable than that between the sensory channel and the feet.

Hand and arm movement coordination is better when these movements are close to the body and symmetrical.

Arm movements that progress forward and/or away from the body are more accurate than arm movements that are directed toward the body. The same is true of leg movements.

Right-handed individuals are more proficient at making clockwise movements than at making counterclockwise movements; left-handed individuals are just the opposite. However, all people make clockwise movements better with the right hand, and counterclockwise movements better with the left hand.

Generally, a person can rotate his or her hand and wrist more precisely in one direction than another, i.e., the right hand clockwise, and the left hand counterclockwise.

Multiple arm and/or leg translatory movements are more efficient when they are similar, i.e., moving the left hand to the left or forward while at the same time moving the right hand to the right and forward.

A person can apply force more accurately to two simultaneously operated controls (i.e., one for each hand) when the controls are located symmetrically with respect to the body and when the directions of movement are similar.

When separate and different kinds of controls are operated simultaneously, there is a high probability that the operation of one or another control will suffer in terms of operator input efficiency.

Combined movements often result in "cross talk" e.g., trying to push and precisely rotate a control at the same time almost invariably introduces inaccuracy in one of the motions.

Excessive disparity between two manual operations (e.g., patting one's head while at the same time rubbing the stomach) often results in a complete breakdown of the sensorimotor response of one of the activities.

Continuous feedback is desirable in order for most control movements to remain optimized, i.e., to maintain an accurate direction, force application, and/or rate of movement.

On the basis of proprioceptive feedback, individuals judge extent of movement more accurately than movement force applied, and force more accurately than duration of movement.

Although visual control is more important while an individual is learning a new perceptual-motor task, as performance becomes habitual, proprioceptive feedback (or "feel") may become the more important feedback resource.

When less than 0.5 s per movement is required, blind movements are as accurate as when using visual positioning.

When the rate of movement is constant, accuracy diminishes as the interval between movements increases.

It takes about 0.04 s to develop maximum tension in the muscles of the human arm.

People tend to compensate for changes in the rate of sensory inputs versus movement rates; i.e., they move a control at a higher rate when the control display ratio is less sensitive.

Friction that is proportional to the rate of movement (viscous friction), in contrast to coulomb friction, is often advantageous in many control systems. Similarly, inertia in the form of a heavy control moved directly by the operator may also improve performance.

The control ratio (i.e., the ratio of control movement to indicator movement) generally has more effect on the time required to make settings within a given degree of tolerance than inertia, friction, or size of the control. Primary adjustment times decrease with increasing distance moved by a controlled marker (pointer or "hook") for each revolution of a control. Secondary adjustment times, however, increase with increases of marker movement per control movement or revolution. Overall time is minimum for gear ratios (e.g., pointer movement in inches per control revolution is between about 1 and 3). Although this ratio is optimum when fine tolerances are required, it may be advantageous to use higher ratios (6:1) in order to arrive at the general, desired setting. For joy-stick controls, the optimum condition for tolerances of 0.10 in (0.25 cm) is about 2½ units of stick movement for one unit of indicator movement.

FINGER DEXTERITY AND CONTROL OPERATION ACCURACY

Although certain people develop considerable dexterity with practice, the average person is able to perform certain types of control manipulations more accurately than others, as the accompanying sketches illustrate.

Rotational manipulation is more accurate than sliding manipulation or movement of thumb or finger wheels, although the latter is more accurate than the sliding manipulation.

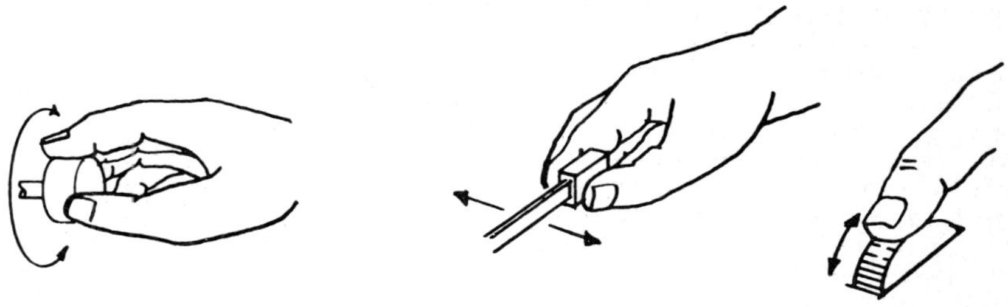

Rotation in the horizontal plane is generally more accurate than rotation in the vertical plane, although the horizontal accuracy depends on the ability of the operator to rest his or her hand on the adjacent surface.

A push button is located and pressed more accurately when it is positioned as shown in the first sketch than when it is positioned as shown in the second sketch.

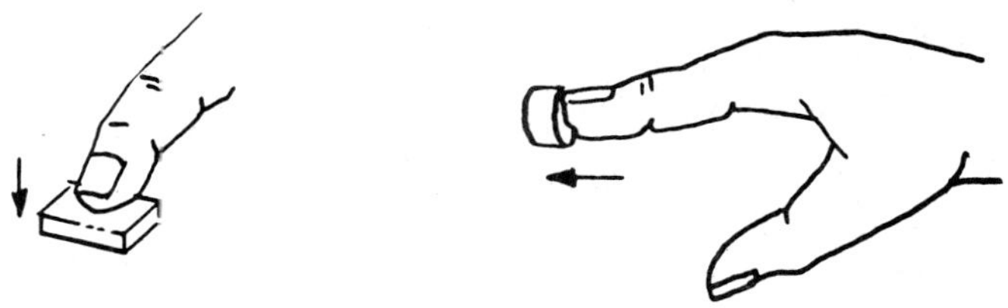

A pencil-sized joy stick is manipulated more precisely than one requiring a full fist grip, and the accuracy is increased significantly if the operator can rest his or her arm on a nearby horizontal surface.

An L-shaped handle is more accurately positioned than a round knob, such as a doorknob.

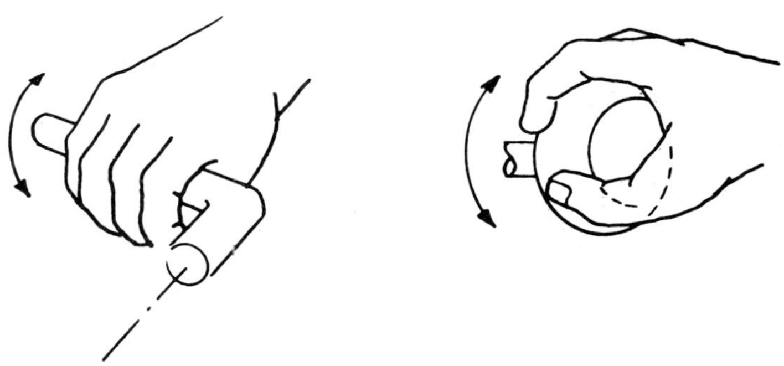

VISUAL POSITIONING ACCURACY

Visual positioning accuracy is a function of motor patterns such as direction and distance of movement. The accompanying chart illustrates the three important components: (A) primary movement time, (B) average reaction time (250 ms), and (C) secondary movement time. Primary movement time increases with distance, but less than in direct proportion to distance. Secondary movement time (e.g., small corrective adjustments) is relatively constant for distance in excess of about 10 cm.

In general and within reason, the relative time required for visual positioning movements is about the same for short and long distances because a higher velocity is typically used as distance increases.

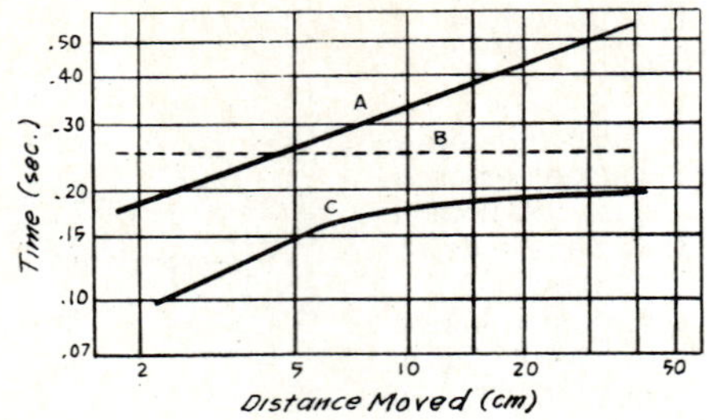

CRANKING MOVEMENTS

Maximum turn or cranking rate is obtained when the cranking radius is between 3 and 5 cm. However, maximum turn rate decreases with increasing load. For example, the curve shown in the accompanying graph is for zero load where the maximum turn rate is about 275 r/min. At 5000 g/cm, the turn rate becomes about 240 r/min.

In general, friction aids the discrimination of small load changes. However, inertia usually provides smoother cranking performance.

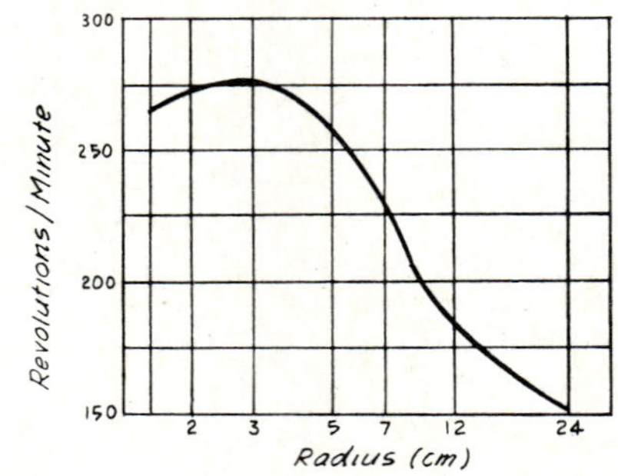

CONTINUOUS-ADJUSTMENT CONTROL RESPONSE

Common continuous-adjustment controlling tasks include steering and controlling speed and altitude. The controlled system may be a plant vehicle. Or, the movement and positioning of an external object may be involved, as when a crane operator acquires, transports, or deposits a package that is attached to the crane cable system.

Control effectiveness depends on at least the following factors:

The ability of the operator to *anticipate* what is going to happen when he or she provides input to the system

The ability of the operator to predict what will happen when he or she makes specific system inputs

Feedback on a timely basis about what is happening as the operator makes control inputs

How much differentiation, integration, and/or algebraic addition the control and display task requires of the operator

How well the specific control and display devices provide compatible relationships between the operator's sensory, perceptual, and motor and physical characteristics

Control effectiveness is degraded by the following factors:

Long delays between inputs, e.g., perceived changes in incoming information, results of operator inputs on system condition, or direct feedback from controller manipulation

Too much noise in the system, e.g., extraneous signals, dynamic disturbances, or mechanical artifacts such as "dead space," "stiction," and force irregularities

Incompatibilities between control and display direction and rate of motion

Controller force requirements that are either too high or too low

Incompatibility of the position, direction, and range of movement of the controller with the operator's musculoskeletal system and operating position

A requirement that the wrong body element be used, e.g., the hand versus the foot, the left hand versus the right hand, or the whole limb versus the hand and fingers

The following general statements about human efficiency with respect to continuous-adjustment control activities should be considered.

Humans are more efficient:

When they can make larger motions, since their own proprioceptive feedback mechanisms impart information about what they are doing.

When the movements they make are in the *same* direction in which the object, system, or displayed element moves.

When the rate of change of their control movement is similar to that of the controlled object or displayed element.

When sufficient information is supplied (naturally or artificially) to allow them to "predict" what is going to happen if they maintain their present control input and/or modify it to some extent. To predict, they must also know the general limits of their control system's response range.

When the control forces are not too high and are approximately equal throughout the controller movement range; i.e., high initiating forces (stiction) require the operator to suddenly compensate, once the controller is put into motion.

When there is some friction in the control system to minimize the effect of external dynamic disturbances and/or inequalities in their own musculoskeletal systems (asymmetry, tremor, fatigue, etc.). However, variable force within certain systems may be desirable to provide the operator with cues relative to the position or condition of the system, e.g., increasingly higher force as a brake pedal approaches "lock-up," toward the limits of a steering-wheel excursion to minimize overcontrolling, or force cues to indicate position of a transmission shift lever.

When they are properly positioned and secured; i.e., a seated operator with appropriate handrests or armrests is less influenced by dynamic disturbance and by problems of maintaining body equilibrium.

In terms of small, accurate inputs, when they primarily rely on hand and finger movement to manipulate a controller.

When they are not required to manipulate too many separate controls in an integrated manner, i.e., when they do not have to perform sequential operation of several hand and/or foot controls while at the same time carrying out a primary continuous-adjustment task (several switching inputs on a joy stick or steering wheel, etc.).

When they do not have to hold a control device (and their arms or legs in suspension) for long periods of time or hold the control in a fixed position for extended periods.

When they are provided appropriate system aids to relieve them of complex mental information processing to differentiate, integrate, extrapolate, or perform algebraic additions during the control tasks.

When the controller-display setup allows them to see what they are doing, e.g., when their hands do not cover up displayed information, and when vehicle elements do not obscure the external view.

Pursuit versus Compensatory Tracking

When performing a tracking task that involves matching a continuously variable input signal with an output of the system, the operator requires information indicating the input to be matched and the output with which it is to be matched. When an operator's control and display system provides separate indications for input and for output, this is referred to as a "pursuit task"; i.e., the operator makes inputs into the control and then to a display element to "follow" a target (system input). On the other hand, when input and output signals are presented to the operator in terms of a difference between the system and the operator's control input, the task is called "compensatory" or "error-reducing."

Pursuit tracking is generally a more "natural" task. In addition, the operator is better able to keep track of input changes and to anticipate them, and the operator can also keep better track of the position of the controlled element. Pursuit tracking is the preferred mode for simple systems, particularly where input frequency is relatively high.

Compensatory tracking is generally less natural (and thus requires training). However, when the dynamic characteristics of the system are complex, compensatory tracking modes minimize the probability that operators will confuse input with the results of their own control actions. Another important advantage is that compensatory tracking information can be displayed in less space because in error-reducing formats the total output range does not have to be displayed.

In general, a pursuit display is best when high-frequency inputs are present, and a compensatory display is best when low-frequency inputs are present. A pursuit display allows the operator to anticipate. It is best for higher-order control systems; the higher the frequency of desired output, the more superior a pursuit mode is to a compensatory mode. With rate controls, a pursuit display is superior as long as the desired output and input are time-varying. With rate controls, a pursuit display is superior except when the cutoff frequency is about 0.1 Hz or less. When the display must be small (because of lack of space) and when the change in desired output is very slow, a compensatory display mode is superior. A compensatory display is best when an "aided system" supplies derivative terms where positional errors are corrected (i.e., the operator performs better when making corrective responses proportional to only the error).

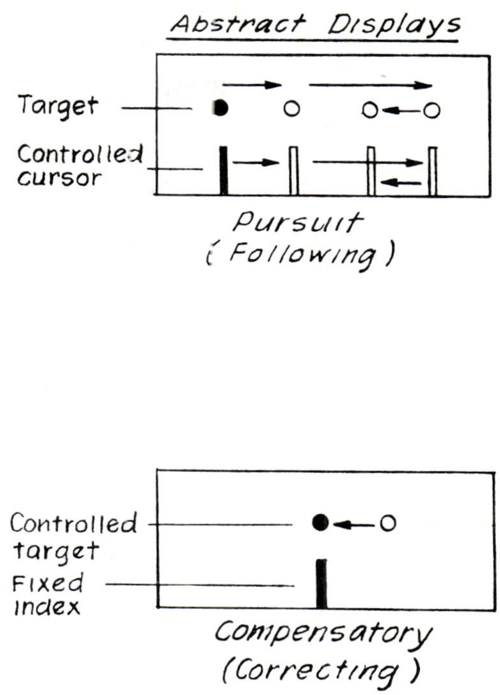

A pursuit display should be used when:

1. The course contains high frequencies.
2. It is a zero-order control system.
3. The operator must know actual output, not just error.

A compensatory display should be used when:

1. The system is aided or quickened
2. The display must be kept small, but output range is large, and/or the precision requirements are high.

Either display can be used when:

1. The course is simple.
2. The display needs to be kept small, but the output range is large, and/or the precision requirements are high.[2]

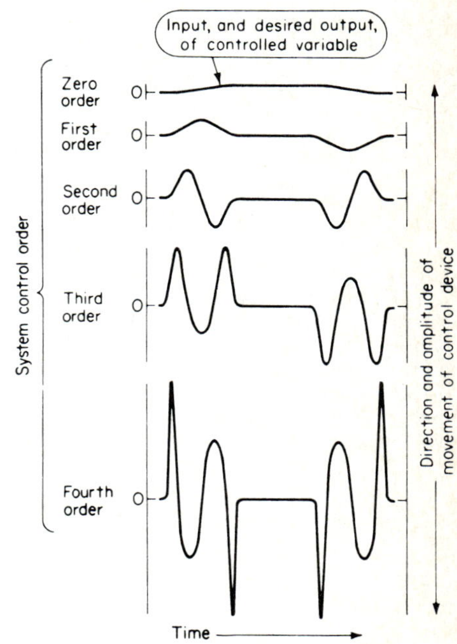

Control order illustrated by changes in a controlled variable and in its first four derivatives. Each line represents the changes over time that would have to be made with control systems of various control orders to make the controlled variable correspond to the input.

[2]For a more complete discussion of human tracking capabilities and limitations and for detailed recommendations regarding the design of closed-loop systems, see E. J. McCormick, *Human Factors in Engineering and Design*, 4th ed., McGraw-Hill Book Company, New York, 1976, p. 220.

LIFTING AND CARRYING

Variables to be considered are the following

1. Task Variables
Location of the object to be lifted (accessibility)
Size and shape of the object
Height from and/or to which the object is lifted
Weight and weight distribution of the object
Relation between the object's c/g and points of grasp
Whether the object is lifted by handles
Working-position characteristics (e.g., awkward rather than normal)
Manipulatory accuracy requirements
Frequency and duration of lifting and carrying movements

2. Human Variables
Age and sex of the person or persons doing the lifting
Body dimensions (stature, arm length, etc.)
Physical fitness
Experience and training

3. Environmental Variables
Thermal (temperature, humidity, and ventilation)
Dynamic (platform motion and vibration)

The most important determinant of lifting force is the distance of the feet from the grasping axis (lifting force is greatest if the lift weight is in the same vertical plane as the body, and it decreases rapidly as the weight moves away from the body). The best height for lifting is approximately fingertip height, with the hands hanging freely at the sides. Above this height, lift capacity decreases very rapidly; below this height, it decreases more slowly. With the load near the floor, capacity is only about 75 percent of best height capacity.

The maximum individual loads carried in either hand by means of handgrips should be about 60 lb (27 kg) for short distances and 35 lb (14 kg) for longer distances. Bulky articles should not weigh more than about 30 lb (13 kg). In general, a package is considered heavy when it reaches about 35 percent of a person's own body weight. Less strain occurs if equal weights are carried in a balanced manner, i.e., half in one hand and half in the other. Carrying one heavy weight a given distance incurs less physiological cost than carrying two weights, each weighing half as much, the same distance in two trips.

Handle size and shape, as well as position, make a considerable difference in the individual's acceptance of maximum weight.

General Guidelines with Respect to Lifting

Assuming that a package is appropriately designed so that it can be picked up, carried, and deposited without damage to the package or injury to the individual, the general weight values shown in the accompanying tables can be used as guidelines in the design of hand-carried packages.

The accompanying sketches are meant to emphasize significant considerations in developing an appropriate package size, shape, and weight configuration. First, one must be able to get hold of the package; thus the designer needs to consider the position the person has to get into and how his hands and fingers can be accommodated to obtain a firm grip. Second, the package should be designed so that the individual can assume a comfortable and safe carrying position (minimum back or arm strain, minimum pressure on the hands or fingers, ability to "see ahead," etc.). Third, the position where the package is to be deposited is important, i.e., if the final resting level is high, the package weight cannot be as great.

Some Typical Industrial Standards

Reasonable Weight Limits for Occasional Lifting

Age, Years	Male	Female
14–16	33 lb	22 lb
16–18	42 lb	26 lb
18–20	51 lb	31 lb
20–35	55 lb	33 lb
35–50	46 lb	29 lb
Over 50	35 lb	22 lb

Source: From the Swiss Accident Insurance Institute.

Maximum Acceptable Weight of Lift for Adult Males

Height of Lift	Percent of Population				
	90	75	50	25	10
Floor level to knuckle height	52 lb	59 lb	66 lb	73 lb	80 lb
Knuckle height to shoulder height	51 lb	56 lb	62 lb	68 lb	73 lb
Shoulder height to arm reach	48 lb	53 lb	60 lb	67 lb	72 lb

Source: From Liberty Mutual Insurance Company.

Lifting Packages without Handles (Young Adult Males Typical of Industrial and Military Populations)[1]

The shape of the package has an appreciable effect on a person's ability to pick it up, lift it, and transport it, as shown by the accompanying sketches. It should also be noted that whether the package is cylindrical or rectangular affects how closely or snugly it can be grasped and/or how effectively it can be manipulated; i.e., at certain points in the handling sequence a rectangular shape provides advantages, and at other points a cylindrical shape provides advantages.

The accompany sketches illustrate a continuing lifting and carrying sequence for a person with a 5th- to a 95th-percentile physique.

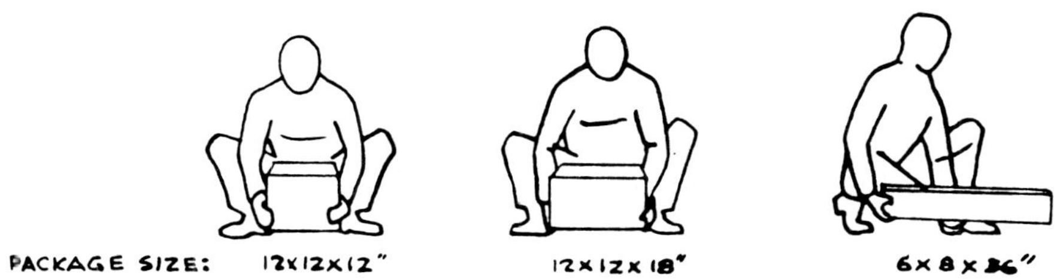

Lifting and Carrying Height above the Floor	A	B	C
36 in	75 lb	95 lb	110 lb
48 in	55 lb	75 lb	95 lb
60 in	40 lb	50 lb	50 lb

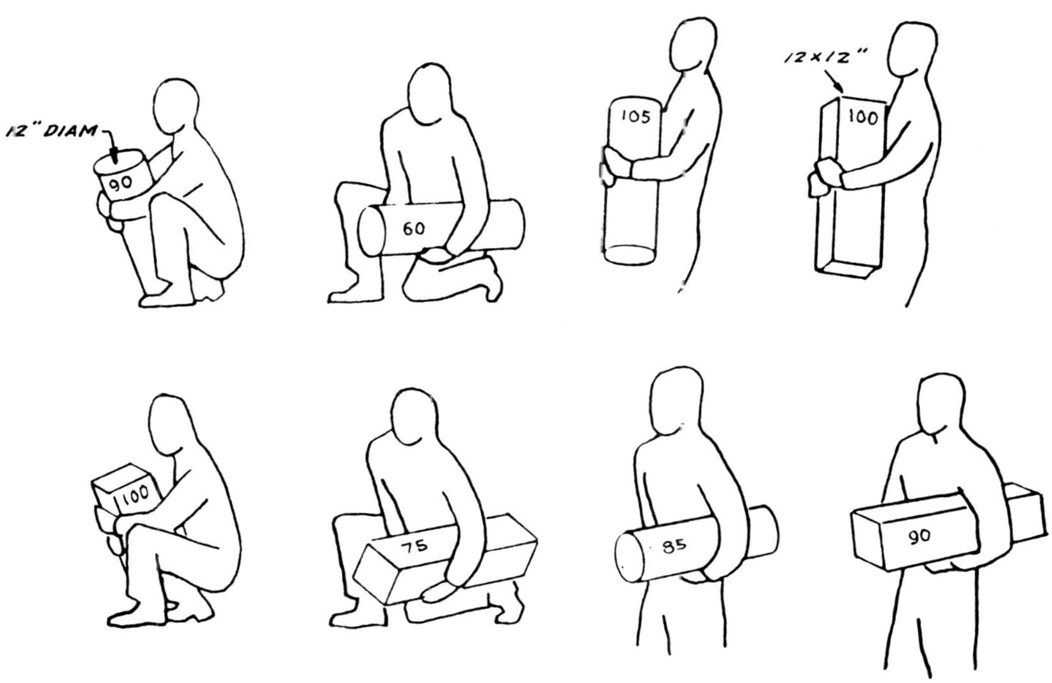

[1] W. E. Woodson and D. W. Conover, *Human Engineering Guide for Equipment Designers,* University of California Press, Berkeley, 1964.

Weight in Relation to Carrying Mode

The ease with which various packages can be carried depends not only on the weight but also on the type and placement of the handles. The accompanying sketches illustrate common situations and suggested upper weight limits for typical package-carrying configurations:

When the package (e.g., a suitcase) has smooth sides, it can be carried against the person's hip or leg without too much stress. The weight should not exceed more than about 45 lb (20 kg) for males and 35 to 40 lb (14 to 18 kg) for females.

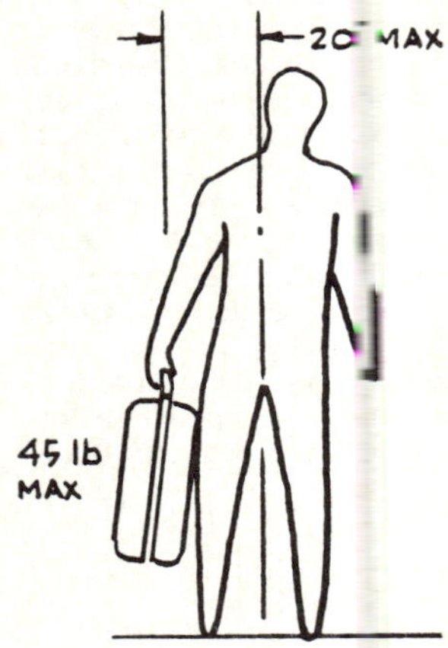

When the package surfaces are not smooth (as is the case with electronic equipment chassis, etc.), limit weight to about 35 lb (14 kg).

Although the typical male can carry an equipment package weighing up to about 60 lb (27 kg) for a short distance, this much weight should involve two persons carrying the package by means of appropriately located dual handles. Two men should not be required to carry 100 lb (45 kg) very far or more than 200 lb (90 kg) more than a short distance.

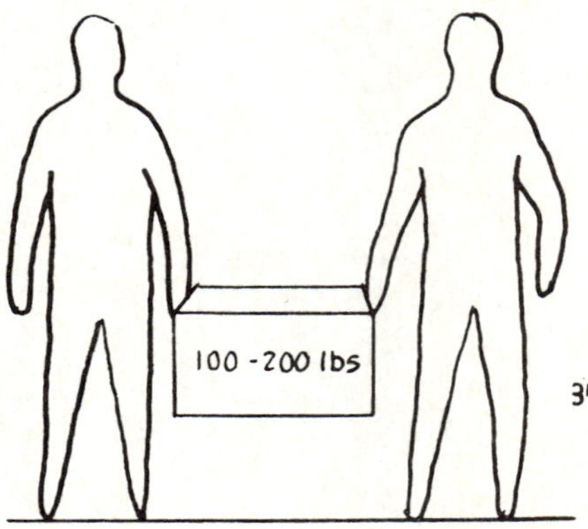

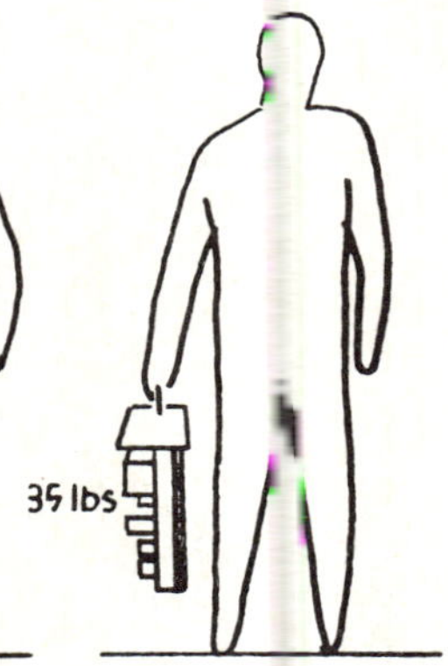

The removal and replacement of equipment units from a rack at different heights require consideration of the suggested maximum package weights in relation to working height, as indicated in the accompanying sketch.

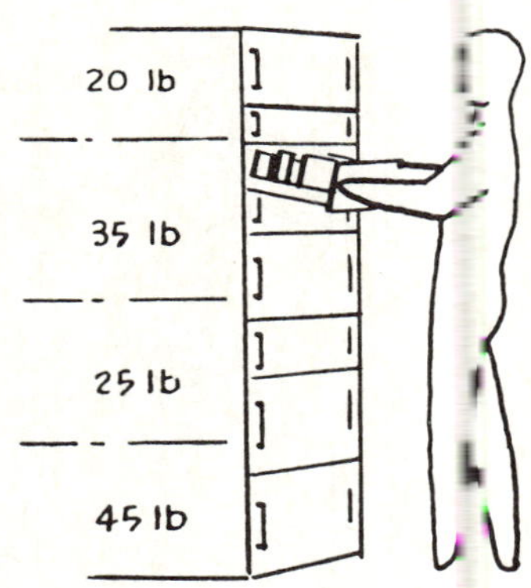

Backpack weight-carrying guidelines are indicated in the accompanying sketch.

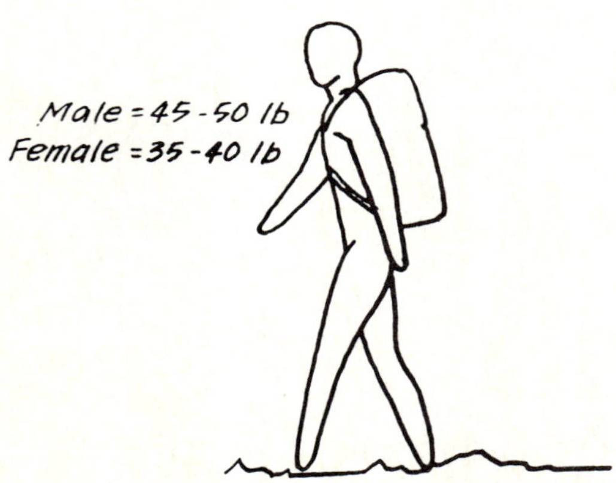

CAPACITY TO MOVE LARGE OBJECTS

The capacity of an individual to move large objects, such as items of furniture, depends on several factors, including the position the individual can assume (see the accompanying sketches). The values shown are general guidelines for each operator-package relationship (forces for women are indicated by the values in parentheses). An individual can move considerably more weight after the object has begun to move; therefore, it behooves the designer to devise appropriate slide or caster systems to reduce the forces required to put the object in motion.

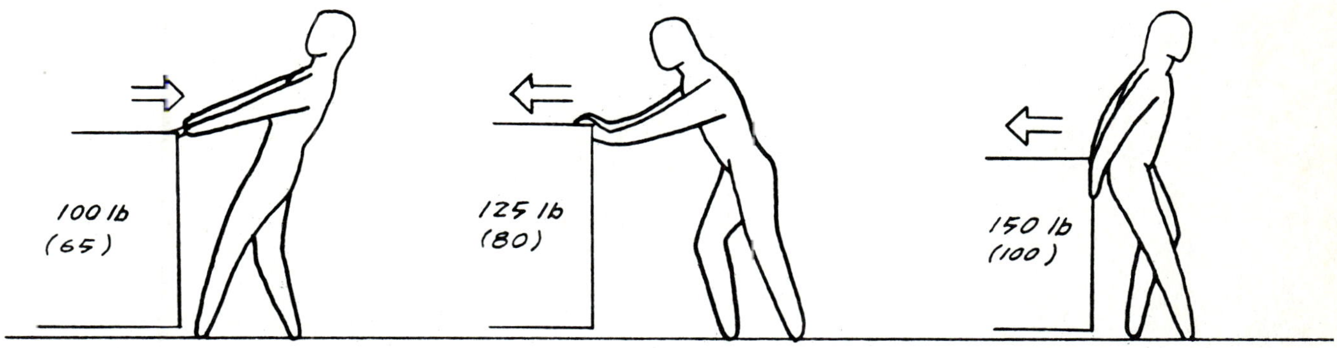

The accompanying tables provide more precise maximal static force capabilities, as determined by Kroemer[2] for adult males, for various application modes. Note that these are for conditions that include structural "push-off" support.

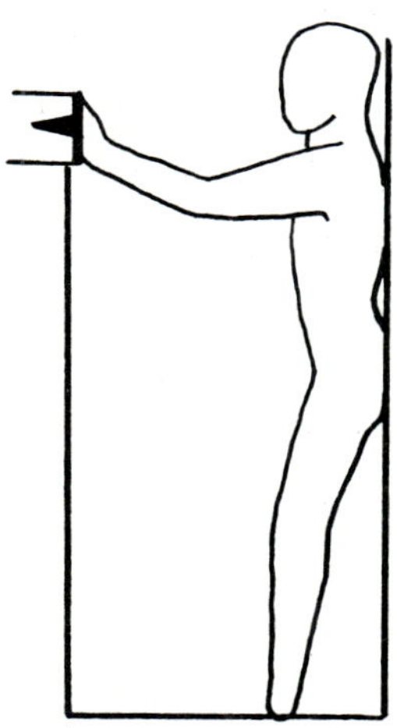

Distance	Force Means	Force SD
	Both Hands	
50	131 lb	32
60	150 lb	36
70	221 lb	61
80	289 lb	90
90	220 lb	68
100	145 lb	57
	Single, Preferred Hand	
50	59 lb	15
60	67 lb	16
70	81 lb	22
80	117 lb	32
90	111 lb	38
100	96 lb	39

[2] K. H. E. Kroemer, *Push Forces Exerted in Sixty-Five Common Working Positions,* AMRL-TR-68-143, Aerospace Medical Research Laboratory, Wright-Patterson Air Force Base, Ohio, 1968.

Distance	Force Means	SD
a. 80	149 lb	40
100	174 lb	48
120	175 lb	37
b. 80	161 lb	36
100	164 lb	52
120	184 lb	31
c. 80	141 lb	33
100	152 lb	44
120	194 lb	2

Note: *a* = 50 in; *b* = 70 in; c = 90 in.; *a, b,* and *c* represent the height of the force plate.

Distance	Force Means	SD
a. 70	171 lb	38
80	192 lb	40
90	178 lb	32
b. 60	130 lb	25
70	157 lb	28
80	163 lb	32
c. 60	117 lb	29
70	139 lb	29
80	143 lb	30

Note: *a* = 60 in; *b* = 70 in; *c* = 80 in.

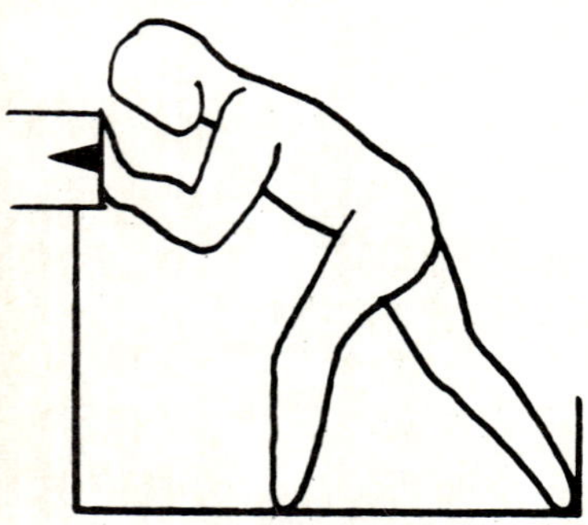

Distance	Force Means	SD
a. 70	140 lb	33
80	155 lb	35
90	132 lb	30
b. 70	123 lb	28
80	122 lb	28
90	120 lb	18
c. 70	97 lb	21
80	101 lb	21
90	109 lb	18

Note: *a* = 70 in; *b* = 80 in; *c* = 90 in.

WORK EFFICIENCY

If one thinks of humans as machines that produce work output, they are much like factories, i.e., they take in raw materials in the form of food, water, and air, and from these they produce materials for maintaining their bodies, power to operate the various body systems, and energy to perform work. The efficiency with which work is performed is a ratio of output to input. The following equation describes the process:

$$\text{Work efficiency} = \frac{\text{work done (e.g., foot-pounds)}}{\text{work done} + \text{heat produced}}$$

It is estimated that human efficiency generally does not exceed about 20 to 25 percent. According to Shilling[3] a typical 154-lb (70-kg) adult male has the fuel and energy requirements for typical activities shown in the accompanying table. The approximate amount of energy required for one day of typical activities are estimated to be as shown in the accompanying table.

Activity	Calories per Hour
Sleeping	65
Resting (awake but lying still)	77
Standing at attention	115
Typewriting rapidly	140
Sweeping the floor (38 strokes per minute)	169
Walking slowly (2.6 mi/h)	200
Performing carpentry, sheetmetal work, or painting	240
Walking moderately fast	300
Sawing wood	480
Swimming	500
Running (5.3 mi/h)	570
Walking up steps	1100

	Calories
8 h of sleep	520
3 h of light work	510
8 h of medium to heavy work	1920
5 h of sitting at rest	500
Day's total	3450

COMPARISON OF ENERGY EXPENDITURES BY MALES AND FEMALES FOR SELECTED WORK TASKS

Occupation	Type of Work	Kilocalories per Day	
		Men	Women
Bookkeeper	Sedentary, light manual	2400	2000
Secretary	Sedentary, light manual	2700	2250
Bus driver	Sedentary, medium manual	3000	2500
Letter carrier	Walking, climbing stairs, and carrying mailbag	3000	2500
Machine operator	Sedentary, heavy manual	3300	2750
Stonemason	Heavy manual	3600	3000
Laborer	Heavy manual, lifting and carrying	3900	3250
Carpenter	Manual, continuous (hammering and sawing)	3900	3250
Ballet dancer	Vigorous, continuous whole-body movements	3900	3250
Coal miner	Standing, extremely heavy body and limb movements	4200	—
Lumberjack	Standing, extremely heavy body and limb movements	4500	—
Athlete	Marathon running, swimming, and football	4800	4250

Energy Expenditure in Relation to Body Position and Activity

Posture and Mobility Involvement	Kilocalories per Minute	Kilocalories per Hour
Sitting	0.3	20.0
Kneeling	0.5	30.0
Crouching	0.5	30.0
Standing	0.6	35.0
Bending over while standing	0.8	50.0
Walking (medium rate)	1.7–3.5	100–200
Climbing (10° grade minus load)	0.75-m climbing height	0–400
Work involving one arm	0.7–2.2	40–120
Work involving two arms	1.5–3.0	80–160

Note: The above values represent general estimates and should not be used except to establish approximations for evaluating the consequence of proposed design or task configurations.

Source: G. Lehmann, *Praktische Arbeitsphysiologie,* Georg Thieme Verlag, Stuttgart, 1953.

Specific Task Output Costs

Activity	Working Conditions	Kilocaleries per Minute
Walking	Without load, level surface (3km/h)	1.7
	(4km/h)	2.1
Walking	With load carried on back (10 kg/4km/h)	3.6
	(30 kg/4km/h)	5.3
	(50 kg/4km/h)	8.1
Climbing	16°/11.5 m/min (no load)	8.3
	(20-kg load)	10.5
	(50-kg load)	16.0
Cycling	16 km/h	5.2
Housekeeping	Cooking	1.0–2.0
	Light cleaning, ironing,	2.0–3.0
	scrubbing floor	4.0–5.0

[3]C. W Shilling, *The Human Machine,* U.S. Naval Institute, Annapolis, Md., 1955.

WORK, HEAT, AND OXYGEN COSTS

The human machine oxidizes its fuel in order to obtain energy. The body oxidizes fuel at temperatures of about 37°C and converts chemical energy directly into work.

Human energy exchange is expressed primarily in terms of heat. When work is performed, 10 to 30 percent of the energy expenditure is applied to the external load, and the rest comes out as heat.

Oxygen costs (e.g., heat production) for typical activities are listed in the accompanying tables.[4]

[4]P. Webb, "Work, Heat, and Oxygen Cost," in J. F. Parker and V. R. West (eds.), *Bioastronautics Data Book* (2d ed., National Aeronautics and Space Administration), 1973.

Everyday Activities

Activity	Oxygen Consumption, l/min	Equivalent Heat Production, kcal/min
Asleep		
Sleeping, men over 40	0.22	1.1
Sleeping, men aged 30–40	0.24	1.2
Sleeping, men aged 20–30	0.24	1.2
Sleeping, men aged 15–20	0.25	1.3
Resting		
Lying fully relaxed	0.24	1.2
Lying moderately relaxed	0.26	1.3
Lying awake, after meals	0.28	1.4
Sitting at rest	0.34	1.7
Very light activity—seated		
Writing	0.36	1.8
Riding in automobile	0.40	2.0
Typing	0.46	2.3
Polishing	0.48	2.4
Very light activity—standing		
Relaxed	0.36	1.8
Drafting	0.38	1.9
Taking lecture notes	0.40	2.0
Peeling potatoes	0.42	2.1
Light activity—seated		
Playing musical instruments	0.58	2.9
Repairing boots and shoes	0.60	3.0
At lecture	0.60	3.0
Assembling weapons	0.72	3.6
Light activity—standing		
Entering ledgers	0.52	2.6
Washing clothes	0.74	3.7
Ironing	0.88	4.4
Scrubbing	0.94	4.7
Light activity—moving		
Slow movement about room	0.50	2.5
Vehicle repairs	0.68	3.4
Slow walking	0.76	3.8
Washing	0.84	4.2
Moderate activity—lying		
Creeping, crawling, prone resting maneuvers	1.14	5.7
Crawling	1.22	6.1
Swimming breaststroke at 1 mi/h	1.36	6.8
Swimming crawl at 1 mi/h	1.40	7.0
Moderate activity—sitting		
Rowing for pleasure	1.00	5.0
Cycling at 8–11 mi/h	1.14	5.7
Cycling rapidly	1.38	6.9
Trotting on horseback	1.42	7.1
Moderate activity—standing		
Gardening	1.16	5.8
Chopping wood	1.24	6.2
Baseball pitching	1.30	6.5
Shoveling sand	1.36	6.8
Moderate activity—moving		
Golf	1.08	5.4
Table tennis	1.16	5.8
Tennis	1.26	6.3
Army drill	1.42	7.1
Heavy activity—lying		
Leg exercises, average	1.50	7.5
Swimming breaststroke at 1.6 mi/h	1.64	8.2
Swimming backstroke at 1.0 mi/h	1.66	8.3
Lying on back, head raising	1.76	8.8
Heavy activity—sitting		
Cycling rapidly, own pace	1.66	8.3
Cycling at 10 mi/h, heavy bicycle	1.78	8.9
Cycling in race (100 mi in 4 h 22 min)	1.96	9.3
Trotting on horseback	1.96	9.3
Heavy activity—standing		
Chopping wood	1.50	7.5
Shoveling sand	1.54	7.7
Sawing wood by hand	1.60	8.0
Digging	1.78	8.9
Heavy activity—moving		

Everyday Activities *(continued)*

Activity	Oxygen Consumption, l/min	Equivalent Heat Production, kcal/min
Skating at 9 mi/h	1.56	7.8
Playing soccer	1.66	8.3
Skiing at 3 mi/h on level	1.80	9.0
Climbing stairs at 116 steps/min	1.96	9.8
Very heavy activity—sitting		
Cycling at 13.2 mi/h	2.00	10.0
Rowing with two oars at 3.5 mi/h	2.20	11.0
Galloping on horseback	2.28	11.4
Sculling (97 strokes/min)	2.52	12.6
Very heavy activity—moving		
Fencing	2.10	10.5
Playing squash	2.10	10.5
Playing basketball	2.28	11.4
Climbing stairs	2.40	12.0
Extreme activity		
Wrestling	2.60	13.0
Marching at double	2.66	13.3
Endurance marching	2.96	14.8
Harvard Step Test	3.22	16.1

Specialized Activities

Activities	Oxygen Consumption, l/min	Equivalent Heat Production, kcal/min
Engineering tasks		
Medium assembly work	0.58	2.9
Welding	0.60	3.0
Sheet metal work	0.62	3.1
Machining	0.66	3.3
Punching	0.70	3.5
Machine fitting	0.90	4.5
Heavy assembly work—noncontinuous	1.02	5.1
Driving vehicles and piloting aircraft		
Driving a car in light traffic	0.26	1.3
Night flying—DC-3	0.32	1.6
Piloting DC-3 in level flight	0.34	1.7
Piloting helicopters	0.36	1.8
Instrument landing—DC-4	0.50	2.5
Piloting light aircraft in rough air	0.54	2.7
Taxiing DC-3	0.58	2.9
Piloting bomber aircraft in combat	0.58	2.9
Driving car in heavy traffic	0.64	3.2
Driving truck	0.66	3.3
Driving motorcycle	0.70	3.5
Moving over rough terrain on foot		
Flat, firm road 2.5 mi/h	0.56–0.98	2.8–4.9
Grass path 2.5 mi/h	0.64–1.02	3.2–5.1
Stubble field 2.5 mi/h	0.80–1.22	4.0–6.1
Deeply plowed field 2.0 mi/h	0.98–1.38	4.9–6.9
Steep 45° slope 1.5 mi/h	0.98–1.38	4.9–6.9
Plowed field 3.3 mi/h	1.56	7.8
Soft snow, with 44-lb load 2.5 mi/h	4.2	21.0
Load carrying		
Walking on level with 58-lb load, trained men		
2.1 mi/h	0.38	1.9
2.7 mi/h	0.58	2.9
3.4 mi/h	0.92	4.6
4.1 mi/h	1.66	8.3
Walking on level with 67-lb load, trained men		
2.1 mi/h	0.46	2.3
2.7 mi/h	0.58	2.9
3.4 mi/h	1.02	5.1
4.1 mi/h	1.68	8.4

(continued)

Specialized Activities *(continued)*

Activities	Oxygen Consumption, l/min	Equivalent Heat Production, kcal/min
Walking on level with 75-lb load, trained men		
2.1 mi/h	0.50	2.5
2.7 mi/h	0.68	3.4
3.4 mi/h	1.04	5.2
4.1 mi/h	1.72	8.6
Walking up 36% grade with 43-lb load, sedentary men		
0.5 mi/h	1.34	6.7
1.0 mi/h	2.46	12.3
1.5 mi/h	3.20	16.0
Swimming on surface		
Breaststroke		
1 mi/h	1.40	7.0
2 mi/h	5.80	29.0
3 mi/h	19.40	97.0
Crawl		
1 mi/h	1.80	9.0
2 mi/h	3.60	18.0
3 mi/h	9.60	48.0
Butterfly		
1 mi/h	2.40	12.0
2 mi/h	5.80	29.0
3 mi/h	15.00	75.0
Walking under water		
Walking in tank minimal rate	0.58	2.9
Walking on muddy bottom minimal rate	1.10	5.5
Walking in tank maximal rate	1.44	7.2
Walking on muddy bottom maximal rate	1.68	8.4
Movement in snow		
Skiing in loose snow 2.6 mi/h	1.62	8.1
Sled pulling—hard snow 2.2 mi/h	1.72	8.6
Snowshoeing (bearpaw) 2.5 mi/h	1.74	8.7
Skiing on level 3.0 mi/h	1.80	9.0
Sled pulling—low drag, medium snow 2.0 mi/h	1.94	9.7
Snowshoeing—trail type 2.5 mi/h	2.06	10.3
Walking, 12–18 in snow, breakable crust 2.5 mi/h	2.54	12.7
Skiing on loose snow 5.2 mi/h	2.92	14.6
Snowshoeing—trail type 3.5 mi/h	2.96	14.8
Skiing on loose snow 8.1 mi/h	4.12	20.6
Measured work at different altitudes		
Bicycle ergometer		
Workload Altitude		
430 kg·m/min 720 mm Hg	1.02	5.1
430 kg·m/min 620 mm Hg	0.98	4.9
430 kg·m/min 520 mm Hg	1.08	5.4
Mountain climbing		
880–1037 kg·m/min 610 mm Hg	1.84–2.20	9.2–11.0
566–786 kg·m/min 425 mm Hg	1.54–1.90	7.7–9.5
393–580 kg·m/min 370 mm Hg	1.28–2.10	6.4–10.5

WORK AND REST

Approximately 4 kcal/min has been established as a desirable upper limit of the average energy cost of work. This requires that there be some rest in order for the body to compensate for the excess above this limit. A formula for estimating the amount of rest required for a given amount of work is

$$R = \text{rest} = \frac{T(K - S)}{K - 1.5}$$

where R = rest (in minutes), T = total working time, K = the average kilocalories per minute of work, and S = kilocalories per minute as the adopted standard. The value 1.5 approximates the resting level in kilocalories per minute.

If an S value of 4 kcal/min is used and we want to calculate R for a 1-h period (e.g., T = 60 min), the formula becomes

$$R = \frac{60(K - 4)}{K - 1.5}$$

Then one can project a curve such as that shown in the accompanying graph. Of course, individual differences in energy considered as an acceptable standard make this formula purely an approximation.

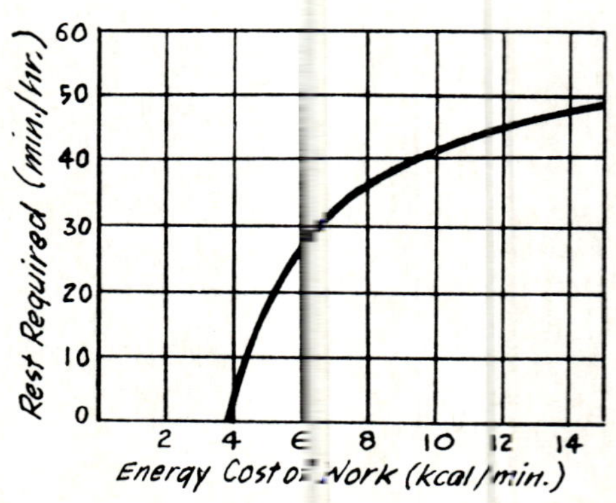

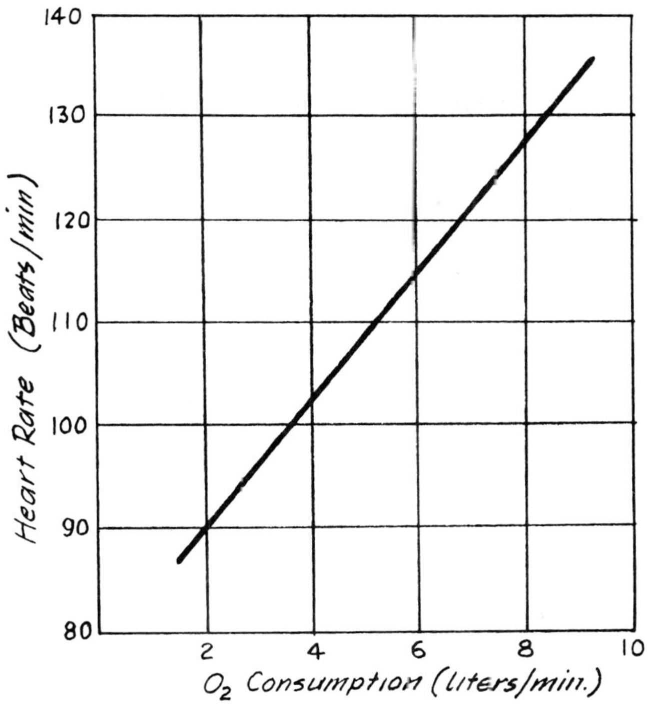

Heart rate versus oxygen consumption.

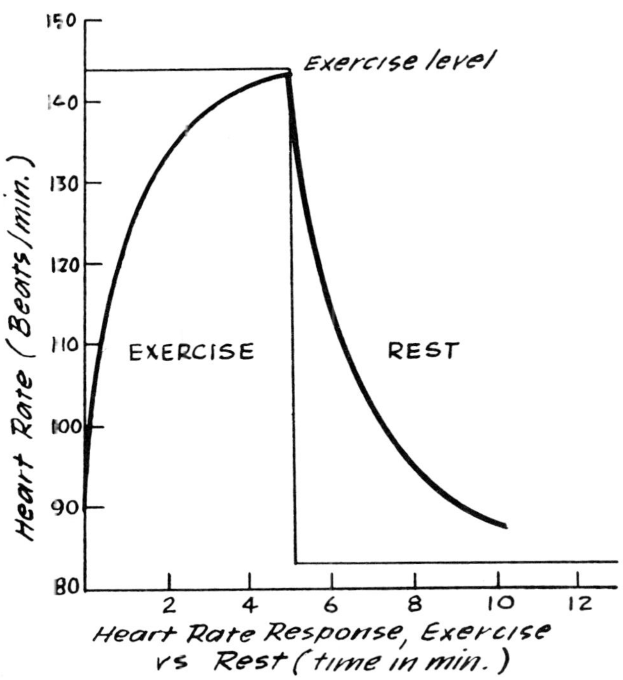

Heart rate: exercise versus rest.

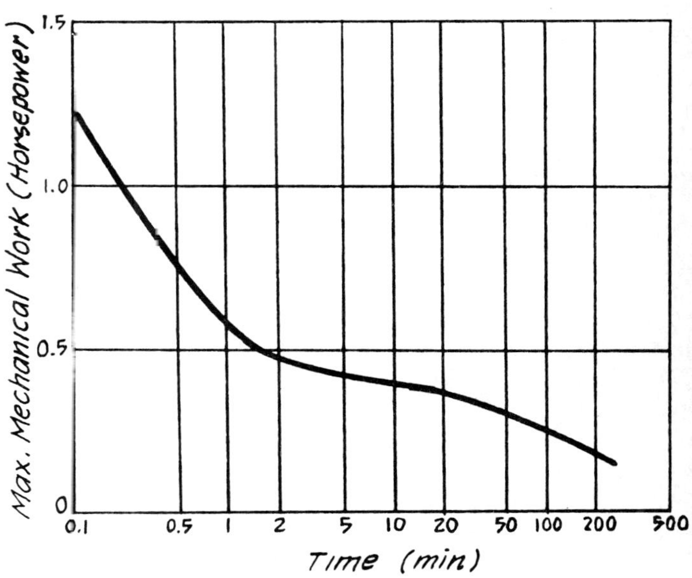

Work output versus work duration.

CALCULATING OXYGEN COSTS

To determine heat output from respiratory data:

1. Calculate oxygen consumption from respiratory ventilation volume and the difference in O_2 concentration between inspired and expired air.
2. Correct volume to O°C, 760 mmHg, dry (STPD).
3. Select heat output corresponding to each unit volume of O_2 by approximating the subject's diet or on the basis of his or her measured respiratory quotient.

The accompanying nomograms help simplify the calculations.

A.

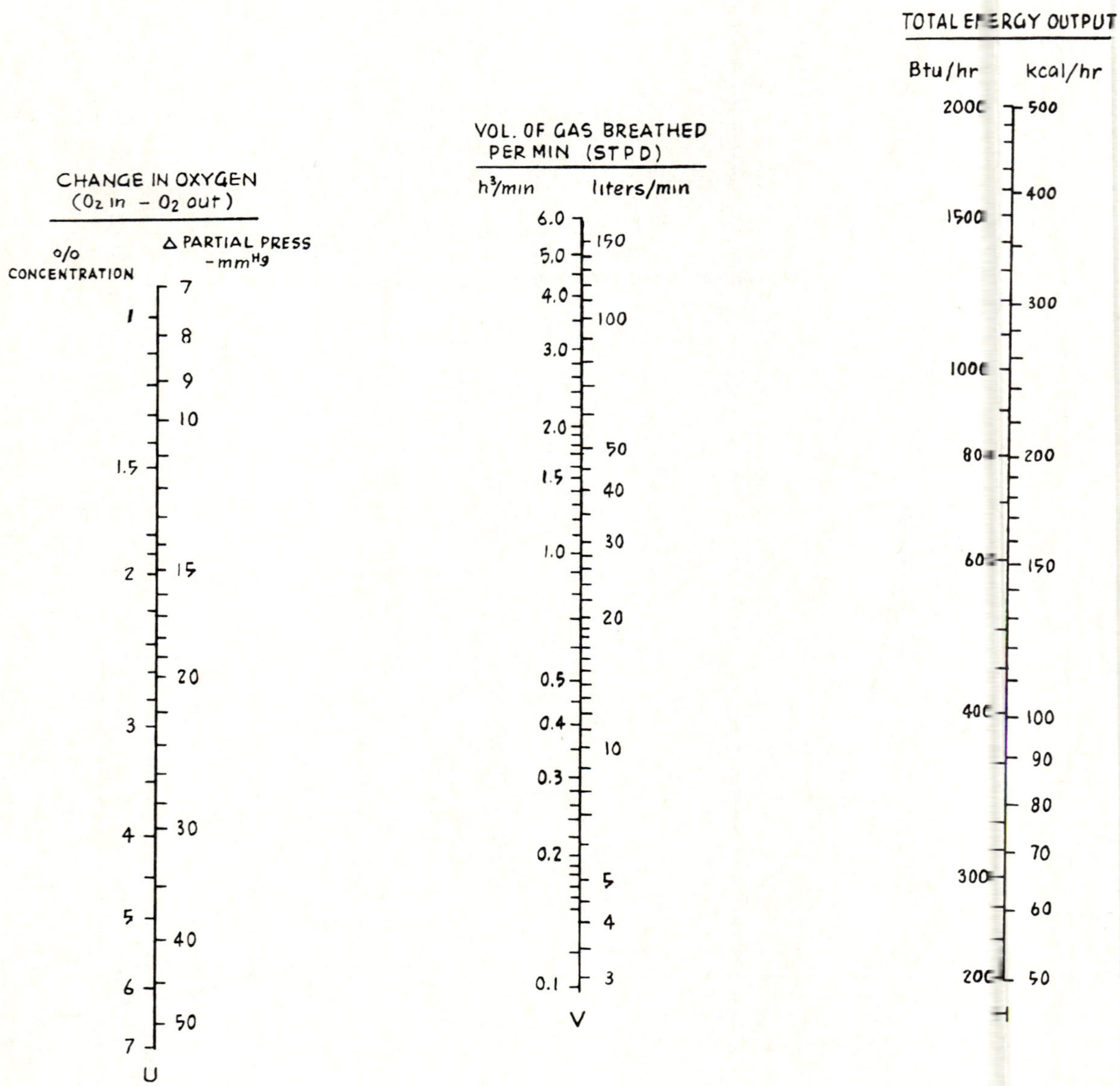

Nomogram A uses the following values: *RQ* = 1.00, and 1 L of O_2 is equivalent to 5.0 kcal, permitting direct calculation of heat output *(H)* in Btu per hour from oxygen uptake *(U)* and ventilation rate *(V)*. Alternatively, *U* can be calculated from *H* and *V*, or *V* can be calculated from *U* and *H*.

B.

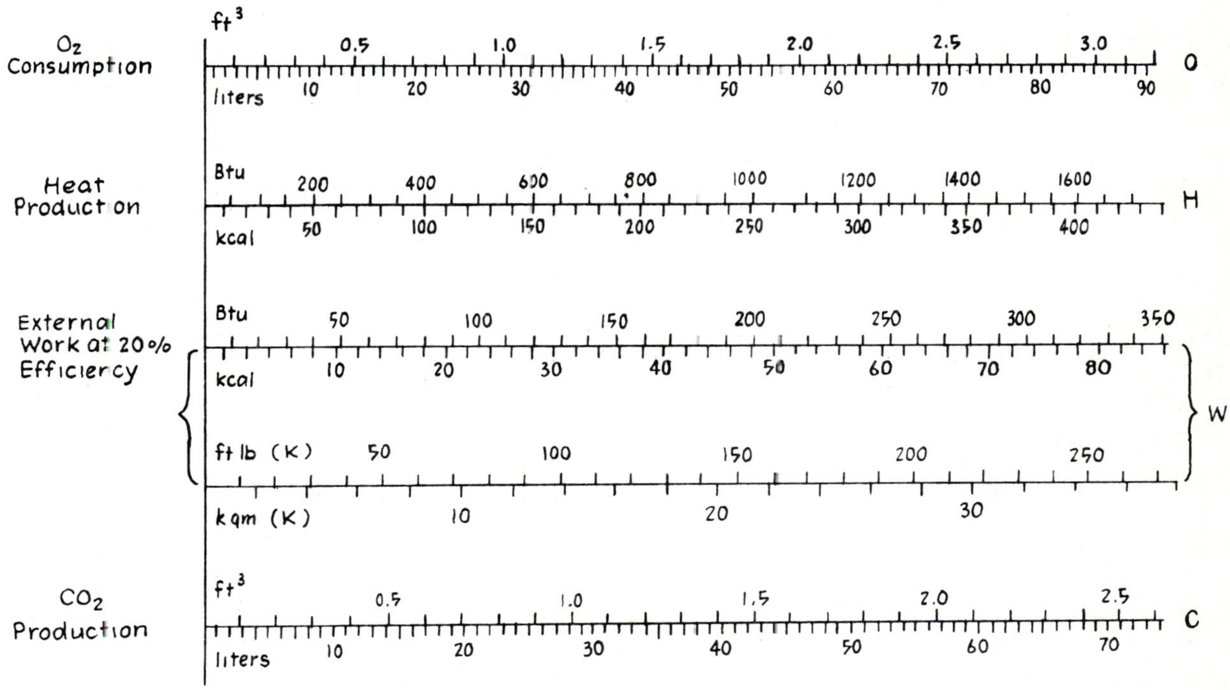

Nomogram B uses the following values: *RZ* = 0.82, and 1 L of O_2 is equivalent to 4.825 kcal. This nomogram allows one to interrelate, by drawing straight vertical lines, the values for oxygen consumption *(O)*, heat output *(H)*, external work output *(W)*, and carbon dioxide production *(C)* at typical conversion rates. *H* may be as much as 3 percent lower or 5 percent higher than the quoted value at any specific oxygen consumption, depending on the *RQ*, which equals 0.7 for a pure fat diet and 1.00 for a pure carbohydrate diet. The values given in the third and fourth lines should be modified if the efficiency changes. Typical ranges are 5 to 25 percent, with an average of 20 percent; thus the listed work output may increase by three-fourths if the task is one that can be performed at high efficiency (e.g., bicycling). Conversely, the true value may be reduced by three-fourths if the function is inefficiently performed (e.g., high-speed walking).

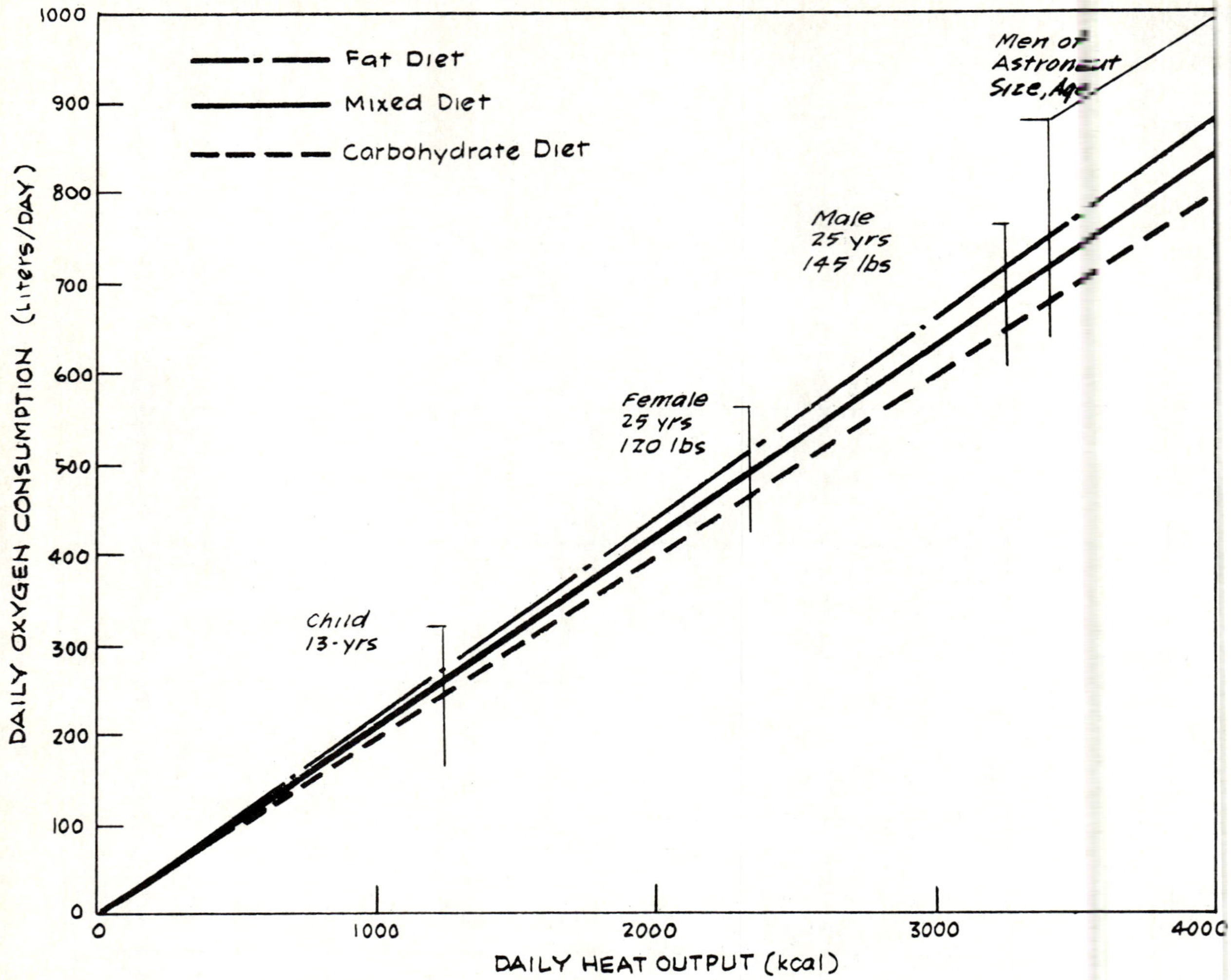

Estimate of total daily O_2 cost and heat produced for normally active people, as affected by body size. (J. F. Parker and V. R. West [Eds.], *Bioastronautics Data Book,* 2d ed., National Aeronautics and Space Administration, 1973, p. 852.)

THERMAL RESPONSE OF THE BODY

Normal body temperatures and cooling requirements for various regions of the body are given in the accompanying table. At 62°F (16.7°C) body cooling begins. Above 78°F (25.5°C) the body gains heat. The body adapts to changes as great as 20 percent within the comfort range through evaporative cooling when it is hot and through exercise when it is cold.

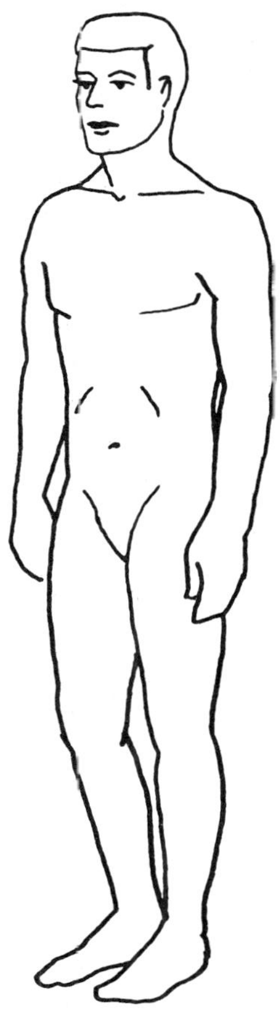

Normal Temperatures, °F		Sweat Rate, g/hr Insensible	
Mouth	98.6	Head	7–11
		Neck	4–5
		Back	1–2
Chest (skin)	94.95	Arm	1–2
		Armpit	0–7
		Chest	1–2
		Forearm	1–2
Waistline	95–97	Abdomen	1–2
		Hand	6–10
Rectum	99.6	Buttock	1–2
		Groin	Variable
		Thigh	2–4
Calf (skin)	80–83	Calf	1–2
Foot (skin)	80–83	Foot	4–7

Region	Preferred Temperature, °F	Heat Loss, Btu/h	Area, ft²	Skin Conductance, Btu/ft²-°F
Head	94.4	15.9	2.15	1.61
Chest	94.4	32.6	1.33	3.87
Abdomen	94.4	17.9	1.29	3.02
Back	94.4	49.3	2.43	4.31
Buttocks	94.4	33.0	1.94	3.70
Thighs	91.4	47.7	3.55	1.76
Calves	87.5	58.0	2.15	2.06
Feet	83.6	39.7	1.29	1.96
Arms	91.4	33.4	1.07	4.10
Hands	83.5	63.5	0.75	6.45

Response in Relation to Skin Temperatures

General Responses

Temperature of Skin, °F	Response
98	Very hot
96	Unpleasantly warm
94	Slightly warm
93	Comfortable
91	Comfortably cool
88	Slightly too cool
86	Unpleasantly cool
84	Very cold

Temperature and Its Effect on Comfort of the Extremities

	Skin Temperature, °F	
	Hands	Feet
Minimum acceptable	68	73
Tolerable	68–59	73–64
Intolerable	59–50	64–55
Numbness sets in	<50	<55

Effects on Skin of Contact with Surfaces of Different Temperatures

Temperature, °F	Sensation or Effect
212	Second-degree burn on 15-s contact
180	Second-degree burn on 30-s contact
160	Second-degree burn on 60-s contact
140	Pain; tissue damage
120	Pain; burning heat
91	Warm; neutral (physiological zero)
54	Cool
37	"Cool heat" sensation
32	Pain
Below 32	Pain; tissue damage (freezing)

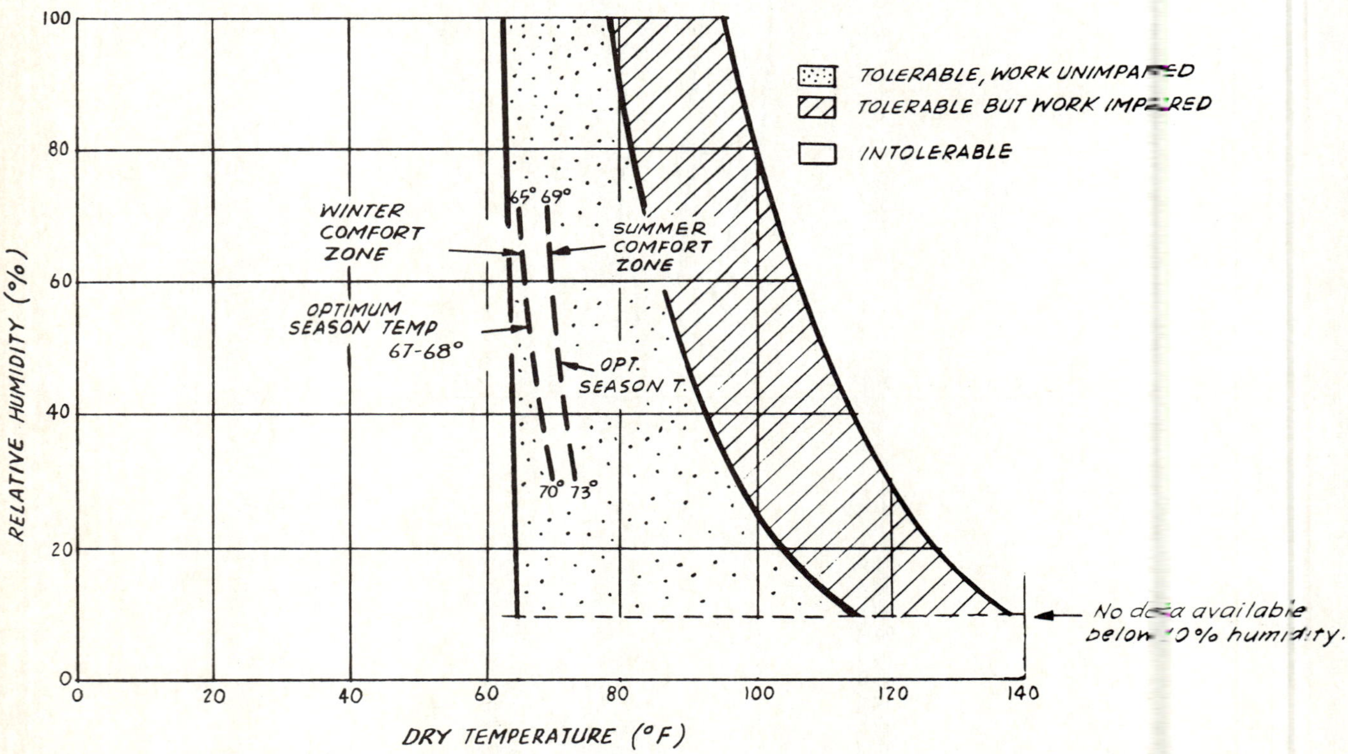

Tolerable temperature-humidity requirements (with conventional clothing).

TOLERANCE TO HIGH TEMPERATURES[1]

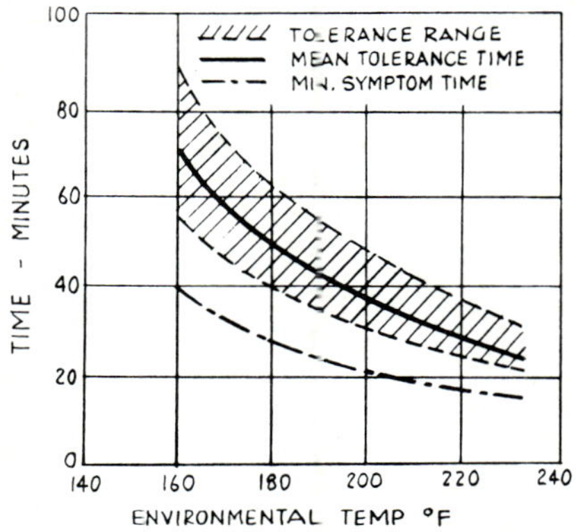

Mean and range of tolerance time and minimum symptom time for high temperature.

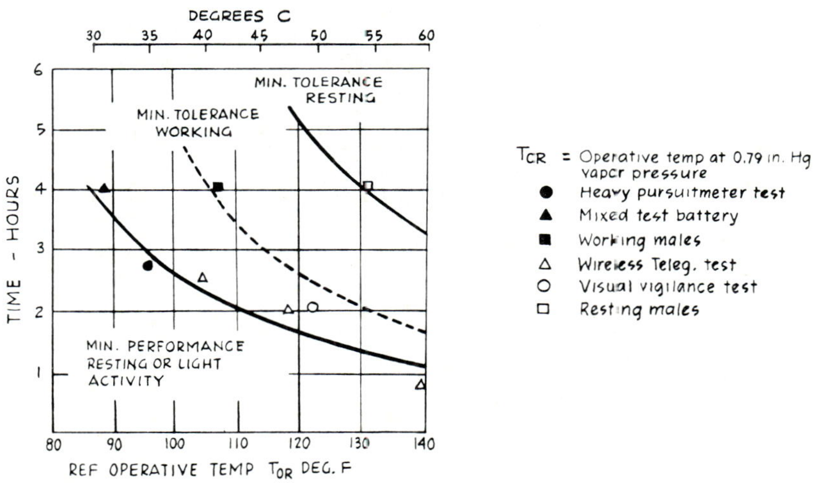

Performance and tolerance limits in the quasi-compensable zone for lightly dressed men.

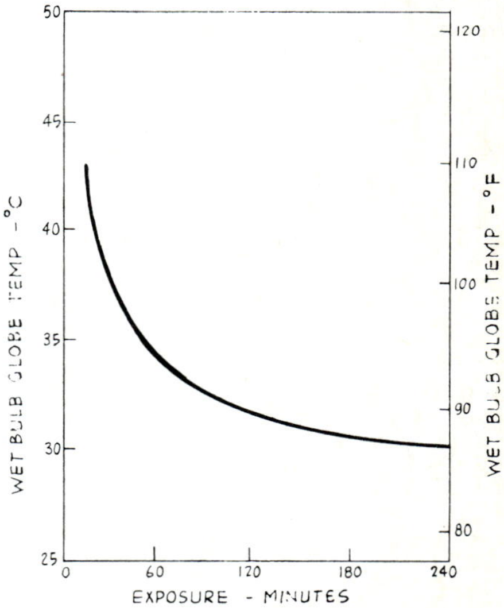

Upper limits of exposure for unimpaired mental performance.

[1] *Criteria for a Recommended Standard-Occupational Exposure to Hot Environments,* National Institute for Occupational Safety and Health, USGPO-HSM72-10269, 1972.

TOLERANCE TO LOW TEMPERATURES

The accompanying graph shows tolerance times for endurance to cold temperatures. Subjects were wearing various types of clothing, were seated, and were performing light work (air velocity, approximately 200 ft/min, 61 m/min; barometric pressure, 1 atm).

A—1-clo (light coveralls)
B—2-clo (woolen underwear, coveralls, and jacket)
C—3-clo (intermediate-weight flight clothing)
D—4-clo (heavy flight clothing)

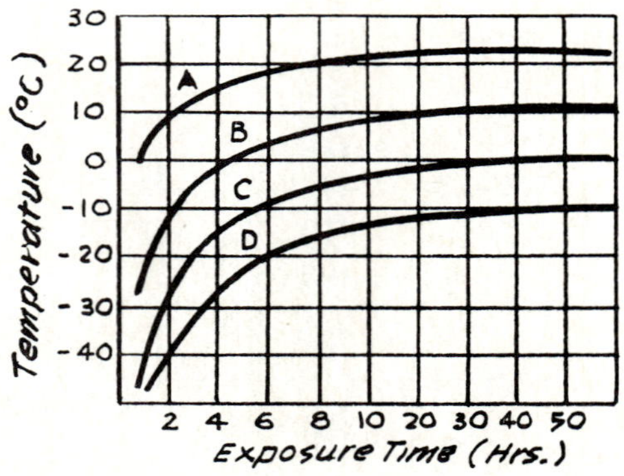

Effects of Cold on Selected Task Performances

A—tactile sensitivity, bare hand
B—simple visual reaction time
C—manual skill

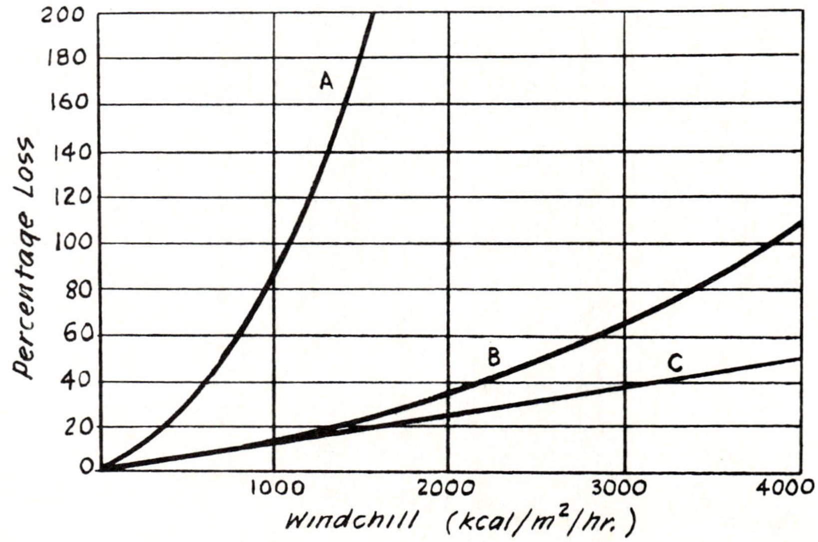

Wind-Chill Nomograph[2]

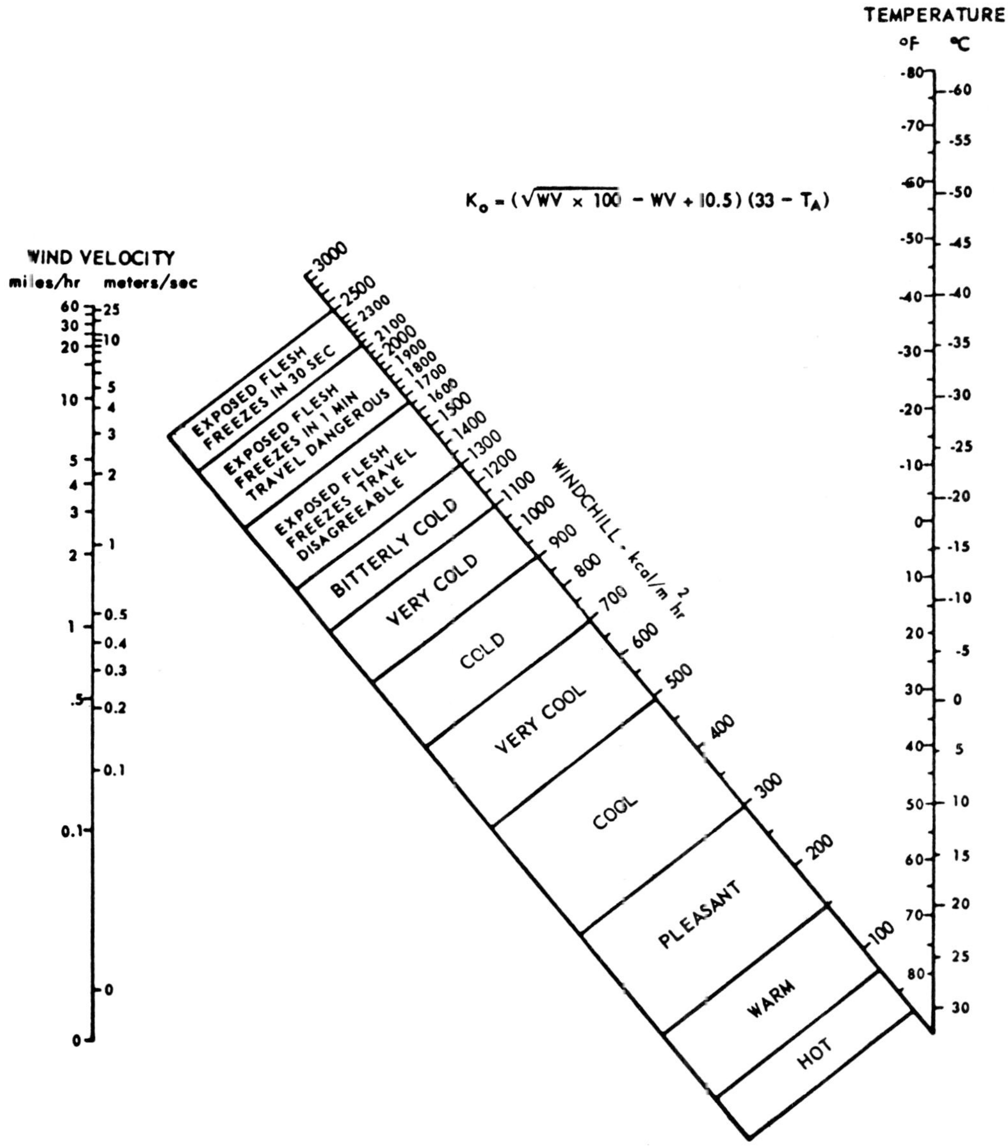

NOTE:

The windchill index does not account for physiological adaptations or adjustments and should not be used in a rigorous manner. It is based on field measurements during World War II of the rate of cooling of a container of water.

[2] J. F. Parker and V. R. West (eds.), *Bioastronautics Data Book,* 2d ed., National Aeronautics and Space Administration, 1973.

TEMPERATURE AND VENTILATION REQUIREMENTS

Ventilation requirements in relation to net airspace and the removal of body odor are shown in the accompanying graph. The graph illustrates how the intensity of body odors in a given area depends on the rate of flow of odor-free air.

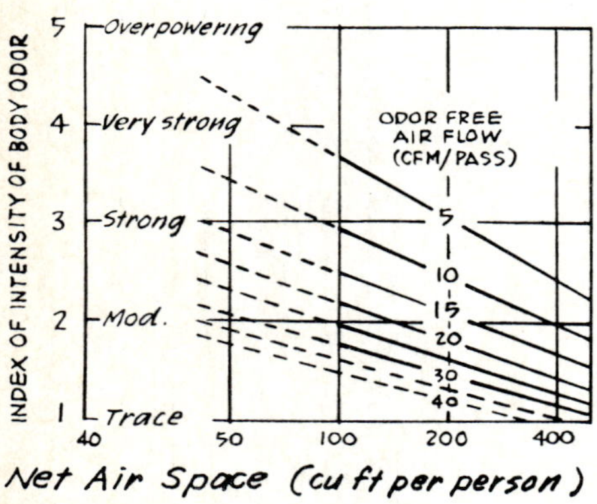

General Minimum Fresh-Air Requirements for Typical Ventilation Needs

Type of Occupancy	Airspace per Person, ft³	Ventilation per Person, ft³/min
Sedentary adults	100	25
	200	16
	300	12
	500	7
Adult laborers	200	23
Grade school children	100	29
	200	21
	300	17
	500	11

Note: These are minimum requirements during the normal heating season, with or without recirculation and without air conditioning. Norms for rooms at sea level range from 5 to 50 ft³/min per person, e.g., 7.5 for theatres; 15 for restaurants, hospital rooms, and apartments; and 30 for private offices, boardrooms, etc.

Source: R. A. McFarland, *Human Factors in Air Transport Design*, McGraw-Hill Book Company, New York, 1946, pp. 119, 121.

HUMAN PERFORMANCE EFFECTS AT VARIOUS TEMPERATURES

Effective Temperature °F*	Performance Effects
90	Upper limit for continued occupancy over any reasonable period of time.
80–90	Expect universal complaints, serious mental and psychomotor performance decrement, and physical fatigue.
80	Maximum for acceptable performance even of limited work; work output reduced as much as 40–50 percent; most people experience nasal dryness.
78	Regular decrement in psychomotor performance expected; individuals experience difficulty falling asleep and remaining asleep; optimum for bathing or showering.
75	Clothed subjects experience physical fatigue, become lethargic and sleepy, and feel warm; unclothed subjects consider this temperature optimum without some type of protective cover.
72	Preferred for year-round sedentary activity while wearing light clothing.
70	Midpoint for summer comfort; optimum for demanding visual-motor tasks.
68	Midpoint for winter comfort (heavier clothing) and moderate activity, but slight deterioration in kinesthetic response; people begin to feel cool indoors while performing sedentary activities.
66	Midpoint for winter comfort (very heavy clothing), while performing heavy work or vigorous physical exercise.
64	Lower limit for acceptable motor coordination; shivering occurs if individual is not extremely active.
60	Hand and finger dexterity deteriorates, limb stiffness begins to occur, and shivering is positive.
55	Hand dexterity is reduced by 50 percent, strength is materially less, and there is considerable (probably uncontrolled) shivering.
50	Extreme stiffness; strength applications accompanied by some pain; lower limit for unprotected exposure for more than a few minutes.

*These temperature effects are based on relatively still air and normal humidity (40 to 60 percent). Higher temperatures are acceptable if airflow is increased and humidity is lowered (a shift of from 1 to 4°); lower temperatures are less acceptable if airflow increases (a shift upward of 1 to 2°).

OXYGEN[1]

The effects of excess oxygen and the lack of oxygen are shown in the accompanying table.

Partial Pressure of Oxygen, mm of Hg	Percent of Oxygen in Dry Air at Sea Level Pressure	Effect
		Oxygen Excess
456	60	Onset of oxygen poisoning after some hours.
167	22	Limit set in British navy to control fire hazard in charcoal filters in nuclear submarines.
		Normal
160	21	Normal atmospheric level
		Oxygen Lack
137	18	Accepted limit of alertness. Loss of night vision. Earliest sign—dilation of the pupils.
114	15	Performance seriously impaired. Hallucinations, excitation, apathy.
100	13	Coordination impaired. Emotional upset.
84	11	Paralysis, loss of memory. Irreversible unconsciousness.
46	6	Death before symptoms apparent.

Note: The effect of falling oxygen is insidious, because it dulls the brain and prevents realization of danger.

The accompanying graph shows that a lack of oxygen results from reduced atmospheric pressure. At altitudes above 2000 m, the fall in atmospheric pressure requires compensation in the inspired air. By the same token, at depths below sea level, similar compensation is required to offset the effects of excess oxygen.

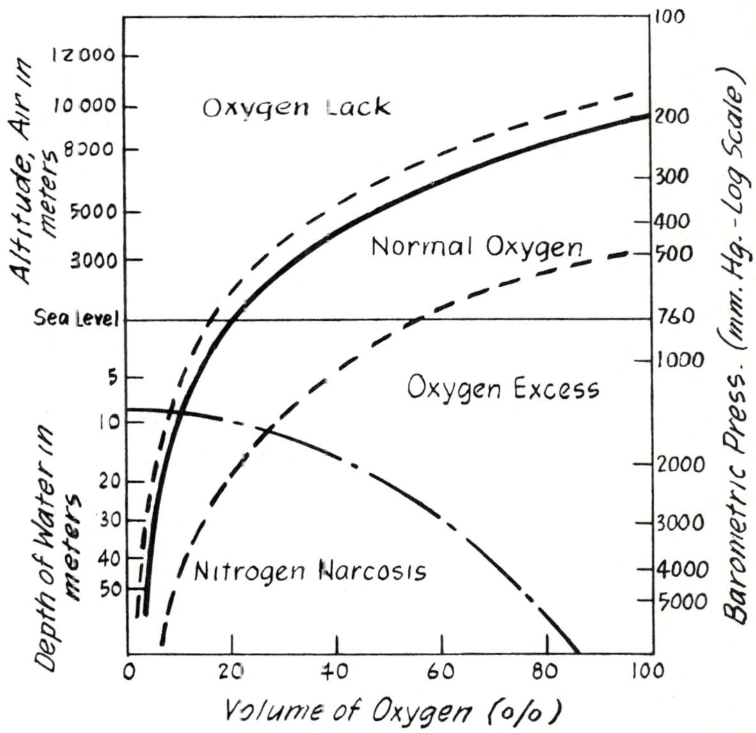

Oxygen concentration, nitrogen narcosis, and atmospheric pressure.

[1] *Human Factors for Designers of Naval Equipment*, Medical Research Council, Royal Naval Personnel Research Committee, 1970, chap. 1.

Oxygen Safety Limits

Chronic oxygen poisoning can occur at or above 460 mm of mercury. Short-term memory deteriorates after about 10 min of breathing pure oxygen.[2]

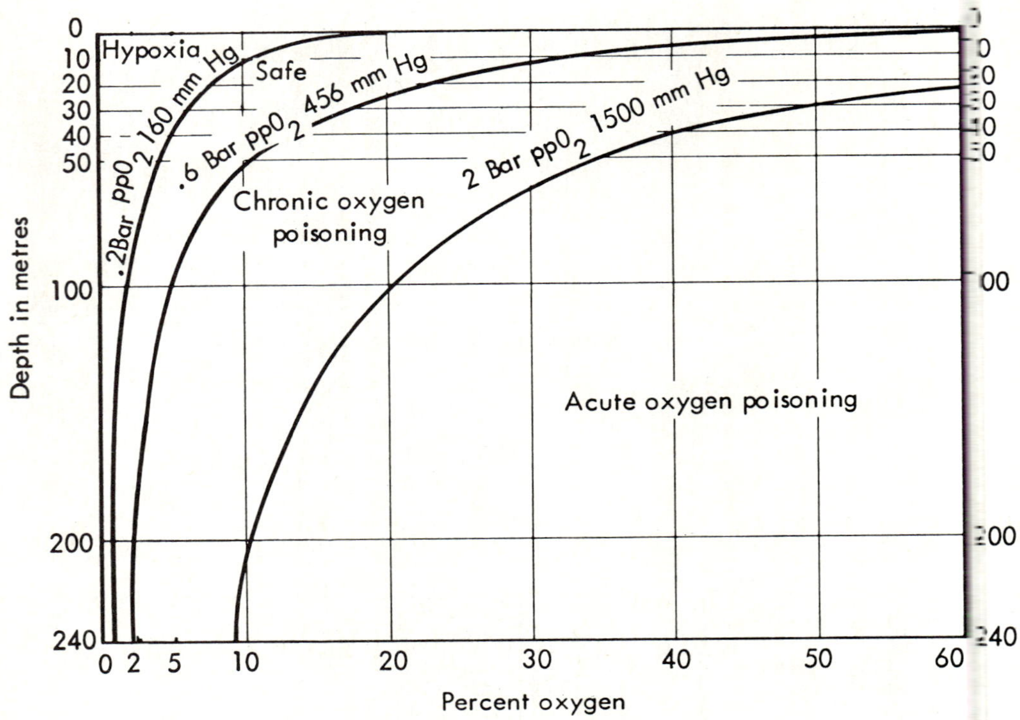

[2]*Human Factors for Designers of Naval Equipment,* Medical Research Council, Royal Naval Personnel Research Committee, 1970, chap. 1.

CARBON DIOXIDE

The air we breathe normally contains about 0.03 percent of carbon dioxide. The concentration of carbon dioxide should be maintained below 0.5 percent. The necessary ventilation rate to maintain this level is illustrated in the accompanying table.

	Oxygen Consumption per Person at Sea Level	Ventilation Rate per Person Sea Level	5000 ft	10,000 ft	15,000 ft
At rest	0.008 ft³/min	1.2 ft³/min	1.4 ft³/min	1.7 ft³/min	2.1 ft³/min
Moderate activity	0.028 ft³/min	3.9 ft³/min	4.7 ft³/min	6.7 ft³/min	6.9 ft³/min
Vigorous activity	0.056 ft³/min	8.7 ft³/min	9.7 ft³/min	11.7 ft³/min	14.5 ft³/min

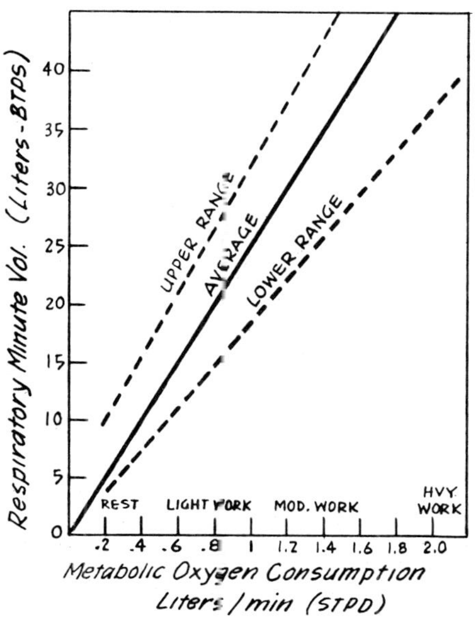

Maximum Limits for Carbon Dioxide Exposure

Maximum Limit, %	Situation	Remarks
0.5	General limit for 8-hr industrial exposure	Threshold limit value (TLV) laid down by the American Conference of Government Industrial Hygienists (ACGIH).
1	90-day limit for continuous exposure in nuclear submarines (to be reduced to 0.5%)	Future design limit for removal machinery is 0.5%.
3	Limit for short-term exposure in conventional submarines	Brain functions deteriorate above this level. Heavy breathing.

Effects of Continuous Exposure to Carbon Dioxide over a Number of Weeks

Level, %	Effect
0.5–1	Probably no significant physiological, psychological, or adaptive changes occur.
1–3	Gradual changes in the composition of the blood and body fluids, which may lead to irreversible changes in the tissues on prolonged exposure.
3 and above	Deterioration in performance, alteration of basic body functions, and irreversible changes in the tissues

CARBON MONOXIDE

The effects of carbon monoxide poisoning are outlined in the accompanying table. Carbon monoxide is particularly dangerous because it is odorless and colorless.

Carboxyhemoglobin levels of over 5 percent have been known to produce intellectual deterioration, and excessive exposure over long periods often leads to arterial and heart disease.

$$\frac{\text{CO in vol/million}}{6} = \text{carboxyhemoglobin \%}$$

Effects of Carbon Monoxide Exposure

Atmosphere, Volumes per million	Carbon Monoxide in Blood, % Carboxyhaemoglobin	Effects
0–60	0–10	None subjectively noticeable, but initial visual and psychomotor impairment is revealed in objective tests.
60–120	10–20	Tightness across forehead, slight headache, flushed complexion.
120–180	20–30	Headache with throbbing in temples, breathlessness from any exertion.
180–240	30–40	Severe headache, weakness, dizziness, dimness of vision, nausea, and vomiting with possibility of collapse.
240–300	40–50	All preceding symptoms with increased pulse rate and respiration and greater possibility of collapse.
300–360	50–60	Loss of consciousness, with increased or irregular respiration, rapid pulse, and possibility of coma with convulsions.
360–480	60–80	Coma, convulsions, depressed heart action, respiratory failure, and possibility of death.

OZONE

Under certain conditions three atoms of oxygen combine to form one atom of ozone, which is poisonous in both short- and long-term exposure. An individual who is exposed for about 10 min to 1 volume per million of ozone is likely to feel irritation in the nose and throat, to cough, and to experience visual degradation; 100 volumes per million may injure body tissues, and 1000 volumes per million produces fatigue, depression, and pain in the head and body. Eventual unconsciousness and death may result.

OTHER TOXIC GASES

The accompanying table[3] shows permissible limits for other toxic gases, excess exposure to which may endanger human health, both temporarily and permanently. The American Conference of Government Industrial Hygienists (ACGIH) has established threshold limit values (TLV) on the basis of an 8-hr working-day exposure. Maximum permissible concentrations (MPC) for 90-day exposures have been established by the British Royal Navy for use in nuclear submarine environments.

This is only a partial list. For more-detailed treatment, see handbooks on toxicology or chemical-plant safety.

Maximum Permissible Concentrations of Toxic Gases

Substance	Formula	Threshold Limit Value for an 8-hr Working Day,* TLV	Maximum Permissible Concentration for a 90-day Exposure,† MPC90
Ammonia	NH_3	25 vpm (to be reduced)	25 vpm
Arsine	AsH_3	0.05 vpm	0.01 vpm
Carbon dioxide	CO_2	0.5%	1% (to be reduced to 0.5%)
Carbon monoxide	CO	50 vpm	25 vpm
Hydrocarbons (not methane)		Not listed as a group	40 mgm/m³ (8 mgm/m³ aromatic)
Hydrogen chloride	HCl	5 vpm	0.1 vpm
Hydrogen fluoride	HF	3 vpm	0.1 vpm
Mercury	Hg	Alkyl (skin effect) 0.01 mgm/m³ Non-alkyl 0.05 mgm/m³	0.05 mgm/m³
Chlorine	Cl	1 vpm	0.1 vpm
Methane	CH_4	Explosive	1.2% (risk of explosion)
Methyl alcohol (including formaldehyde and formic acid)	CH_3OH	200 vpm	10 vpm
Nitrogen dioxide	NO_2	5 vpm	0.5 vpm
Ozone	O_3	0.1 vpm	0.02 vpm
Phosgene	$COCl_2$	0.1 vpm	0.05 vpm
Stibine	SbH_3	0.1 vpm	0.05 vpm
Sulphur dioxide	SO_2	5 vpm	1 vpm
Benzene	C_6H_6	25 vpm	1 vpm
Toluene	C_7H_8	100 vpm	4 vpm
Xylene	C_9H_{12}	100 vpm	4 vpm

Note: Vpm—volumes of gas or vapor per million of air. Mg/m³—milligrams of solid material, particles, or dust per cubic meter of air.

*Standard of the American Conference of Government Industrial Hygienists.

†MODN Book of Reference 1326, 1973.

[3]After *Human Factors for Designers of Naval Equipment,* Medical Research Council, Royal Naval Personnel Research Committee, 1970, chap. 1.

RADIATION AND IONIZATION[4]

The accompanying chart shows wavelengths of electromagnetic radiation and how they are used, indicating the region in which damage to living tissue can occur.

Nuclear radiation may damage the tissues of the body. An acute dose of 500 rad of penetrating ionizing radiation is likely to kill about half of an exposed population within 30 days. However, if the dose is distributed slowly over 10 years, there is only an increased risk that an exposed person will develop cancer. An individual who receives an acute dose of 200 rad of penetrating ionizing radiation equally distributed over his or her body is likely to feel sick for a day or two before recovering. With doses of over 5000 rad, the brain is directly affected; the person may collapse within minutes and die within hours.

Ultraviolet light, after repeated exposure, may produce localized cancer of the skin. Laser beams, microwaves, and very intense radar waves may produce temporary tissue damage.

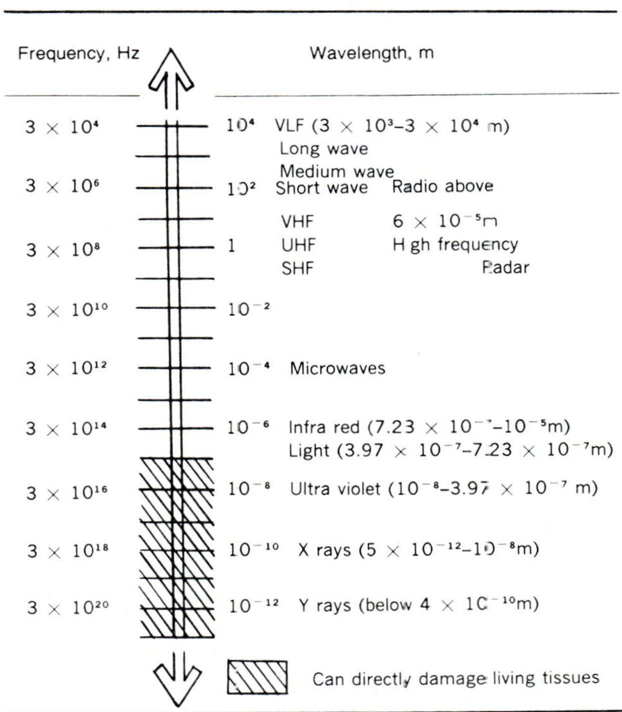

[4]*Human Factors for Designers of Naval Equipment,* Medical Research Council, Royal Naval Personnel Research Committee, 1970, chap. 1.

U.S. BASIC RADIATION PROTECTION STANDARDS

Exposed Parts	Permissible Accumulated Dose
Whole body Head and trunk Active blood-forming organs Gonads Lens of the eye	1.25 rem in any calendar quarter 5 (*N*-18) during a lifetime (*N*-present age)
Skin of whole body Thyroid	7.5 rem in any calendar quarter 30 rem in any calendar year
Hands and forearms Feet and ankles	18.75 rem in any calendar quarter 75 rem in any calendar year

The accompanying table shows the effects of acute, whole-body, external radiation exposure and the probable therapy required.[5] (A rem is the dosage of an ionizing radiation that will have the same biological effect as 1 part R of x-ray or gamma-ray dosage.)

0.25 rems	25–100 rems	100–200 rems	200–300 rems	300–600 rems	600 rems or more
			Immediate Effects		
No detectable clinical effects	Slight transient reductions in lymphocytes and neutrophils.	Nausea and fatigue with possible vomiting above 125 rems.*	Nausea and vomiting on first day.	Nausea, vomiting, and diarrhea in first few hours.	Nausea, vomiting, and diarrhea in first few hours.
	Disabling sickness not common; exposed individuals should be able to proceed with usual duties.	Reduction in lymphocytes and neutrophils with delayed recovery.	Latent period up to 2 weeks or perhaps longer.	Latent period with no definite symptoms, perhaps as long as 1 week.	Short latent period with no definite symptoms in some cases during first week.
			Delayed Effects		
Delayed effects may occur	Delayed effects possible, but serious effects on average individual very improbable.	Delayed effects may shorten life expectancy in the order of 1%.	Following latent period, the following symptoms appear but are not severe: loss of appetite, and general malaise, sore throat, pallor, petechiae, diarrhea, moderate emaciation.	Epilation, loss of appetite, general malaise, and fever during second week, followed by hemorrhage, purpura, petechiae, inflammation of mouth and throat, diarrhea, and emaciation in the third week.	Diarrhea, hemorrhage, purpura, inflammation of mouth and throat, fever toward end of first week. Rapid emaciation and death as early as the second week, with eventual death of up to 100% of exposed individuals.
			Recovery likely in about 3 months unless complicated by poor previous health or superimposed injuries or infections.	Some deaths in 2 to 6 weeks. Possible eventual death to 50% of the exposed individuals for about 500 rems.	
			Therapy		
	Reassurance is probably the only therapy needed.	Patients should be kept under hematological surveillance.	Antibiotics should be administered as indicated.	Antibiotics and blood transfusions should be administered as indicated.	Bone marrow transplantations may be tried. Electrolyte balance should be maintained.

*Occurs on the first day or so following irradiation, followed by a "latent period" up to 2 weeks or more, during which the patient has no disabling illness and can proceed with his regular occupation. The usual symptoms, such as loss of appetite and malaise, may reappear, but if they do, they are mild.

[5]*Design Handbook, Human Factors Engineering,* AFSC DH 1-3, Ser. 1-0, Department of the Air Force, Headquarters Aeronautical Systems Division, Wright-Patterson Air Force Base, Ohio.

EFFECT OF ELECTRICAL CURRENT ON THE HUMAN BODY

Body resistance to electrical current varies over a wide range, making it difficult to define dangerous potentials exactly. Hand-to-hand resistance varies from 1000 to 4000 Ω even with good contact. Hand-to-foot resistance is slightly less. With dry skin, resistances as high as 250,000 Ω are possible. Under ordinary working conditions when the hands are sweaty, a resistance of 5000 Ω is commonly expected. Safety recommendations suggest 1000 Ω as the highest resistance value. The accompanying table identifies some of the potential effects that can be expected for various current levels.

Current, mA	Effects
Less than 1	Usually not felt, i.e., no sensation.
1–2	There is a sensation of shock, but not pain. The individual can let go because muscular control is not lost.
3–15	Painful shock occurs, but the individual can still let go.
15–20	Painful shock occurs, and the individual may not be able to let go because control of the immediately adjacent muscles is affected.
20–50	Very painful shock occurs, plus severe muscular contractions. Breathing typically becomes difficult.
50–100	Ventricular fibrillation (a heart condition that results in instant death) is very likely.
100–200	Same as above, except that the results are certain.
200 and over	Severe burns occur as well as muscular contractions so severe that the chest muscles stop the heart during the duration of the shock.

SKIN REACTION TO CHEMICAL SUBSTANCES

The accompanying table presents a partial list of chemical substances and their reaction on the skin. For more-detailed treatment, see more-specialized handbooks on toxicology or chemical-plant safety.

Agent	Reaction
Acids:	
Acetic	Dermatitis and ulcers
Carbolic	Irritation and erosion, eczema, and anesthesia
Chromic	Ulcers (chrome holes on the skin), inflammation, and perforation of the nasal septum
Hydrochloric	Irritation and ulceration
Hydrofluoric	Severe burning, erosion ulcers, and blisters
Lactic	Ulcers (if strong solution)
Nitric	Severe burns and ulcers
Oxalic	Local caustic action on the skin
Sulfuric	Corrosive action on the skin and severe inflammation of the mucous membranes
Alkalis:	
Calcium cyanamide	Irritation and ulceration
Calcium oxide	Dermatitis, burns, and ulcers
Potassium hydroxide	Severe burning, persistent ulcers, and loss of fingernails
Sodium hydroxide	
Sodium silicate	Thickening of the skin and ulcers on the fingers
Sodium or potassium cyanide	Blisters and ulcers
Salts:	
Antimony and its compounds	Irritation and eczematous eruptions
Arsenic	Skin darkening, perforation of the nasal septum, eczema around the mouth and nose, and possible loss of nails or hair
Barium	Eczema and cyanosis of skin
Bromine	Brownish stains and skin erruptions
Chromium (hexavalent compounds)	Chrome holes on the skin, perforation of the nasal septum, and eczematous eruptions
Mercury compounds	Corrosion and irritation and mercurial eczema
Sodium	Burns and ulcers
Zinc chloride	Ulcers of the skin and nasal septum
Solvents:	
Acetone	Dry (defatted) skin
Benzene	Dry (defatted) skin
Carbon disulfide	Dry (defatted) skin
Chlorinated phenols	Severe eruptions
Petroleum distillates	Acne and epithelioma
Trichlorethylene	Dry, cracked skin
Turpentine	Red, blistered skin and eczema
Dyes:	
Chlorinated compounds	Blisterlike eruptions
Dinitrochlorobenzene	Blisterlike eruptions
Nitro and nitroso compounds	Red skin and eczematous eruptions
Phenyl hydrazine	Blisterlike skin eruptions
Insecticides:	
Chlorophenols	Red skin, and blisters
Creosote	Pustular eczema, warts, and epithelioma
Fluorides	Severe burns and dermatitis
Pyrethrum	Red skin, blisters, and pimples
Rotenone	Red skin and blisters
Resins:	
Coal tar, pitch, and asphalt	Acute dermatitis, acne, inflammation, epitheliomatous cancer, eczema, and ulcers
Synthetics, e.g., phenol-formaldehyde	Extremely red and itchy skin
Synthetic waxes, e.g., chloronaphthalenes and chlorodiphenyls	Dermatitis and acne

Source: Julian B. Olishifski and Frank E. McElroy (eds.), *Fundamentals of Industrial Hygiene,* National Safet Council, Chicago, 1971.

GENERAL BEHAVIORAL EXPECTANCIES DATA TABLE FOR DESIGNERS

Behavior Pattern	Safety Implications	Design Considerations
Built-in habits and/or natural behaviors and associations:	Some appearances, operations, and procedures seem more natural than others to people; hence their first impulse is to interpret, operate, or use a device according to these expectations.	If the product is designed so that it either is or appears to be contrary to natural expectations, the user frequently makes mistakes, some of which could lead to hazardous situations.
1. People tend to assume that an object, device, or package is small enough to get hold of and also light enough to pick up.	If an object actually is very heavy, users may suffer severe strain or lose their balance and fall.	Either the user should be warned or the design should include some type of fastening to prevent the object from being lifted.
2. People assume that a vehicle is designed to "fit" their physical characteristics and thus try to adapt to it, whether they can do so effectively or not.	Lack of proper fit often causes users to assume a precarious position or attempt a reach effort that not only makes their situation insecure but also forces them to assume a fatiguing position or apply uncontrolled forces and manipulatory patterns.	The vehicle should be adjustable when it cannot be made to fit all users by a single configuration.
3. People have learned (and therefore expect) that water faucets and liquid valve handles rotate counterclockwise to increase the flow of a liquid, gas, or steam.	If such valves are designed contrary to expectations, when such valves have to be turned off rapidly (in an emergency), individuals frequently will turn them in the wrong direction first and then make a correction—which could be too late.	Valve handles should be marked and mounted according to standard practice. No liquid valve should be incorporated into a design that does not operate according to this standard.
4. Electrically powered control switches are expected to move upward, to the right, or clockwise to effect electrical continuity, i.e. to turn power on.*	If such switches are designed contrary to expectations, people may turn an electrical device on when they think they have turned it off, thus exposing themselves or others to a functioning system which they assume is off.	Design or select electrical switches and control handles that are compatible with the expected directions of motion.
5. People expect that when they move a vehicle control to the right, or clockwise, the vehicle will move in that direction.	Incompatibilities with user expectations result in confusion and often in loss of control.	In the design of vehicles, vehicle control motions should be similar. If the operator must turn around to face a control, the vehicle-control relationships become confusing. Avoid such arrangements if at all possible.
6. People expect that when they actuate a knob or switch that causes an instrument pointer and/or some part of a machine to move, the directions of motion of both the control and the pointer (or part) will be the same.	Incompatibilities between the motion of the instrument pointer (or part) and that of the control element lead to erroneous dial settings and movements of subelements, often causing delays to correct the error and sometimes considerable danger to the user.	Knob and switch motion directions should be consistent with motion compatibility rules, and no control relationship should be used that does not produce a consistent natural response.†
7. People have become conditioned to certain color meanings (e.g., red for "danger," "fire," and "hot"; green for "OK," "go," and "acceptable"; yellow or amber for "caution," "look out for," and "yield"; and blue (in the case of electrical devices) for "cold" or "cool."	Misuse of colors may cause observers to ignore a caution or danger signal or to assume the presence of one that does not actually exist. Either condition creates confusion and delay.	Observe current color-coding standards (which may differ for various product categories).
8. People's attention is drawn to bright and vivid colors, bright lights, loud noises, flashing lights, and repeated and undulating sounds.	People may be distracted by an inadequately selected display and fail to attend to an important task detail, or an important task may not be given attention when needed because the device, signal, or sound was not designed to be conspicuous. When their	Use stimuli of adequate intensity when attention needs stimulation; do *not* use high-intensity stimulation when distractions should be minimized.

(continued)

Behavior Pattern	Safety Implications	Design Considerations
	attention is drawn to a bright light, individuals may disturb their own visual adaptation level to the point where they cannot see necessary visual detail (as when they face oncoming automobile headlights while driving).	
9. Dark, dull colors appear to advance toward the observer; lighter, brighter colors appear to recede.	People may react instinctively (i.e., "duck") to miss a dark object above their heads, even though there is ample clearance.	Observe normal brightness and color impression conventions in selecting colors for interior surfaces.
10. People expect to face auditory signals in order to corroborate and/or reinforce reception.	Placing an auditory warning signal behind an individual could cause the person to displace his or her eye reference as a result of turning the head or body. If this occurs at the wrong moment, it could cause a vehicle operator to miss seeing something in front of the vehicle.	Locate auditory warning signals so that they appear to come from in front of the operator and also locate them close to head height.
11. People assume a relationship between objects on the basis of their spatial proximity (i.e., they assume that things are related somehow when they are located or grouped together).	If a label or control appears to be associated with a display or another control, the individual's inclination is to use that control to effect a modification of the nearby display, whereas it may be the wrong control.	Careful positioning and/or arrangement of controls, displays, labels, or instructions is required to ensure the proper associations.
12. People generally regard products as being safe; that is, they: a. Assume that a system is *ready* to go and hence do not check. b. Grab a handrail assuming that it is strong enough to support them. c. Assume that a system is not on and hence proceed to touch or manipulate without caution. d. Assume that no one else has turned the system on and therefore do not check to make sure.	Although the assumption of "safe" should not be made, the fact that it is tends to preclude users from *thinking* safety, thus reducing the likelihood that they will anticipate possible hazards and make the necessary check.	Try to design the product so that it cannot be used improperly. If that is impractical, provide guards or other means to cause the user to think about the hazards. Finally, if neither of the above is practical, take steps to warn the user by means of labels or instructions, appropriately coded to imply caution or warning.
13. When about to lose their balance or fall, people instinctively reach for and grab the nearest thing.	Sometimes the things most convenient for grasping are not strong enough to support the individual, or they may be hot or slippery or cause an electrical shock.	Configure the design to provide appropriate emergency supports, or else make sure that hazardous components that might be used for support are not within reach.
14. People instinctively use their hands first to test or explore.	Many injuries occur because users touch an object with their hands or fingers, even when it was intended that the task be performed by means of an intermediate device of some kind.	Either make the product suitable for handling or else make it quite clear that its use requires a device supplied to obviate the necessity for using one's hands.
15. People expect a seat height to be at a level they are used to; a stair angle or tread pattern similar to one with which they are familiar and equally spaced in terms of riser height and tread depth; and a handhold or railing of a size and shape that can be gripped securely.	Individuals tend to adjust their posture, weight, mass, and gait to what they expect. When their expectations are not fulfilled, they are apt to lose their balance, take a wrong step, or trip.	Adhere to recognized standards and guidelines so that furnishings and architectural features are consistent with learned use patterns and are compatible with human physical dimensions.

(continued)

Behavior Pattern	Safety Implications	Design Considerations
Lack of knowledge and/or experience:		
1. Across the broad population there is a lack of knowledge about, and experience with, the technological features of many products, e.g.: a. Structural integrity b. Mechanical relationships c. Electricity d. Thermal characteristics e. Gaseous and combustible materials and conditions leading to explosion or fire f. Implications regarding center of gravity and balance g. Acceleration, velocity, and inertial factors	Since people generally are not prepared either to anticipate or to analyze conditions or the possible results of incompatibilities among physical elements and phenomena, they inadvertently initiate events that lead to improper and hazardous interactions. And because many do not know what makes structures collapse or tip over, what causes materials to ignite or explode, or what inertial forces will do either to others or to themselves, they do not approach the product prepared to deal with these hazards.	Analyze the potentially hazardous interactions between the naive user and the various product elements to make sure that every means possible of precluding misuse has been designed into the product, that misuse has been warned against, and/or that the potential severity of the hazard has been minimized.
2. Most people know very little about their own physical, sensorimotor, or cognitive characteristics or limitations. Thus they do not recognize: a. How they might cause muscle strain, sprains, or skeletal fracture b. How easily they can be thrown off balance c. How vertigo is induced d. How momentum and centrifugal forces affect their balance or ability to reach or apply force e. What is required for them to see effectively and reliably, hear effectively, and feel (touch) accurately f. The importance of adequate sensory feedback in order to control their actions properly or how various internal and external stresses can modify their interpretation of the feedback phenomena	Product users often do not recognize that they are not getting adequate inputs through their sensory channels or that they are reacting to illusory phenomena, upon which they are basing their decisions, rather than to the true informational input.	Learn more about how the human body works, and about its basic characteristics and limitations and then design in such a way as to minimize overstressing, creating illusory stimuli, or creating sensory contradictions and uncertainties.
3. The average nonspecialist is not familiar with much of the technical terminology used for identification	Typical users do not understand the terminology, or they may misinterpret it and thus proceed on an erroneous basis to misuse the product.	Avoid highly technical terms (i.e., use standard nomenclature and abbreviations) and/or test new terms to make sure that they are clear to nonspecialist individuals.
Inattention:		
1. Inattention is a phenomenon exhibited by everyone at one time or another, either as a natural trait or because of external and internal stresses, fatigue, boredom, or lack of motivation.	Lapse of attention often results in failure to observe a hazardous condition, read a label or instruction correctly, understand an instruction fully, see a warning light, or watch where one's feet, hands, head, or other body and limb components are going and whether they are clearing obstructions.	Anticipate the possibility of inattention in conceptualizing product design; i.e., think of the design in *use* terms. Where the possibility of inattention may be critical to the proper operation and safety of the product-user system, steps must be taken to preclude misuse and/or to provide means for attracting the user's attention.
Distraction:	Distraction from the primary product use task may cause users to miss seeing an important element or event, to fail to observe the path of a part of their bodies or the	Analyze the product design and potential use procedures to eliminate distractions and/or to provide a stimulus to maintain user procedural continuity and reliability, or else design to
1. Many people are easily distracted, either by certain aspects of the product's features or by the procedures for its use.		

(continued)

Behavior Pattern	Safety Implications	Design Considerations
	bodily position they have assumed, or to break their train of thought, causing them to omit a step or fail to remember what they have completed.	minimize the possible effects of procedural errors. The environment within which the product will be used must be known, and affective influences must be guarded against.
Hurrying: 1. Most people tend to hurry at one time or another.	When individuals hurry, they tend not to observe or to react carefully; they may not have time to think out or arrive at decisions based on a full analysis of a given situation. Thus, oversights may lead to inadvertent omissions, inappropriate responses, or even inaccuracies in sensorimotor performance.	Recognize the probability that a user will be in a hurry and design the product so that the use procedures help pace the user's response and/or so that the possibility of a hazardous consequence of an error or omission is minimized.
Deliberate risks and shortcuts: 1. Although risk taking and frequent shortcutting of proper procedures are more typical of certain users (e.g., young people), almost everyone takes some risk or shortcut at one time or another.	When a calculated risk cannot be based on a full evaluation of the potential failure modes and consequences, a product use failure probably will occur. Usually the user has neither the knowledge nor the time to make such an evaluation; hence the probability of failure is high.	Anticipate the possible risk-taking and shortcutting behaviors and design the product in such a way as to preclude shortcutting or make the consequences of omissions less serious.
2. Many people refuse to look at and read signs or other visual warnings.	Failure to see or read labels, instructions, or signs leaves the individual uninformed and therefore more prone to error.	Make critical signs and signals conspicuous, legible, visible, and understandable and place them where the user is expected to be looking.
Complacency, neglect, and overconfidence: 1. Because of overconfidence, individuals often proceed without thinking, observing, checking, or reading instructions or labels. 2. Complacency tends to dull people's sensitivity and alertness. 3. Some people fail to exercise caution because of a neglectful and disorganized approach to life.	When people assume that they know how something should be operated, they often make faulty judgments that lead to errors. In some product use situations, alertness is vital to safe performance.	Anticipate possible lack of interest or overconfidence on the part of the user and design in such a way as to alert and stimulate the user to possible critical aspects of product use and operation. The design should be made as simple as possible, and potential hazards should be conspicuously identified and brought to the user's attention. The design should preclude the operational steps' being performed out of sequence.
Forgetting: 1. Although everyone occasionally forgets things, some people by their very nature seem to forget more regularly than others or to forget because of stress or fatigue.	Forgetting a critical step, for whatever reason, can have disastrous consequences. People also can forget about something they ordinarily know is dangerous or potentially injurious.	Anticipate the possibility of forgetfulness on the part of the user and design out critical features where such procedural omissions could lead to injury. Where necessary and possible, the design should provide some technique to monitor whether the user has forgotten a critical step.
Confusion: 1. People are easily confused by things that are unfamiliar.	Lack of familiar features not only lengthens the time required for a user to respond (perhaps making the response too late for safety) but also may freeze the individual into inaction, resulting in an unsafe condition.	Avoid designing products which are so entirely new that nothing about them (shapes, colors, nomenclature, etc.) is familiar to the user.
2. People are easily confused by complexity (in numbers, sizes, shapes, arrangements, concomitant operations, etc.).	Complexity places considerable demand on human perceptual-motor and cognitive processes, and under these circumstances any outside stress could lead to a collapse of the human system response.	Simplify all operator involvements; i.e., complex decisions should be made by the designer of the product.
3. Most people are confused by	Not only will proper identification	Do not assume that people can

(continued)

Behavior Pattern	Safety Implications	Design Considerations
lack of identification of operating interface elements and by the inability to see or find a display or control or to locate the spot where the hand should be placed in order to pick up the device.	improve understanding, but it will also reduce the time necessary to figure out what to do and how to do it. Too much time and/or faulty deductions can lead to misuse or other human errors affecting operator and equipment safety.	recognize product features and understand them without labels or instructions (often labels or instructions are not included because of aesthetic design objectives or cost economies).
4. Many people are unable to perceive mechanical relationships, geometric relationships, or pictorial representations and to interpret these correctly.	If the designer assumes that all people have his or her skill in interpreting relationships and graphics and if the designer uses these to communicate with the typical user, the result will be confusion and misinterpretation by the user, resulting in possible misuse.	Avoid using special graphics or depending on relationships that are familiar only to specialists (i.e., design for the nonspecialist user).
Lack of manual dexterity, skill, or practice: 1. Many people either lack the innate capacity for dextrous manipulation or control of their bodies and limbs or else have not developed (through practice) the manual skills needed for many tasks involving the fingers, hands, arms, legs, or feet.	When the use of a product (and the ultimate safety of the operation) is highly dependent on manual dexterity or skill and practice, there is a great opportunity for error at one time or another—error that could lead to an unsafe manipulation or operation.	Try to design products in such a way that they do not require great dexterity, precision, or speed or highly sensitive response to various feedbacks. Designers should understand the innate characteristics of the human sensorimotor servosystem and design so that lags in the human system are taken into account.
2. Some manipulatory and other manual skills are especially degraded when the specific human-product relationship is not optimum (e.g., the position of the operator in relation to the task, the extent and direction of movement, and the rate and rate of change of movement).	Increased physical and mental strain reduces the ability to coordinate body, limb, hand, and finger movements and to make precise direction, rate, and force inputs, and it reduces attention and perceptual awareness of errors.	Human-product design relationships should be "natural," convenient, minimally demanding, and not subject to potential misuse.
3. Some manipulatory skills are especially degraded by gloves and clothing.	Clothing may snag and preclude completion of movement; bulk may cause inadvertent contact.	Provide *ample,* not just minimum, clearance between components.
Spurious autonomic responses: 1. Fear, anxiety, and uncertainty often lead to irregular, uncontrolled motor responses. 2. Panic often leads to unexpected responses (e.g., withdrawal, inaction, and violent and uncontrolled input to a device or control). 3. Pain leads to automatic recoil in most cases, although the opposite may be true of small children.	When responses cannot be predicted under these unusual conditions, many undesirable reactions may occur that have adverse consequences in terms of safety; e.g., a driver may freeze to the wheel or accelerator or suddenly exert forces or movements opposite to those which should be exerted, or a person may jump back after receiving an electrical shock, a burn, a sharp cut, or a blow to the body, twisting his or her body suddenly in an effort to recover from a shift in body mass, etc.	Be careful about deciding how users may react to very sudden disturbances in normal situations; these usually do not allow time for thinking, since they are the result of completely involuntary and unpredictable reactions. The design should preclude the possibility or probability of such situations and provide "wiggle room" for the possible reactions so that they do not set off a chain of hazardous events.

TYPES OF SPACE

Besides needing enough space in order to move about and perform various tasks, people react to space in a variety of ways. Several researchers have defined the space surrounding the individual in terms of the limits within which people categorically respond (see the accompanying sketch). *Intimate space* is that area in which a person tends not to allow anyone to intrude unless intimate relationships are expected. *Personal space* is that area within which a person allows only selected friends or fellow workers with whom personal discussion is mandatory. *Social space* is that area within which the individual expects to make purely social contacts on a temporary basis. And, finally, *public space* is that area within which the individual does not expect to have direct contact with others. Obviously, the more intimate the spatial relationship becomes, the more people resist intrusion by others. Personal space factors are important in establishing the privacy requirements for architectural design.

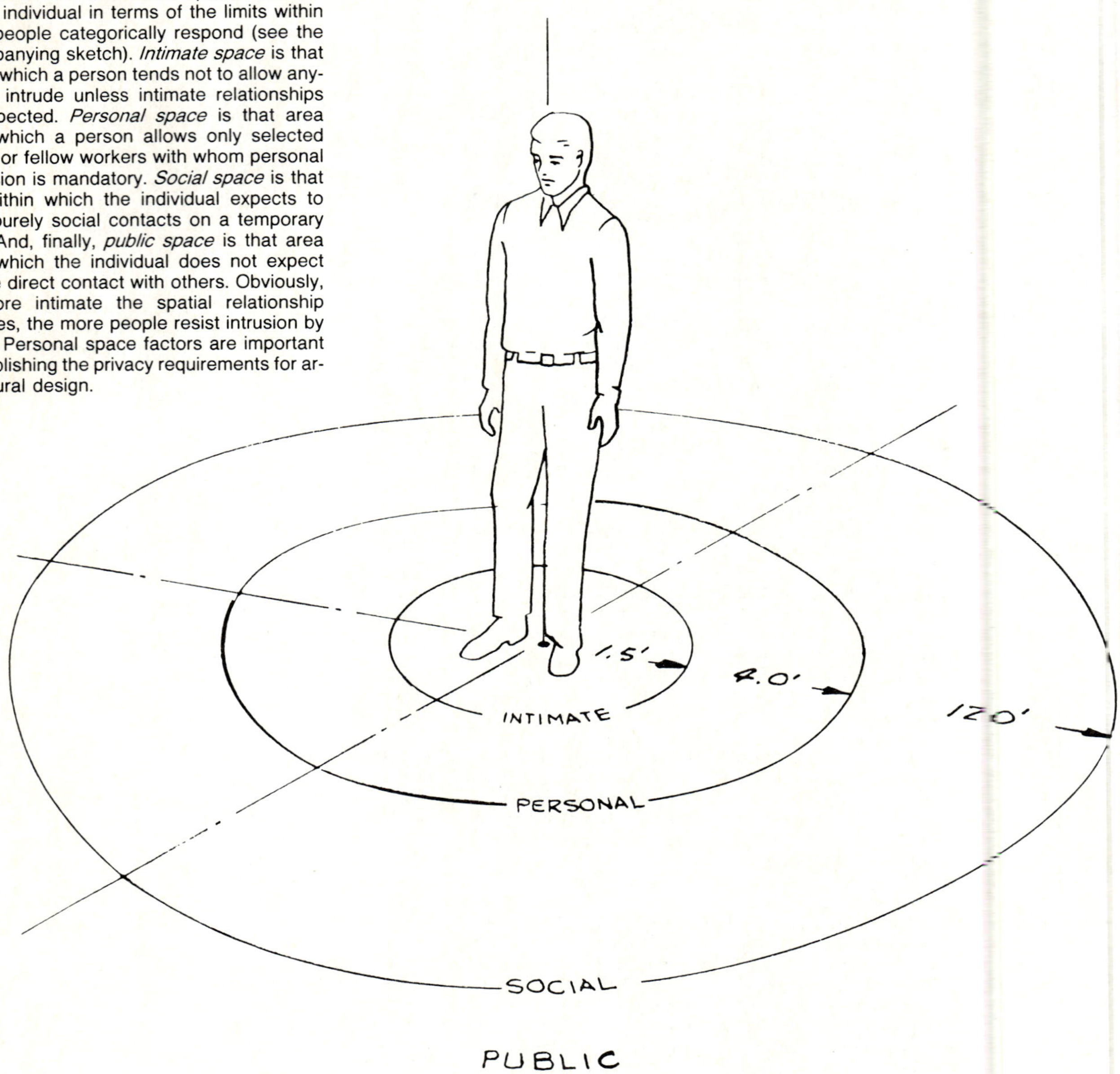

TYPICAL SUBJECTIVE RESPONSES TO SELECTED SPATIAL FEATURES

Although few research data have been generated with regard to how people respond to specific spatial factors (at least in terms of being able to prescribe precise, quantitative guidelines), it is important for the designer to reflect on potentially negative reactions that often result when a given space is not made compatible with what the user expects in terms of the size, shape, organization, color, and illumination of a particular space. The considerations listed in the accompanying table are suggested as a checklist for the designer.

Space Characteristic	Probable Response
Size (generally volume)	If the space is too small for the number of people, furnishings, equipment, or other objects that occupy it, people will consider it to be crowded. Although they may accept a crowded condition on a temporary basis, they will object to living or working in such a space for extended periods of time. If the space is too large for the people, furnishings, equipment, or other objects that occupy it, people will consider it "unfriendly," inconvenient, and/or overly demanding in terms of communicating, travel distance, maintenance, etc. Although they may accept the "barnlike" atmosphere for temporary periods, they will object to living or working in such a space for extended periods of time.
Shape (generally proportion)	If the space is out of proportion (too narrow, wide, long, high, etc.) for the intended use, people will consider it awkward and often distracting or oppressive. Although they may accept proportional distortion on a short-term basis (i.e., as they pass through briefly), they will object to living or working in such a space for extended periods of time. If the space contains such distortions as all curved surfaces, acute wall junctures, and too many projections or surface changes, people will consider it confusing and difficult to maneuver in and/or furnish. Although they may accept such distortions (or even consider them interesting) on a temporary or one-time basis, they will object to living or working in such a space for extended periods of time. When a ceiling is extremely high relative to the lateral dimension of a space, people feel as though they are working in a pit and that the walls are closing in on them. When a ceiling is extremely low and the space in front of the observer is very long, people feel as though the room is "endless" or as if they will hit their heads unless they duck.
Color and illumination	If a space is dark (unless this is required for a particular operation, such as a motion picture presentation), people tend to become lethargic and less active or they may feel anxious. As a rule, the less bright a room is, the less cheerful it seems. A small space will seem even smaller. If a space is too bright, people will feel overly exposed, or they will complain of glare or thermal discomfort (even though actual glare in terms of accepted light levels or inappropriate thermal conditions for comfort are not present). If there are too many different colors, too large expanses of very saturated color, or too many and too "busy" patterns of color within a space, most people become irritated after more than a brief exposure to the space. If there is too little color, no visual pattern, or no other decorative "break" in the visual environment, people will find the space monotonous, boring, and eventually irritating to the point of wanting to escape. Although isolated points of highly reflective surface provide interest, all-metallic and highly reflective surface treatments create both subjective and directly objective interference for most people who have to work in the space.
Windows	Generally, most people do not like to live and work in a space that is devoid of windows. First and foremost, people seem to need visual contact with the outside world. Too many windows, on the other hand, can cause the following possible negative reactions: too much glare, too much exposure (fishbowl effect), lack of protection from outside elements, true anxiety (caused by floor-to-ceiling glass at high elevations).

(continued)

Space Characteristic	Probable Response
Space organization	The internal components within a space and the traffic corridors and entrance and exit locations will seem either well organized or badly organized. The furnishings, partitions, decorative objects, etc., will appear as being either organized or disorganized, depending on the observer's ability to comprehend what things are and where they are with respect to his or her vantage point. Key behavioral response issues are: apparent capability to find one's way to specific locations, apparent ease for interacting and communicating with others with whom the individual must associate, apparent privacy provisions necessary to perform individual tasks. Although these are sometimes conflicting needs, the people who use a space will perform on the basis of how well each of these factors has been executed for *them,* not for the designer or the boss. The organization of internal space components obviously interacts with all the other space characteristics; i.e., the individual perceives and reacts to the combined effects of size, shape, color and illumination, windows, and organization simultaneously. A significant behavioral response will be an individual's interpretation of whether sufficient options are available for local modification of his or her own portion of the space. Even though people may never require a modification, they react to their own space in terms of permanently established restrictions that eventually elicit the feeling that the space is too small, the wrong shape, too dark, or isolated from the rest of the world, for example.
Furnishings	As a general rule, people are sensitive to improperly proportioned furniture, i.e., furniture that is too large, too small, or the wrong shape for the space in which it is placed. Although the designer normally tries to select furnishings that are properly proportioned for the space he or she has created, this may ultimately restrict the efficiency of the individual (e.g., a desk or storage cabinet may be too small). Thus, although the general visual proportions of furniture in relation to space must be taken into account to avoid negative observational responses, shortchanging the individual in terms of specific furniture and use requirements soon stimulates an even stronger negative response.

Unique Responses to the Individual's Location within a Space

Observation and study of what people do when they occupy spaces provide several important thoughts with respect to where a designer expects to place people. For example:

People in auditoriums will not sit next to a side wall, especially if the wall is very high.

People generally dislike sitting facing a wall that is too close, unless they can look out a window.

People generally prefer to sit in a position from which they can observe the entrance to the room and also observe other people, but they do not like to sit so that they are directly observed by others (or think they are).

People dislike having to face other people (in a waiting room, airplane, train, etc.) as they come out of a rest room.

People coming into an auditorium from the rear will tend to take the back seats, not those farther toward the front. Similarly, they will take the seat nearest an aisle, rather than one toward the center of a pew or row of seats.

In a conference situation, most people will take a seat at the corner of the conference table, not an end position or one halfway along the side of the table.

People generally will sit facing a light-colored rather than a dark or highly (saturated) colored or patterned wall.

When there is a row of chairs or stools, most people will seek out one that is not next to one that is occupied.

PSYCHOLOGICAL MOOD

Many attempts have been made to define basic human needs in relationship to designing an environment that is satisfying to the eventual customer, user, or occupant of living and working spaces. The accompanying model illustrates one such attempt to formalize human needs.

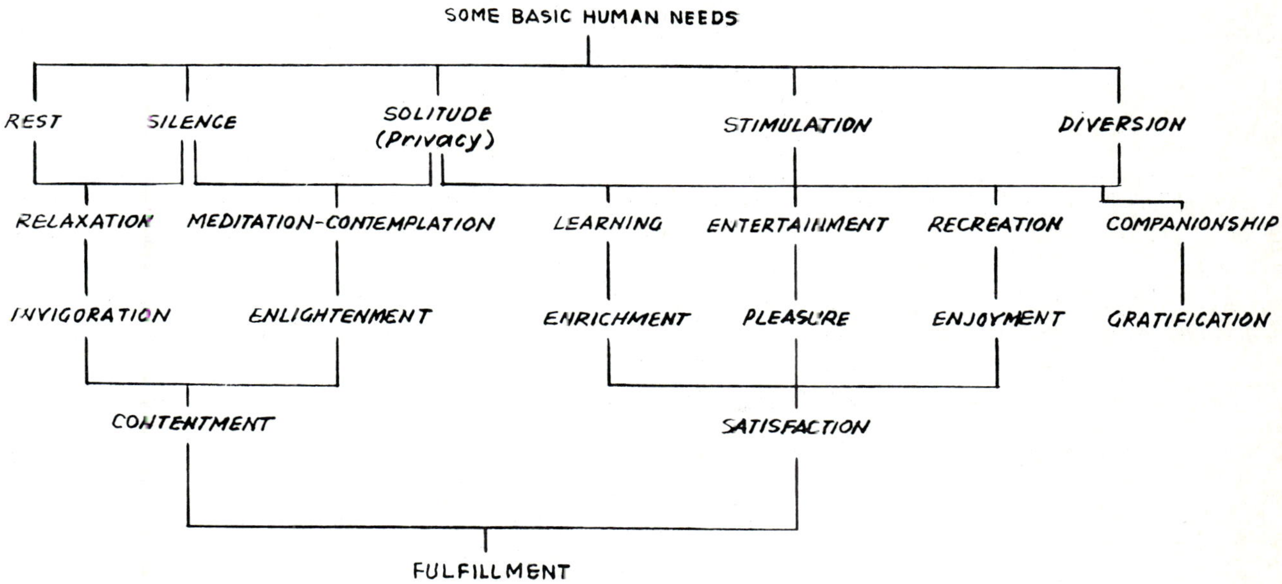

Complete fulfillment probably is beyond the expectation of most people. However, certain negative factors may seriously affect moods, attitudes, and eventual response to the environment. Some of these factors are the following:

Boredom
Apathy
Anger
Fatigue
Depression
Distress
Frustration
Anxiety
Pain

Although any or all of these may not be controllable through design, they can be influenced by it, i.e., by the nature of the task, by the environment within which the task is performed, and by how much stress or lack of stimulation the task involves.

EXPECTANCIES WITH RESPECT TO PRODUCT APPEARANCE

Although there are apparent relationships between various aspects of a product's appearance and the attitudes individuals may express toward it (especially in terms of what people decide to buy), there is little quantitative information to assist the designer in knowing what to do to make a product more acceptable. Key issues regarding product appearance include the following:

Size, shape, and proportion
Color
Texture
Apparent functionality
Compatibility with contemporary concepts (and/or with the style of a particular period, e.g., early American)

It becomes obvious that each of these characteristics hinges upon the particular visual environment within which the product will operate. For example, the size of furnishings needs to be compatible with the size of the space within which the furniture will be placed; the color of the product should be compatible not only with its general surroundings but also with the colors of the other products within these surroundings; the texture of the product should fit with the other textures used in the operating environment; the product must appear to have a purpose within the given environment; and, finally, the product should not appear outdated, unless it is obvious that it is meant to look like an antique, for example.

There is some validity to the theory that an attractive product will be treated with more respect than one that looks "cheap" or poorly designed. However, such concepts are difficult to quantify and write specifications for. On the other hand, it is important to recognize that most people soon neglect products that are hard to maintain in terms of good appearance.

A product should have the appearance of being substantial and sturdy; i.e., it should not appear too fragile and easily damaged.

HUMAN FATIGUE

The problem of fatigue is very important to the worker, and therefore it should be of concern to the designer; i.e., at least a design should be such that it does not result in activities that are unnecessarily fatiguing in terms of operation or maintenance.

A design may create two different types of fatigue: physiological fatigue, in which the operator's muscles are overstressed, and psychological fatigue (or mental fatigue), which may be caused by design-induced stress, i.e., complexity, high accuracy demands, or environmental implications, such as noise.

MUSCLE FATIGUE

Muscle fatigue occurs because of certain biochemical reactions within the muscle. Nerve impulses acting on a muscle fiber initiate a series of chemical reactions which result from, and contribute to, muscle contractions. Adenosine triphosphate breaks down under the influence of enzymes to create adnosine diphosphate, which in turn releases energy to enable the muscle to perform work. The triphosphate has to be regenerated before another contraction can occur. Energy required for this to happen is supplied by a breakdown of glycogen to lactic acid. Lactic acid is a poisonous by-product that must be removed by oxidation to carbon dioxide and water. The oxygen thus removes the by-products of the energy-producing reaction, which may continue some time after the muscular activity has ceased. The energy for muscular activity thus comes from a reaction which does not depend primarily on the presence of oxygen and which allows work to be done even when the immediate supply of oxygen is insufficient. This often permits the body to make a sudden extreme effort that might otherwise be impossible if the energy always had to be obtained directly from the oxidation of some substance within the muscle fiber.

Oxygen may come either from that stored in red muscle fiber or from the blood supply. As long as the supply is adequate to preclude lactic acid buildup, work is classified as "aerobic." If, however, the rate of work exhausts the reserve of oxygen, the work is said to be "partly anaerobic," and the muscle builds up an oxygen debt; i.e., there is an accumulation of lactic acid in the muscle and bloodstream, causing pain or muscular fatigue.

PSYCHOLOGICAL FATIGUE

The exact nature of psychological or mental fatigue is less clearly understood since it is extremely difficult to isolate and quantify causes and effects. Mental fatigue also confounds the researcher because opposite conditions may produce essentially the same mental or psychological aberration; i.e., one can become just as fatigued by boredom as by overwork. Unfortunately, we know of many instances in which individuals have been able to summon some reserve during a crisis, and thus we tend to believe that, with the proper stimulation, most people can continue mental activity indefinitely. Considerable evidence to the contrary, however, shows us that when an individual works too near mental capacity for long periods, almost any emergency that suddenly occurs may push the individual beyond his or her capacity to cope, the result often being a complete collapse or disorientation. Finally, the question of the threshold of psychological fatigue is further confounded by the fact that an individual's threshold may be stressed to within a few degrees of tolerance by preoperating conditions (prior activities), with the result that he or she has no tolerance to cope with an overly demanding mental task or situation.

Design Implications Relative to Minimizing Potential Muscle Fatigue

Avoid the following:

- Designs that require operators to apply near-maximum force capacities over many cycles and for long periods of time
- Designs that require continuous, rapid, repetitive muscle contractions for long periods, e.g., pounding, tapping, cranking, or push-pull cycling
- Designs that force operators to "hold" some device in a fixed position for long periods without intermittent rest periods
- Designs that require operators to maintain an upright posture for long periods without adequate body support (as in the case of a seat)
- Designs that require operators to make very long reaches, frequently and for extended periods of time
- Designs that require operators to stand or sit in an awkward position and to hold their arms above their heads for a long period of time
- Designs that require operators to work in a "bent-over" or squatting position or in a position on their stomachs or backs, with the accompanying stress of holding the head and arms in a strained position
- Designs that require operators to bend over and straighten up frequently and over a long period of time
- Workplace layouts that require many steps, repeated again and again over a long period of time
- Workplace layouts that require operators to sit "askew" (in a twisted position) in order to watch a display and at the same time operate some control (especially a foot control)
- Workplace layouts that require operators to hold one foot above a foot control (between pedal depressions) for long periods of time
- Workplace layouts that require operators to continuously move their heads from side to side or up and down
- Workplace layouts that require operators to step up and down frequently for long periods of time
- Equipment designs that require operators to hold and push against a work surface or component to maintain contact pressure
- Tool designs that require operators to hold a very heavy tool in a precise position for a long period of time
- Tool or other equipment designs that require operators to maintain a very tight grip to keep the tool in place (especially if the grip must be maintained for a long period of time)

POSTURE IN RELATION TO FATIGUE

Incorrect posture produces both physical and mental fatigue. As the lower sketches illustrate, the least fatigue occurs when the body can be kept in balance, since that is when there is the least demand on the muscular system to keep the body upright. The musculoskeletal system is most nearly balanced in the standing position. Even in this position, the various sets of muscles require fairly constant flexing to minimize fatigue; i.e., if the body were maintained in a "stiff," erect position, as when a person is at attention, the muscles would be in a state of continuous tension, and thus considerable fatigue would result. On the other hand, a generally erect and balanced position in which the muscles are constantly flexing a slight amount uses up the least amount of energy and produces the minimum amount of fatigue.

When an individual sits, the same principle of muscle balance and minor flexing is desirable. However, in order to create this balance, the body needs to be positioned so that the head mass is easily erected in a generally vertical axis above the torso and buttock masses.

As the accompanying sketches illustrate, the best and least fatiguing posture is one in which it is easy to position the head approximately in a vertical column above the torso and buttock masses (in a seat). If a seat, for example, is sloped too far back, as shown in the second sketch, the individual will naturally pitch his or her head forward to try to reestablish the columnar effect. If the seat is at 90°, as shown in the third sketch, the individual naturally pitches his or her head back to reestablish the columnar effect. Either of the latter conditions produces considerable strain on the neck and back muscles; this is why it is important to provide seats that have the proper seat-pan and seat-back angles.

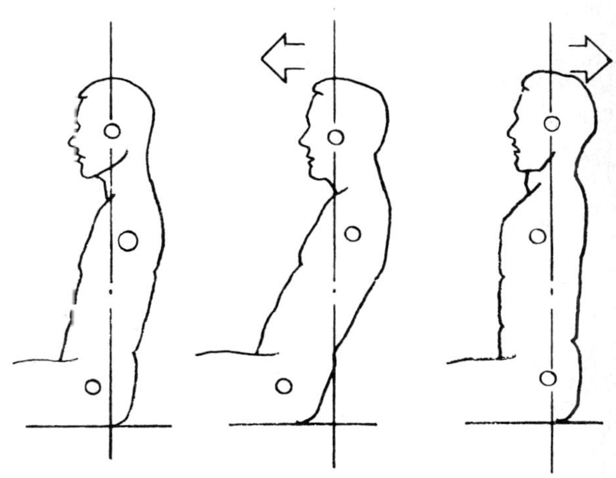

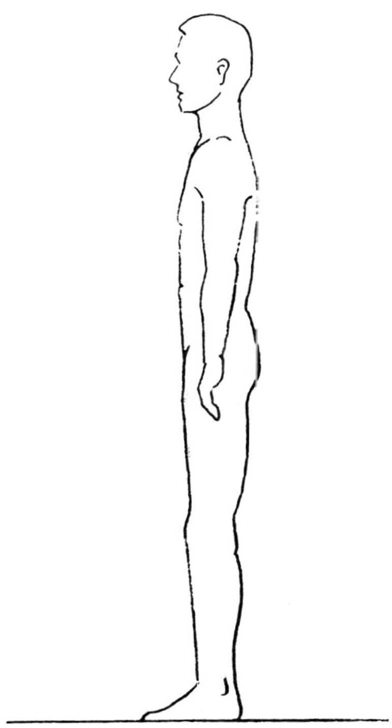

Common Fatigue-Producing Postures

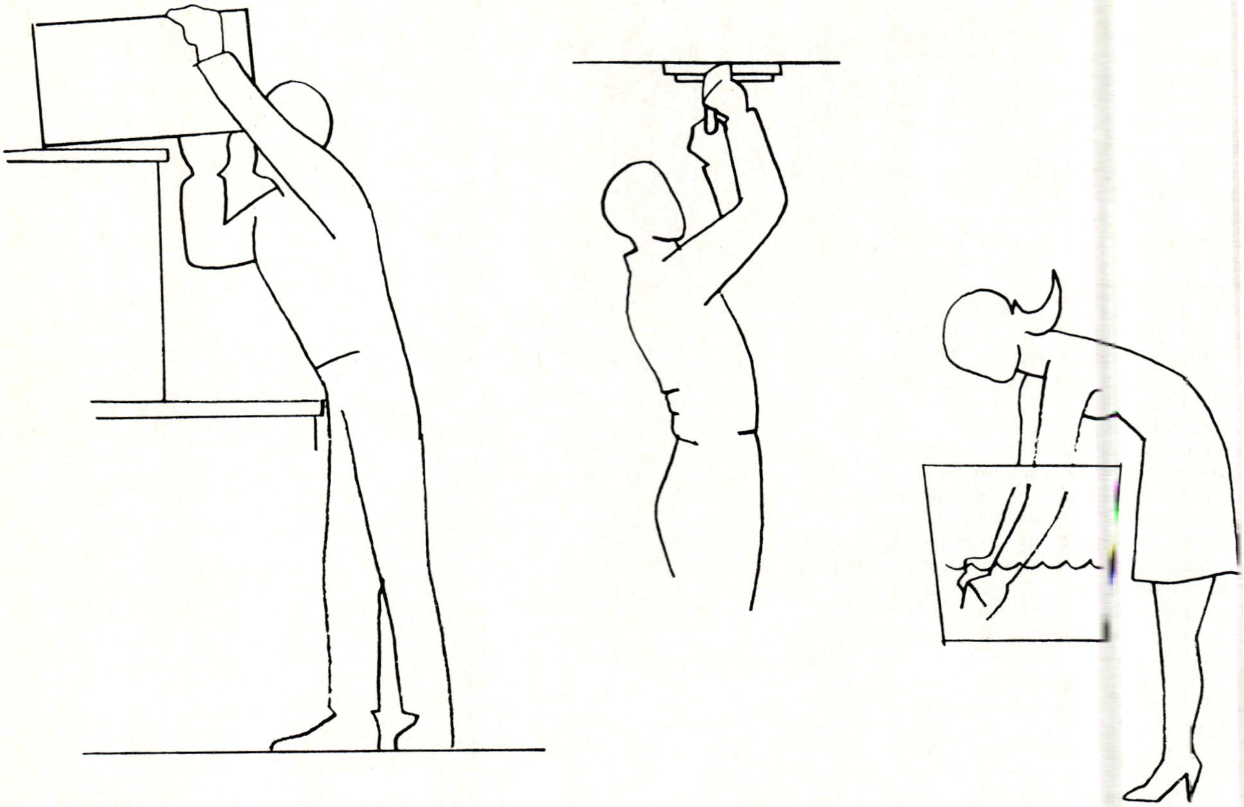

Repetitive lifting to an awkward position. An extended work task located above the worker's head. Bending over for extended periods.

Lack of attention during work-station design planning often results in the worker's being subjected to severe strain, especially if his or her task requires assumption of these poor postures for a considerable length of time. Neural and skeletal disorders often can be traced to the workplaces illustrated.

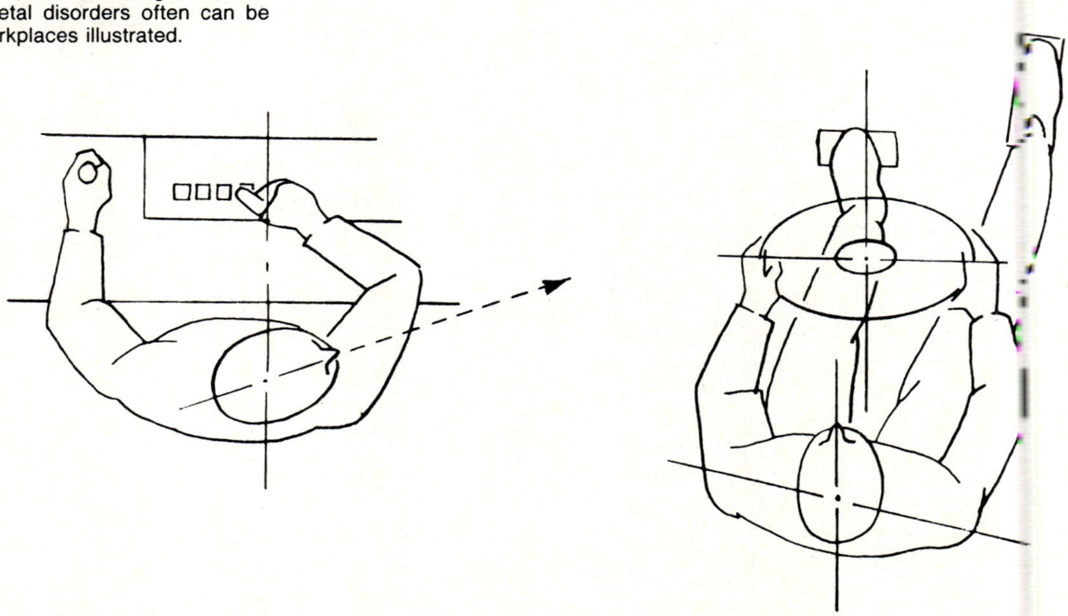

Sitting for long periods in a skewed position.

EQUIPMENT DESIGN AND OPERATING FEATURES KNOWN TO CONTRIBUTE TO MENTAL FATIGUE

1. Too many separate visual displays have to be monitored simultaneously.
2. Visual display formats require extrapolation rather than providing directly usable information.
3. Visual display detail is considerably greater than required; e.g., there are more scale marks than are warranted by the inherent accuracy of the instrumentation.
4. The legibility of visual display details is borderline, requiring unnecessarily close scrutiny in order to detect, recognize, and interpret what is being displayed.
5. Visual displays vibrate because they are not properly shock-mounted.
6. Visual displays are not adequately illuminated, or there are uncontrolled glare sources within the critical viewing envelope.
7. Simultaneous audio communications and/or excessive background noise is present.
8. Overly precise control adjustments are required.
9. Poor control dynamics, in terms of force-displacement, control-display, direction-of-motion, and/or movement ratio incompatibilities, are present.
10. There are long delays in informational feedback, i.e., long periods between signals or changes in equipment status.
11. There is a lack of timely indication of whether the equipment is functioning properly.
12. Continuous manual monitoring or control tasks are required that could just as well be automatic, with periodic operator alerting.
13. There is a lack of standardization among various similarly operated pieces of equipment, thus requiring operators to shift their point of reference.
14. The control panel layout is poorly organized, making it necessary for the operator to search for appropriate panel elements.
15. The workplace environment is inadequately controlled in terms of:
    a. Lighting, temperature, humidity, ventilation, noise, vibration, acceleration, pressure, etc.
    b. Support furnishings (standing platform, seating, writing surface, reference storage, restraint system, etc.).
    c. Space, e.g., clearance.

WORK OVERLOAD VERSUS BOREDOM

Mental stress and/or fatigue may be due either to an overload condition or to sheer boredom. Generally it is rather obvious when a design creates an overload condition, i.e., when the operator is required to do too many things at the same time. However, boredom is more difficult to cope with in terms of what can be done to a particular design. It has been suggested that one of the primary methods for dealing with the problem of potential boredom is to make sure that the human is given tasks that are more suitable for his or her unique capabilities, namely, planning and decision making. Alternatively, the machine or equipment should be given those tasks which are more difficult for the human and/or which are monotonous and unchallenging. The following general guidelines are suggested:

To prevent work overload:

Sequence tasks rather than creating overlaps.
Make individual tasks short.
Minimize task precision requirements.

To prevent boredom:

Provide task variety.
Distribute tasks equally throughout the work period.
Assign the operator only significant tasks and make it clearly evident that the operator rather than the machine is in control.

TASK ANALYSIS

As the term implies, task analysis involves the examination of what a particular design requires of an operator, maintainer, or product user who must perform a monitoring, control, or maintenance task. The sooner one can examine and define the task implications of a proposed task, the sooner he or she will know whether the design is compatible with the user's capabilities and limitations.

Task analyses and task descriptions serve two main purposes, among others. First, as noted above, the task analysis tells us not only whether the design is feasible but also whether it is optimum from the user's standpoint. Second, the task analysis provides important information for the personnel subsystem specialist who must plan and devise personnel acquisition and training programs for the system.

Preliminary task analyses should be performed as soon as possible during or directly following the initial conceptual phases of a development program; thereafter, they should be regularly refined and updated until the task description becomes a procedural guide.

The accompanying materials are meant to provide guidance in task analysis through the several examples of descriptive materials and documenting formats. Though the examples do not relate specifically to process plants, the principles are broadly applicable. It will be observed that various analysts have taken different approaches to documenting their analysis and have used different formats for describing tasks. Obviously, there are no set rules, and therefore these examples merely illustrate the types of information that various analysts felt were necessary to provide useful documentation—for assisting in design review and/or for developing manning and training requirements.

Example: Preliminary Task Analysis

Function 4.1.2:
Load xx projectile into x-1 breech block

PRELIMINARY OPERATOR FUNCTIONAL REQUIREMENTS WORKSHEET

SYSTEM/OP CONSTRAINTS: Loading rate = 2/Minute; Sea State 4

TASK DESCRIPTION: Monitor ammunition shaft delivery signal, open door, position projectile carriage, transfer projectile to carriage, move loaded carriage to loading position, open gun breech, transfer projectile to breech, check position, adjust fuse, close breech, notify fire control officer "ready".

PHYSIOLOGICAL CONSIDERATIONS: Maximum lift capability for one man is 60 pounds; higher weights require special sling or method for direct roll-transfer of projectile.

PSYCHOLOGICAL CONSIDERATIONS: Visual codes for identification and special calibration markings must be legible as well as visible under battle (red) lighting environment.

SAFETY CONSIDERATIONS: Automatic door closure on ammunition delivery system and possible effects of ship roll require special indoctrination and drill to prevent accidents (e.g., closure of door on personnel) or loss of control of projectile.

Example: Task Analysis Documentation Format Showing One Analyst's Approach for Defining Task Characteristics at Both a Gross and a Refined Level

Gross Task Analysis

Task	Display: Description	Display: Critical Values	Decisions	Subtasks Listed	Characteristic Errors or Malfunctions
A. Set up before first run	Position of brake	Not in 9 o'clock position	Release brake—move to 9 o'clock position	1	
	Recall of last link	Copy light since last inking (est.)	Measure ink—add if necessary	18, 19, 20, 21, 22	

Analysis of Subtasks

Subtask or Task	Display-Control Descriptions	Control Action	Indication of Response Adequacy	Objective Criterion of Response Adequacy	Characteristic Errors or Malfunctions
A. Release brake brake	Brake	Turn clockwise (up)	Brake stop in 9 o'clock position		
B. Attach the stencil	Wheel	Turn	Stencil head clamp available		
	Stencil head clamp	Lift left end	Stencil head clamp loosens		

Function	Human Functional Element	Task	Subtask
Activate gyros and other components	Make power ON input	1. Turn on ac and dc missile guidance power	*a.* Check to assure that all circuit breakers are "in" *b.* Position ac missile power guidance switch at ON *c.* Position dc missile guidance power switch at ON
Activate test instruments	Make power ON input	2. Turn on missile preflight analyzer power	*a.* Check to assure that test patchboard is inserted and locked *b.* Position analyzer master power switch at ON *c.* Insert test tape in tape reader and lock in place
	Make analyzer self-test initiating input	3. Start analyzer self-test test sequence	*a.* Position test switch at SELF-TEST
	Monitor self-test readout		*b.* Check test result window for indication of analyzer self-test result
	Provide corrective inputs and monitoring when self-test yields a "no-go"	Alternative procedure: isolate and replace malfunctioning analyzer module	*a.* Check code displayed in malfunctioning module window against modular code table *b.* Remove module from analyzer and take to storage *c.* Obtain replacement module from storage; take to analyzer and insert it *d.* Repeat task 3

Suggested Information Requirements for Expanded Task Analyses

Item	Description
Operator and maintainer position	Position-type title or specialty rating that it is believed the task should be assigned.
Job operation	Job operation performance description, i.e., operation performed in support of a project function, usually performed at a single location as a unit of work, in the sense that it has a well-defined beginning and ending.
Duty	Larger units of work under a job operation assigned to an individual in execution of a position. Duties are made up of operationally similar tasks within a given operator or maintainer position. An example might be "carries out overall subsystem test on XX radar."
Task title	A short title that indicates what an individual does (in functional terms).
Task index number	Task-identifying number, usually corresponding with OSD numbering system.
Data sources	Major sources of data useful for later verification (equipment, location, performance criteria, etc.).
Work-area location	Where task is accomplished.
Task description	Includes: Equipment used Displays used Controls used Actions (detailed, sequential description of each task step) Support equipment and aids used Feedback re individual's actions
Type of task	Qualities or characteristics of task (fixed procedure, variable procedure, motor skill, system analysis, circuit analysis, display interpretation, etc.).
Frequency of performance	Once, twice, etc.; hourly; per shift, daily, weekly, monthly.
Performance time	Estimated hours and minutes job or task will take; cite maximum permissible time, etc.
Criticality	Effect of failure to perform on the success or failure of job or project; potential for personal injury.
Newness	Extent to which the task is new to the using agency.
Other positions	Other tasks whose performance depends on, or interacts with, the performance of other personnel position types (indicate functional interrelations).
Safety factors	Any known or suspected hazards, sources, or preventive requirements.
Tools and equipment	Tools, test equipment, protective gear, reference manuals, etc.
Skills and knowledge	Types and level of skill or knowledge pertinent to the selection and training of individuals for the job or tasks.
Physical characteristics	Special physical requirements necessary for effective task performance (e.g., body size, strength, dexterity, coordination, physiological tolerance to specific and anticipated stresses).

Example: Analysis to Break Out Tasks and Subtasks for Human Functions

Behavior Classification for Task Analysis

Processes	Activities	Specific Behaviors
Perceptual processes	Searching for and receiving information	Detect Inspect Observe Read Receive Scan Survey
	Identifying objects, actions, and events	Discriminate Identify Locate
Mediational processes	Information processing	Categorize Calculate Encode Compute Interpolate Itemize Tabulate Translate
	Problem solving and decision making	Analyze Calculate Choose Compare Compute Estimate Predict Plan
Communication processes		Advise Answer Communicate Direct Indicate Inform Instruct Request Transmit
Motor processes	Simple, discrete tasks	Activate Close Connect Disconnect Join Move Press Set Raise Lower Hold
	Complex, continuous tasks	Adjust Align Regulate Synchronize Track Transport

Source: Adapted from C. Berliner, et al., "Behaviors, Measures and Instruments for Performance Evaluation in Simulated Environments," paper presented at a Symposium and Workshop on Quantification of Human Performance, Albuquerque, N.M., 1964.

Requirements Allocation Sheet

Requirements Allocation Sheet	Functional Diagram Title and Number		Equipment Identification		Personnel and Training Equipment Requirements				
	Nomenclature and Number of CEI								
Function Name and Number	Design Requirements	Facility Requirements	Nomenclature	Detail Spec, CEI, Master Control Number	Tasks	Time Required	Performance Required	Training and Training Equip. Required	Procedural Data Requirements
3.3.1.3 Electrical continuity check	Line voltage-0–150 Ω	None	Line tester	1.3	Attach clips. Turn on-off and SW and knob. Adjust Monitor	Estimated 120's	Read to ± 5 Ω	Actual tester	*Note:* Instructions in maintenance management
3.3.1.4 Hydraulic line press check	Line Pressure-350 lb/in²	None	Hydraulic press tester	2.4	Remove protection cap. Attach quick-connect line connectors. Open line valve. Read Meter	Estimated: 325's	Read to ± 2 lb/in²	Hydraulic pressure tester	See instructions in checkout manual *Note:* Be sure line valve is in closed position before making line connection.

Example: Task Analysis Worksheet

1. Job Operation:	2. Task Title:	3. Location of Task:	4. Source Reference:	5. Operational Site:
Prepare Swimmer Delivery Vehicle (SDV) for Launch	Replenish Breathing Gas Supply	On Dock or in Water	SDV Specification	World Wide

EQUIPMENT DATA / PERSONNEL DATA

6. Equipment Required	CSE Required	7. Source Ref.	8. Number of personnel	9. Recommended Position Title	10. Applicable Rating
Breathing gas supply pump & containers, press gauge	Breathing gas supply containers	none	2	Vehicle Maintenance Man	BM

TASK PERFORMANCE DATA

11. Time Required — Hours 0, Minutes 25

12. Frequency — Day / Week / Month: unknown

13. Probable Error — Minimal on land, moderate in water

14. Speed
- ☐ 1. Not critical
- ☑ 2. Moderately critical
- ☐ 3. Highly critical

15. Positioning/Handling Equip, Special care
- ☐ 1. Little
- ☑ 2. Moderate
- ☐ 3. Considerable

16. Manipulating Controls
- ☐ None
- ☑ Hand Valves
- ☐ Toggle Switches
- ☐ Selector Switches
- ☐ Pushbuttons
- ☐ Cont. Audio Feedback
- ☐ Cont. Visual Feedback

17. Source of Special Dangers
- ☐ None
- ☐ Mechanical
- ☐ Electrical
- ☑ Explosive
- ☐ Temperature
- ☐ Volatile Fuels
- ☐ Toxic Substance
- ☐ Other

TRAINING DATA

18. Nature of Procedure
- ☐ Fixed
- ☑ Variable
- ☑ Motor Skill
- ☐ System Analysis
- ☐ Circuit Analysis

19. Technical Manual Function
- ☐ Primary
- ☑ Secondary

20. Training Difficulty
- ☐ Easy
- ☑ Moderate
- ☐ Hard

21. Training Equipment
- ☐ Complex
- ☑ Simple
- ☐ None

22. Technical Manual Title: Life Support System, Maintenance & Replenishment of SDV-TR-013

23. Training Course Title: Life Support System, Repair and Replenishment of SDV

24. Task Performance Date
- ☐ Estimated
- ☑ Tried in mockup
- ☐ Tried, hardware

25. Equipment Development Status
- ☐ Design Plan
- ☐ Preliminary Design
- ☐ Mockup
- ☑ Prototype
- ☐ Production Design
- ☐ Production Equipment

26. Narrative Task Information; Remove skin cover plate over life support quick-connect fitting. Attach gas supply nozzle to boat, open supply valve, and monitor flow rate. Shut off valve when pressure reaches 2800 psi. Disconnect supply and check for leaks at nozzle. Replace skin cover.

Example: Providing Inputs to Design from Task Analysis

TASK CRITICALITY This is not an independent output; rather, it is deduced from the nature of the task and is presented to the engineer with the task description in the form of a note to the task description.

There are three major steps in the derivation of task criticality:

1. *Identify the potential errors which can be made in performance of the task.* This is largely a matter of considering the elements of the task and the perceptual, motor, and decision-making demands imposed on the operator. Thus, in the case of a simple task which involves *(a)* reading a pressure gauge regulating the internal pressure of a rocket and *(b)* stopping a pump at a specified pressure, errors may manifest themselves in two ways:
   a. Failing to stop at the prescribed point
   b. Stopping the pump before the prescribed point
2. *Identify the effect of each potential error on system operation.* In the example above, failing to stop the pump at the prescribed point may result in overpressurization and bursting of the rocket being pressurized. Stopping before the prescribed point will result in underpressurization, which will cause certain sensitive instruments requiring a pressurized atmosphere to function erratically.
3. *Estimate the relative criticality of the potential errors.* Criticality may be scaled in terms of categories such as loss of personnel, destruction of the system, mission failure or abort, mission degradation, and mission delay. In these terms, overpressurization may be more critical than underpressurization, since it may result in explosion of the rocket and destruction of the rocket and launch pad as well as loss of life, while underpressurization is less critical, since the mission may (though not necessarily will) be degraded.

Pointing out a task as being critical to the engineer "flags" that task as one requiring special consideration in design. Among the solutions which are possible (certainly the list is not exhaustive) are the following:

1. Replacing the human with an automatic means of accomplishing the function if the desired level of correct performance cannot be achieved in any other way
2. Providing means to reduce the probability of error, e.g., assigning a special feedback device to warn the operator when the task is being performed incorrectly
3. Assigning the task to only highly skilled personnel

Task criticality is highly related to specification of task difficulty and error likelihood. Since the engineer thinks in terms of physical effects on the system, it is preferable to flag the task as being critical without indicating that it also has a high difficulty or error-likelihood index. The provision of quantitative indices of a highly precise nature, such as probability of operator error to four figures (e.g., .0013), is not advised, since the engineer cannot interpret the quantitative values in design-relevant terms. A gross categorization of task difficulty, such as the following three-part scale, is as much precision as the engineer can handle in design terms:

1. Simple, routine
2. Somewhat difficult
3. Very difficult

Moreover, it is necessary to apply the above scale only to those few very critical tasks which merit design attention, not to all tasks.

TASK DURATION Task duration should be considered in two ways:

1. As a system requirement, i.e., the time within which the task must be performed in order to accomplish a given system function
2. As an anticipated human performance capability, i.e., the time within which the operator can actually perform the task

Item 1 is a criterion against which item 2 can be evaluated as satisfying or failing to satisfy system time requirements.

As a system requirement, a task may have to be accomplished in so short a time either that the operator cannot physically perform it or that the probability of the operator's making an error will be substantially increased because of the time loading. In either case, special attention must be drawn to such a task. If the system time requirement is inflexible, it may be necessary to automate the function involved (to eliminate the operator) or else to redesign the manner in which the task can be performed or the equipment is to be operated or maintained.

Task duration is, of course, not critical unless the system's required response time is also critical to the successful accomplishment of the mission. Hence, it is necessary to analyze the mission segment in terms of its time demands before examining any individual task duration. Information required in order to perform task duration analysis will include:

1. System performance time requirements
2. A description of the tasks to be performed by personnel in each mission segment
3. The estimated time required to perform the task

Of these informational requirements, the last is the most difficult to secure because it requires data on the performance time capability of personnel, such as the time required to hook up an umbilical connection. That information can be secured from previous comparable systems in which similar or identical tasks have been timed or from the body of general human performance data in the literature. Neither of these sources is readily available.

Example: Task Duration Analysis for SM-X "Assemble Missile" Segment

Function or Task	Output		Duration		Input		Remarks
	Not Earlier Than	Not Later Than			Earliest	Latest	
A. Transport W/H, etc.							① Output cannot be made earlier than time missile clears track bed.
2. Position crane	—	T + 1	1 min		—	T + 0	
5. Rotate nose cone, forward	①	T + 3	1		T + 2②	T + 2.5	
6. Position truck ahead of transporter erector	T + 3	T + 4	1		T + 2.5	T + 3	② Time of initiation depends upon rate at which a second technician hoists nose cone.
B. Mate missile and nose cone							③ Output can start as soon as components are mated.
1. Use crane and truck to mate components	—	T + 6	2		T + 3.5	T + 4.5	
C. Attach nose cone to missile							④ This task is also performed by a second technician, who exerts a pacing effect.
1. Install bolts to attach components	③	T + 10	4④		T + 5.5	T + 7	

Source: D. Meister, et al., *The Impact of Manpower Requirements and Personnel Resources Data on System Design,* AMRL-TR-68-44, Air Force Systems Command, Wright-Patterson Air Force Base, Ohio.

TASK DIFFICULTY (ERROR LIKELIHOOD)[1]

The human factors specialist is especially concerned about task difficulty because this, in turn, may lead to higher error probability, with its attendant effects on project accomplishment. Task difficulty arises because system requirements are incompatible with and overload the skill capabiliy of the individuals assigned to perform the task.

Task difficulty is not the same as error likelihood. A difficult task need not automatically have a higher error probability if personnel of higher skill are available to compensate for the increased task difficulty. The significance of task difficulty is intensified when the task is also critical to the accomplishment of the mission. Such difficult and critical tasks automatically demand redesign because their attendant error probability cannot be accepted. As in the case of maximum task durations which the operator's performance cannot meet, it may be necessary to automate the performance of the task, relax the accuracy requirement (thus implicitly accepting a higher error probability), or redesign the task to simplify it.

The determination of task difficulty must be made by analyzing the individual task in terms of the inputs which initiate the task (e.g., a verbal message) and the outputs which accomplish the task (e.g., a switch action). The human factors specialist will look for the following characteristics, which may (not necessarily will) indicate an excessively difficult task:

1. The input which initiates the task requires excessively precise visual discriminations or fine motor responses.
2. The operator's response to the initiating inputs must be performed so quickly that he or she has problems keeping up with the initiating inputs.
3. The accuracy demanded of the operator in responding to the initiating inputs is excessive (e.g., heading error must be within 0.5°).
4. The task must be coordinated extremely precisely with other tasks performed by other personnel.
5. The environment in which the task must be performed tends to degrade task performance (e.g., high levels of noise or acceleration).
6. Information from multiple sources (e.g., several displays on a control panel) must be integrated by the operator in order to make a decision.
7. There is less than the desirable amount of information available on the basis of which a decision must be made or an action taken.
8. The task is composed of many subtask elements, the correct performance of which is necessary to task performance, but the amount of feedback provided (knowledge of the correctness or incorrectness of subtask accomplishment) is inadequate.
9. Short-term memory requirements for task performance are excessive (e.g., memory for long sequences or target coordinates).

The design solutions available for reducing task difficulty include the following:

1. Providing additional training or selecting more highly skilled personnel
2. Simplifying the task by such means as combining information sources, providing additional feedback, subdividing the task among several operators, or changing the manner in which the task must be performed
3. Reducing system requirements by accepting a higher error probability, longer response time, etc.

Example: Task Analysis versus Training Requirements Definition

JOB: ______________________

TASK ANALYSIS/TRAINING OBJECTIVES & CRITERIA WORKSHEET Page No. 1 By: ________ Date: ________

TASK NUMBER	TASK, SUB-TASK, STEP, ELEMENT DESCRIPTION	INSTRUCTIONAL OBJECTIVES/CRITERIA PERFORMANCE	CONDITIONS	CRITERION	CRITERION ITEM
0.000	Describe duties, tasks and task elements in the	Write specific statement	Specify important condi-	State standard of	Write test question
	order of their occurrence and in abbreviated form.	of intent, i.e., what	tions under which per-	acceptable performance.	that matches ob-
	Arrange generally to show input (display, informa-	learner is expected to	formance is to occur,		jective in perform-
	tion received), mediation (diagnosis, decision),	be able to do and how you	tools he will be given,		ance and conditions.
	and output (action/feedback).	will know when he is able	job aids, prescribed		
		to do it (i.e., terminal	procedures; also anything		
	Note any unique constraints that would affect task,	behavior).	learner will be denied.		
	such as: environmental factors (noise, illumina-				
	tion, temperature), stress (time limits, etc.),				
	physical (location considerations that may add to				
	task difficulties, etc.).				
	Indicate whether order of task occurrence is				
	critical or non-critical.				
	Make sure task descriptions are stated in <u>behavioral</u> terms!				
	- - - - - - - -	- - - - EXAMPLE - - - -	- - - - - - - -	- - - - - - - -	- - - - - - - -
	<u>Corrective Maintenance: Radar Bright Display Equipment</u>				
3.0	Determine if isolation amplifier is performing its				
	design function.				
3.1	Compare input and output signals of amplifier to	Be able to connect test	Given a Sperry synchro-	In accordance with pro-	Given a Sperry
	ensure no objectionable noise or gain distortion	equipment to power	scope. With reference	cedures prescribed by	synchroscope, con-
	introduced.	source and test points,	manual but without	TM 7-864 and Instructio	nect it to power
		read voltage values and	assistance or super-	Manual 8-60. Perform	supply, check volt-
		... alignment	vision ...	...ks in corre...	...ordance

[1] D. Meister et al., *The Impact of Manpower Requirements and Personnel Resources Data on System Design,* AMRL-TR-68-44, Air Force Systems Command, Wright-Patterson Air Force Base, Ohio.

Example: Personnel Requirements Documentation

PERSONNEL REQUIREMENTS INFORMATION

Job Category:______________________ Duty:______________

Operation:______________________ Task:______________

Date:______________

Task Analysis Information

Task Performance Frequency:______________ Performance Time:______

Task Requirements:
- Perceptual Level:______________
- Decision Level:______________
- Stress Conditions:______________
- Motor Skills:______________
- Physical Requirements:
  - Height:________ Strength:________ Weight:______ Other______

Recommended Changes in Task Structure and Human Requirements:______
____

Personnel Used in Task Verification:______________

Company Procedures Knowledge Task Requirements:______________

Recommended Personnel Selection Requirements:

- General Aptitudes:______________
- General Training/Education:______________

Recommended Pre-selection Testing:

- Performance:______________
- Intelligence:______________
- Interests:______________
- Personality:______________
- Special:______________
- Scoreable Checklist Performance:______________

In-Plant Training Recommendations:

- Requires No Training:______________
- Requires Familiarization:______________
- Requires Special Training:______________

Action Taken by Personnel Department:

Date Report Forwarded:______________ Date Report Returned:________

Example: Quantitative/Qualitative Personnel Requirements Inventory Documentation

Position Title: Liquid Fuel Systems Maintenance

Specialist/Technician — Air Force Specialty Code (AFSC) 56850B/70B

Duties and Tasks	Performance Reliability	Skill Level	Skill Type and Proficiency Availability	Training Requirements
1.0 Connect propellant transporter(s) to fixed-base propellant transfer system Possible Human Errors Cross-connection of hoses Misalignment of coupling 2.0 Prepare to transfer propellant from transporter(s) to ready-storage vessels (RSV's) Valves in system left open System not completely bled RSV vents not opened 3.0 Transfer propellant from transporter(s) Ready Storage Vessels (RSV's) Leaks in system due to improper connection or maintenance Failure to close valves when transporter is empty	.9975 .9960 .9978	5 5 5	1. Position Summary The LFMS/T is responsible, under the direction of the Fuel Specialist/Supervisor, for the conduct of propellant loading and unloading operations at the launch site, including emergency or damage control activities if required. He decontaminates the surface area, the launch duct or the sump, or the rocket as required. He directs emplacement and removal of mobile PTS equipment and seals fixed portions of the PTS. LFMS/T's assigned to maintenance teams perform unscheduled maintenance on PTS fixed equipment at the launch complex. Any LFMS/T's performing PTS procedures at the launch complex should know both the emergency and normal loading and unloading procedures. 2. Environment a. Launch Site (1) The LFMS/T controls traffic into and from the launch complex security area. He directs positioning of propellant transporters and the interconnecting of PTS elements.	The training should be designed to train selected Air Force personnel to: 1. Operate control panels to determine malfunctions in the propellant transfer system. 2. Perform interconnections of the electrical and mechanical mobile equipment. 3. Operate and maintain the propellant transfer system components and the various items of GOE/GSE including the propellant transporters, propellant holding tanks, nitrogen storage tanks, and propellant transfer control unit. 4. Perform emergency unloading procedures for the PTS. 5. Inspect storage tanks, disconnects, valves, piping, pressure regulators, filters, pressure gages, and temperature gages for leakage, corrosion, damage, and wear; also solenoids, relays, switches, chassis of the transfer control panel, sensors, amplifiers, motors, terminals, interconnecting cabling and instrumentation for proper operation. Instruction should include removal and replacement procedures for the above listed items.

Example: Maintenance Training Requirements Analysis

Maintenance Level	Maintenance Task Category	Support Equipment (AGERD)	Skills and Knowledges Required to Do the Job	Category of Skills or Knowledges to Be Taught or Strengthened	Equipment and/or Training Recommendations
Organizational Unscheduled Sub-system Troubleshooting (Continued)				f. Oral Verbalization: Communications procedures between maintenance personnel during task performance.	
Intermediate Unscheduled LRU Mechanical and Electrical Repair	4. LRU (Mechanical and Electrical) Repair: involves disassembly in varying degrees, visual inspections, running of tests and measurements (e.g., clearance tolerances, pressure loss and leakage tests, torque tightening checks). Certain minor servicing functions such as lubrication or dehydrator recharging is accomplished. Repair consists mainly of parts	a. Regulated air supply. b. Power supply. c. Valve and flange caps. d. Pressure and height gauges. e. Depth and thickness micrometers. f. Flowmeter. g. Replacement parts. h. Standard electronic AGE. i. Cleaning materials. j. Anti-G valve test	a. Knowledge of appearance, location, nomenclature and fitting sequences of LRU parts. b. Knowledge of part functioning principles and relationship to LRU and fitting principles. c. Be able to use precision mechanical tools, and other associated AGE. d. Knowledge of visual appearance of acceptable part condition. e. Know how to perform each step of procedures.	a. Nonverbal Identification: Identification of LRU parts configuration. b. Nonverbal Detection: Spatial judgments in parts fitting tasks. c. Learning facts: Knowledge of part appearance, location nomenclature, fitting sequence, visual cues of acceptable condition, gauge scale readings, test setup. d. Learning Principles: Parts fitting, parts and LRU	1. Transparencies (animated and static): For presentation of facts and principles on LRU and precision measuring tools. 2. Self-Teaching Programmed Texts: Using appropriate illustrations to produce identification skills of parts and nomenclature knowledge to aid assembly and disassembly tasks. Also used to teach the

(continued)

Example: Maintenance Training Requirements Analysis

Maintenance Level	Maintenance Task Category	Support Equipment (AGERD)	Skills and Knowledges Required to Do the Job	Category of Skills or Knowledges to Be Taught or Strengthened	Equipment and/or Training Recommendations
Organizational Unscheduled Sub-system Troubleshooting (Continued)				f. Oral Verbalization: Communications procedures between maintenance personnel during task performance.	
	replacement. Some external surface damage repair. Assembly follows with running of functional "quality assurance" tests.	set (0500).	f. Be able to make minor repair decisions regarding extent of repair required from test results. g. Be able to make minor maintenance decisions.	functioning, and relationships. e. Learning Procedures: Testing, inspection, and test setup procedures. Disassembly and assembly procedures. f. Using Principles: Evaluating test results. Apply principles in performance of tasks. g. Positioning Movement: Using precision tools in making measurements, fitting parts together.	concept and use of scales and interpolating readings of scales on precision instruments. 3. Real Precision Tools: For active student use to permit perceptual and motor skill practice and learning on appropriate LRU's and setups (for task specific tools only). 4. Bench Setups with Real LRU's and Support AGE: The nature of mechanical work requires the development of perceptual and motor skills, spatial.

Acceptable Task Verbs

Activate: To initiate system operation by movement of a control.
Actuate: To move a control switch (push button, rotary selector or round knob, toggle, lever, etc.).
Adjust: To reset a control as a result of new or nonprogrammed information. (Refers usually to nondetented control hardware.)
Advise: To notify others verbally.
Align: To bring control and display elements into coincidence or to match physical hardware for the purpose of assembly or checking of assembly.
Assemble: To join major assemblies and subassemblies. (Includes all attach actions plus mating of component interfaces.)
Calculate: To compute numerically (either mentally or physically).
Calibrate: To check accuracy, deviation, or variation by comparison with a standard.
Carry: To transport, except without mechanical assistance.
Check, test: To carry out those activities associated with verifying operational readiness.
Clean: To remove dirt, grease, or other contaminants.
Connect: To couple matched connection hardware (electrical, hydraulic, or pneumatic lines or nonflow mechanical or structural supporting interfaces).
Decide: To select a course of action on the basis of previous analysis.
Depress: To actuate a push-button control.
Disassemble: To disjoin major assemblies and subassemblies. (Includes all detach actions.)
Disconnect: To decouple matched connection hardware (see *connect*).
Drain: To empty liquid contents from tanks, reservoirs, lines, etc.
Evaluate: To analyze information for the purpose of establishing its validity and significance.
Fasten: To secure common fastener hardware (nuts and bolts, latches, Zeus fasteners, screws, etc.).
Fill: To introduce fuel, oil, hydraulic fluid, etc., into tanks or reservoirs.
Flush: To introduce fluids into tanks, reservoirs, or lines for the purpose of decontamination.
Insert: To put into (e.g., a bolt into a hole).
Inspect: To make a programmed, critical visual examination of equipment for obvious defects.
Install: To carry out the initial activity of attaching components or subcomponents.
Interpret: To analyze information for the purpose of establishing its meaning.
Listen: To carry out an auditory activity for the purpose of extracting specific information.
Measure: To make a qualitative and/or quantitative determination of the operating condition of a component or subcomponent.
Monitor: To carry out those activities associated with maintaining cognizance of a system's status (e.g., observing panel status lights).
Note: To carry out a perceptual activity (visual, auditory, etc.) for the purpose of extracting specific information from display hardware and/or other operational phenomena.
Observe: To carry out a visual activity for the purpose of maintaining an awareness of a general condition.
Operate: To carry out those activities associated with the *use* of consoles and test equipment.
Place: To move a toggle switch or lever arm in a specific direction.
Position: To physically orient an object for the purpose of accomplishing a specified task.
Read: To carry out a visual activity for the purpose of extracting specific information from display hardware.
Record: To fill out checklists, log sheets, or other required forms.
Remove: To detach components or subcomponents, usually in connection with maintenance activities; to move from the original position (e.g., raising an engine cover prior to start).
Repair: To carry out those activities associated with restoring damaged, worn-out, or malfunctioning equipment to an operable condition.
Replace: To attach components or subcomponents, usually in connection with maintenance activities.
Secure: The reverse of *operate.*
Service: To perform the scheduled cleanup, lubrication, and replenishment necessary to prepare equipment for use.
Set: To place a rotary control pointer in a predetermined position.
Transfer: To carry or transport something (with the aid of equipment) from one location to another.
Transport: To move an assembly, unit, component, etc., from one place to another with the assistance of a mechanical aid; includes steering or guiding the mechanical aid.
Troubleshoot: To isolate a fault within a system, subsystem, or component.
Unfasten: To release common fastener hardware (see *fasten*).
Verify: To confirm that an expected condition exists.
Withdraw: To retract from (e.g., the reverse of insert).

LINK ANALYSIS

Link-Analysis Procedure

Link analysis is a technique for developing a best arrangement of components on a control panel, functional elements within a work station, elements and traffic flow within a work area, or humans and machines in a system. This technique is used only after decisions have been made regarding what items are to be on the panel, what equipment will be used in the system, how many people will be used, etc. Thus, link analysis is used to place the system components, be they items on a panel or people and machines in a workroom, in the "best arrangement" on the basis of criteria (such as minimum distance between workers or fewest movements between workers and equipment) important for the system under evaluation.

The term "link" as used here refers to any connection between a human and a machine or between one person and another. If one person must talk or physically contact another (i.e., hand that person a message), this is represented by a link. If a person must see a display or reach a control on a machine, he or she has a link to that machine. Ordinarily, any links between machines can be neglected unless the link possesses some quality which might cause the system to operate inefficiently (i.e., excessive length of a hardwire connection between two machines might result in high line loss in power). In these cases, links between machines are also included in the link analysis.

A typical example of a link analysis for a system involving four operators and four pieces of equipment is shown below.

STEP 1 Draw a circle for every person in the system and label it with a code number:

①

Draw a square for every item of equipment and label it with a code letter:

[A]

STEP 2 Determine the type of link between each of the equipments and operators, equipments and equipments, and operators and operators. The different link types should be coded, such as:

———— control links
━━━━ visual links
. . . . . talk links

The three codes are fairly standard for link analysis use. Other codes required can be made up by the individual doing the analysis.

STEP 3 Establish the "link value" for each link. The link value is based on two factors: (1) the importance of the link in accomplishing the mission assigned to the system under evaluation and (2) the number of times (frequency) the link is used in completing the mission. A number, between 1 and 3, is assigned for both frequency and importance, with the higher number representing maximum importance and frequency. As an example, if a particular control movement on the panel was very critical in fulfilling the equipment use (3) and it was frequently activated during the equipment use (3), the link value for that particular control link would be 9 (3×3). If, on the other hand, a link was infrequently used (1) but was critical when it was needed (3), the link value for that particular link would be 3 (1×3). Thus, the link value is found by multiplying the frequency rating by the importance rating.

Selection of the frequency and importance ratings are usually based on past experience, either of the individual doing the analysis or someone familiar with similar systems already in use. In addition, the engineer responsible for the equipment design should be of assistance.

STEP 4 Prepare an analysis chart (see illustration on page 968) of the link values established for the system under evaluation. For each operator, show all the links (with each link value) associated with that operator. Do the same for each piece of equipment. For each operator and equipment item, add up the total of the link values and write this number to the right side. This provides an idea of the priority of equipment use and the operators most active in the system operation.

STEP 5 Prepare a schematic diagram (or series of diagrams) of possible arrangements. It is preferable to make this schematic to scale by cutting components out of paper and laying them on a scaled drawing of the space available (panel face, work-space floor plan, etc.). Starting with the operator or machine with the highest total link value, place the remaining components around it, moving them as necessary to minimize link crossings and shorten links, especially those with high link values. If conflicts occur between links, it may be necessary to reassess the original link values. The evaluation and rearrangement continue until the "best fit" solution is obtained. (See the final layout of the sample system on the bottom of page 141.) It should be emphasized that additional changes may be required in system layout once full-scale mockups or early hardware make actual system layout and evaluation possible.

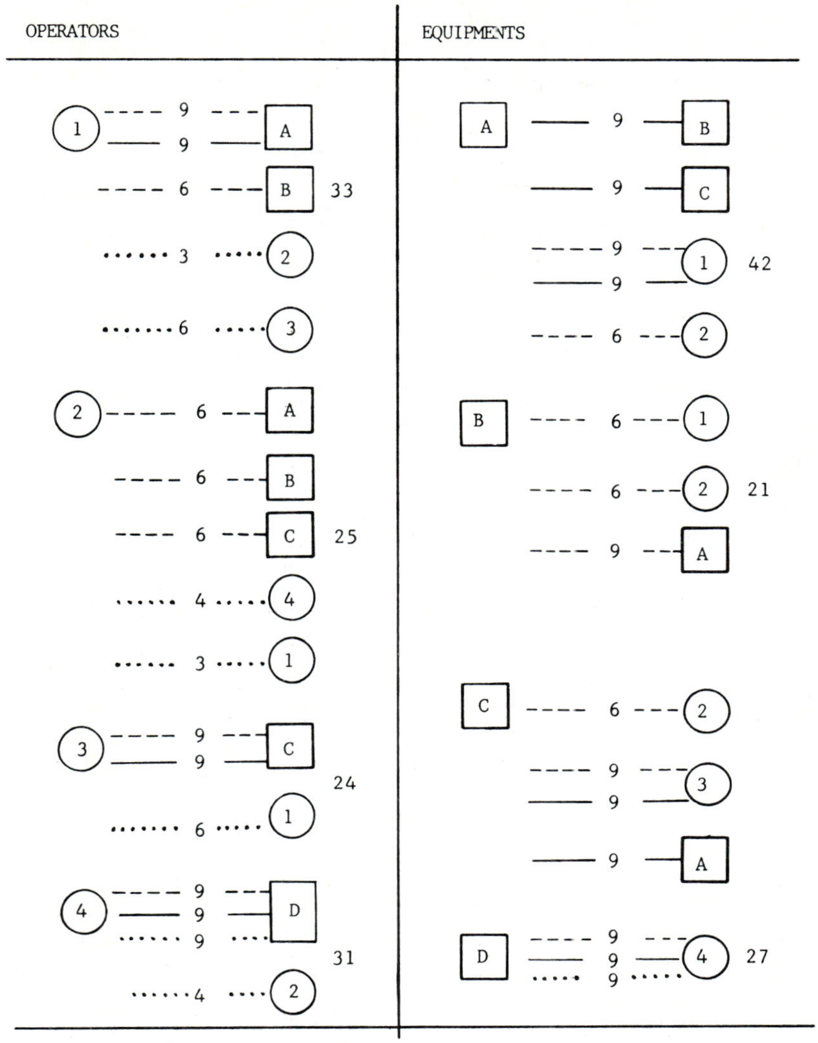

Sample Analysis Chart for Link Values

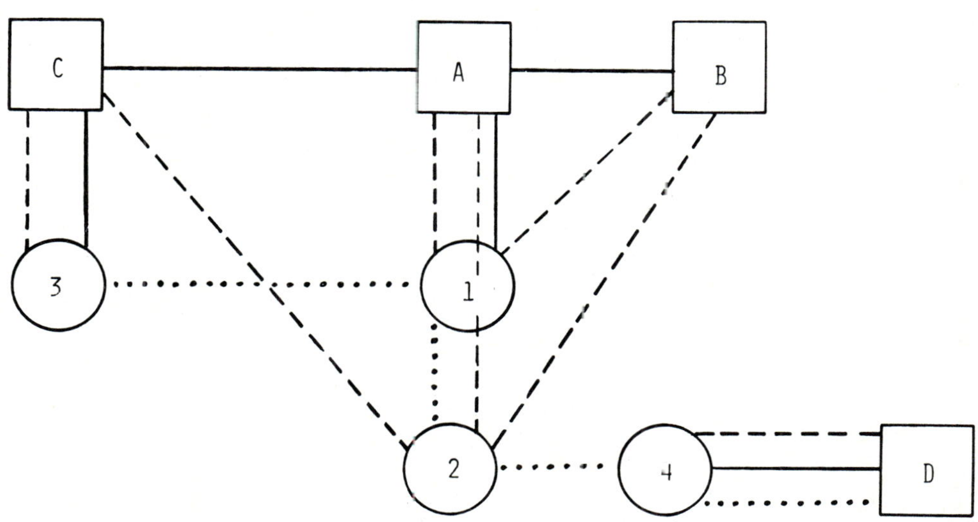

Schematic of Final System Layout

HUMAN CAPABILITIES AND LIMITATIONS

Comparison of Human Capabilities with Machine Alternatives

Human	Machine
Can recognize and use information redundancy (pattern) in the real world to simplify complex situations.	Has limited perceptual constancy and is very expensive.
Has high tolerance for ambiguity, uncertainty, and vagueness.	Is highly limited by ambiguity and uncertainty in input.
Can interpret an input signal even when subject to distraction, high noise, or message gap.	Performs well only in a generally clean, noise-free environment.
Is a selecting mechanism and can adjust to sense specific inputs.	Is a fixed sensing mechanism, operating only on that which has been programmed for it.
Has very low absolute thresholds for sensing (e.g., vision, audition, and touch).	To have the same capability, becomes extremely expensive.
Has excellent long-term memory for related events.	To have the same capability, becomes extremely expensive.
Can become highly flexible in terms of task performance.	Is relatively inflexible.
Can improvise and exercise judgment on the basis of long-term memory and recall.	Cannot do this; is best at routine, repetitive functions.
Can perform under transient overload; performance degrades gracefully.	Stops under overload; generally fails all at once.
Can make inductive decisions in novel situations; can generalize.	Has little or no capability for induction or generalization.
Can modify performance as a function of experience; can "learn to learn."	Is not characterized by trial-and-error behavior.
Can override own actions, should the need arise.	Can do only what it is built to do.
Is reasonably reliable; can add reliability to system performance by selection of alternatives.	Is reliable only at the expense of increased complexity and cost, and then only for routine functions.
Complements the machine, i.e., can use it in spite of design failures, can use it for a different task, or can use it more efficiently than it was designed to be used.	Has no such capability.
Complements the machine by aiding in sensing, extrapolating, decision making, goal setting, monitoring, and evaluating.	Has no capacity for performance different from what was originally designed.
Can acquire and report information incidental to the primary mission.	Cannot do this.
Can perform time-contingency analyses and predict events in unusual situations.	Does very poorly at this.
Is relatively inexpensive for corresponding complexity and is generally in good supply, but must be trained.	Is more limited in terms of complexity and supply by cost and time.
Is light in weight and small in size for function achieved for most situations.	To have functional equivalence of the human, requires more weight, power, and cooling facilities.
Is relatively easy to maintain; demands a minimum of "in-task" extras.	Maintenance problems become disproportionately serious as complexity increases.

Comparison of Human Limitations with Machine Alternatives

Human	Machine
Is a poor monitor of infrequent events or of events which occur frequently over a long period of time.	Can be constructed to reliably detect infrequent events or events which occur frequently over a long period of time.
Has a limited channel capacity.	May have as much channel capacity as can be afforded.
Is subject to coriolis effects, motion sickness, disorientation, etc.	Is not subject to these effects.
Has extremely limited short-term memory for factual material.	May have as much short-term (buffer) memory as can be afforded.
Is not well suited to data coding, amplification, or transformation tasks.	Is well suited to these tasks.
Performance is degraded by fatigue and boredom.	Performance is degraded only by wearing out or by lack of calibration.

(continued)

Comparison of Human Capabilities with Machine Alternatives *(continued)*

Human	Machine
Performance is degraded by long duty periods, repetitive tasks, and cramped or unchanged positions.	Is less affected by long duty periods and performs repetitive tasks well; some may be restricted by position
Becomes saturated quickly in terms of the number of things that can be done and the duration of the effort.	Can do one thing at a time so fast that it seems to do many things at once, for a long period of time.
May introduce errors by misidentification, reintegration, or closure.	Utilizes these processes.
Expectation or cognitive set may lead an operator to see what he or she expects or wants to see.	Does not exercise these processes.
Much human mobility is predicated and based on gravity relationships.	May be built to perform independently of gravity.
Is adversely affected by high g forces.	Is unaffected by g forces.
Can generate only relatively small forces and cannot exert large forces for very long or very smoothly.	Can generate and exert forces as needed.
Generally requires a review or rehearsal period before making decisions based on items in memory.	Goes directly to stored information for a decision.
Has a built-in response latency of about 200 μs in a go–no-go situation.	Has no response latency.
Is not well adapted to high-speed, accurate search of large volumes of information.	Computers are designed to do just this.
Does not always follow an optimum strategy.	Will always follow the strategy designed into it.
Has physiological, psychological, and ecological needs.	Has only ecological needs.
Is subject to anxiety, which may affect performance efficiency.	Is not subject to this factor.
Is dependent upon the social environment, both present and remembered.	Has no social environment.
Diurnal cycle imposes cyclic degradation of behavior.	The machine cycle may be whatever is desired.
Is subject to stress as a result of interpersonal problems.	Is not affected by such problems.
Great differences exist among unselected individuals.	There are no unselected machines.

Typical Functions Assigned Either to Human or Machine[1]

Some Comparisons of the Limitations of Machine and Human Functions, in Relative Terms

Conditions for function	Limitations: Human	Limitations: Machine
Sensing		
Display	Limited to certain ranges of energy change affecting human senses	Range extends far beyond human senses (X-rays, infrared, etc.)
	Sensitivity: very good	Sensitivity: excellent
Filtering	Easy to reprogram	Difficult to reprogram
Identifying		
Display	Can be varied over relatively wide range of physical dimensions	Can be varied only in very narrow range of physical dimensions
	Channel capacity: small	Channel capacity: large
Filtering	Easy to reprogram	Difficult to reprogram
Memory	Limits to complexity of models probably fairly high, but not precisely known	Potential limits of complexity very high
	Limits to length of sequential routines fairly high, but time-consuming to train	Potential limits of length of routines very high
Interpreting		
Display	Same as Identifying	Same as Identifying
Filtering	Easy to reprogram	Difficult to reprogram
	Highly flexible, i.e., adaptable	Relatively inflexible
	May be reprogrammed by self-instruction following input changes contingent on previous response (dynamic decision making)	
Shunting	Can be readily reprogrammed to lower levels of functioning	Difficult to reprogram
Memory	Limitation to rule storage not known	Limits of rule storage ("logic") quite high
	Speed of reinstatement of rule sequences relatively low (as in computing)	Speed of using rule sequences high (e.g., computing)
	Use of novel rules possible (inventing)	Limited in use of novel rules

[1]Gagne, R. M., *Psychological Principles in System Development,* Holt, Rinehart & Winston, New York, 1962.

SYSTEM CLASSIFICATIONS[2]

The Designing of Man-Machine Systems

Kind of System and Its Mode of Operation	Components	Couplings between Components	Examples
1. Manual system—Operator directed, flexible	hand tools or aids	one human operator	cook plus utensils, craftsman plus tools, singer plus amplifying equipment
2. Mechanized system—System directed, rigid	powered mechanical subsystems	on-line human operators, tracks, conduits, etc.	railway system, assembly line
3. Automatic system—Preset, programmed, or adaptive	powered mechanical subsystems	cables, pipes, conduits, levers, etc., forming a control circuit	clock, telephone exchange, digital computer
4. Collaborative human machine system—Exploratory and flexible	one or more human operators, one or more complete automatic systems	complex displays and controls	multiple-access computers
5. Mechanical subsystem—Operator controlled and inflexible	highly interdependent physical parts forming indistinguishable components and couplings		engine, automobile, machine tool
6. Administrative system—Goal directed and hierarchical	human operatives with tools or aids	rules, messages, human administrators, and informal contacts	army of foot soldiers, a business, a school
7. Voluntary system—Self-rewarding and collaborative	any number of persons each of whom is also a biological system and some of whom also act as administrative subsystems	affection, shared aims, laws, customs, managers, physical presence, mutual aid, common language, ancestry, etc.	family, religious order, club, society, (university)
8. Environmental system—Permissive of a range of human activities and contacts: prohibitive of others	inhabitants and facilities within an environment, the outside world	spaces and the barriers between and around the components	occupied building, city, or region
9. Biological system—Homeostatic, adaptive, evolutionary, growing, differentiating, and self-reproducing	cells, organs, subsystems, all of which are also physical systems	nerves, glands, chromosomes, etc., past experience and environment	cells, plants, animals, "human operators"
10. Physical system—Dynamically stable but subject to eventual decay	elementary particles, planets, seas, land, etc.	gravitation, electrical forces, radiation, physical motions, and forces	solar system, molecule, crystal, cloud, strut, tie, beam, shell
11. Symbol system—Semantic, analogous, ambiguous, or precise	words, signs, symbols, numbers, etc.	syntactical rules	languages, mathematics, codes, etc.

Note: Systems classified according to mode of operation and the physical nature of their components and couplings.

[2]J. C. Jones, "The Designing of Man-Machine Systems," in W. T. Singleton et al. (eds.), *The Human Operator in Complex Systems,* Taylor & Francis Ltd., London, 1967.

2

The Workplace

Minimum Toilet Fixture Requirements (New York State Labor Code)

No. of MEN	Water Closets	Urinals	No. of WOMEN	Water Closets	No. MEN or WOMEN	Wash Basins
1-9	1	0	1-15	1	1-20	1
10-15	1	1	16-35	2	21-40	2
16-40	2	1	36-55	3	41-60	3
41-55	2	2	56-80	4	61-80	4
56-80	3	2	81-110	5	81-100	5
81-100	4	2	111-150	6	101-125	6
101-150	4	3	151-190	7	126-150	7
151-160	5	3	191-240	8	151-175	8
161-190	5	4	241-270	9	176-200	9
191-220	6	4	271-300	10	201-225	10
221-270	6	5	301-330	11	226-250	11
271-280	7	5	331-360	12	251-275	12
281-300	7	6	361-390	13	276-300	13
301-340	8	6	391-420	14	301-325	14
341-360	8	7	421-450	15	326-350	15
361-390	9	7	451-480	16	351-375	16
391-400	10	7	481-510	17	376-400	17
401-430	10	8	511-540	18	401-425	18
451-450	11	8	541-570	19	426-450	19
461-430	11	9	571-600	20	451-475	20
481-520	12	9	601-630	21	476-500	21
521-540	12	10	631-660	22	501-525	22
541-570	13	10	661-690	23	526-550	23
571-530	14	10	691-720	24	551-575	24
581-630	14	11	721-750	25	576-600	25
631-640	15	11	751-780	26	601-625	26
641-660	15	12	781-810	27	626-650	27
661-700	16	12	811-840	28	651-675	28
701-720	16	13	841-870	29	676-700	29
721-750	17	13	871-900	30	701-725	30
751-760	18	13	901-930	31	726-750	31
761-810	18	14	931-960	32	751-775	32
811-820	19	14	961-990	33	776-800	33
821-840	19	15	991-1020	34	801-825	34
841-880	20	15			826-850	35
881-900	20	16			851-875	36
901-930	21	16			876-900	37
931-940	22	16			901-925	38
941-990	22	17			926-950	39
991-1000	23	17			951-975	40
					976-1000	41

WASH FOUNTAINS REQUIRED

Number of Fixtures	Persons Accommodated By			
	54" CIRCULAR (8 each)	54" SEMI-CIRCULAR (4 each)	36" CIRCULAR (5 each)	36" SEMI-CIRCULAR (3 each)
1	1-175	1-80	1-100	1-60
2	176-375	81-175	101-225	61-125
3	376-575	176-275	226-350	126-200
4	576-775	276-375	351-475	201-275
5	776-975	376-475	476-600	276-350
6	976-1175	476-575	601-725	351-425
7		576-675	726-850	426-500
8		676-775	851-975	501-575
9		776-875	976-1100	576-650
10		876-975		651-725
11		976-1075		726-800
12				801-875
13				876-950
14				951-1025

FACTORIES/WAREHOUSES

Employee Facilities

Both the quantity and the quality of the product depend not only on the good design of the plant machines but on the proficiency, pride, and fitness—both mental and physical—of the personnel. The development of plant design in recent years has become more and more concerned with creature comforts for the employees.

The facilities should be near the work space, so that no time is lost getting back and forth, but they should be sufficiently insulated from the sights and sounds of the work area itself so that a real change of scene is provided. If a pleasant outside view is available, it should obviously be used.

A clear distinction should be made between quiet lounging places and recreation and cafeteria areas. The problems are interesting, the solutions may be various, but the reigning criteria seem to be constant—cheerfulness, comfort, and durability.

The areas in this category include the following:

- Cafeteria and kitchen
- Coffee lounges
- Recreation areas (indoor and outdoor)
- Quiet lounges
- Plant men's and women's lockers and toilets
- Office men's and women's lockers and toilets
- Meeting rooms
- First Aid and nurses station

From J. De Chiara and J. H. Callender, *Time-Saver Standards for Building Types*, McGraw-Hill Book Company, 1973.

Offices

The office facilities may include any or all of the following:

- General office
- Message center
- Billing office
- Cashier
- Telephone room
- Foreman's office
- Office manager
- Terminal manager
- Operations manager
- Salesmen's room
- Record room
- Heater room
- Central checking
- Drivers' locker room
- Transportation department
- Dormitory
- Cafeteria
- Drivers' ready room

Other Facilities

- Maintenance shop
- Fueling area (near shop)
- Weighing area
- Truck and trailer parking area (two parking spaces per dock stall recommended)
- Employee and visitor parking area

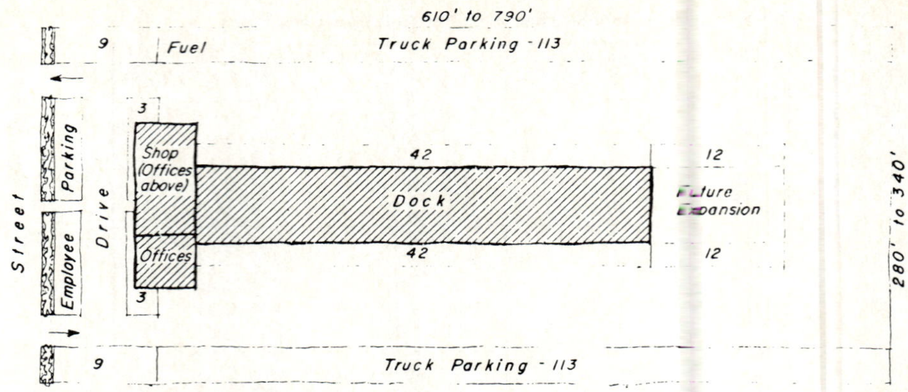

Truck terminal with short side to street.

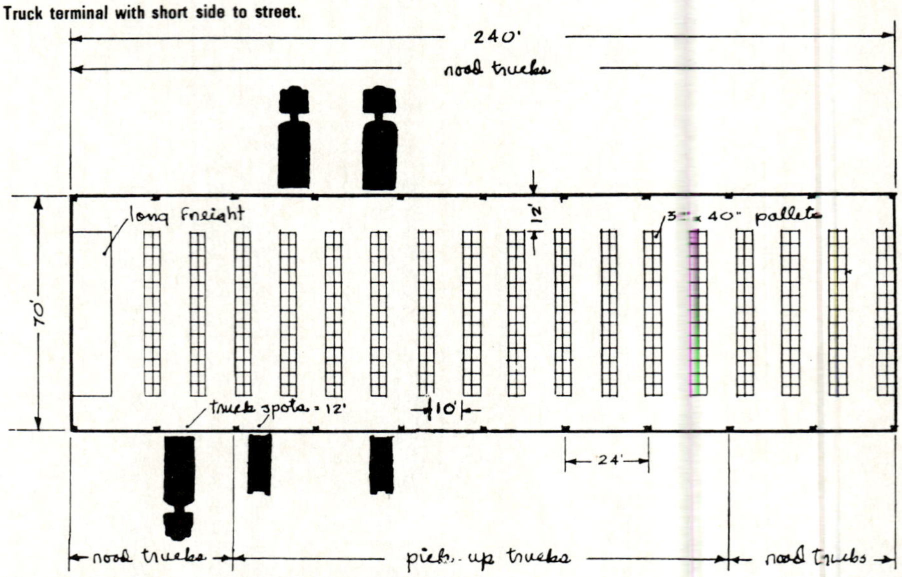

Truck terminal dock plan for fork-lift truck and pallet storage.

Minimal Parking Space or Dock Approach Length and Width (Width – 12 ft)

Overall length of tractor-trailer, feet	Apron length, feet	Dock approach length, feet
40	43	83
45	49	94
50	57	107
55	62	117
60	69	129

Percent of Grade for Material Handling Equipment

Type of equipment	Allowable percent of grade*
Powered handtrucks	3
Powered platform trucks	7
Low-lift pallet or skid trucks	10
Electric fork trucks	10
Gas fork trucks	15

* Contact manufacturer and check manufacturer's specifications before operating beyond allowable percent of grade.

Most standard truck dockboard lengths range from 6 to 10 ft. For most applications, dockboards should be 6 ft wide. Use 7-ft wide dockboard for loading or unloading unit loads with fork truck.

8. *Provide area for access to trucks.* A minimum area measured inside the plant from the edge of the dock should be kept clear and unobstructed for the movement of freight and materials handling equipment. The depth of the area must allow for maneuverability of materials handling equipment in and out of vehicles and for two-way cross traffic behind the dock.

From J. De Chiara and J. H. Callender, *Time-Saver Standards for Building Types,* McGraw-Hill Book Company, 1973.

Calculation of Storage Space (Area Utilization)

Gross warehouse area = inside total square footage of warehouse

Net storage area = actual area occupied by inventory, not including aisles plus space for empty pallets plus shipping and receiving areas plus allowance for "honeycombing" plus special inventory (inspection, etc.)

Interference = irregularities due to columns, odd corners, etc.

Miscellaneous = space for offices, equipment, etc.

Gross warehouse area = net storage area plus aisles plus interference plus miscellaneous

Rule of thumb (for general package warehousing):

$$\frac{\text{gross warehouse area}}{\text{net storage area}} = \frac{3}{2}$$

or Net storage area + (50% net storage area) = gross warehouse area

This rule is accurate for average warehouses, but actual analysis of the layout is recommended.

"Honeycombing" is a warehouse term used when space is not fully occupied because of partial withdrawal of inventory. Maximum honeycombing factors are in the range of 75 to 90 percent of maximum capacity, depending upon the activity, number, and quantity of items stored.

Straight-Line Flow or Assembly-Line Principle Straight-line flow is inherently efficient and usually is adopted in warehouses adjacent to production areas. Conveyors and pallet systems illustrate typical methods of efficient handling.

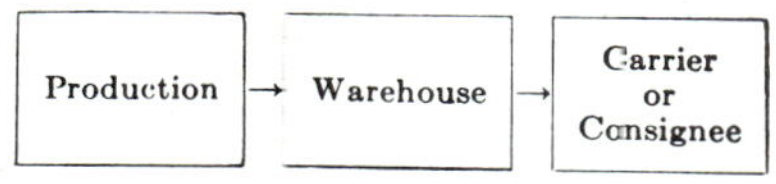

Order pick lines in a grocery warehouse are characterized by the assembly-line principle.

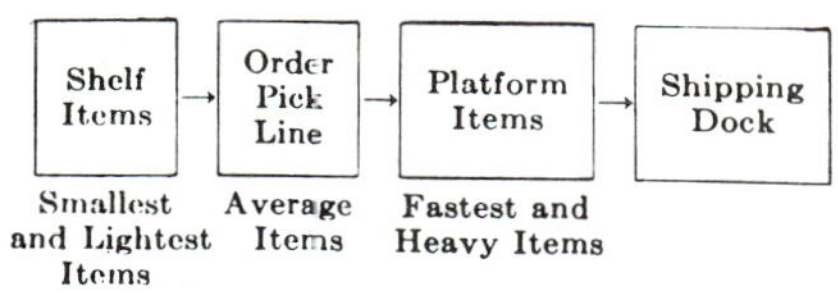

Shipping and Receiving Areas The receiving area of a warehouse should be located adjacent to incoming rail or truck facilities and as convenient as possible to the storage area. The receiving dock is usually separated from the shipping area if possible to minimize cross traffic and possible confusion. The number of unloading positions required is dependent upon the volume of receipts or the maximum number of cars or trucks spotted at the same time. The light weight of portable aluminum or magnesium dockboards is desirable when power equipment is not available for positioning units of conventional steel construction.

Weather protection at the unloading positions permits continuous handling operations. Loading platforms located outside the warehouse building can be designed for one-way or two-way traffic where required.

The proper control, checking, and sorting of inbound materials is important for the prompt and efficient delivery of outbound shipments. The size of the receiving area is determined by the analysis of the temporary storage lag needed to perform the necessary inbound handling and inventory control operations.

A shipping area (or dock) receives materials for outbound shipment after selection and transfer from storage. The preassembly of orders according to plan requires sufficient room to perform packing, packaging, or preparation operations prior to shipment.

The size of the shipping area is dependent on the makeup time of filling orders and the quantity of simultaneous loading operations during peak periods.

Utilizing centrally located short-lot and bin-storage operation, coordinated by use of dragline conveyors

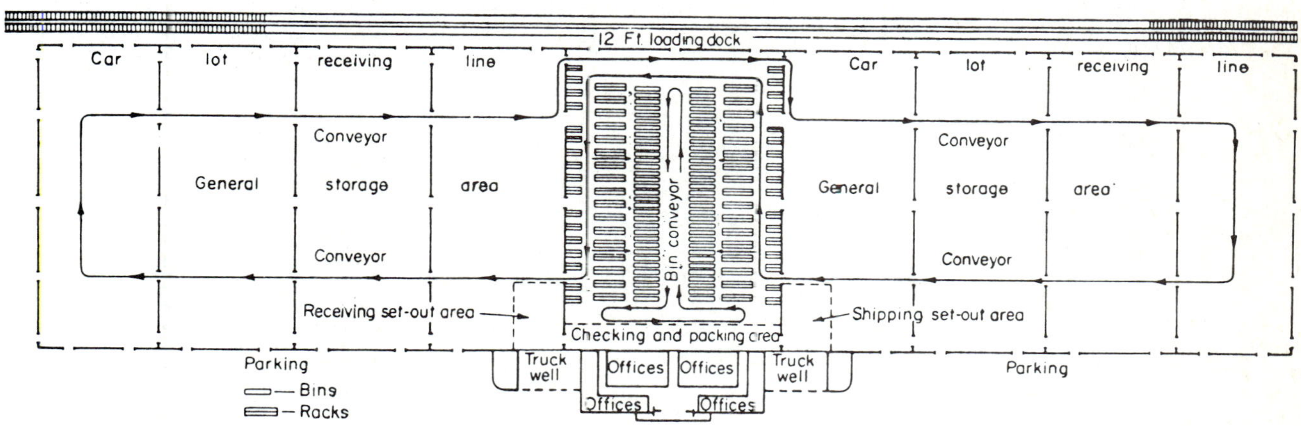

This bin layout emphasizes a direct-flow replenishment and stock selection operation

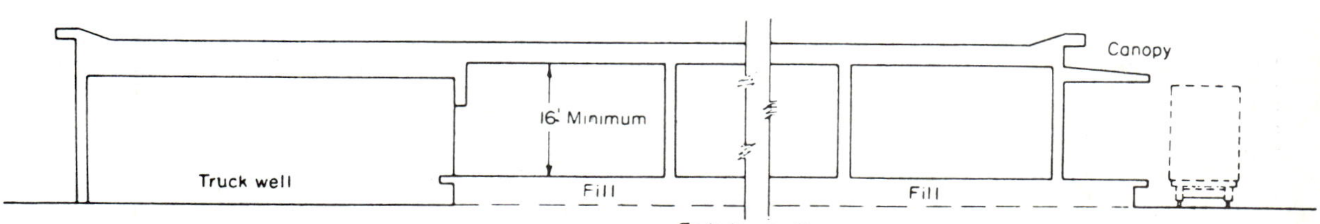

End view-section

Typical large warehouse layout.

From J. De Chiara and J. H. Callender, *Time-Saver Standards for Building Types*, McGraw-Hill Book Company, 1973.

Hypothetical Relationship of Parking Area Requirements to Location

Location	No. of employees at peak shift overlap	Percent as drivers or auto psgrs.	Number of autos to be parked*	Approx. site, sq ft	Area, acres
Urban	1,600	60	740	222	5.0
Suburban	1,600	80	990	297	6.8
Rural.	1,600	95	1,180	354	8.0

* Assuming car occupancy of 1.3 persons per car.

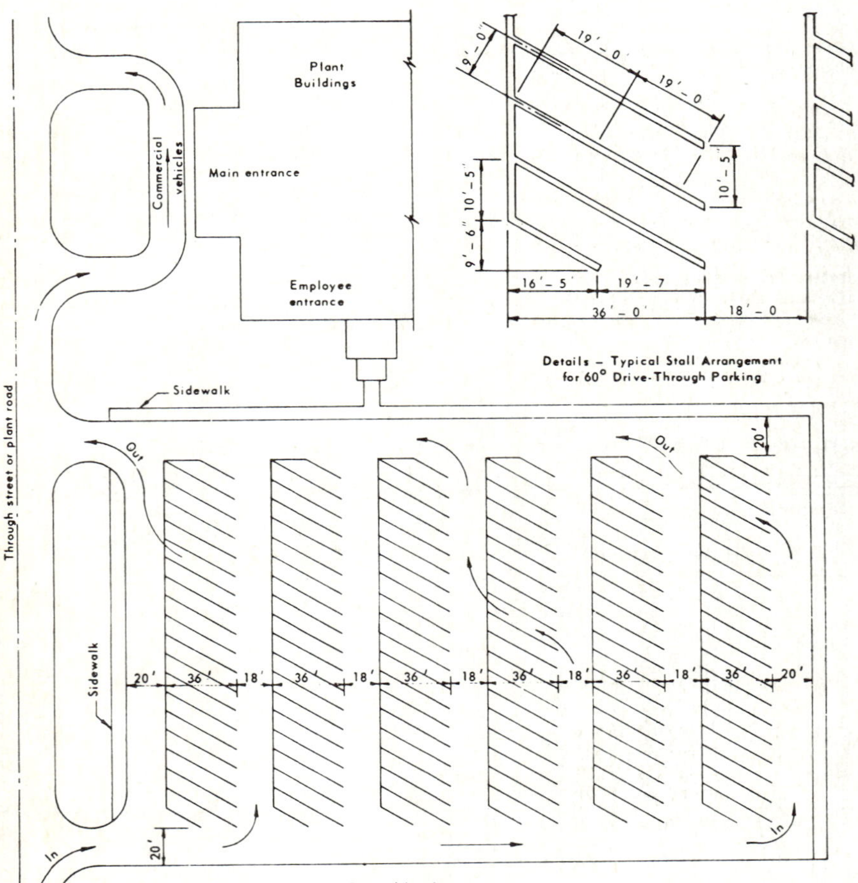

Details – Typical Stall Arrangement for 60° Drive-Through Parking

Typical Lot Layout

DESIRABLE STALL AND AISLE DIMENSIONS FOR DRIVE-THROUGH PARKING

Angle of Parking	Width of Stall	Depth of Stall Perpendicular to Aisle	Width of Aisle	Unit Parking Depth	Width of Stall Parallel to Aisle
°	W	L	A	UPD	W
90	10'	38' – 0"	24'	62' – 0'	10' – 0'
90	9'	38' – 0"	26'	64' – 0'	9' – 0'
60	9'	36' – 0"	18'	54' – 0'	10' – 5
53	9'	35' – 10"	18'	53' – 10'	11' – 3

Drive-through lot layout.

Parking Facilities for Industrial Plants, Institure of Traffic Engineers, Washington, D.C., 1969.

Employee Automobile Parking

Stall Arrangements Decisions about the choice of angle and the layout of aisles must be based on individual site conditions. The placement and number of entrances and exits, and the site shape and contour are the major controls. At large plants, blocks of parking by groups of three to five hundred cars are preferable to larger aggregations. Pedestrian-vehicle conflicts can be reduced, and assigned parking for different shifts and employment groups can be better controlled, through the use of such relatively small blocks.

The following general practices are desirable: use natural grades to facilitate drainage; provide for counterclockwise traffic aisle flow, since left turns are easier than right turns for drivers; have parked vehicles face downhill rather than uphill, to allow for stalled vehicles or winter weather conditions.

This report includes layout details for only one type of parking. The figure illustrates a stall arrangement and an aisle design that have not been widely published—the drive-through double stall pattern, usable in either 90° or acute angle parking layouts.

In general, angle parking is preferred for large industrial parking facilities. First, properly designed angle parking can employ space as effectively as right-angle parking. Second, it virtually forces one-way movements, thereby simplifying control, reducing conflicts, and ensuring that daily parking practices conform to the established design. Third, it provides for easier turning movements into and out of stalls.

Drive-through angle parking design offers the further advantages of minimizing backing out of stalls and directing all aisle travel in the same direction. It conserves space more effectively than other angle parking designs. Typically, the angled drive-through layout requires 36 ft for the double stall and an 18-ft aisle (to permit passing stalled vehicles), for a unit parking depth of 54 ft. Compared with 90° parking, the space loss along the length of the bay—eight spaces in 500 ft according to the example—will be compensated for by the reduction in unit parking depth, from 62 or 64 ft to 54 ft, if enough bays can be used.

The disadvantage of this design of drive-through parking is that it increases the travel distance and time of a search pattern if the lot is nearly full. It also is imperative to keep the end circulation aisles two-way so that a driver will not be forced out of the lot in order to return to another parking aisle.

The drawing also gives dimensions for angle parking at 53° 8′, an angle which has the layout convenience of being a 3-4-5 triangle. Other angles commonly used for parking are 45, 55, or 60°. However, any angle smaller than the 3-4-5 configuration tends to be wasteful of space, without offering any significant advantage.

Where two-way aisle flow may be desirable, as in visitor parking lots, 90° parking is more appropriate. Site dimensions sometimes may be such that 90° unit parking depths are most appropriate regardless of other circumstances. The desirable 90° parking depth is 62–64 ft. When unit parking depths are less than desirable, shortened stall lines (10–15 ft long) may encourage drivers to pull all the way into stalls.

From J. De Chiara and J. H. Callender, *Time-Saver Standards for Building Types,* McGraw-Hill Book Company, 1973.

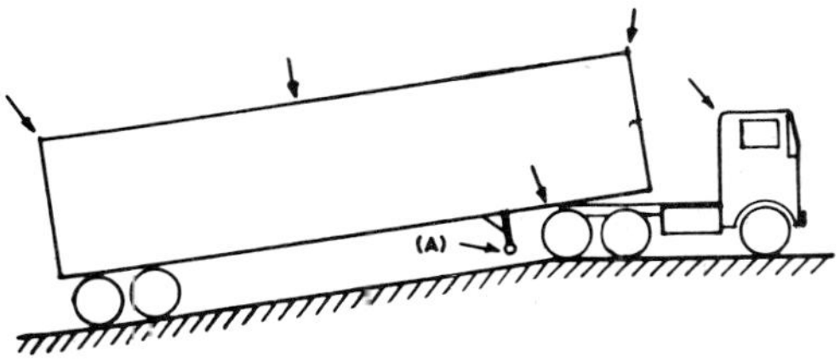

When designing for ramps, dips, or crowns in the terminal area, special care must be taken to provide clearance at the points indicated in the diagram. Actual dimensions must be obtained. Cab clearances are more critical when the combination is jackknifed. Landing gear height (A) may be as low as 10 in.

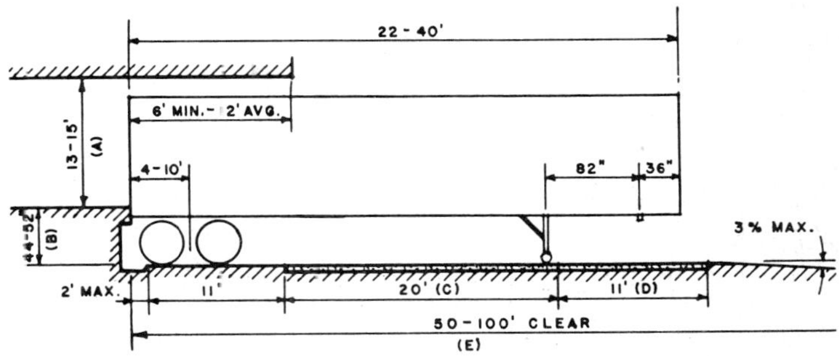

NOTES:
(*A*) *Should be at least 6 in. over legal height for level area, more for slope.*
(*B*) *Dock height, 48 to 52 in. for road trailers, 44 in. for city trucks.*
(*C*) *Concrete apron of the dimensions shown will accommodate trailers from 22 to 40 ft long.*
(*D*) *Additional slab length recommended to support tractor wheels.*
(*E*) *General rule for distance required: total length of tractor-trailer times 2.*
Trailer width—8 ft
Trailer stall width—10 ft minimum, 12 ft recommended.

Recommended dimensions and clearances for truck loading docks.

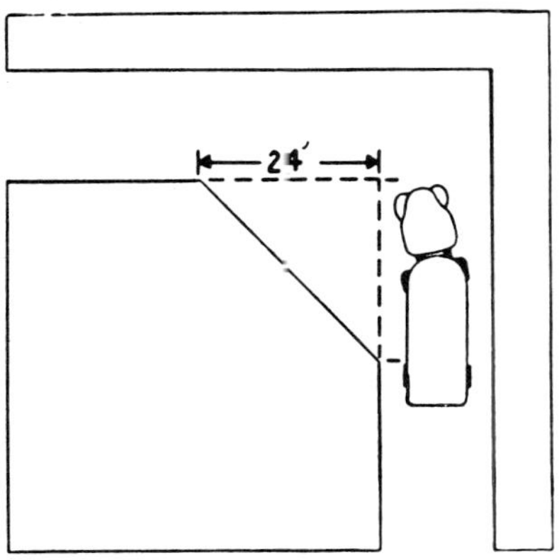

Turning clearance for driveway.

Site

Location: In selecting a site, consider the following factors:

1. Proximity to pickups, deliveries, and connecting carriers
2. Accessibility to main traffic arteries
3. Obstructions such as bridges, underpasses, and railroad crossings
4. Zoning
5. Urban and regional plans; future growth pattern of city
6. Transportation facilities for employees
7. Utilities

Grade: Site should be approximately level: maximum slope 3 percent; minimum slope for drainage, 1 percent. Storm drains recommended 60 to 75 ft on centers, 100 ft maximum.

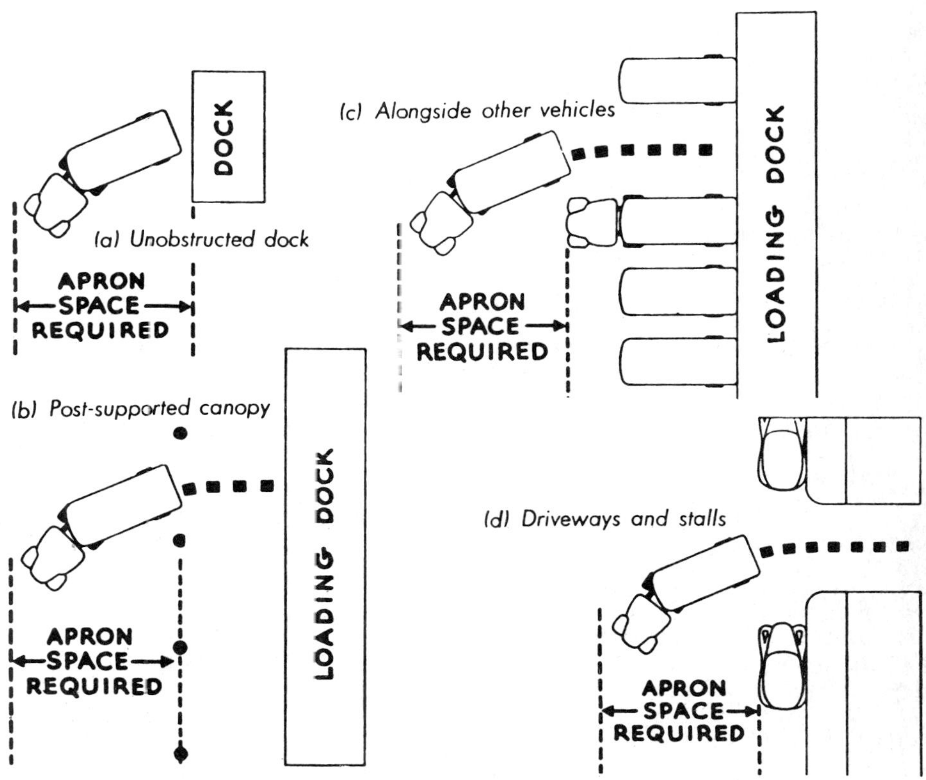

Apron space required for various conditions.

From J. De Chiara and J. H. Callender, *Time-Saver Standards for Building Types,* McGraw-Hill Book Company, 1973.

EXTERIOR SIDEWALKS

Outdoor sidewalks should always be laid out so that they are relatively level (except for sufficient slope to cause rainwater to drain away).

Sidewalks should be set back from the street. This prevents pedestrians from having to walk precariously close to vehicular traffic and provides sufficient space so that driveway and/or wheelchair ramps do not cross the primary walkway, which is sometimes difficult for elderly persons to negotiate without turning an ankle.

Sidewalks should be at approximately the same level as adjacent lawns so that someone who accidentally steps off the walkway will not turn an ankle.

Curbs should not be more than 10 in (25 cm) high (this is the accepted maximum for a single step).

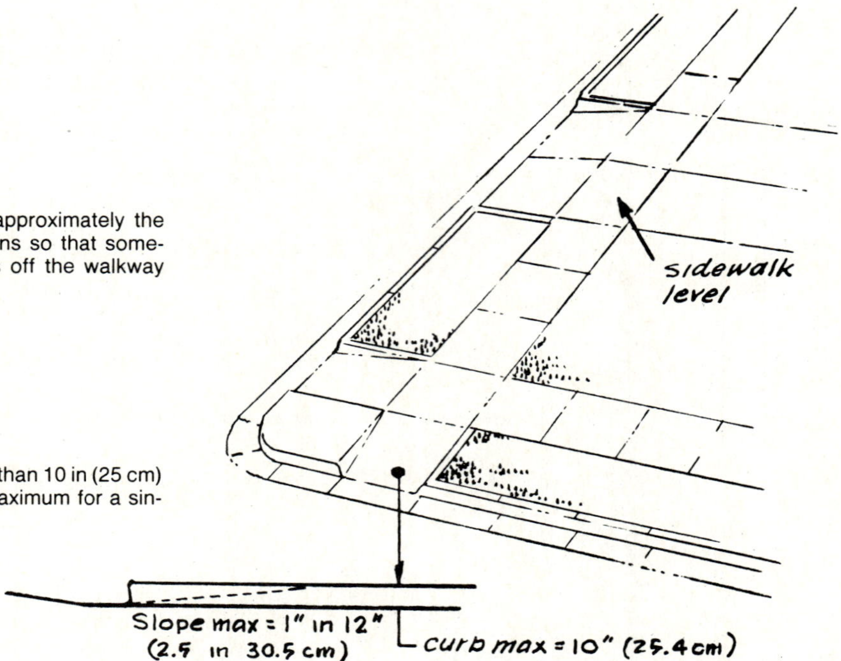

Whenever a walkway is more than 10 in (25 cm) above the ground, a suitable railing should be provided. The railing should contain some type of intermediate grille or other barrier to prevent small children from falling off the walkway.

Walkways should have nonskid surfaces, especially at ramps, so that pedestrians will not slip if the surface is covered with water or other liquid that might have been spilled on the walkway.

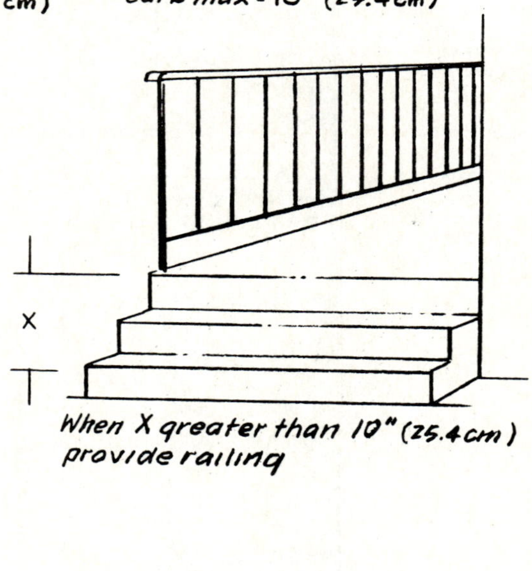

Avoid placing a single step along a walkway; people tend not to notice single steps. If such a configuration seems necessary, mark it conspicuously for daytime warning and make sure that it is properly illuminated at night.

If decorative walkways are planned, examine the layout in terms of possible hazards, such as stepping-stones that are too far apart or wooden planks that are irregular or full of rough splinters that may cause a person to trip.

Design Specification for Wheelchair Ramps between Sidewalk and Street

The criteria shown in the accompanying illustration will provide not only optimum passage of the wheelchair but also tactile cues for blind pedestrians and visual cues that will help prevent other pedestrians who do not notice sidewalk irregularities from stumbling.

Each of the special features has been derived from a number of attempts to create the optimum compromise for all users.

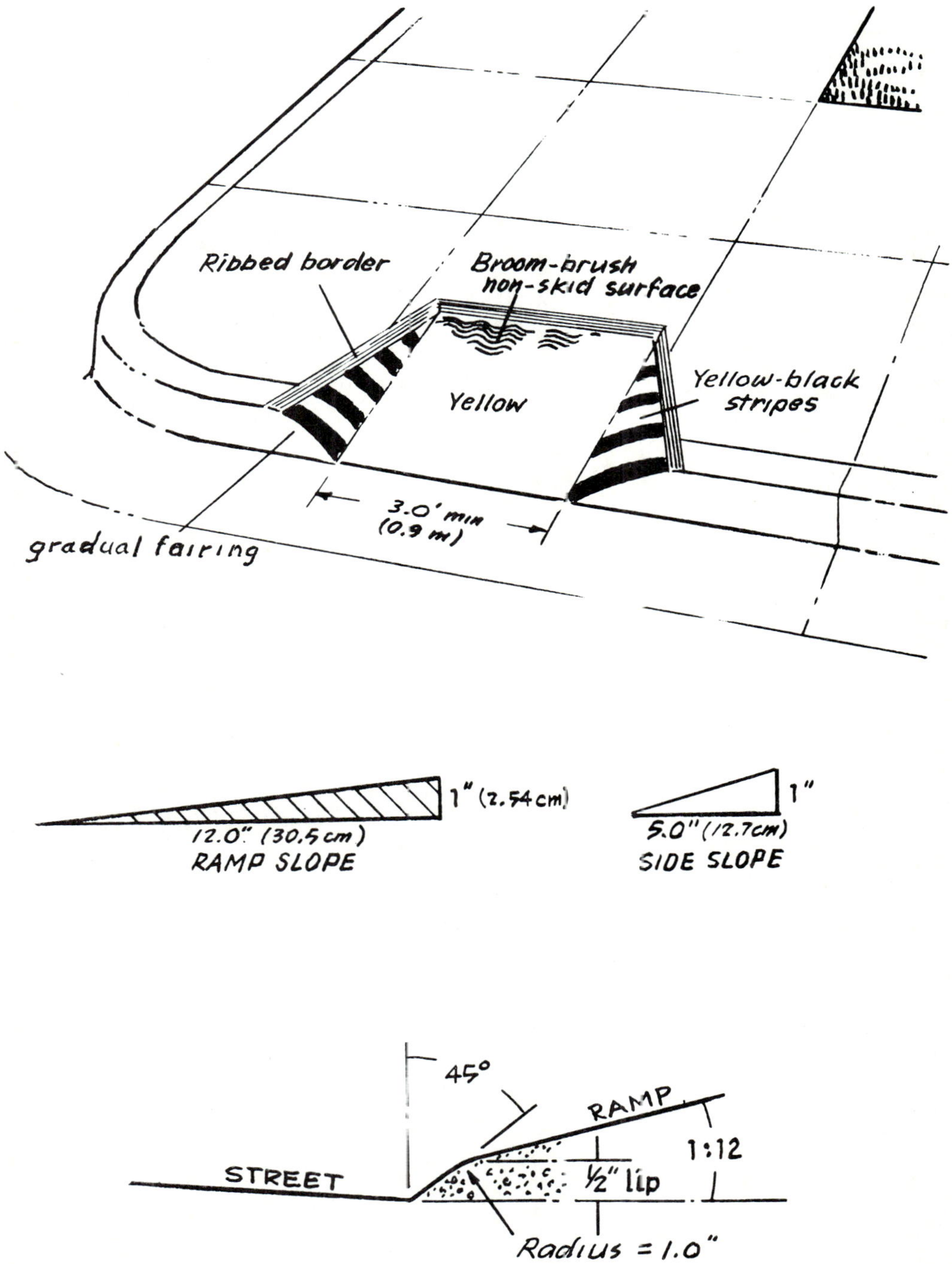

Recommended wheelchair ramp design.

INTERIOR HALLWAYS AND AISLES

Interior hallways, usually being defined by walls on both sides, should be wide enough so that people do not have to walk carefully in order to keep from bumping into a wall, into other people, or into equipment that is attached to the wall or is being carried or moved on wheels. Although this may sound like an unnecessary caution, most standards for hallway widths do not account for all these factors. For example, a hallway standard is often applied without consideration of the fact that a drinking fountain, a public telephone, or a fire extinguisher will later be hung on one of the walls.

The accompanying sketches reflect preferred dimensions to ensure lack of conflict during passage of persons and/or equipment through typical hallway configurations:

1. Although a minimum of 30 in (76 cm) is wide enough for one person, designers should always plan for the person in a wheelchair situation, which requires more space. The extra width will also accommodate a person on crutches.
2. A person should not have to step aside or crowd against a wall to let another person pass.
3. When mobile equipment can be identified, the additional space required for moving such equipment past people who are using the hallway should be determined and designed for accordingly.
4. Aisleway widths for in-plant vehicular transportation systems should not be narrowed to the point where people who are using the aisle bump into and annoy fellow passengers. If it is expected that the transportation vehicle will be accommodating persons in wheelchairs, be sure that the aisle is widened so that wheelchairs can pass freely, or lay out the vehicle so that persons in wheelchairs can park their wheelchairs near the door (e.g., in a bus).

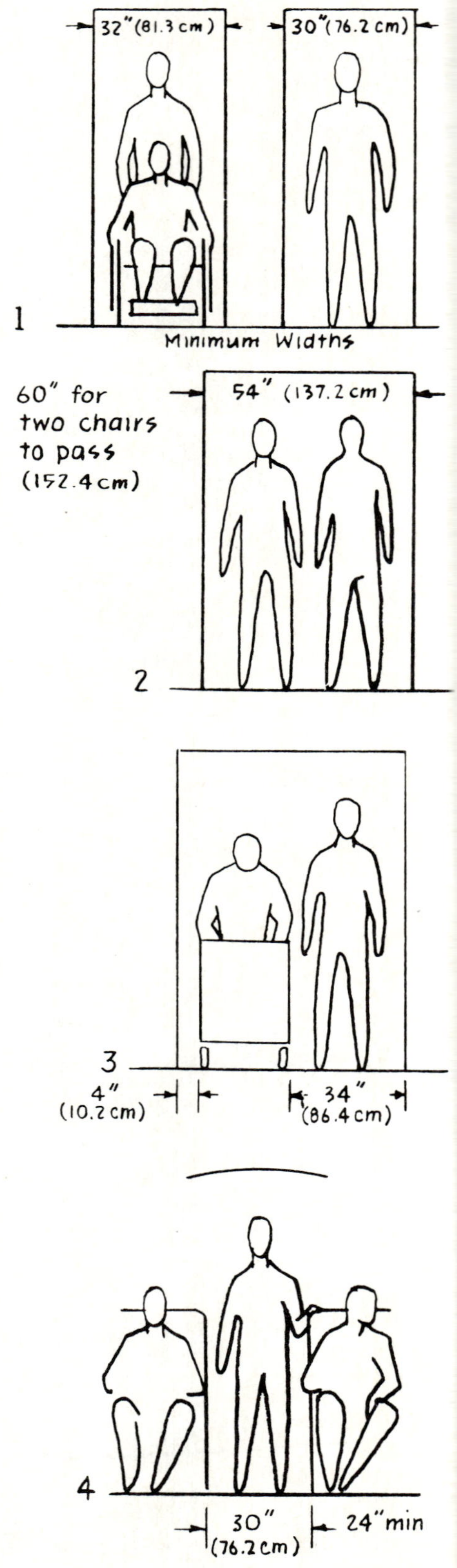

Hallway Path Definition

A defined pathway limit slightly apart from the walls of a hallway tends to keep people from wandering or absentmindedly cutting the corner at a change of hallway direction. Avoid the use of large-radius cove design at the juncture of the wall and floor; people tend to trip or step on the radiused area and fall.

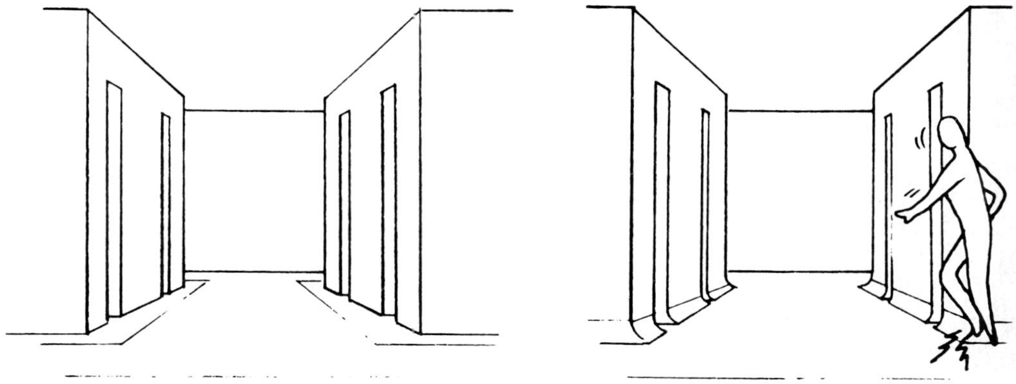

PREFERRED HAZARDOUS !

In laying out hallways, do not place doorways directly adjacent to corners where the hallway junctures with another hall and/or the single hallway merely turns the corner. The doorway should be at least 5 ft (1.5 m) from the corner; more distance is preferred.

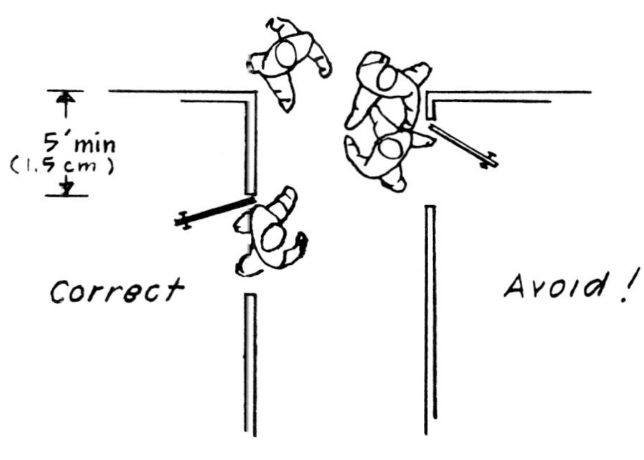

Make sure that hallways are adequately lighted both during the day and at night. Natural light during the day should come from the side of a hallway, not from the end, where it tends to contrast with the less light exterior and blind the person who is facing the bright sunlight.

If a building is not ordinarily occupied at night except by security personnel, plan the hallway lighting circuits so that only some of the fixtures will burn.

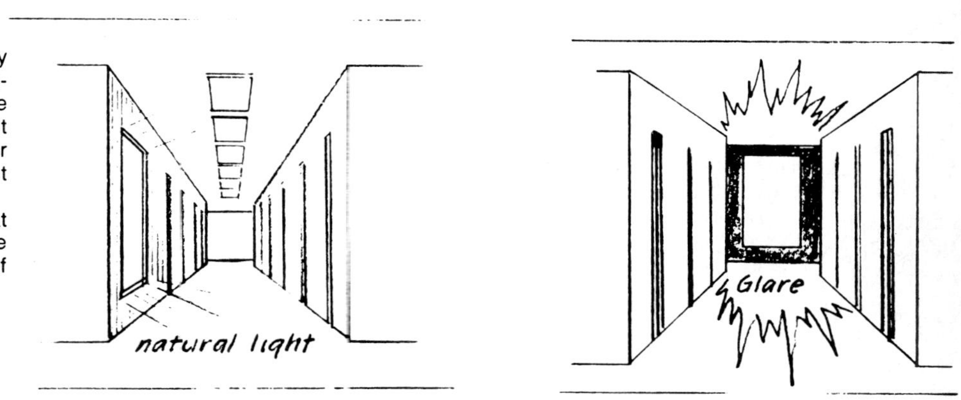

Correct Poor !

Design-Generated Confusion

Hallway paths should be laid out in a rectilinear pattern so that people make approximately 90° turns left or right. Although radiused corners often are desirable because they tend to reduce conflicts between people who are approaching the corner at the same time, large curved hallways are not desirable; people tend to get lost and find themselves in doubt as to which direction they are going in.

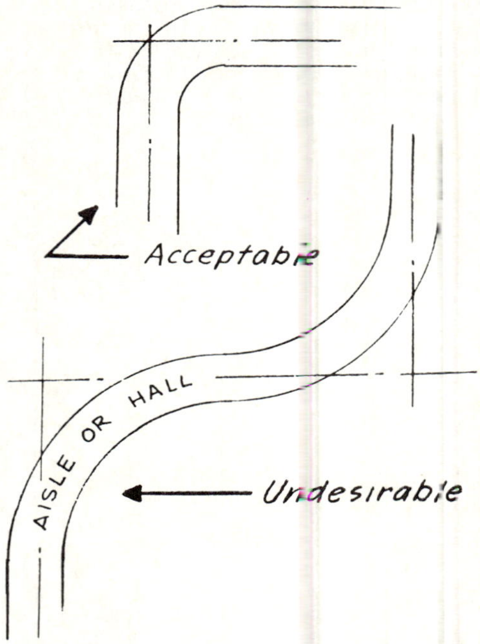

Hallways should have smooth, level floors (i.e., avoid inclines and single step-ups). If a change of grade is necessary, provide at least two steps, and make sure that the steps are well lighted and conspicuously defined. If the installation serves persons in wheelchairs, add a suitable ramp to one side of the hall. This should have a railing separating it from any steps so that nonhandicapped persons will know which is the ramp aisle and which is the stairway.

Hallway flooring should be made of nonskid materials. If carpet is used, make sure that it is properly laid so that it will not slip or bunch as people shuffle or drag their feet on it.

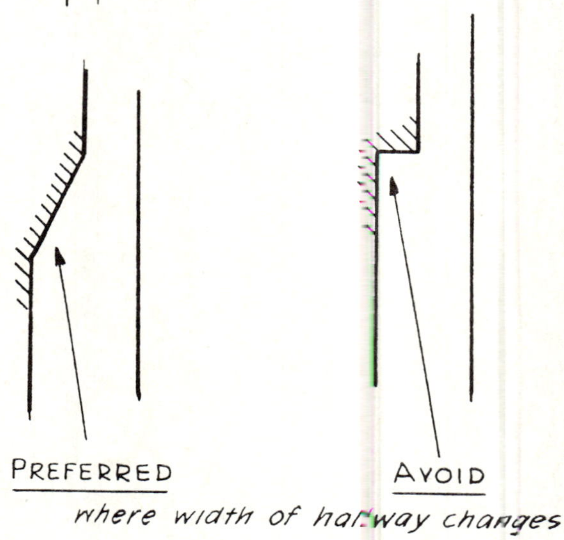

Floor patterns should be chosen that will not create unusual visual illusions, as shown in the accompanying illustration.

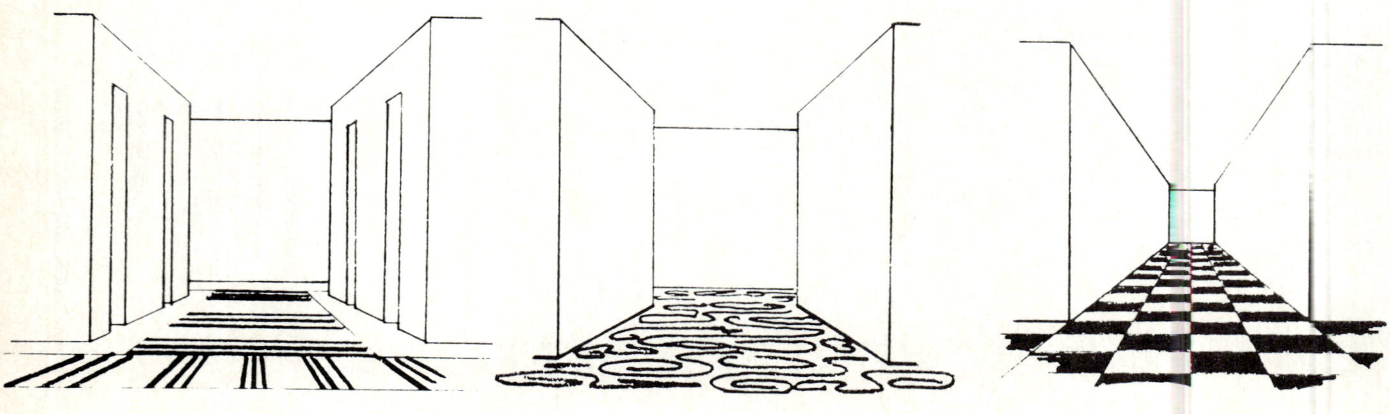

Although there have been no studies of the psychological effects of hallway proportion on user performance, it has been observed that certain kinds of proportional exaggeration (e.g., very long halls with low ceilings and very narrow halls with very high ceilings) tend to produce negative reactions.

In some instances, low ceilings have been used on long halls to accommodate overhead air-conditioning ducting. In other cases, however (especially in home design), very low hallway ceilings have been used purely for effect.

Because the hallway walls tend to create a "confined" appearance, care should be exercised not to use very low ceilings for more than a few feet (6 ft, 1.8 m or less). Otherwise, people may feel that they have to duck to keep from hitting their heads. No ceiling should be less than 7 ft (2.1 m).

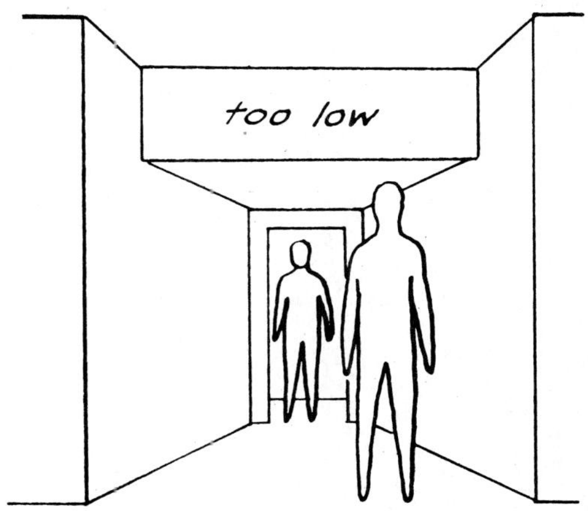

Very high ceilings in a long hallway tend to create the impression of being in the bottom of a chasm.

When low ceilings are used, make sure that the ceiling color is light. When high ceilings are used, the color should be darker and more subdued.

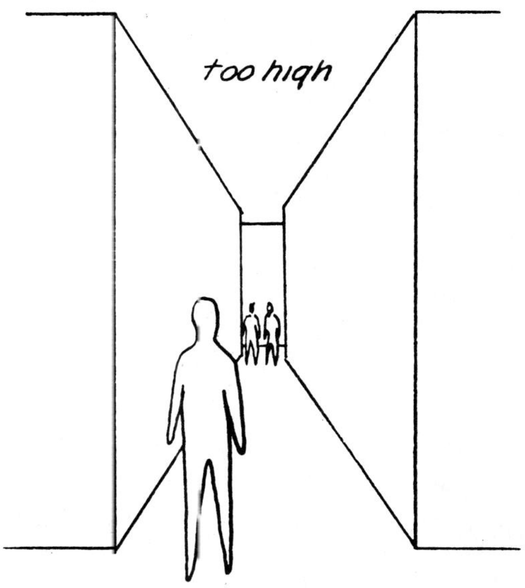

SPECIAL WALKWAYS AND PASSAGEWAYS

The accompanying sketches illustrate a few of the special types of walkways and passageways that often are associated with industrial applications. The accompanying sketch shows the minimum passage envelope for the adult male, walking at a slow and careful pace. Passageways of this type should be laid out in a straight line whenever possible, with no intervening obstacles which people have to step over, or which might snag their clothing.

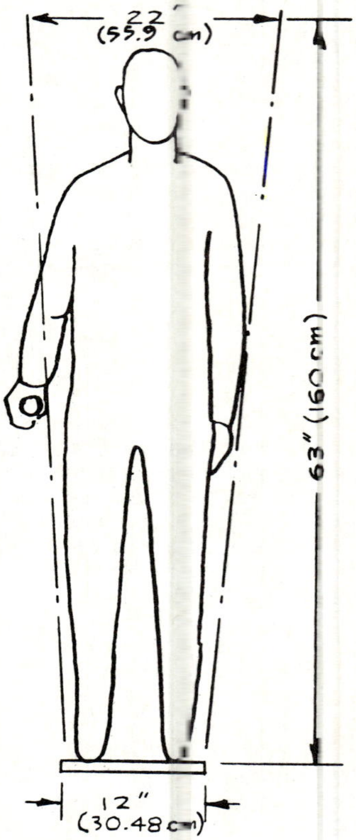

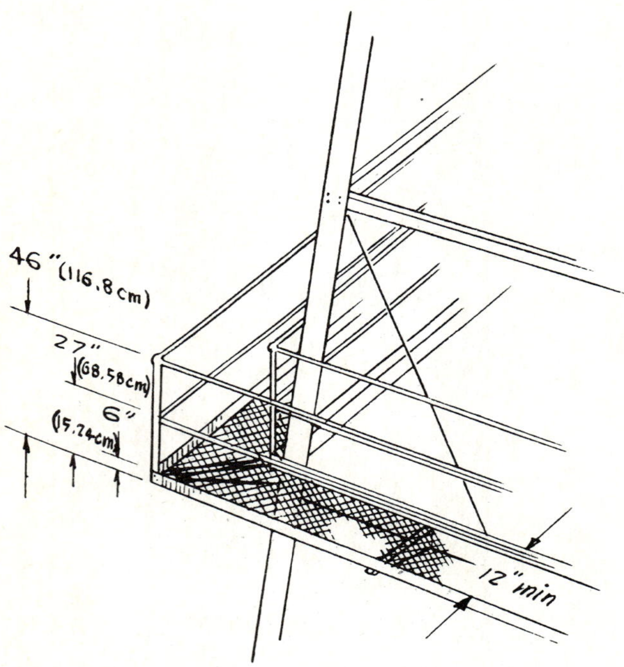

Walkways and catwalks should have nonskid surfaces, especially when they are outdoors and when they are extremely high above the ground-floor level indoors.

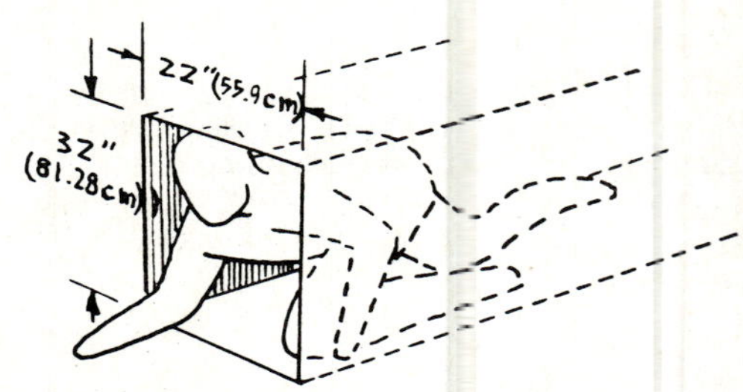

CRAWL - THROUGH

COMMUNICATION TUNNEL WITH MANUAL TROLLEY SYSTEM

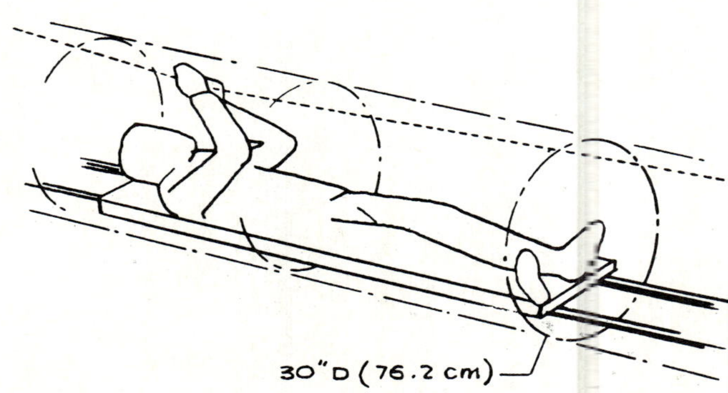

Catwalks should have handrails on both sides, a midrail, and a toe board for safety.

Rough edges, bolt ends, and other protrusions must be avoided. The interior surfaces of crawl spaces and tunnels should be painted a light color.

RAMPS

The primary purpose of a ramp is to provide a smooth change of elevation for wheeled vehicles, mobile equipment, and wheelchairs. Unless a ramp is quite shallow, it is not (as some people think) preferred over a properly designed staircase for many applications. The principal considerations for deciding on the use of a ramp as opposed to a staircase are the following:

1. Wheeled vehicles, mobile equipment, or wheelchairs must be accommodated.
2. There is sufficient space to meet the ramp angle guidelines.
3. The ramp can be positioned where opening windows or doors will not interfere with it.

It should be noted in considering ramps that, although more energy may be required to go up a ramp, the major hazards occur when people are going down a ramp. Although healthy, nonhandicapped individuals can negotiate a fairly steep ramp (up to about 15°), the main criteria for designing most ramps should be based on the needs of the elderly or handicapped user. The guidelines shown in the accompanying sketches are generally accepted by those who design for the handicapped person, who must travel by wheelchair.

A slope of 5 percent is the maximum allowed for a ramp before it is required that a handrail be added. When the slope becomes greater, it is very taxing for most wheelchair users to "pull the hill." Thus this upper limit is a preferred slope guideline for all ramps, whether they are for indoor or outdoor applications.

No ramp should exceed about 8 percent slope under any circumstances.

If a ramp slope is created that is over the 5 percent grade, handrails should be provided to assist individuals who are weaker and/or less steady on their feet. There should be a separate railing for both sides of such ramps.

Nonskid surfaces should be provided on all ramps.

Ramps should not extend more than 30 ft (9.2 m) before an intervening level "resting platform" is provided. There should also be similar flat platforms both at the bottom of the ramp and at any point at which the ramp system changes direction. The accompanying guidelines are best for wheelchair users but are equally valid for all users.

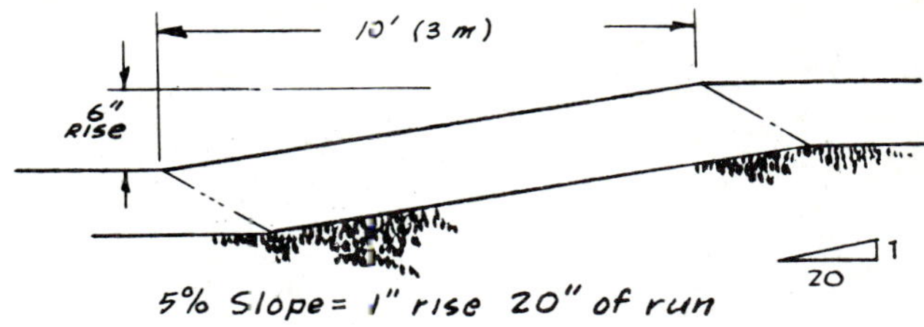

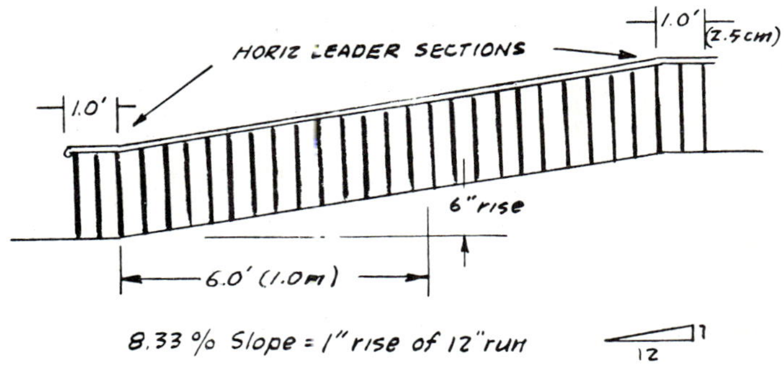

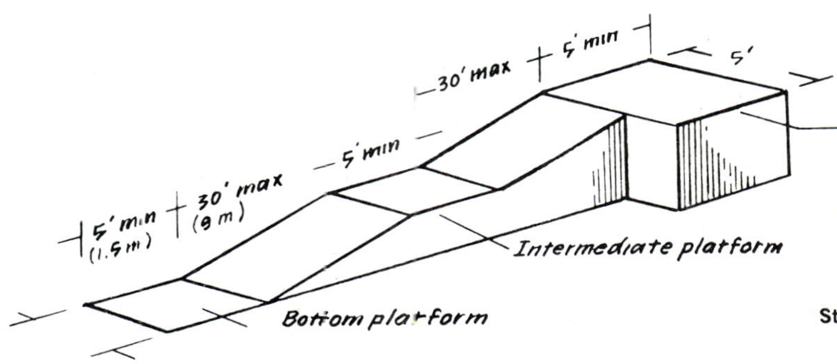

Straight run ramp.

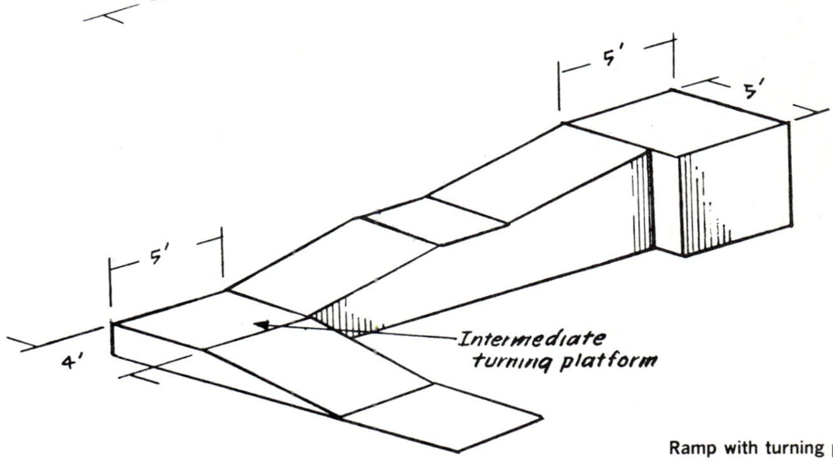

Ramp with turning platform.

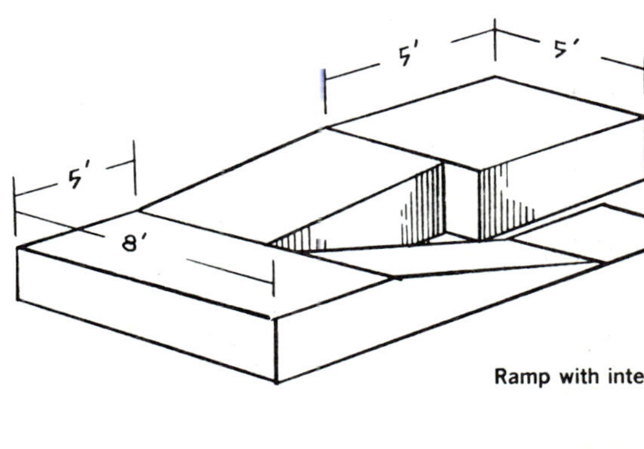

Ramp with intermediate switch-back platform.

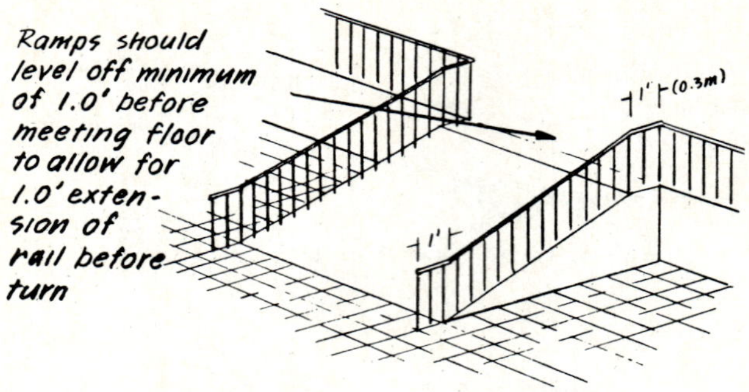

In order to improve traffic flow in busy public buildings, consider a combination ramp and staircase system, which allows the nonhandicapped to "move through" without waiting for someone who is using the ramp (who is usually slower). This scheme prevents a handicapped person from trying to hurry to get out of the way, which could cause an accident.

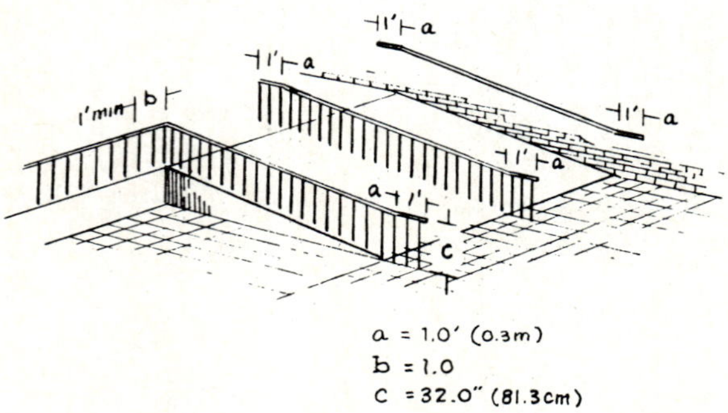

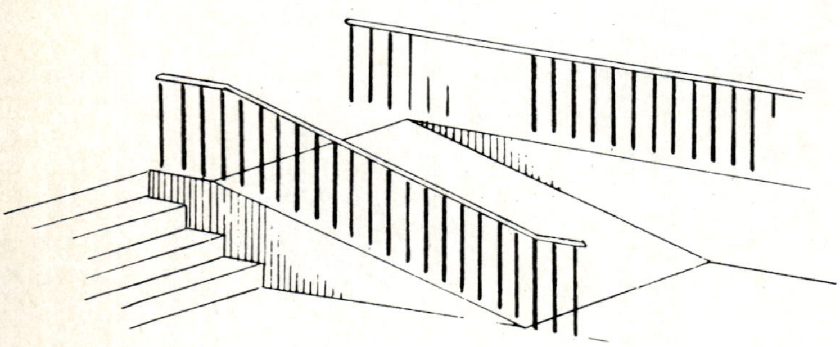

Special Ramp Situations

Sloping (Ramp) Floors in Assembly Halls and Theatres

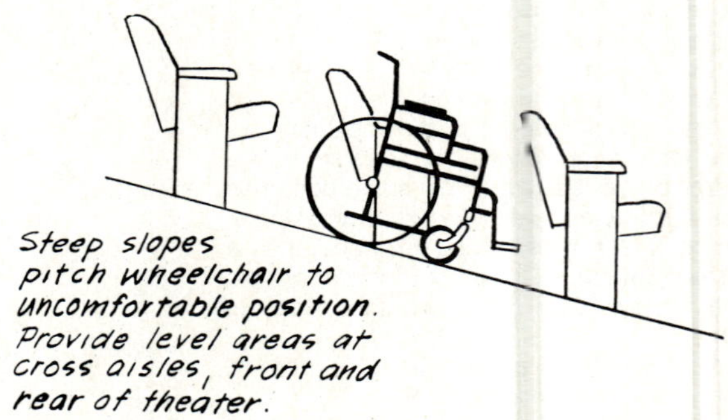

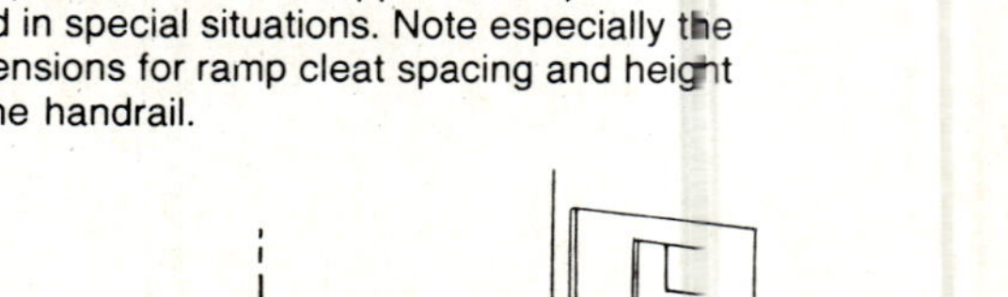

Industrial Sites

It is sometimes desirable in certain industrial plants to provide a combination pedestrian–mobile equipment ramp. The design shown in the accompanying illustration (although not acceptable for handicapped users) can be used in special situations. Note especially the dimensions for ramp cleat spacing and height of the handrail.

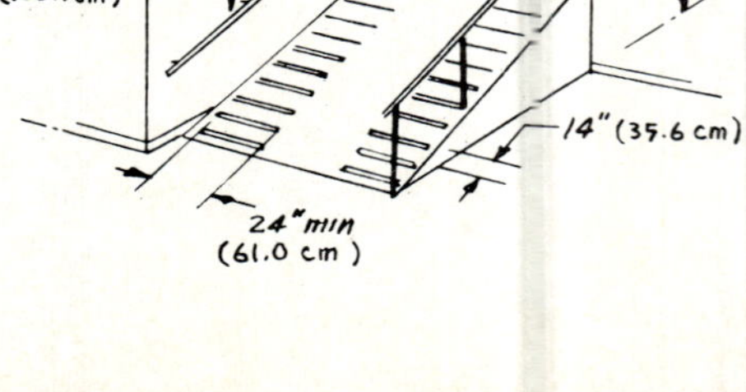

STAIRS

Optimizing Stair Design

Studies have shown that a staircase slope of approximately 30° requires the least expenditure of energy on the part of the average adult. Such a slope requires a tread depth of approximately 11 in (28 cm) and a riser height of about 6.7 in (17 cm)

This, of course, does not adequately take into consideration the two groups who have perhaps more stairway accidents than any other, i.e., children and the elderly. Unfortunately, it is probably impractical to design separate staircases for everyone (except in special cases that might be identified as custom installations). Therefore, it is recommended that all staircases (both indoors and outdoors) be designed with the following characteristics:

Tread 11.0 in (27.9 cm)
Riser 6.5 in (16.5 cm)
Nosing 1 to 1.5 in (2.5 to 3.8 cm)

As indicated in the accompanying sketches, certain stair features are undesirable, principally in regard to the "nosing." The reverse taper configuration is preferred over the "lip" because the latter tends to trip older people and the handicapped, who are not always able to lift their feet to clear the nosing. The "open-face" configuration creates a similar problem, and, in addition, some people are confused by the "see-through" staircase.

Other important optimizing features include, obviously, the proper width (a minimum width between handrails of 30 in, 76 cm, for one person at a time and of 52 in, 132 cm, for two persons at a time), the overhead clearance, and the height of the handrail (all stairs should have at least one handrail, and preferably two).

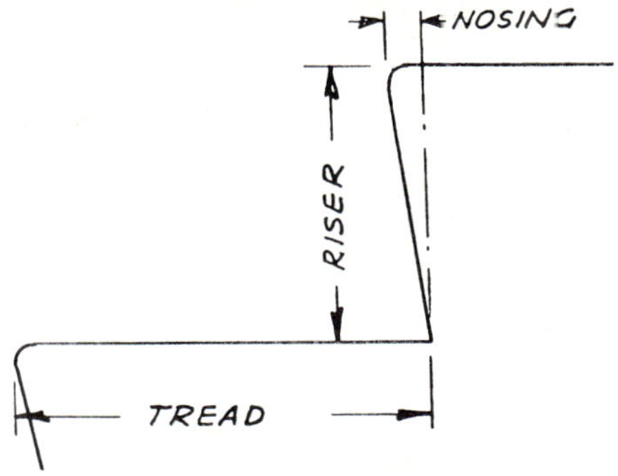

PREFERRED DESIGN

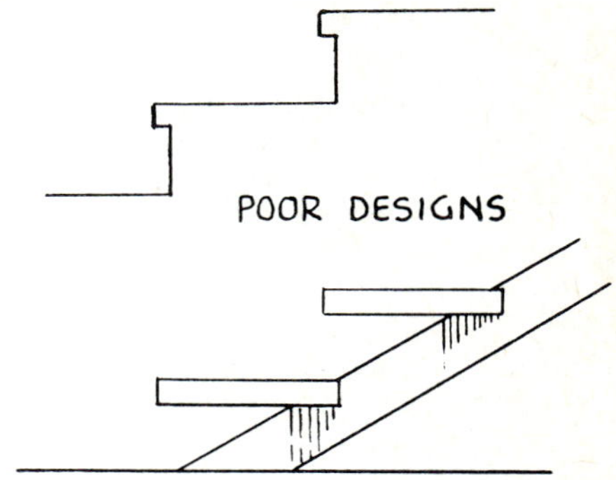

POOR DESIGNS

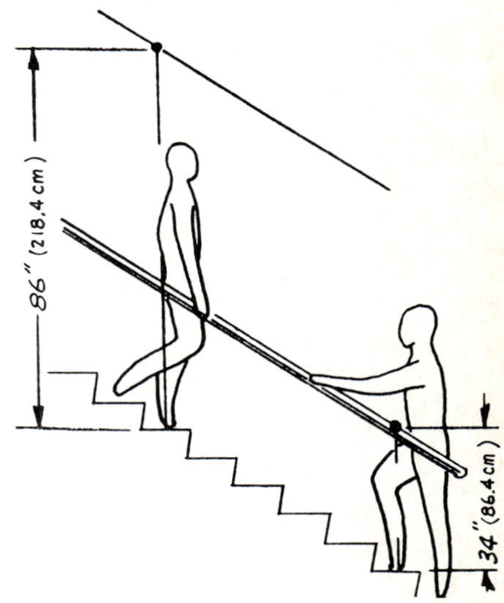

Stair Design in Relation to Equipment and Furniture Handling

Although equipment and furnishings can often be transferred between the floors of a structure by means of an elevator, sometimes the only avenue is by the staircase.

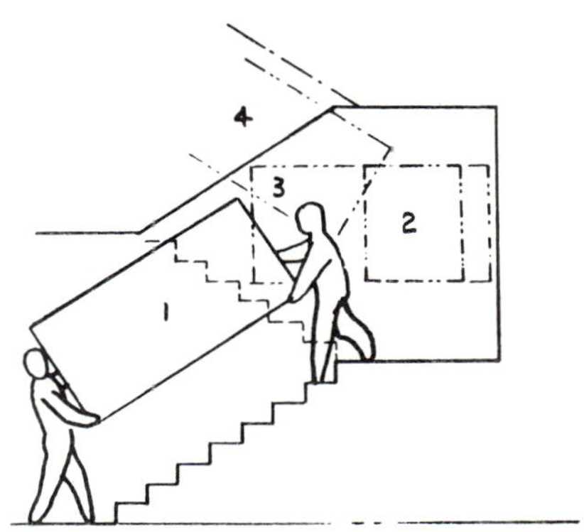

The recommended spacing between stair rails for safely moving the typical large piece of furniture or equipment is shown in the accompanying sketch.

It should also be noted that windows located at staircase landings are vulnerable when furnishings are being manipulated within the landing; therefore, the designer should consider the possibility of eliminating such windows.

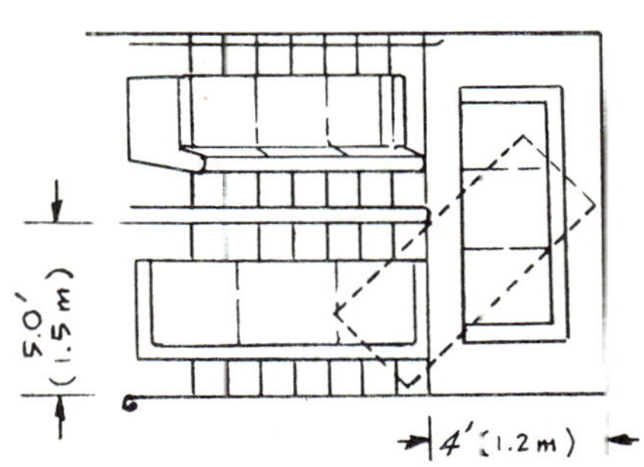

Special Safety Considerations in Stair Design

Spiral staircases should not be used as a general rule. The varying tread depth (the inside and the outside dimensions differ) often leads to missteps and accidents.

All main stairways within a single structure or dwelling should be alike in terms of tread-riser configuration.

Do not leave the backside of a staircase open, or people (especially the blind) may walk into the stair structure.

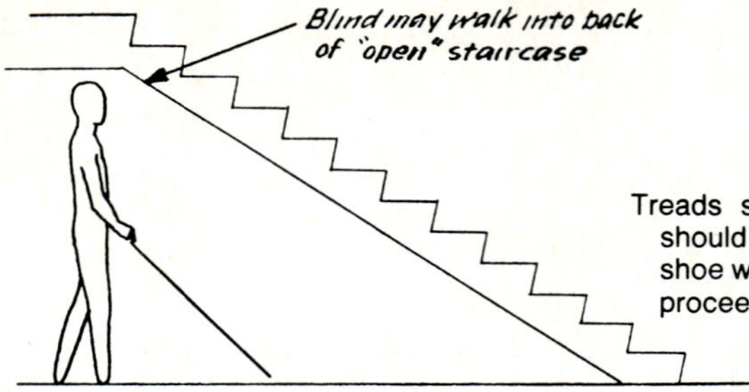

Treads should be of nonskid material but should not be so rough that the sole of the shoe will not slide slightly, especially as one proceeds up the stairs.

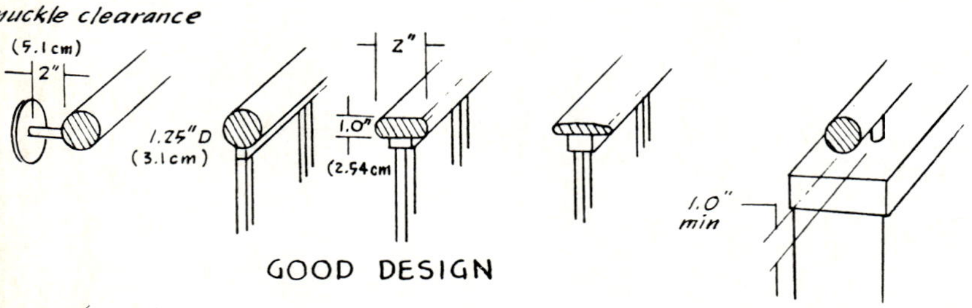

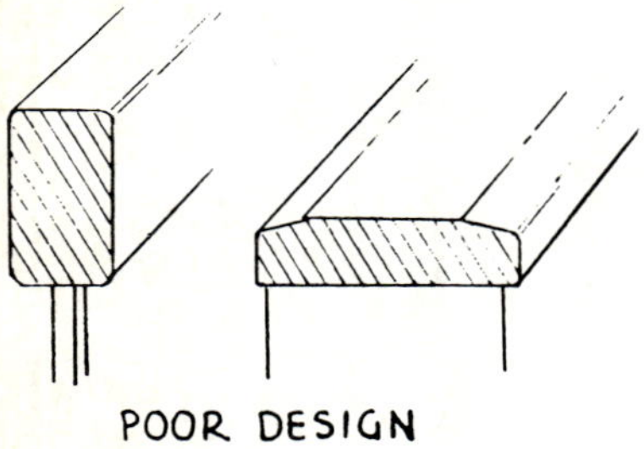

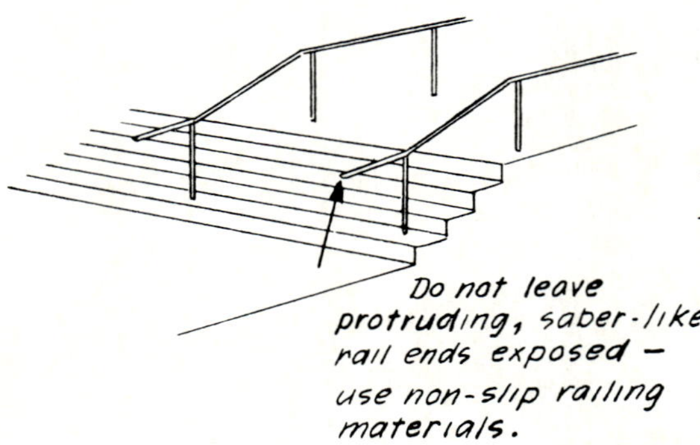

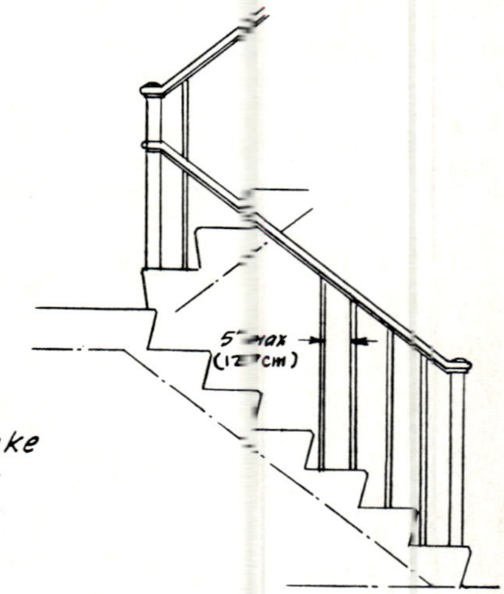

Select handrails that people can grasp easily.

Protruding, "saberlike" rail ends should not be left exposed, and nonslippery handrail material should be used.

Open stair rail designs should not be used because it is too easy for people to fall through. Equally important is the use of a closure design that is in itself not hazardous.

If vertical post designs are used, the spacing between the posts must be such that a child cannot get his or her head caught between them. Spacing should not exceed about 5 in (13 cm).

Other fancy, filigree-type supports should also be examined before they are used to determine whether they have sharp projections or edges and/or possible entrapment potential.

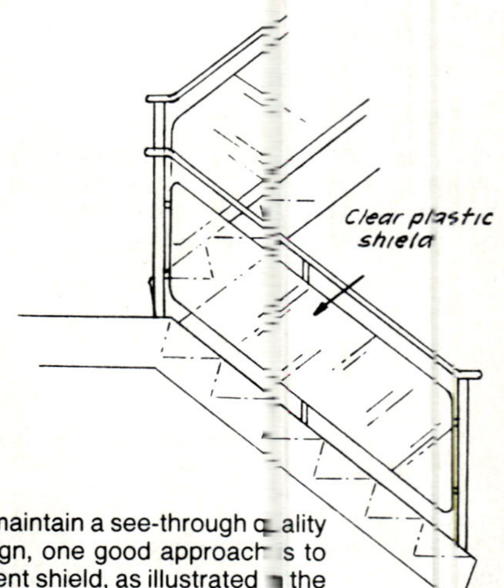

If one wishes to maintain a see-through quality about the design, one good approach is to use a transparent shield, as illustrated in the accompanying sketch.

Special Industrial Stairs

When specialized stair applications require some compromise of the preferred slope, tread and riser, and other previously noted features, one should still consider the possible hazards that may be associated with stair use.

The accompanying sketches illustrate some of the considerations that should be addressed:

The slope should never be steeper than 50°; otherwise, the user cannot decide whether to walk down head first or back down.

Always provide a safety screen behind open stair designs.

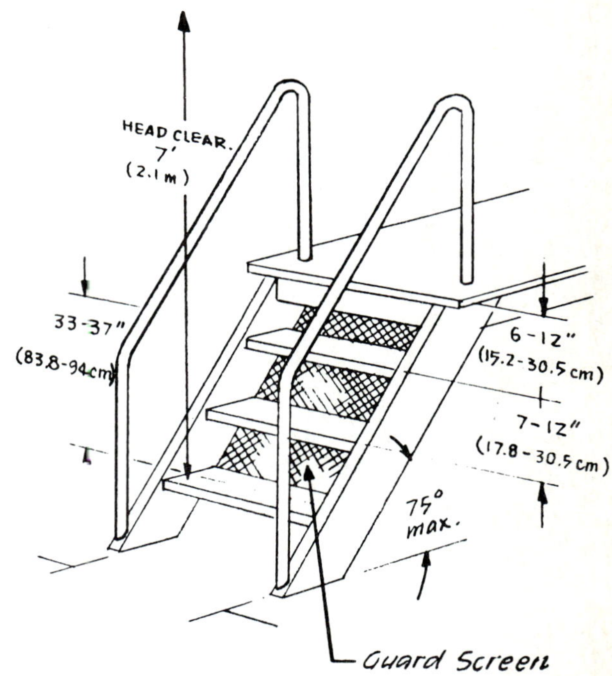

Use similar screening at landings.

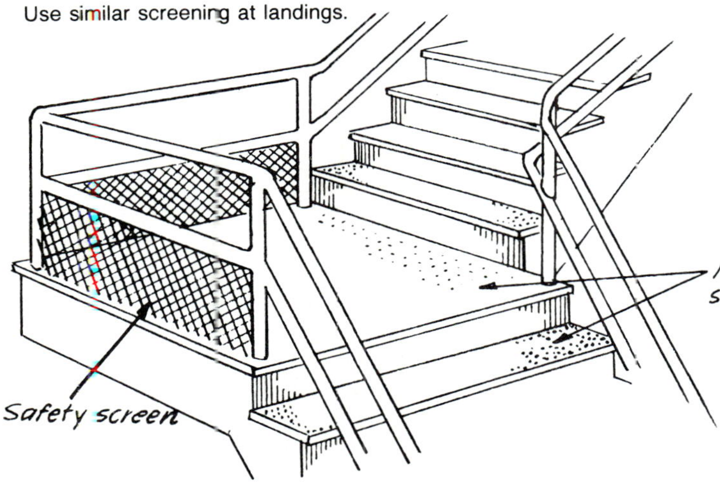

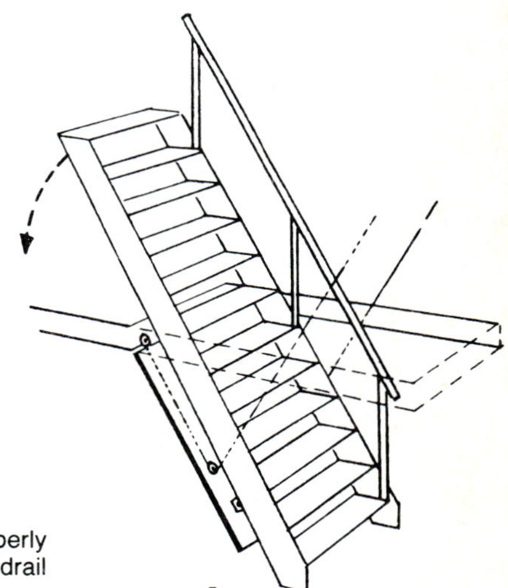

Foldaway stair systems must be properly counterweighted, and at least one handrail should always be provided.

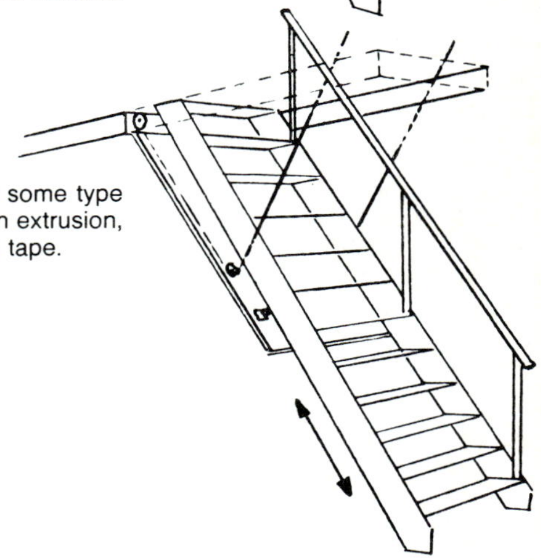

When using metal treads, provide some type of nonslip surface, such as open extrusion, raised bead, or nonskid overlay tape.

LADDERS

Once the incline requirement reaches about 50°, the designer is no longer within acceptable stair concepts; i.e., the device should be conceived as a ladder which should be employed only where it is expected that agile adults will be using the system. Ladder angles between 50° and 75° ordinarily should be designed with flat treads at least 4 in (10.2 cm) deep (with at least 6 in, 15 cm, of toe clearance behind the tread) and with a handrail that starts and ends at a point 36 in (91 cm) vertically above the bottom tread and similarly above the top tread and/or platform.

When a ladder becomes steeper than 75°, a railing should no longer be used, and rungs should replace the flat treads; this is done because at these steeper angles, the user should not be encouraged to come down the ladder frontward.

A, degrees	*R*, in (cm)
55	10.0 (25.4)
60	10.5 (26.7)
65	11.1 (27.9)
70	11.5 (29.2)
75	12.0 (30.5)

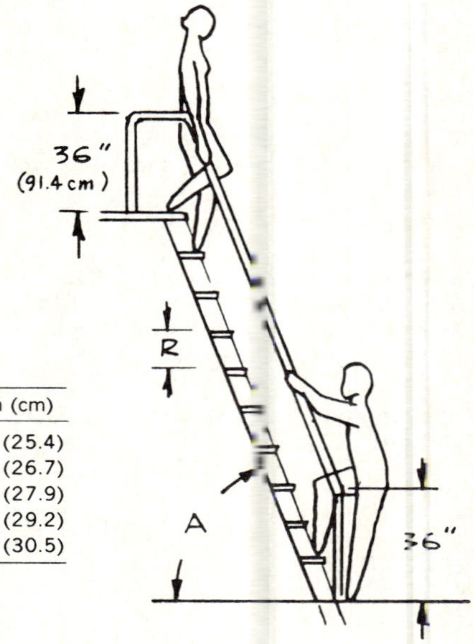

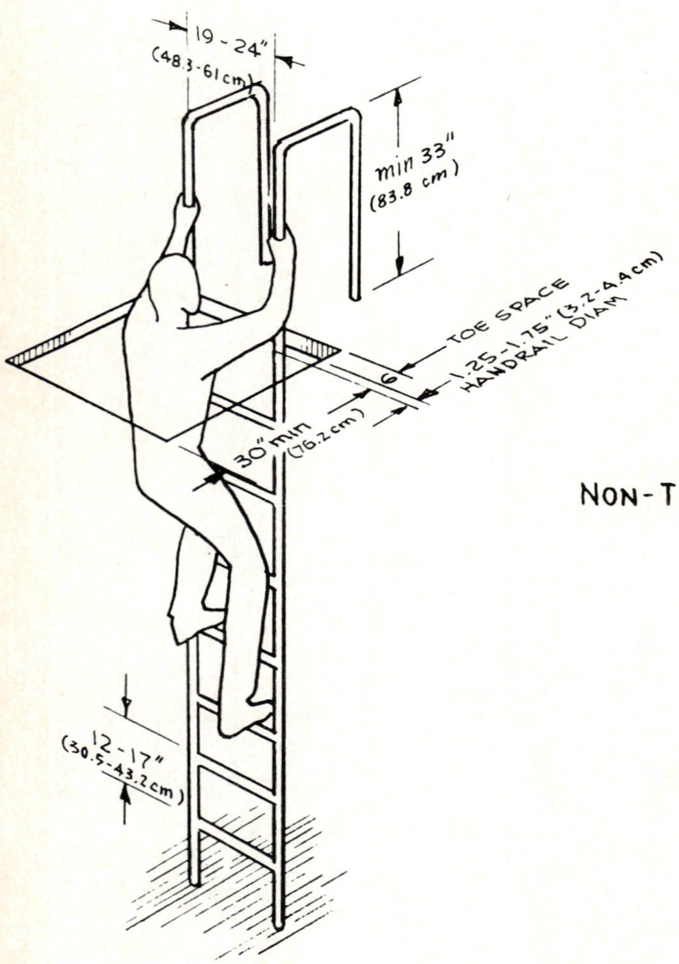

NON-TREAD LADDERS

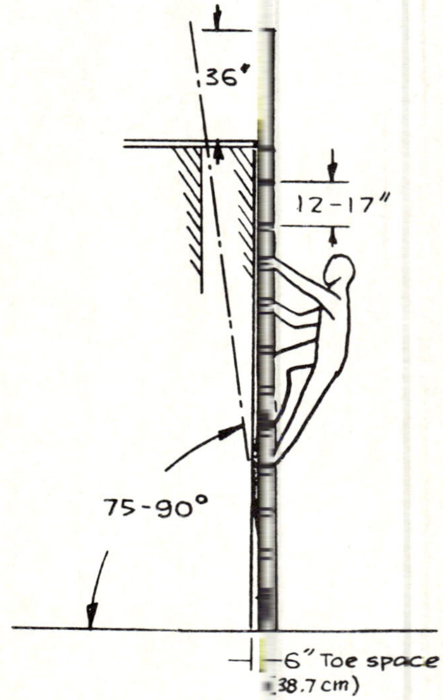

Common Design and Application Errors with Respect to Ladders

The most common ladder design and application errors occur relative to the final interface between the user's hands and feet. Several of these are shown in the accompanying illustrations:

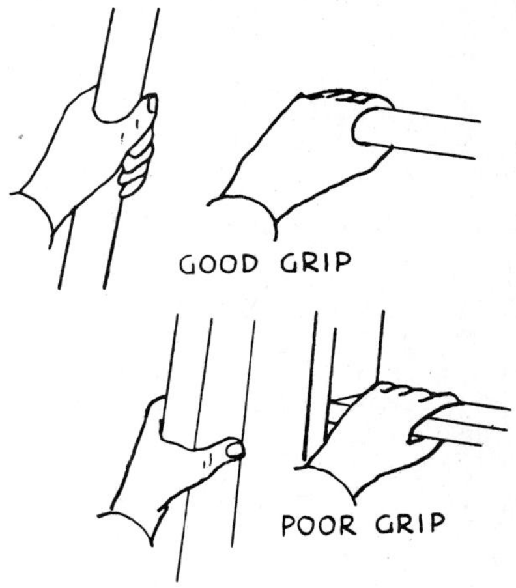

1. The rail or rung is either too large or too small for the user to obtain a secure hold on these structural elements. A rectangular cross section is harder to grasp than a round one.

2. Vertical rails are often either too close or too far apart, making it awkward to maintain a comfortable hold on either the vertical rails or the rungs.

3. Flat steps invite the user to think that the ladder is negotiable by going down frontward. When the ladder is too steep, the user usually discovers partway down that he or she cannot safely proceed in this fashion, at which point it is difficult to change modes.

4. To be safe, a flat tread must always be horizontal. Often the ladder is set up so that the flat tread is not horizontal.

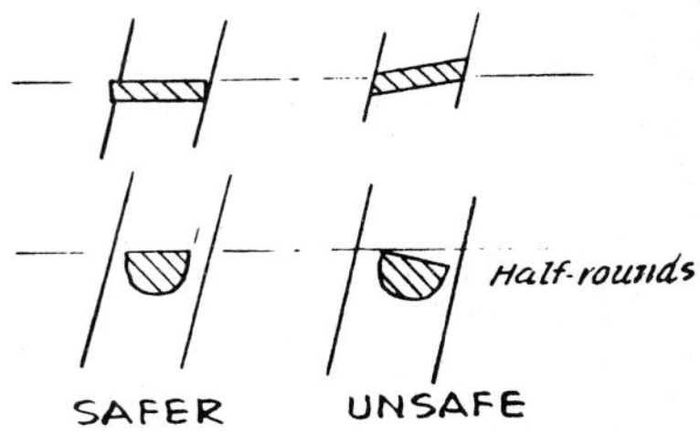

Ladders that are erected 75° to 90° should always have round vertical and tread features to ensure maximum flexibility, obviousness of use mode, and safety.

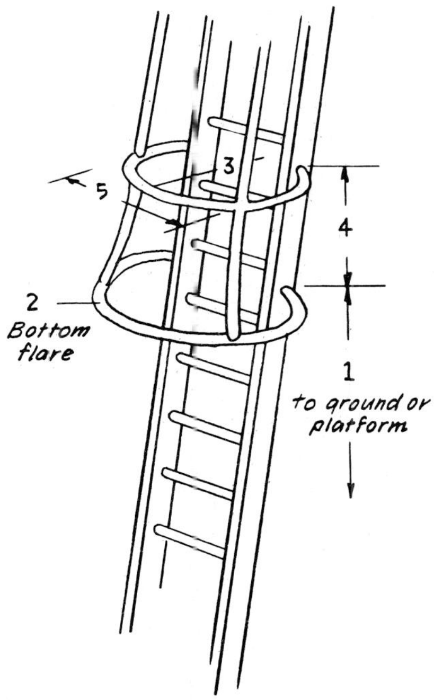

A safety cage should be provided on very high ladder subsystems, both when the ladder extends for a considerable vertical length and when the ladder is fairly short but is positioned high above the ground. The following guidelines should be considered in such installations:

1. The height of the cage from the ladder base should be about 7 ft (2.1 m).
2. Approximately a 32-in (81-cm) "flare" should be provided at the bottom of the cage in order to ease the user's entry and exit.
3. The depth of the primary cage from the center of the ladder should be about 28 in (71 cm).
4. The maximum distance between cage ribs should not exceed 18 to 20 in (46 to 51 cm).
5. The width of the cage should be about 27 in (69 cm); i.e., the cage should have an oval shape.

LOCKER-ROOM GUIDELINES

The accompanying illustrations show two examples of locker-room arrangement and critical dimensions. Key considerations should also include the following:

1. Good ventilation in toweling and dressing areas
2. Water control curbs
3. Nonskid floor surfaces throughout
4. Good lighting
5. Ample accessories such as mirrors, clothing and towel racks and hooks, waste receptacles, drinking fountains, and lockable, ventilated lockers
6. A first-aid room directly adjacent to the locker-room facilities

Note: In communal shower rooms, adjustable-height, nonclogging shower heads at least 4 ft (1.2 m) apart should be provided. There should be both individual and master controls.

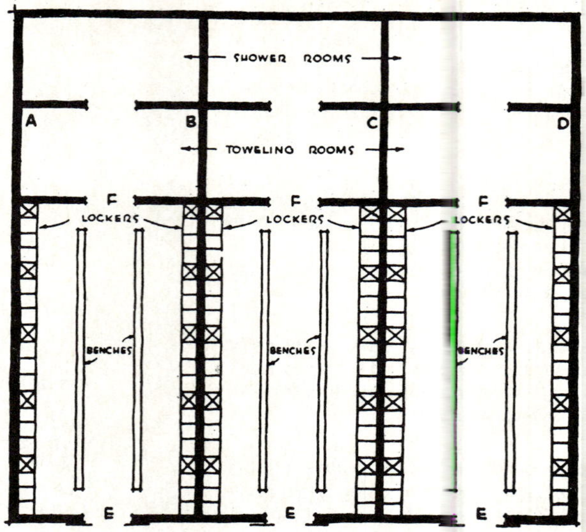

GENERAL PRINCIPLES FOR CONCEPTUALIZING SPECIAL WORKPLACES

Working Position

Workers or operators should stand when:

1. The task they must perform requires reaching beyond a point that typically can be reached comfortably while seated.
2. They have to move frequently from one type of task or workplace to another.
3. Certain tasks require lateral reaching beyond a specific task site, i.e., when it is easier to step to another position.
4. They frequently interface with others who are standing.
5. Several persons have to work jointly at a large display area such as a diagram.
6. They have to reach long distances to touch, transport, or adjust controls.

Workers or operators should be provided an alternative sit-stand configuration when:

1. They normally remain at a task more than 30 min, but others frequently must observe their operation.
2. They need the advantages that seating provides for their major task, but often have to move to another position for a temporary standing task.

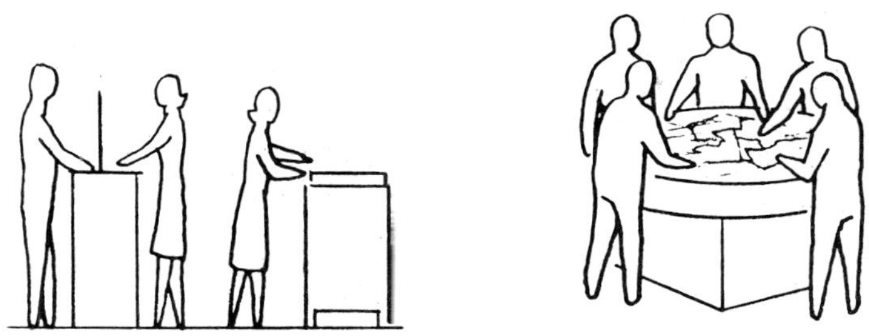

Have workers stand if tasks require frequent position change and tasks are short.

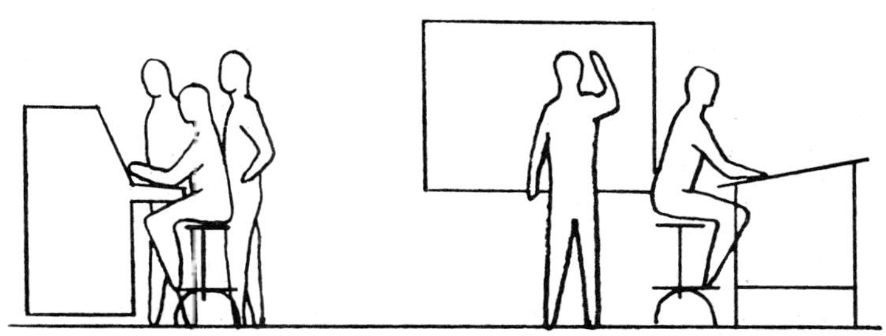

Provide sit-stand configuration if primary task is long, but frequently interspersed with standing operations - getting up and down is fatigueing.

Workers or operators should sit when:

1. They must work at the same task for an extended time period (30 min or more).
2. Task precision demands steadiness both for them and for the devices being used.
3. They need to be restrained to prevent their displacement by dynamic environmental forces.
4. They have to use their feet to operate controls.
5. They are required to apply maximum force to operate a control.
6. They have to perform extensive writing tasks.

Note: Although the above principles apply generally, it is obvious that each decision must be based on a thorough analysis of all the interacting factors, e.g., restraint, steadiness, mobility, interoperator coordination, and potential fatigue. For example, a draftsman usually can perform fairly precise drawing movements either standing or sitting. On the other hand, a vehicle operator generally cannot perform the typical hand and foot control operations as well when standing as when seated.

In addition, the nature of the equipment configuration may require that elements be widely spaced, in which case excessive reaching can be minimized by providing a more mobile operator facility.

Above all, avoid the tendency to repeat traditional workplace configurations without making sure these are compatible with all the task demands.

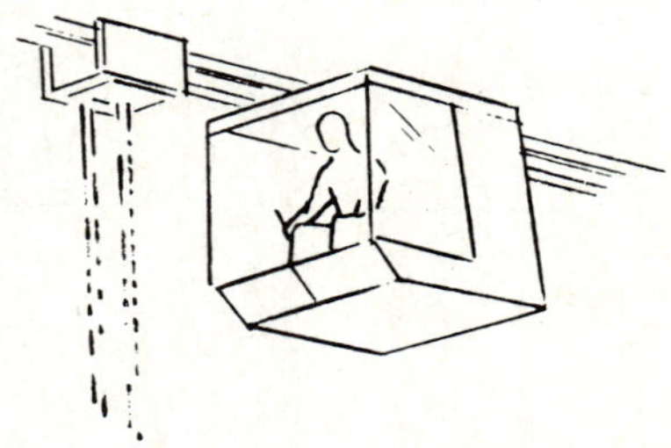

General Worker and Operator Positioning Considerations

1. Avoiding Awkward Positioning

a. Minimize the necessity for operators to lean; i.e., arrange the workplace so that operators can maintain a more or less normal alignment of their bodies, especially when standing.

b. Operators should not have to use their maximum reach capability; i.e., they should be able to shift their bodies to a better position.

c. Operators should not have to sit or stand for long periods with their head, torso, or limb positions skewed; i.e., they should be able to keep their heads, necks, torsos, and limbs in a symmetrical relationship as much as possible.

d. Operators should not be forced to work frequently or for long periods with their hands and arms above normal elbow level.

e. Avoid positioning operators in supine or prone positions.

f. If the limbs must be extended for long periods, provide some kind of limb support.

g. Configure the workplace so that operators can see what they are doing without having to assume an awkward head or body position to see around their hands or the control device.

2. Creating Arrangements That Are Compatible with Normal Limb Movement Patterns and Reach Limitations

a. Minimize the need for operators to move their torsos during continuous control operation, except where a ballistic advantage is desirable.

b. Repetitive arm and leg motions should be in the direction that limbs articulate normally about typical shoulder, elbow, wrist, hip, knee, or ankle pivot points.

c. When both hands or both feet are involved simultaneously, create a configuration in which the motions are opposite rather than in the same direction; i.e., the right hand should rotate in the opposite direction to the left, and the right foot should push forward while the left relaxes.

d. Minimize the necessity for operators to shift their position in a seat in order to look at a display or operate a control. In certain situations, however, it may be desirable to provide seat movement, i.e., fore and aft, laterally, or rotating.

e. Avoid workplace arrangements that force operators to stand too close to, or lean across, a hazardous element, such as a moving belt or chain or a hot component.

f. Take into account the extra clearance requirements and/or reduced mobility caused by special apparel that operators may be required to wear and make sure that there are no features within the mobility envelope that could prevent proper limb movement, i.e., components which might be bumped, which limit full limb movement, or upon which clothing might become snagged.

3. Force Application

a. Place manipulable objects or controls in positions that are compatible with the best geometric relationship for applying pushing, pulling, or rotating force by operators whenever such forces cannot be minimized.

b. Provide an appropriately positioned support against which operators can minimize counterforce effects, e.g., an armrest, handrest, or backrest.

c. Arrange force-demanding controls in positions where operators can apply the necessary force without disturbing their normal body position, especially when this might interfere with a primary visual or control activity and/or with necessary body-referencing posture.

4. Minimizing Fatigue-Producing Workplace Arrangements

a. Provide backrests for seated operators.

b. Provide armrests.

c. Provide handrests when operators are using a continuously operated controller, such as a desk-mounted joy stick or roll-er-ball controller.

d. Provide footrests.[1]

5. Minimizing Potential Safety Hazards

a. The operator's posture should be such that his or her body is in a position that can accommodate dynamic force inputs, such as the buffeting of a vehicle as it impacts against rough terrain.

b. Provide body restraints for vehicle operators which will protect them from impacting the interior structure or equipment and which will not allow them to slip into a position in which the restraint itself, such as a seat belt, becomes a hazard.

c. Make the workplace as safe as possible in terms of locating and designing components so that they will not penetrate an operator's skin upon contact, i.e., so that they will spread the energy rather than concentrate it.

Key Factors to Consider for Each Type of Operator Working Position

1. Standing Operators

A smooth, level surface should be provided.

There should be sufficient surface area for operators to establish an adequate spread of their feet, to move when necessary, and to brace their feet when required.

A nonslip surface should be provided if a platform is subject to movement, as in a vehicular environment.

A resilient surface should be provided if operators must stand all day.

Visually observed workplace elements should be arranged so that they can be seen without excessive movement, so that there is no parallax for reading instruments, and so that reading distance and size of display are compatible with readout precision.

Manipulative tasks should be arranged and/or oriented to be compatible with respect to reach convenience, required motion patterns and excursions, force application requirements, precision demands, and response speed. Special attention should be given to the implications of eye-hand coordination.

2. Sit-Stand Operators

The eye reference for both seated and standing operators should be approximately the same.

The arrangement of visual displays that have to be monitored by both seated and standing operators should be such that there is the same level of readout accuracy from both positions; i.e., operators should not make reading errors because of parallax.

If both seated and standing operators may have to use the same control device, place this in a position that will minimize interoperator interference.

Provide a footrest for seated operators.

3. Seated Operators

Provide a seat that ensures optimum working posture for the tasks being performed.

Arrange visual displays and controls so that an operator's hand will not cover critical displays and so that the displays are normal to the expected viewing line of sight.

If the primary viewing task is directed outside the workplace (vehicle control), position the most important internal displays so that they can be viewed without excessive eye movement from the nominal exterior line of sight.

Position all controls so that, in manipulating them, operators do not appreciably move their nominal eye reference and possibly miss seeing important events occurring outside or on the principal internal display.

4. Miscellaneous Considerations

a. Organizational standardization versus individual adjustment: When similar workplaces are to be repeated, a compromise should be sought between the benefits of a standardized workplace (which permits operators to shift from one workplace to another without confusion) and the benefits that accrue from allowing individual operators to modify their workplace to fit each type of activity more conveniently (i.e., to rearrange tools, materials, or other aids so that they are more accessible).

b. Illumination: Do not assume that a generalized type of illumination will satisfy all task-seeing requirements; i.e., the type and location of light fixtures should be determined according to the varying needs of the operator. Especially important is the creation of seeing conditions that minimize glare and reflection problems.

c. Storage: Almost all workplaces require the provision of storage space for books, files, tools, or writing and drawing equipment and materials. Equally important is the provision of storage space for communications devices, discarded items, and personal articles such as glasses or eye shields.

[1]Refer also to the discussion of problems of awkward positioning.

d. Service and maintenance: Careful consideration should be given to ensuring that, in arranging the workplace and/or packaging of equipment, access to critical elements for service or other maintenance is provided.

e. Design for handicapped operators: Whenever a workplace may be used by a person in a wheelchair, a blind person, or a deaf person, special care must be taken not only to dimension the workplace properly but also to provide those special features necessary for the handicapped operator to perform more effectively within his or her limitations (e.g., tactile features for the blind and special visual signals for the deaf).

f. Environmental considerations: Be aware of the possibility that a particular individual workplace may reduce the effectiveness of general ambient thermal, ventilation, and noise control systems, thus requiring additional features to ensure that the operator does not become overheated or is not subjected to respiratory contamination or to interfering or annoying noises that may be produced within his or her particular workplace.

g. Manipulative clearance for large materials: Often large materials must be maneuvered into place before they can be worked on. Provide adequate clearance so that such materials can be maneuvered into place easily.

BASIC TASKS IN RELATION TO WORKPLACE DIMENSIONING

1. Writing: The writing surface should be at approximately elbow level and should be relatively horizontal (maximum 5° slope); greater slopes make it difficult to keep a pen or pencil from rolling off the surface.
2. Typing: The center of the keyboard should be at about elbow level.
3. Data entry keyboard: The center of the keyboard should be at about elbow level. An associated visual display should be positioned so that the operator's line of sight is perpendicular to the face of the CRT.
4. Drawing: A drawing board used for precision work should slope approximately 3° to 4°; this provides the best compromise between reaching, viewing, and precise instrument manipulation. For less precise drawing, the drawing surface should be adjustable, so that the worker can match the size of the work with the best position for drawing. A sit-stand arrangement is most desirable for average mechanical drafting work since it allows the worker to reach all areas of a large drawing easily

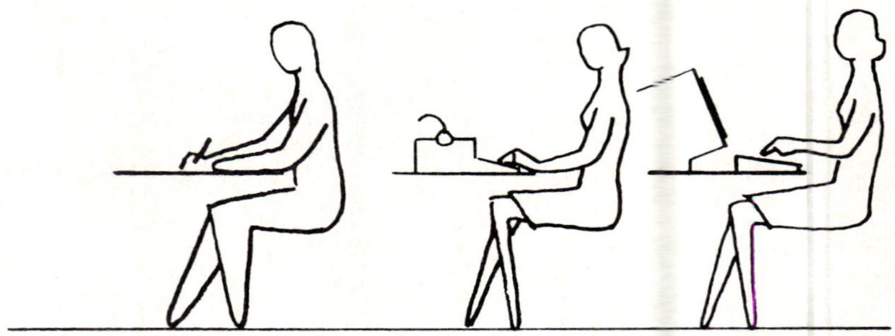

Writing Typing Data entry

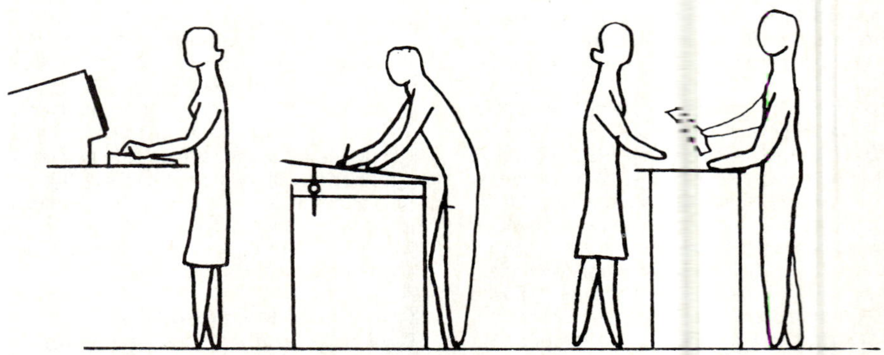

Data entry (stand) Drawing Communication across counter

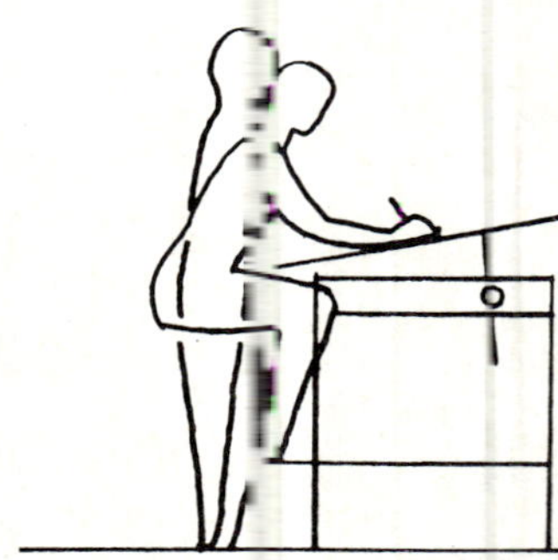

Large layout work

5. Storing and retrieving shelves: Design and/or select storage shelving that allows the worker to reach and grasp objects to be stored or retrieved easily and without fear of dropping them. Avoid deep, high shelving since it is easy for small objects to be pushed to the rear of a shelf, where they are neither visible nor manually accessible.
6. Office machine use: Either select machines that have been properly designed (including special stands) or select or build tables that assure proper positioning of the principal work level of the machine. The work level should be at approximately the operator's elbow height.
7. Tool operation: Various tools require that the work level be carefully chosen, especially if there is a requirement for precision and/or the application of controlled force.

Storage-retrieval

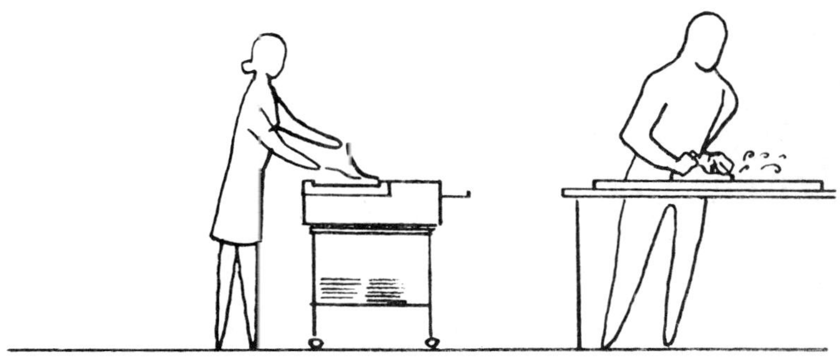

Office machine operation — Shop work

Nominal Guideline Dimensions

Although the accompanying general dimensional suggestions are recommended to approximate the optimum, one should be careful not to assume that they necessarily apply exactly to a specific problem. Use of a test mockup to verify these is recommended.

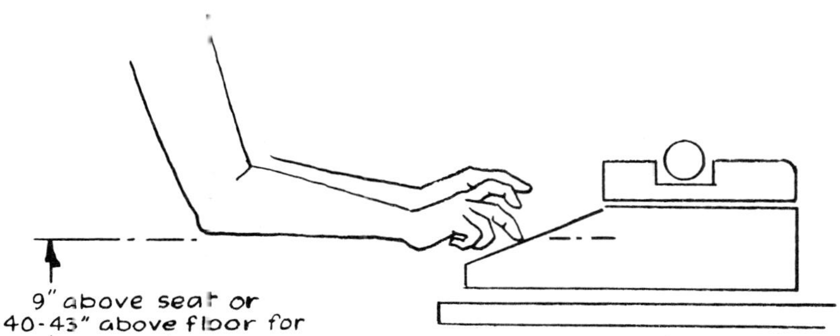

Proper height for typewriters

Writing space

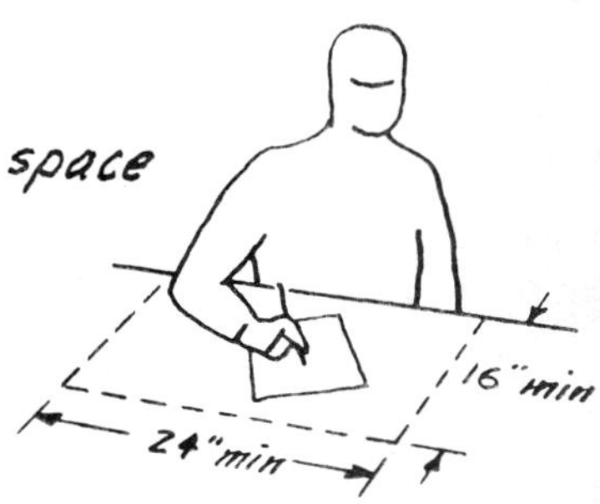

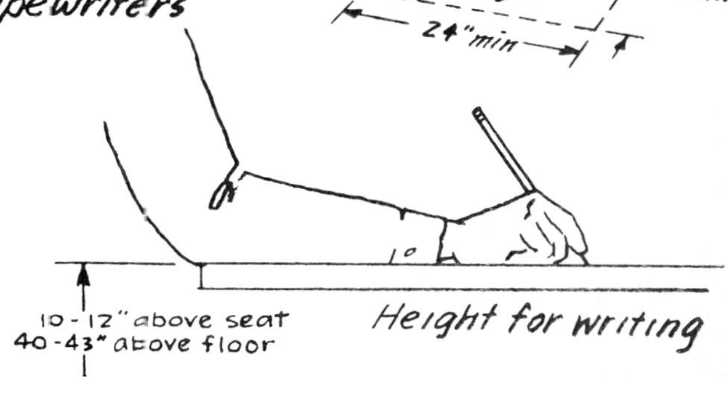

Height for writing

USE OF OPTICAL INSTRUMENTS Both eye alignment and control manipulation must be considered in determining the proper working height for optical instruments. Because of the multiple points of articulation in an operator's back and neck, it is difficult to anticipate exactly how operators may have to position their bodies and heads in order to align their eyes with the eyepiece of a particular instrument. The most practical approach is to obtain the actual instrument that is to be used and then mock up a mounting geometry that has sufficient adjustment for operators of various sizes to try out. In some cases it will not be possible to "fix" the work height so that it is acceptable for all operators; an adjustable worktable must then be provided.

Nominal Guidelines for Tool Use

Assuming the typically available tool and clamping devices obtainable from the average hardware store, a standard workbench height of 36 in (91 cm) is generally acceptable for most applications.

More difficult is the selection or positioning of machine tools that require precise visual inspection and control adjustment. Especially important is removal or positioning to avoid visual blocks.

Table heights must be lower for ease of manipulating or working on large equipment.

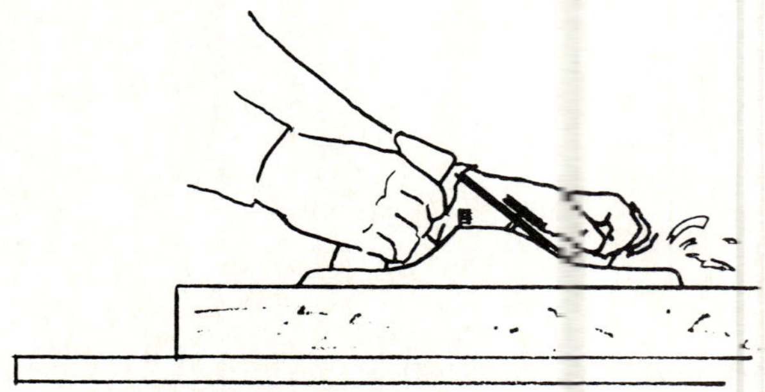

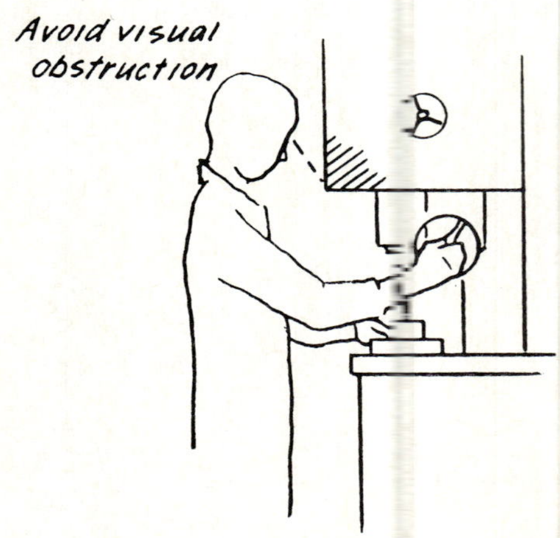

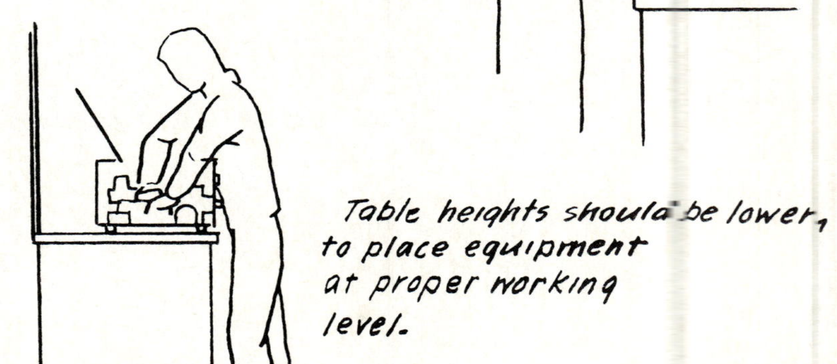

Eye-Positioning Path for Aligning on an Optical Device[2]

The accompanying diagram illustrates the approximate eye-positioning path of a person sitting in a fixed seat and using an articulating optical device. (The designer is cautioned not to consider this diagram a final recommendation; rather, it is useful only in terms of preparing a mockup test bed for further evaluation with an expected range of users.)

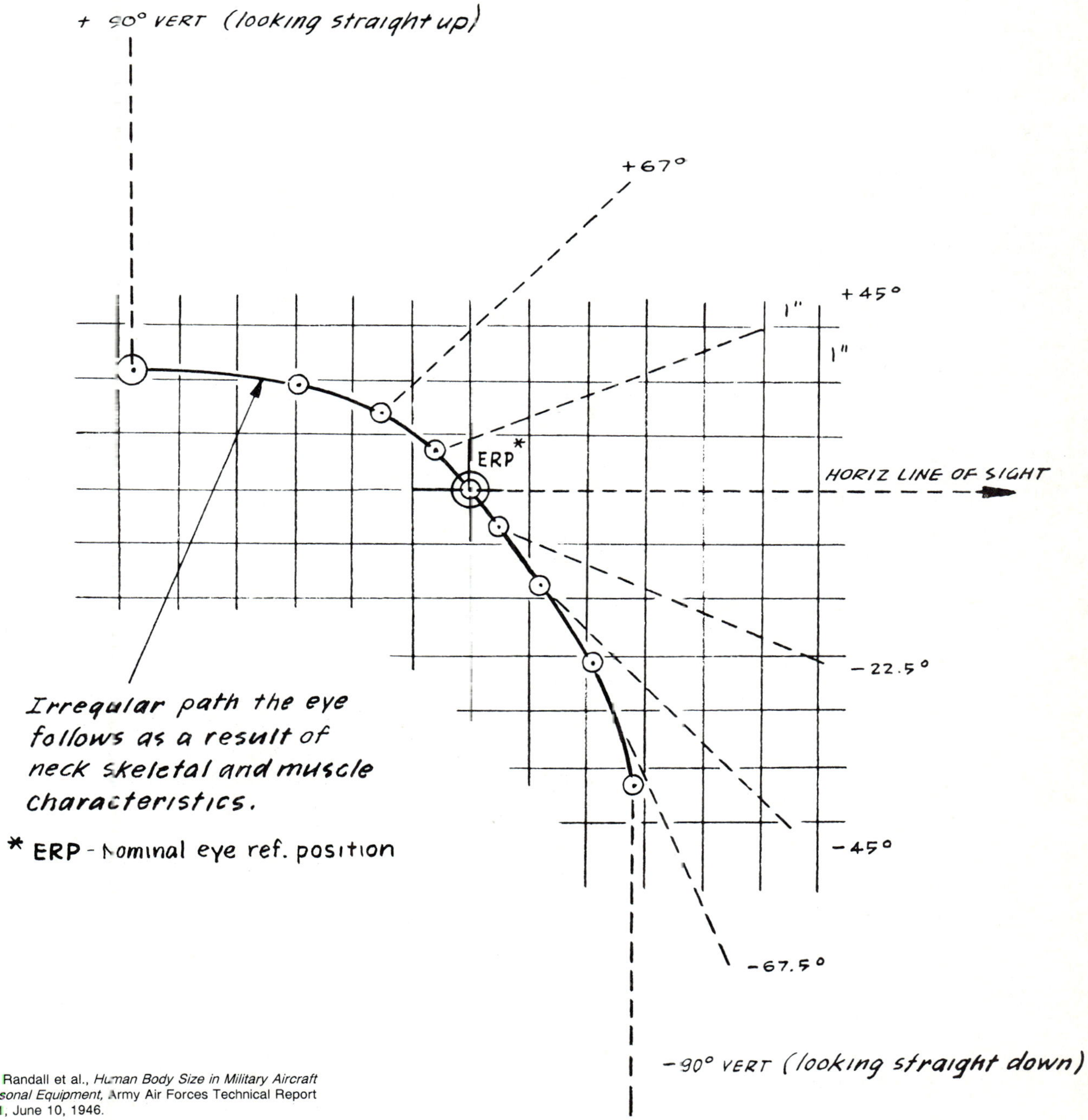

[2]F. E. Randall et al., *Human Body Size in Military Aircraft and Personal Equipment*, Army Air Forces Technical Report no. 5501, June 10, 1946.

ELECTRONIC CONSOLES AND RACK-MOUNTED PANELS

1. Seated Operator Consoles
Operators who are seated at an electronic console are necessarily limited in terms of how far they can reach. In addition, their viewing angles may be critical in order that display faces are reasonably perpendicular to their normal lines of sight.

In general, consider the suggestions illustrated by the accompanying sketch:

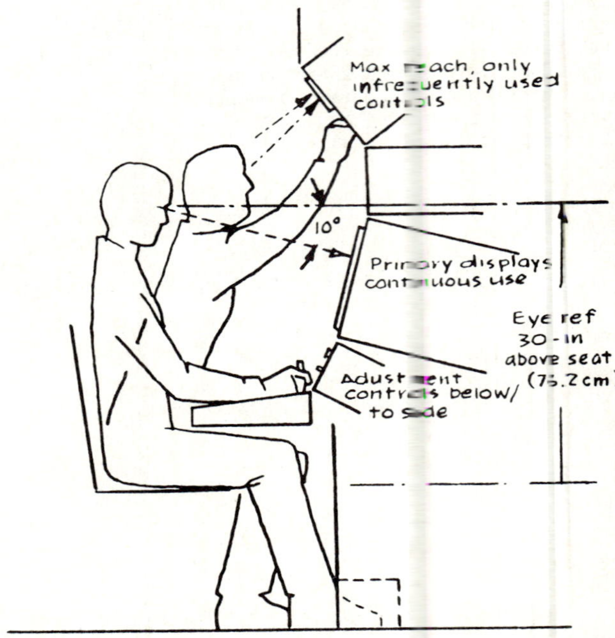

a. Position the primary display—say, a large CRT—slightly below the operator's eye level, but sloped to match his or her normal head slump of about 10°.
b. Adjustment controls that are used most frequently should be positioned at approximately elbow level.
c. If a finger joy stick or roller-ball controller is required, place it nearer the back edge of the desk, not near the outer edge, where the operator will not have a resting surface for his or her arm.
d. Displays or controls that are used infrequently can be located in less convenient positions, as long as they are not "detached" from related functional monitoring areas.

2. Standing Operator Cabinets
See the accompanying sketch.

3. See-Over Consoles
If the operator needs to see over the console to monitor other displays, reduce the console height so that the operator is not required to stretch his or her torso or neck.

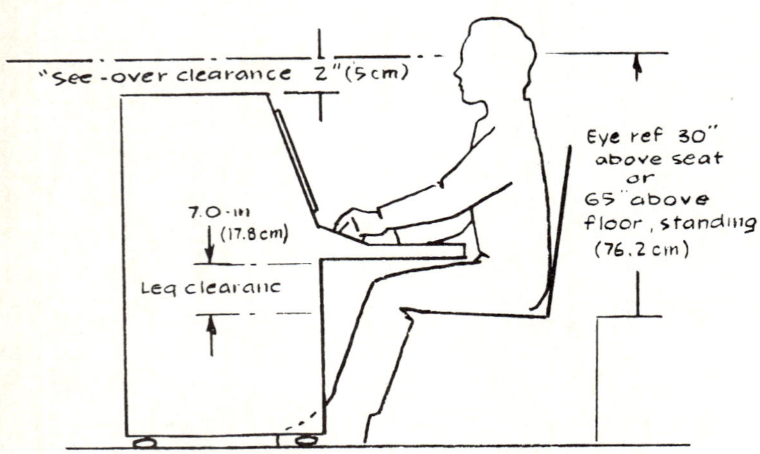

4. Extreme See-Over Configurations
When it is necessary for an operator to "see down" (as illustrated in the accompanying sketch), one should examine the total facility architecture in conjunction with the design of the operator's control station. As the illustrative example implies, the operator's position for viewing objects outside the workplace has to reflect consideration for *(a)* what the operator has to see, i.e., the required down angle; *(b)* the operator's position relative to the structural obstruction; and *(c)* the operator's position in relation to the shape and size of the control package so that the package configuration (at least) allows the operator to take full advantage of whatever clear view the structure allows. In some cases it may be desirable to use a sit-stand configuration in order to take advantage of the fact that the operator is freer to lean over the console.

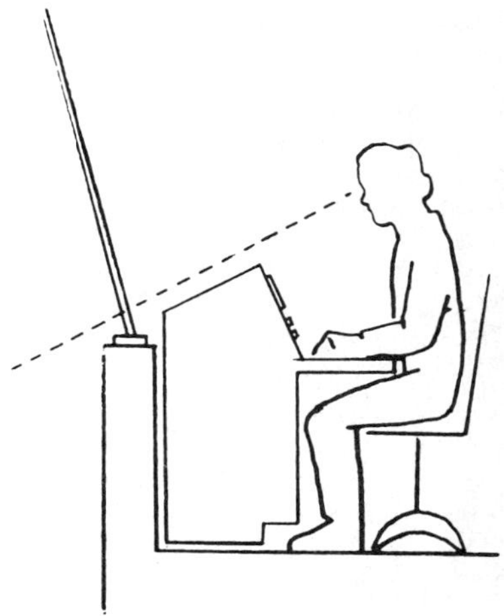

5. Large Plot Subsystems
The two accompanying sketches illustrate general dimensional guidelines for large plot subsystems; these are based on *(a)* limitations created by knee clearance if the operator sits and *(b)* convenient use area when the operator stands.

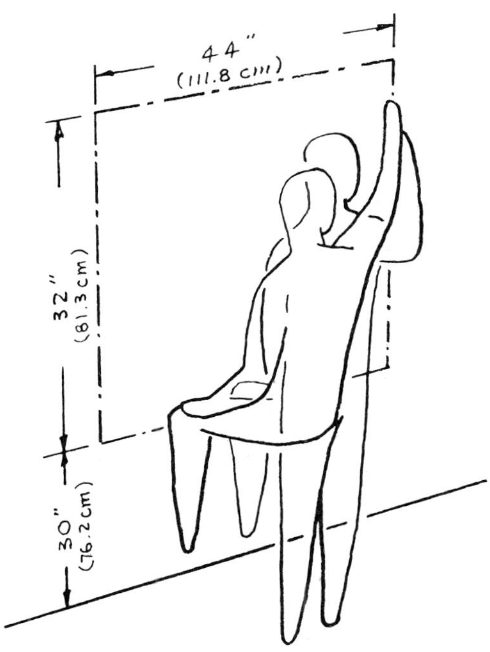

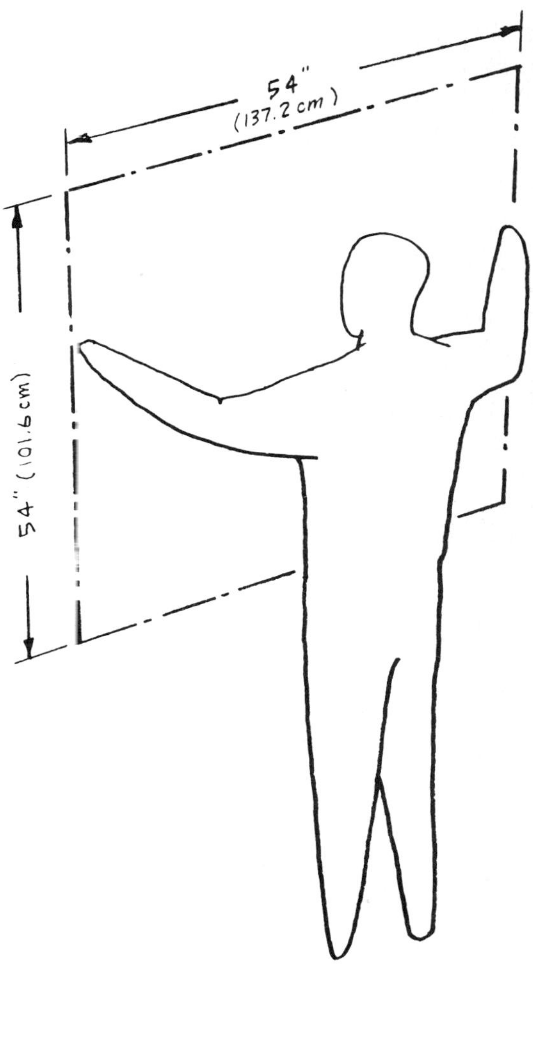

Other Considerations

As a general rule, panels for either seated or standing operation are easier to use when the displays are laid out for horizontal, sequential scanning. This is due to the fact that an operator's eye motions and head movements are less tiring in the horizontal direction. Studies also have shown that operators can respond slightly faster when they attempt to read instruments or other visual information from horizontally arrayed data than when this is presented in vertical arrays.

On the other hand, standing operators, because they are free to walk backward and to the side, can more easily adjust to control and display arrangements that are organized in vertical patterns.

Horizontal arrangement for seated operation.

Display Positioning Relative to Illumination and Reflection or Glare

Large CRTs and/or glass-covered instruments are subject to interference by ambient lighting conditions. Whenever possible, one should attempt to control both the positioning of displays and the ambient lighting conditions. Although an important objective usually is to position a display so that the face is perpendicular to the operator's normal line of sight, it may be necessary to compromise in order to eliminate undesirable reflections from artificial and/or natural light sources.

When control of ambient lighting is not the responsibility of the designer, other steps must be taken to minimize the effects of possible glare or reflection. The glare shield illustrated in the accompanying sketch is typically used to accomplish this.

Horizontal and vertical for standing operator.

Arranging for Added Convenience

Although the typical console or equipment operator has considerable reach capability, it is desirable to minimize the necessity for reaching, especially if it may be required on a frequent basis. The so-called wraparound concept is helpful as long as care is taken to ensure that unusual interferences are not introduced.

The wraparound console is especially helpful for the seated operator, as shown in the accompanying illustration. If a movable seat is acceptable, the arrangement can be expanded, taking advantage of a chair with casters.

A sit-stand operation (see the accompanying sketch) that includes both worktable and panel rack operations can be accommodated best if the rack can be made movable.

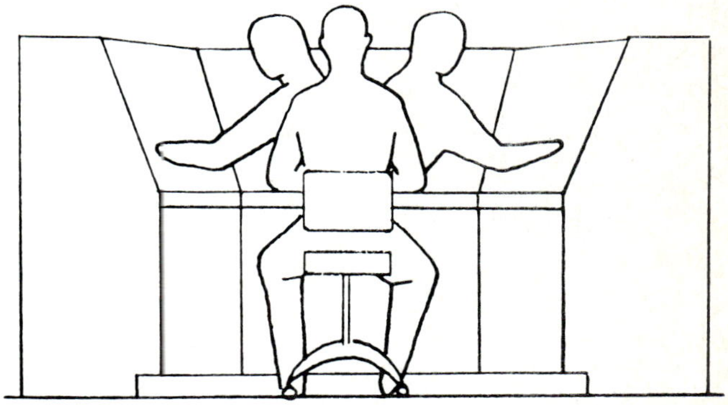

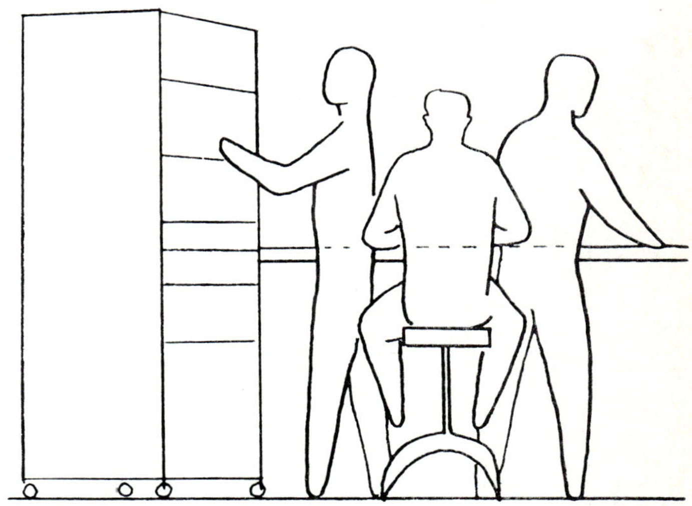

Start with basic single-operator requirements

expand for additional operators

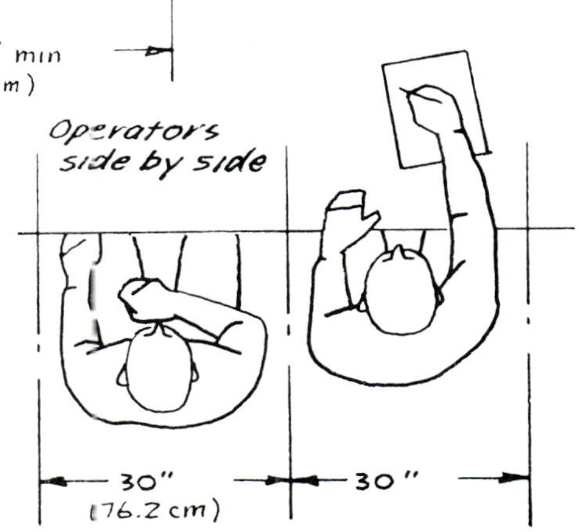

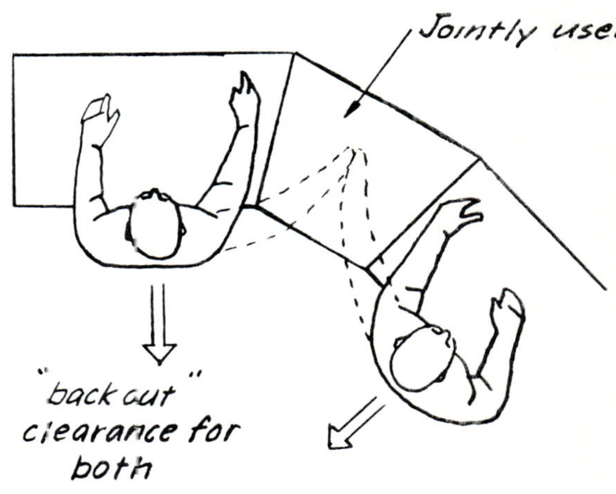

LARGE GROUP DISPLAY AND OBSERVER ARRANGEMENT

When a number of operators must view a large visual display in conjunction with console operation, the large group display should be arranged so that the various observers can see the display from their normal working positions without having to peer around their neighbors or their neighbors' equipment. Many alternative arrangements are, of course, possible, and sometimes a compromise must be made because of facility space constraints.

The accompanying illustrations and guidelines are helpful in making sure that all the sight-line requirements have been evaluated.

As the two accompanying sketches illustrate, angle of view may be improved by seating the operators further apart.

Operators and their consoles can be staggered in order to maintain the recommended screen-observer line-of-sight limit of 60°, as shown in the accompanying illustration.

When the recommended clear line of sight cannot be accomplished by lateral spacing, consider placing the operators in the rear on platforms so that they can see over the heads of those in front.

Possible but awkward

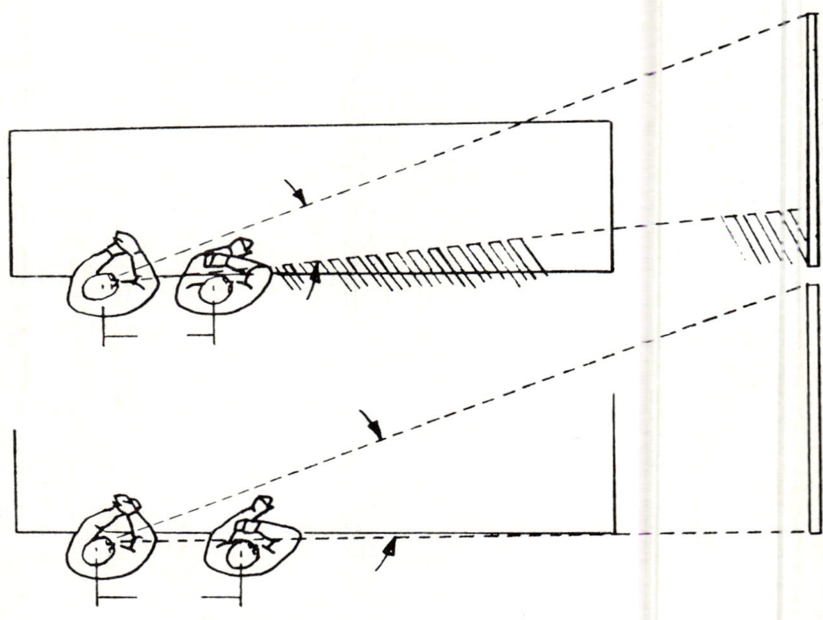

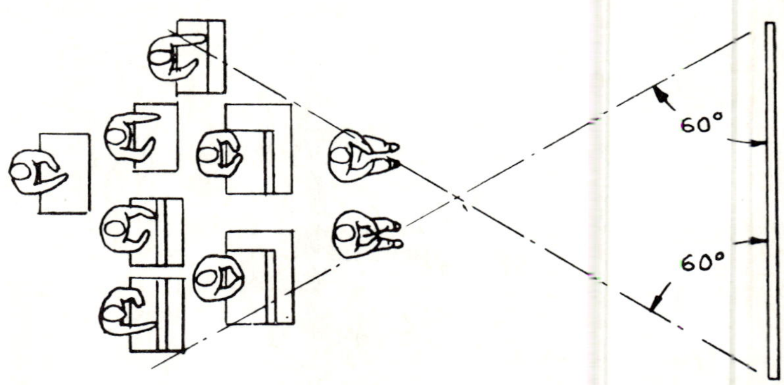

If, because of space constraints, the large display cannot be placed far enough away from a group of observers, consider breaking up the display so that the edges can be rotated for a more suitable angle of viewing.

When the large display is created by a projector, take care to arrange the projector and the observers so that intervening observers do not block the projected image.

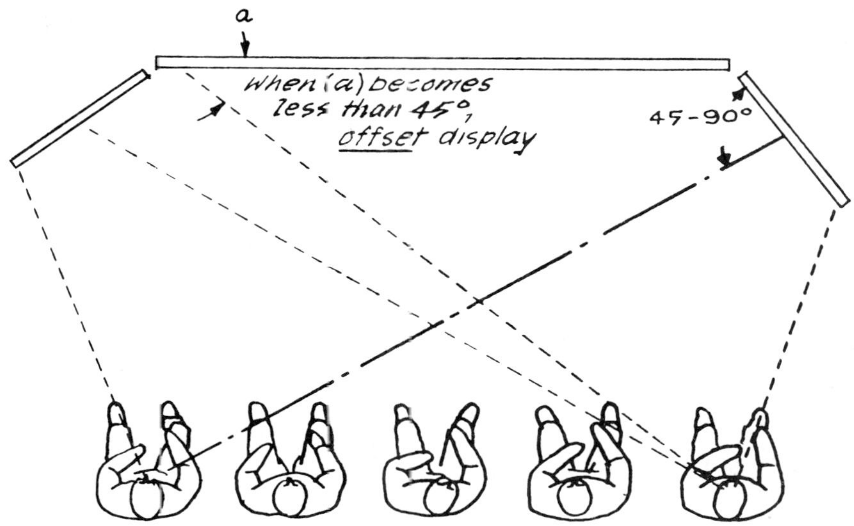

ROW

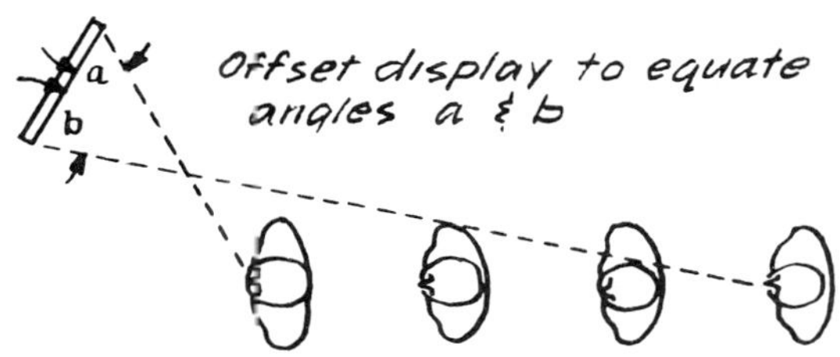

COLUMN

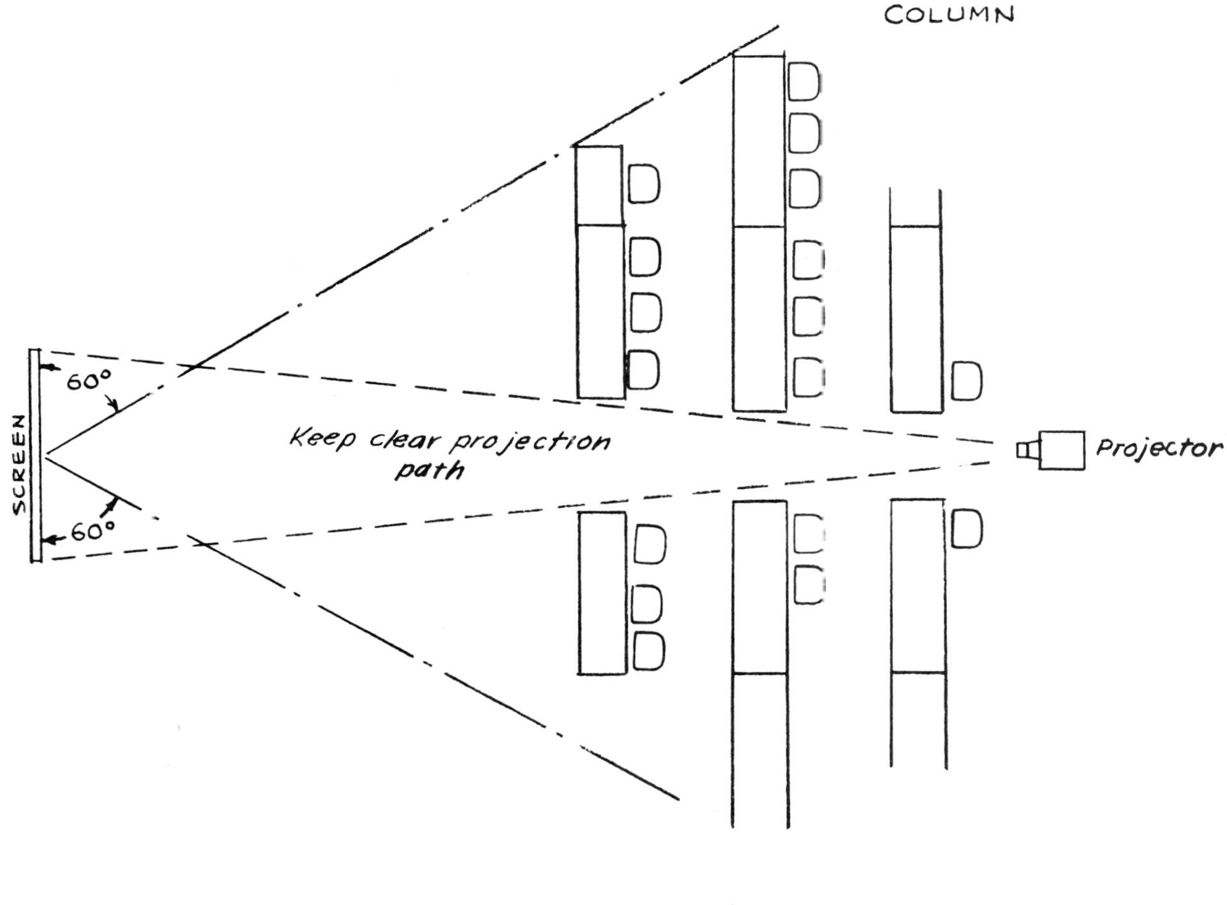

Typical Operator Dimensional Guidelines[3]

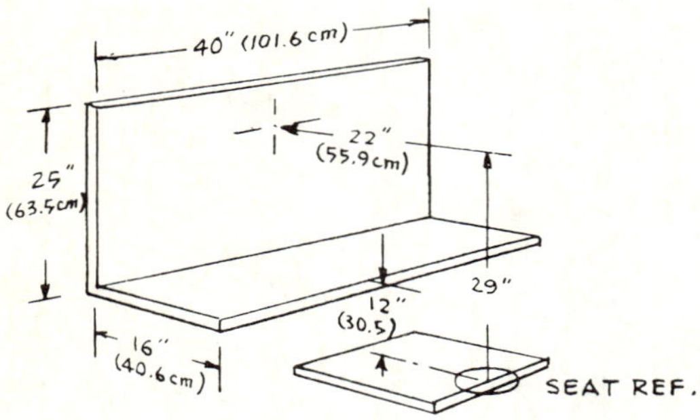

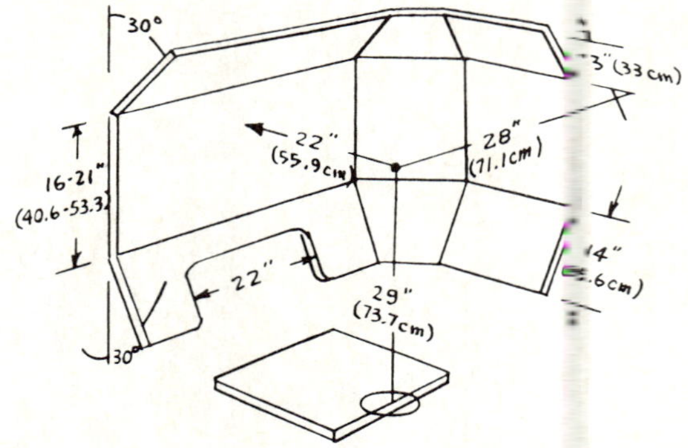

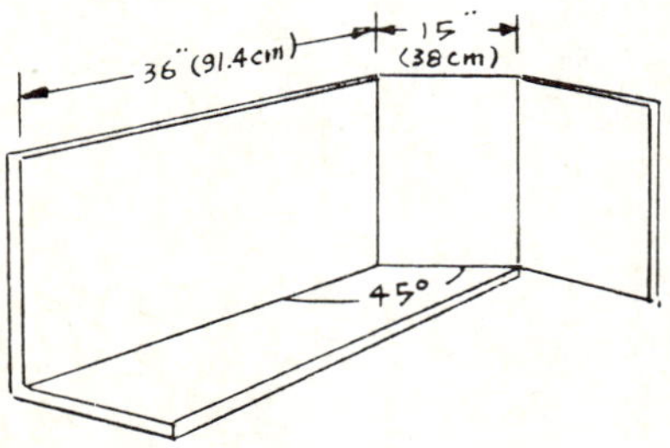

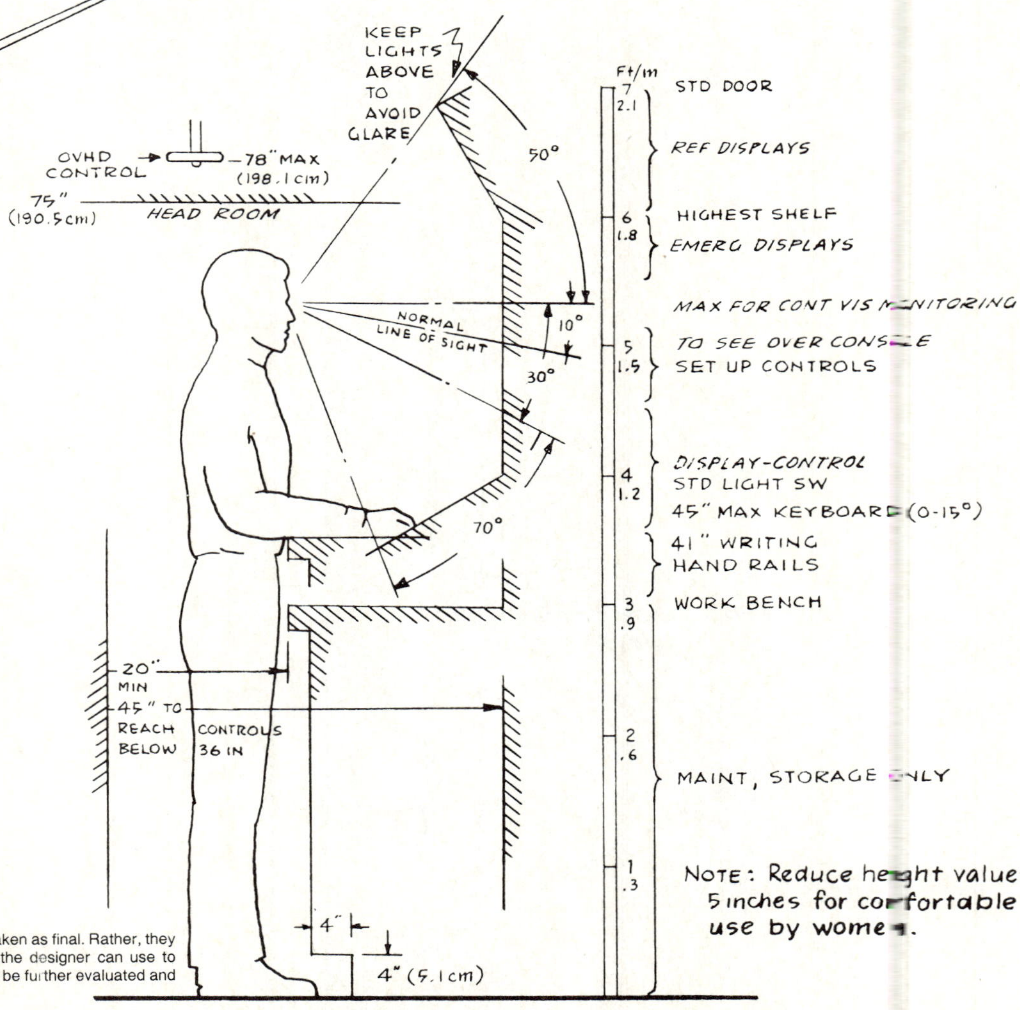

[3]These guidelines should not be taken as final. Rather, they provide "ball-park" estimates that the designer can use to create a mockup, which then should be further evaluated and adjusted if necessary.

GUIDELINES FOR POSITIONING CRT-TYPE DISPLAYS

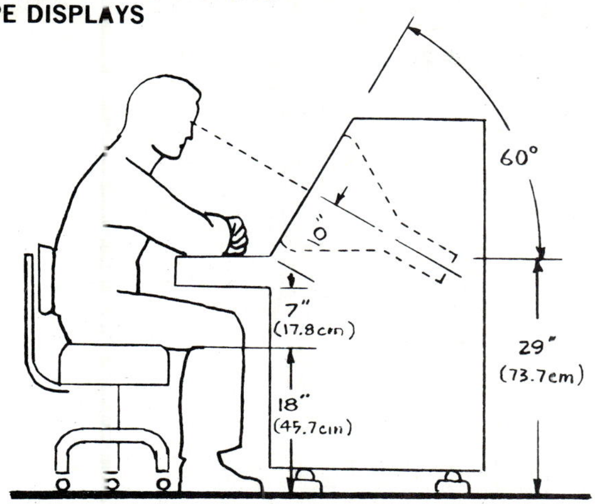

Seated operator.

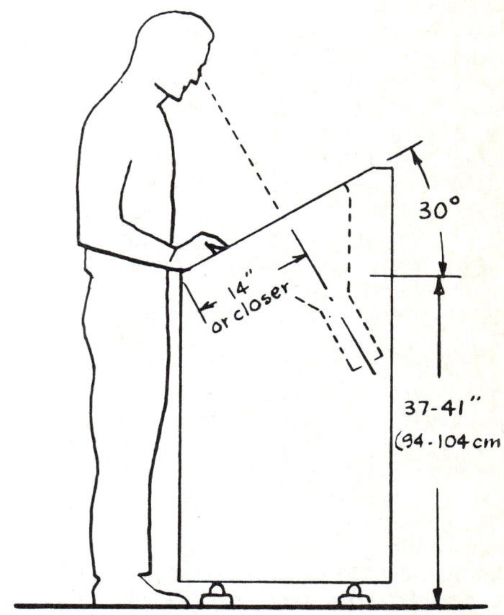

Standing operator.

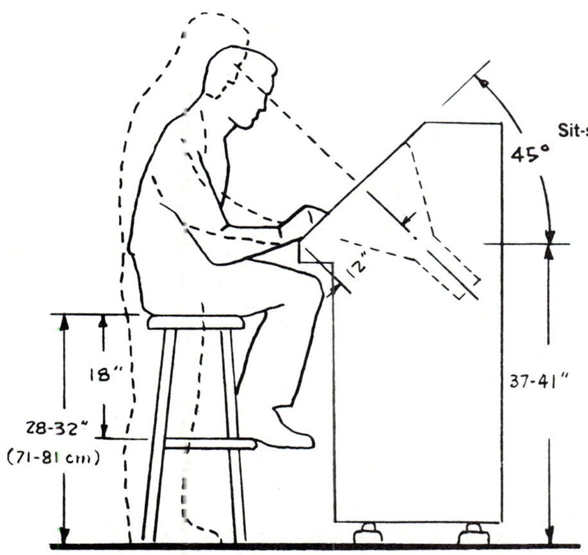

Adjustable height stool suggested

Sit-stand, joint operation.

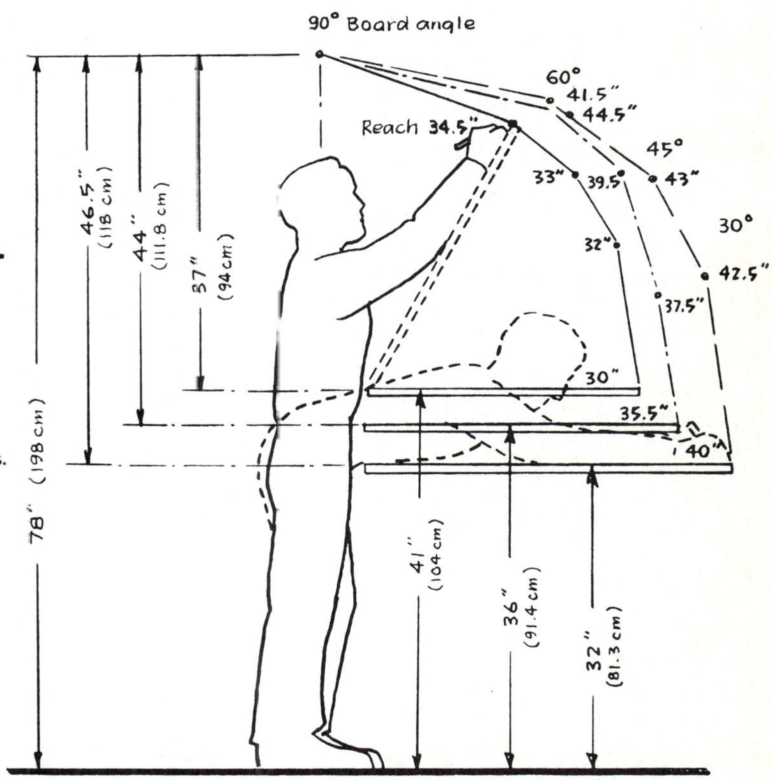

Drawing and plotting reach limits.

NOTE: Reach criteria should be reduced approx 10-15 o/o for women.

Miscellaneous Console Considerations

Avoid the tendency to make all elements of a console operator's operation fixed. That is, it is sometimes desirable to modularize portions of the console so that the modules can be rearranged to be more compatible with the particular activity. One typical example is the separate packaging of a keyboard. The operator may perform several different operations at the console, only one of which requires the keyboard. If the keyboard is fixed in a central position, it is obvious that that prime space cannot be used in any other manner.

When a pencil joy stick or track-ball controller is involved in the console use, be sure that the control position is far enough forward so that the operator has a place on the desk for his or her arm.

CRT viewing angle is important in that when the device is used for long periods, the operator should not have to hold his or her head in an awkward position. Although we often try to set the CRT angle in a so-called compromise position, this may still create problems for either the shortest or the tallest operator. One obviously desirable approach is to make the CRT package adjustable, as shown in the accompanying sketch. Not only does this allow operators to find the best vision angle to reduce neck fatigue, but it also allows them to adjust the face of the display to minimize any annoying reflections of themselves or of objects behind or above them in the room.

A convenient and effective workplace can be created using independent modules, as shown in the accompanying illustration.

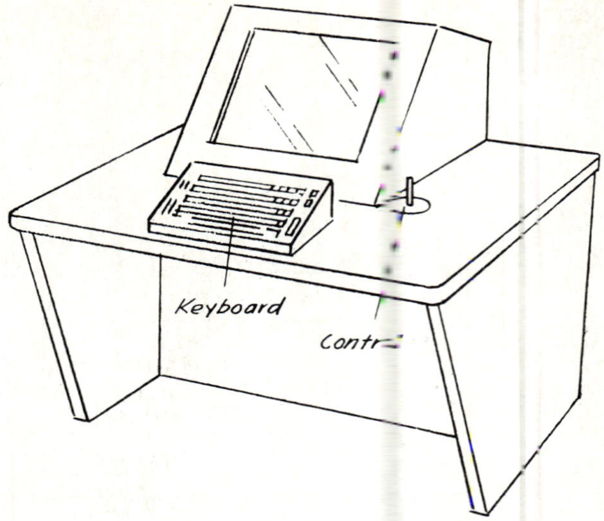

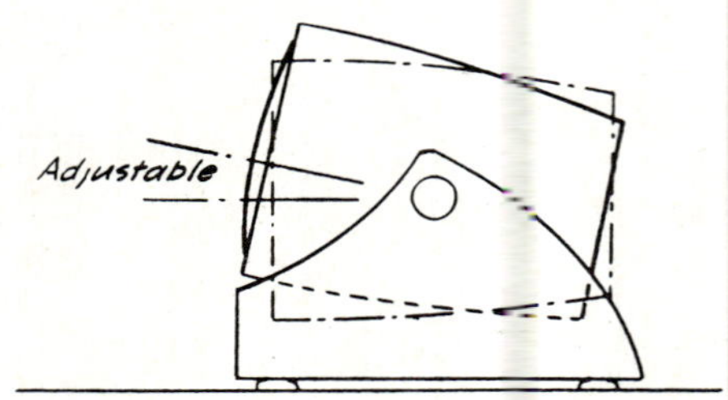

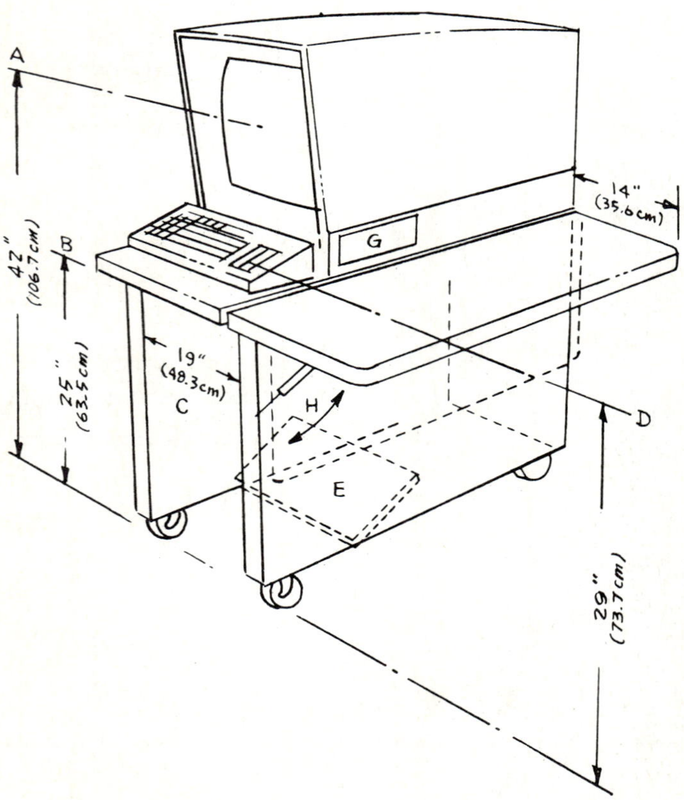

A—The center of a vertically oriented CRT should be approximately as shown. Note that the case extends over the tube to minimize glare from overhead ambient light.
B—Vertical knee clearance minimum.
C—Lateral knee clearance minimum.
D—Center height of keyboard.
E—Footrest.
F—Minimum width for fold-down writing surface.
G—CRT controls (under cover).
H—Supporting bracket for fold-down desk. Casters with locks allow the unit to be moved around to fit the user's needs.

Television Viewing

Screen Size*	Minimum Viewing Distance	Maximum Viewing Distance
17 in (43 cm)	4 ft 11 in (1.47 m)	14 ft 9 in (4.43 m)
19 in (48 cm)	5 ft 1 in (1.53 m)	15 ft 2 in (4.58 m)
21 in (53 cm)	6 ft 4 in (1.89 m)	19 ft (5.7 m)
23 in (58 cm)	6 ft 6 in (1.95 m)	19 ft. 4 in (5.8 m)
25 in (64 cm)	7 ft 6 in (2.25 m)	21 ft. 9 in (6.5 m)

*Diagonal.

For arranging observers in direct TV viewing situations, use the guidelines shown in the accompanying table. For projected television viewing, use the following guidelines:[4]

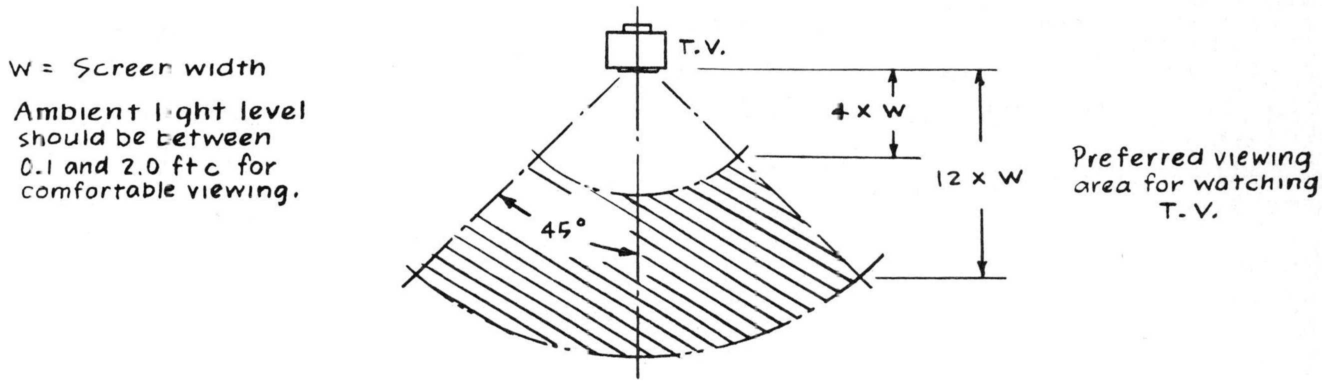

1. Screen brightness 2 to 20 ft (0.6 to 6.1 m) L
2. Brightness ratio 2:1—excellent
   3:1—very good
   10:1—acceptable
3. Contrast ratio 100:1 for pictorial images
   25.1 for printed characters
   5.1 for white characters on black background

[4]A minimum contrast ratio of 30:1 is for the poorest conditions, e.g., the observer who is furthest away, high ambient room lighting levels, and poor materials.

MISCELLANEOUS GUIDELINES FOR OTHER TYPICAL OPERATOR WORKING CONDITIONS

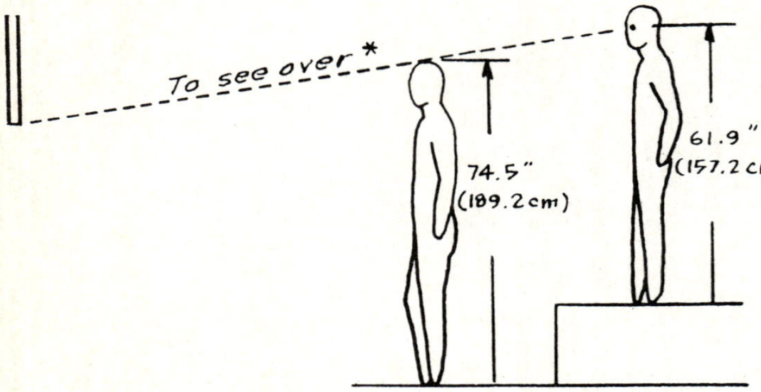

* NOTE: *Values for male observers. For mixed male/female situations, consider tall male in front, short female in rear.*

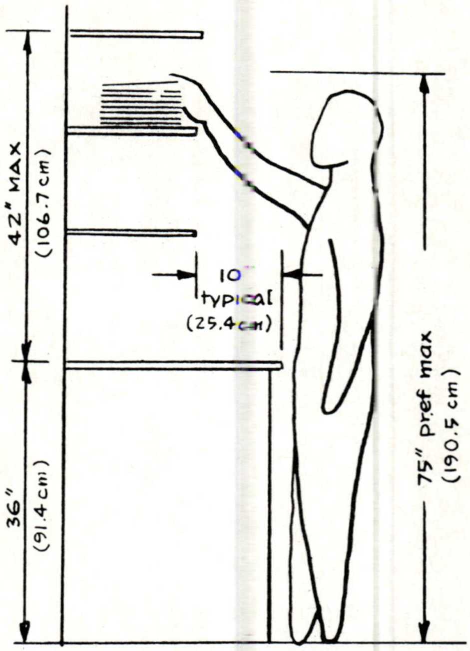

For mixed male/female use

Manual accessibility and clearance.

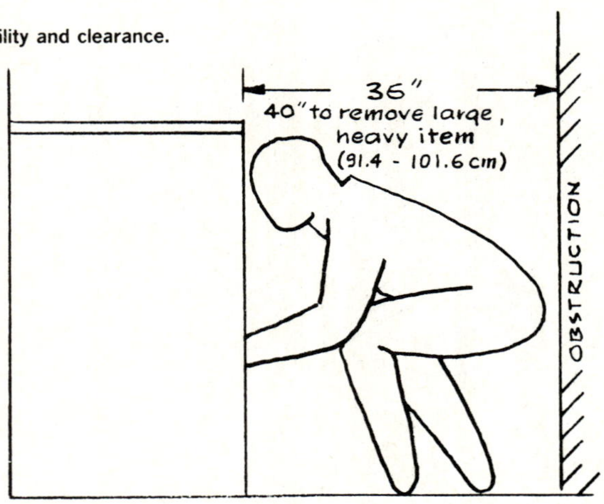

Visual accessibility.

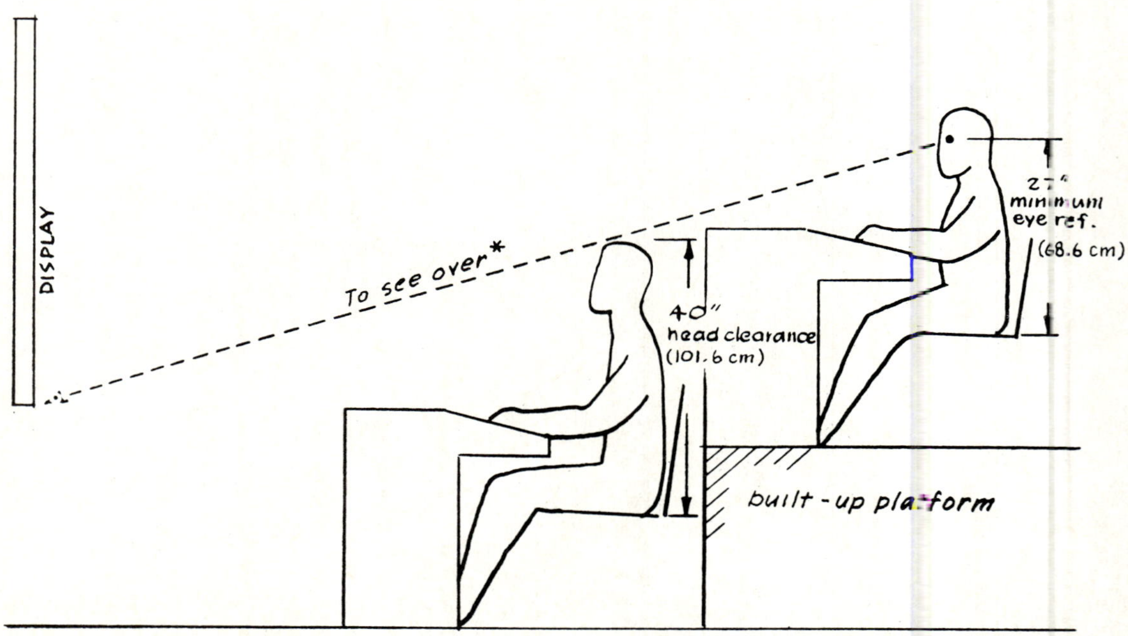

When purchasable workplace furnishings are to be accommodated, their use-dimension characteristics should be examined carefully so that the general workplace geometry that is created (and/or the proposed arrangement in an existing facility) will provide adequate clearance.

The typical examples shown in the accompanying illustrations emphasize the importance of relating drawer-pull distance and operator or worker body clearance requirements.

The worst situation occurs when the worker has to bend over to see and reach into a lower drawer or compartment. Although it is obvious that the worker could step aside and work from one side of the drawer, this position makes it awkward to scan printed or written material, and the worker is forced to assume difficult and tiring body, neck, and head positions.

When several workers may use the workplace, the designer must consider the fact that one worker may be larger or smaller than the others. Although the general dimensions shown in the accompanying illustrations provide a reasonable compromise in most situations, the designer is urged to make a mockup of the proposed workplace and to check it out using several test subjects of different sizes.

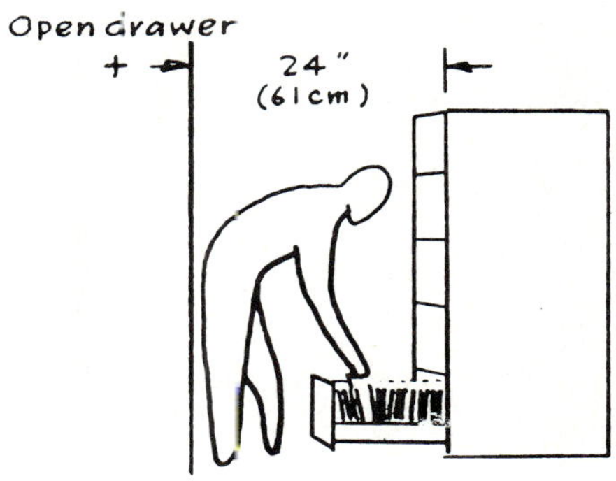

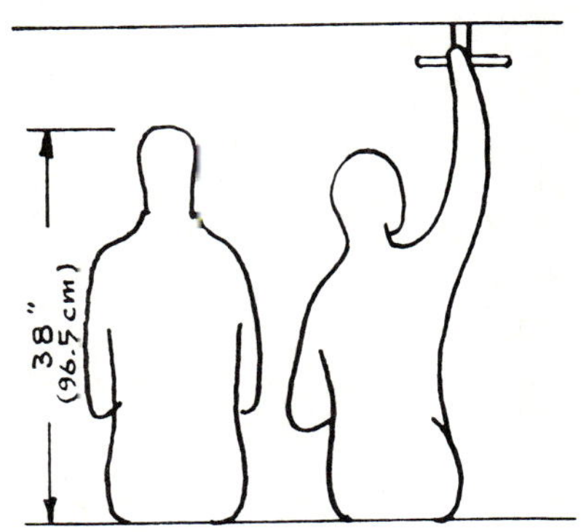

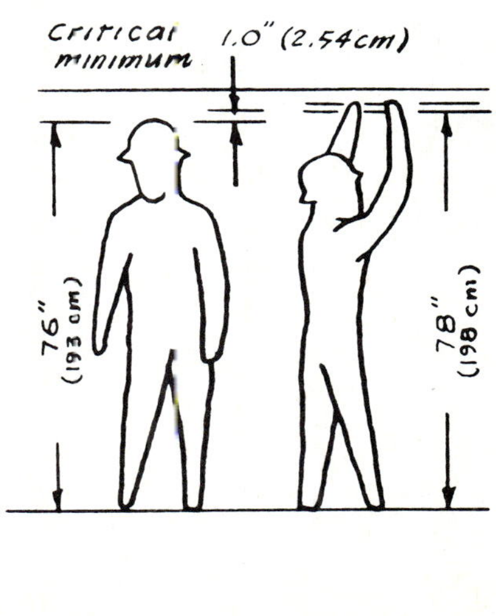

WORKING IN CLOSE QUARTERS As a general rule, avoid creating situations in which workers have to work in close quarters and/or in the awkward positions illustrated in the accompanying sketches.

When this cannot be avoided, consider the general limiting dimensions provided in these illustrations. Critical factors are the following:

1. There must be sufficient vertical clearance for the worker to enter and leave the environment.
2. There must be sufficient horizontal and lateral clearance for the worker to perform the task (considering the nature, size, and configuration of any tool being used).
3. In the supine position, it is critical that the worker's eyes do not have to be too close to the specific work point; otherwise, the worker's eyes cannot adjust for clear viewing.
4. Openings must be large enough to accommodate both the tool package and the operator's hand, finger, or arms, as shown in the accompanying illustrations.

Note: These opening dimensions are for shirt-sleeve conditions. If heavier clothing may be worn, the openings have to be at least 1 in (2.5 cm) larger on all sides (1.5 in, 3.8 cm, is better).

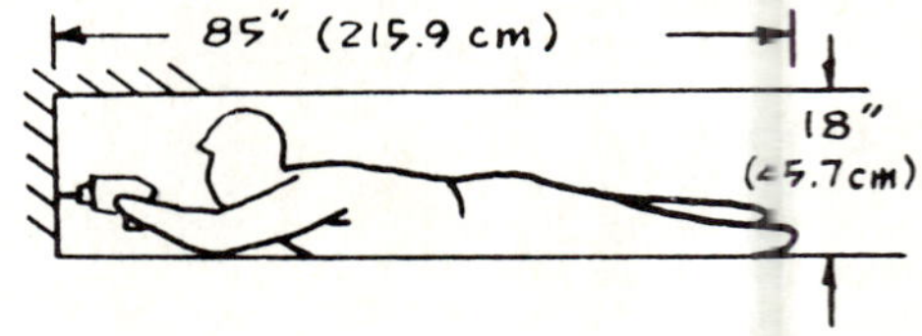

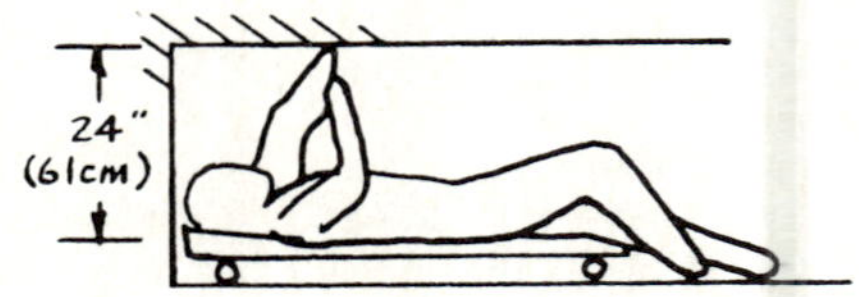

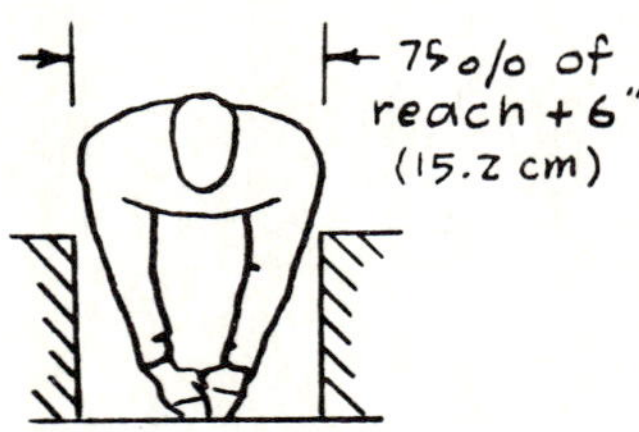

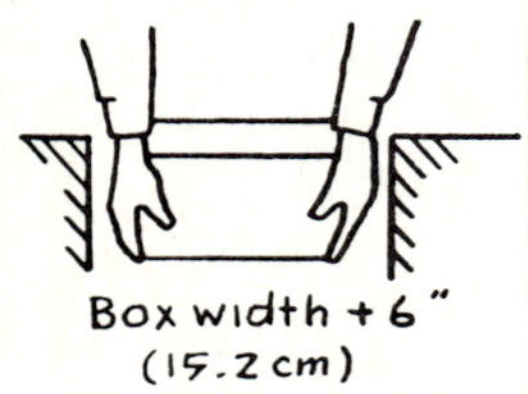

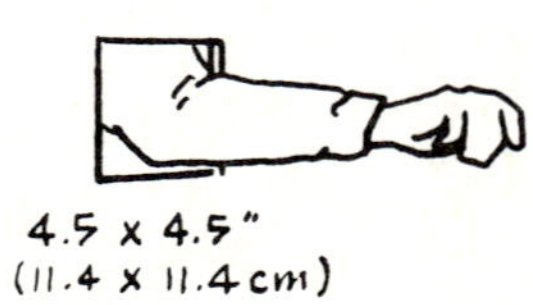

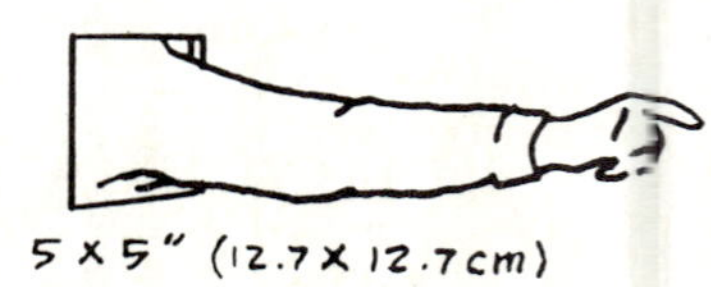

Flat hand

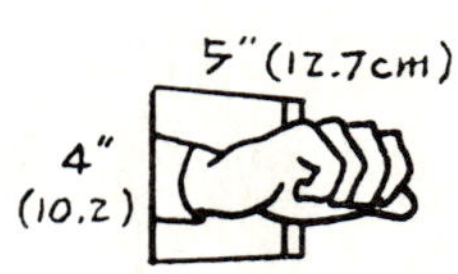

Clinched

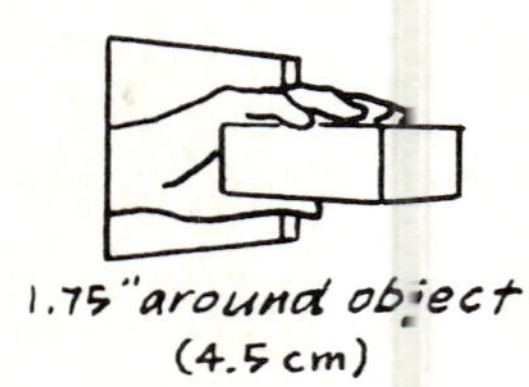

Push button

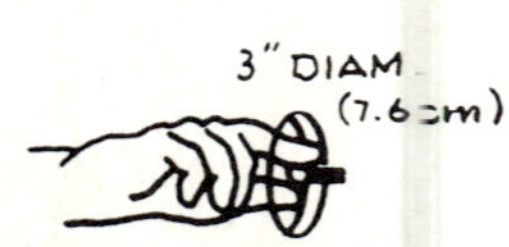

Two-finger twist (or knob diam + 1.5"; add 1.0" for gloves)

MAINTENANCE WORKPLACES

Although equipment maintenance often is not considered as important as the operator's interface, design of the maintenance interface is important to the maintenance technician and therefore should be considered at the same time that operator interfaces are being designed. Several typical examples of maintenance problems, created by lack of proper attention, are illustrated in the accompanying sketches.

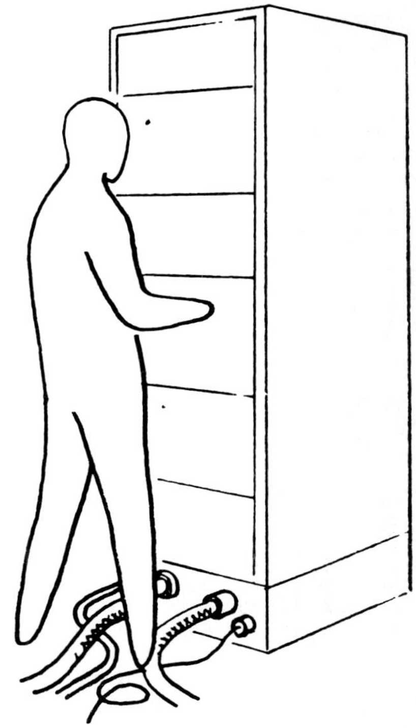

Electrical cables in the way.

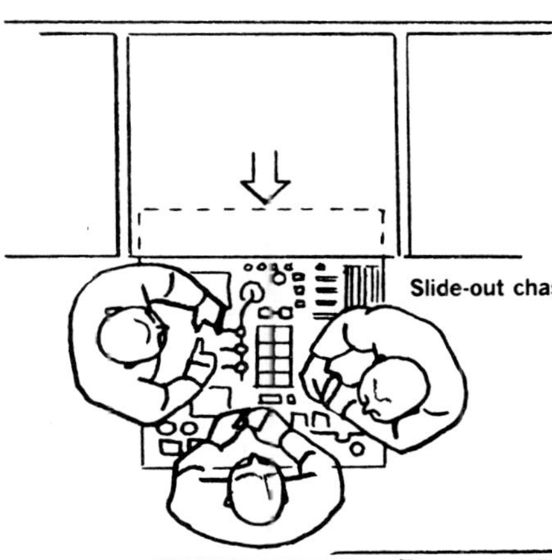

Slide-out chassis provides access on all sides.

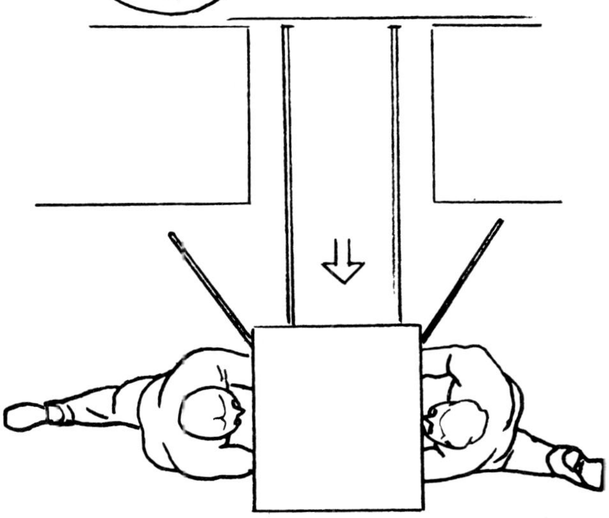

Cabinet tracks allow the cabinet to be removed from among a row of cabinets that block access for maintenance.

Cables should exit the back or top of equipment.

Different equipment packaging concepts either aid or hinder the maintenance technician, as illustrated by the accompanying sketches:

Access is better if the case can be lifted off the chassis. More freedom to work is provided, and there is no cover to prop up.

Avoid designs in which the chassis has to be lifted out of the case or in which a cover has to be removed or propped out of the way.

Design covers so that the minimum number of fasteners have to be removed.

Fasteners should be captive so that they do not become mislaid or lost.

When high voltages are involved, removal of the case or cover should automatically shut the power off. A secondary switch or jumper should be used to reactivate the power when maintenance is being performed.

Correct

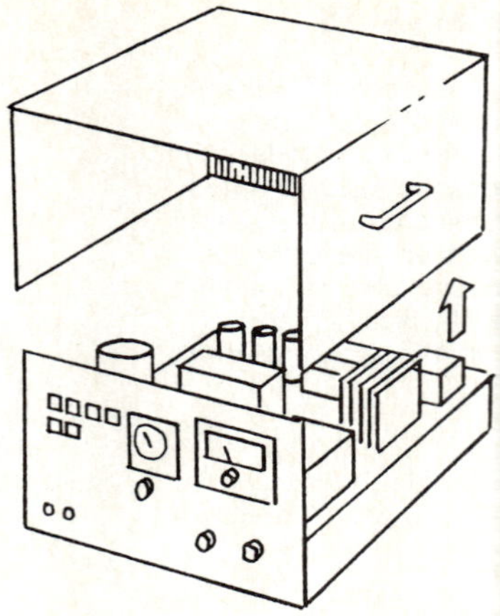

Poor

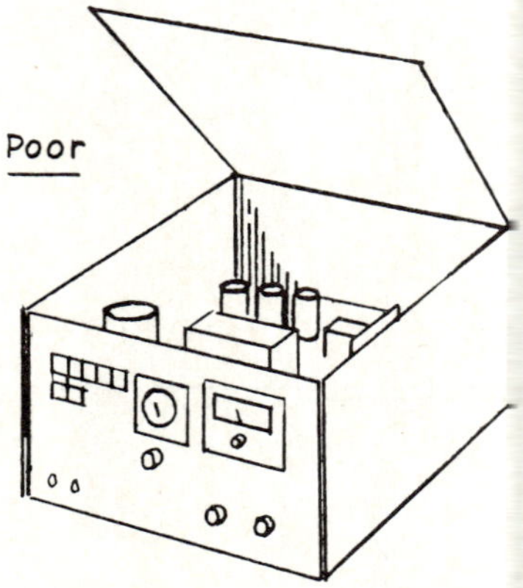

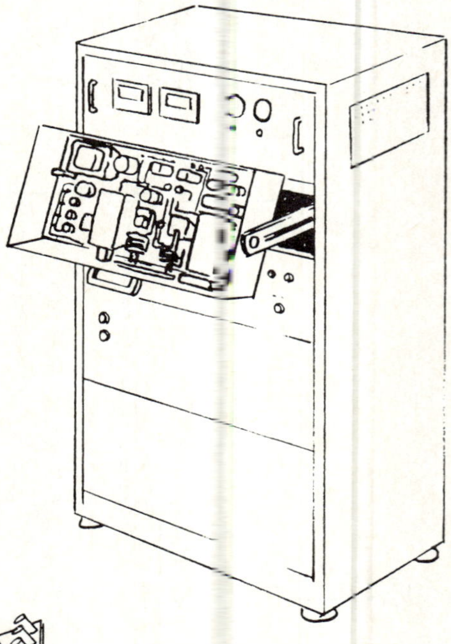

A rotating chassis makes both the top and the bottom accessible.

The fold-out package makes the entire inner assembly accessible.

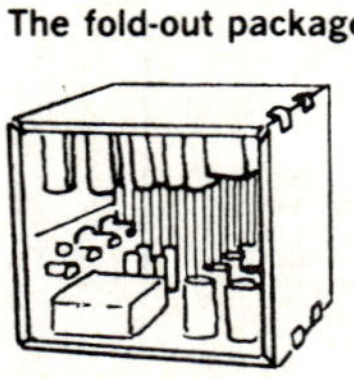

Closed

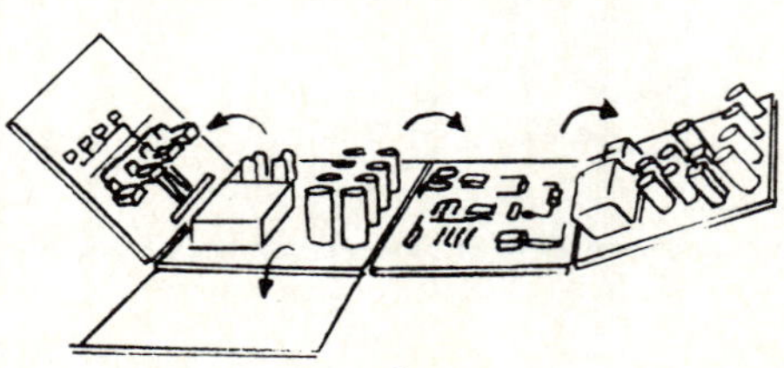

Open

Although this is not always possible, try not to create maintenance working situations in which the technician or mechanic has to work in awkward positions for extended lengths of time. Often this happens because sufficient planning for maintenance was not done before a design was frozen.

The most serious and most common difficulties are illustrated in the accompanying sketches.

Working overhead with arms outstretched is extremely tiring, and there is the added possibility that debris or parts may fall into the technician's eyes.

Working close to the floor or ground is tiring, and it also may make it difficult to get one's eyes in the proper position to see what needs to be done. Even the common gas water heater used in homes illustrates how designers tend to ignore this problem. To light, adjust, or simply inspect the working parts of the heater, one almost has to stand on one's head. In addition, while a person is peering into the darkness to find a pilot light, his or her eyes and face are in the direct line of an explosion, should one occur.

The accompanying sketch illustrates a common problem of most car owners and mechanics. Too little thought was given to the reach distance for engine components that are frequently serviced or replaced. Most serious is the possibility that the mechanic will reach across hot parts, moving belts, etc.

Poor working positions!

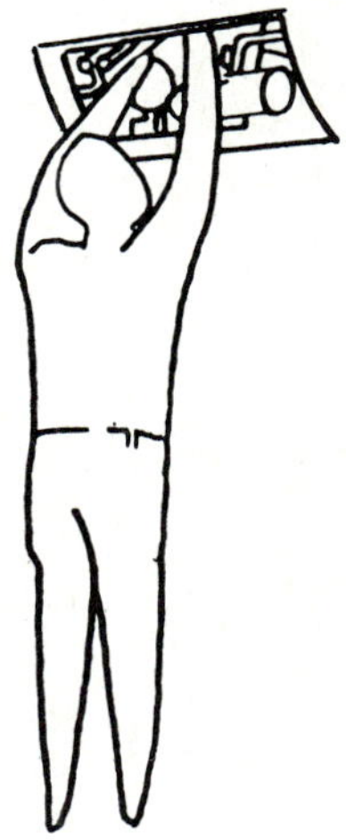

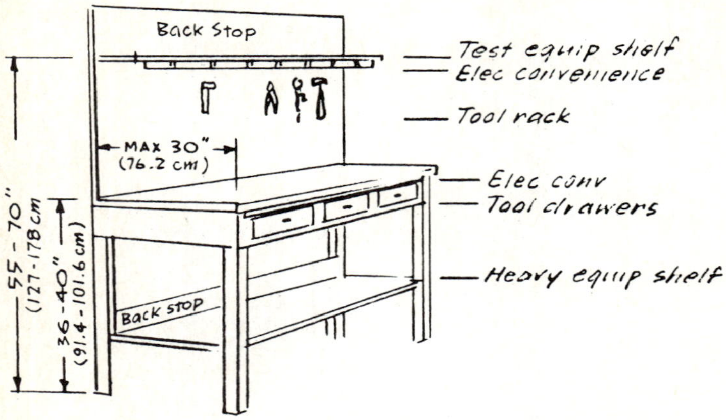

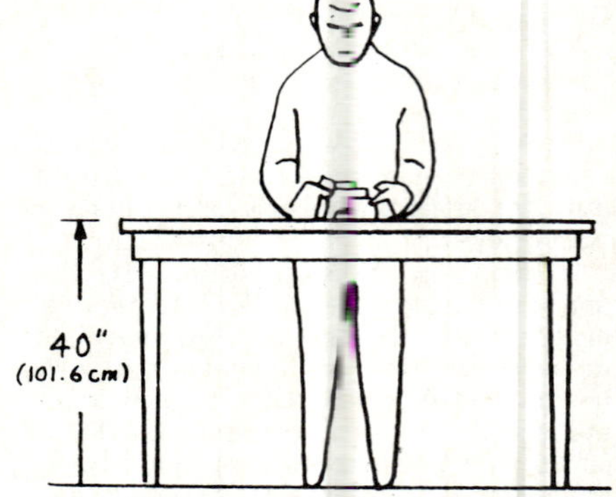

High precision assembly

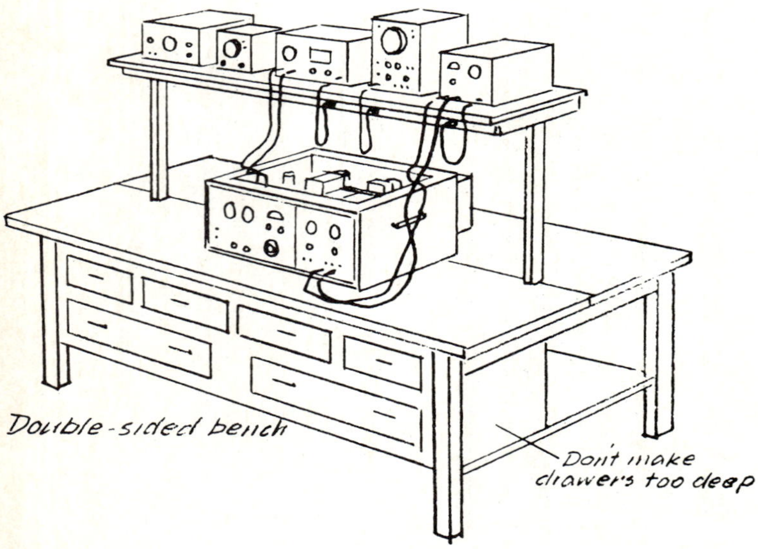

Double-sided bench

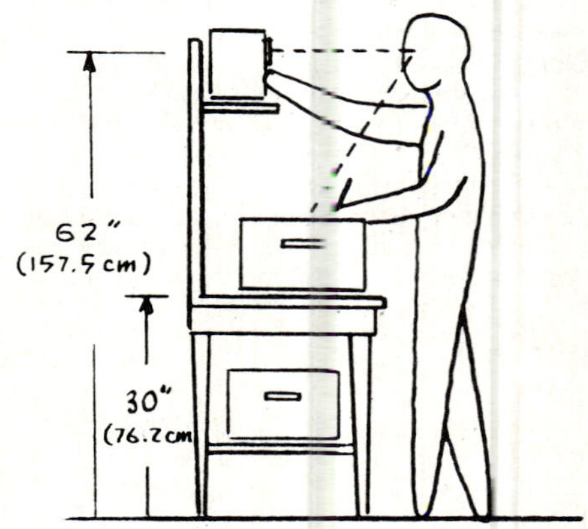

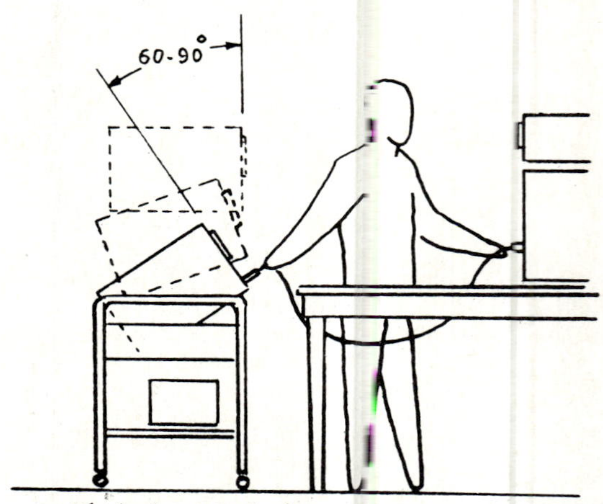

Mobile cart to work site

Complete flexibility

Shop workplaces should be designed for maximum flexibility so that a variety of tasks can be accomplished in the most convenient manner.

OPERATOR CONTROL PANEL LAYOUT

Basic Principles

Four basic principles should be considered in planning the layout of an operator workplace or, more specifically, the equipment control interface (usually a panel containing controls and displays). These are discussed below.

1. Sequence of Operations
The order in which operators will normally use controls and displays is important in terms of their being able to find what they want and being able to go through the sequence with a minimum possibility of missing a step and also in terms of reducing unnecessary motion due to "backtracking." Although some operations may not have a single sequential pattern, there are usually one or two elements that proceed according to a fairly standard sequence of steps.

2. Functional Grouping
Certain functions within almost any operation are related to each other, and/or certain controls are related to certain displays. Related functions should be arranged together so that their association is readily apparent to the operator. When such functions are not grouped, operators tend to get lost and to wander from one point to another trying to find what they want. Although there may be a few elements within the workplace that cannot be placed together for obvious reasons (i.e., a foot pedal obviously cannot be placed where an associated visual display would be), an attempt should be made to avoid scattering controls and displays anywhere on the panel without regard to functional associations.

3. Frequency of Use
Some controls or displays will probably be used more frequently than others. These should be located so that they are maximally convenient; i.e., visual displays should be near the normal sight line, and controls should be near the hand when the arm is in a normal position. For example, a frequently used visual display should not be located where the operator must continually turn his or her head or body in order to see it, and a control should not be placed where the operator continually has to reach a long distance for it.

4. Emergency Use
Once in awhile there will be a control that is critical under emergency situations; i.e., it may be vital that the operator get to it and operate it successfully in the shortest possible time. Obviously, if such a control is hidden, out of reach, or otherwise difficult to get to and operate, the operator may not be able to reach it in time.

Note: Each of the above principles must be considered in light of a particular operator interface problem; there is no set priority that can be all-prevailing.

Positioning Panel Elements Relative to Line of Sight

Visual displays, including indicator lights, instruments, panel and control labels, etc., are useless to the operator if he or she cannot see them. This requires that no control (joy stick, steering wheel, etc.) be placed between the observer's eye and other displays. In addition, it requires that care be taken in the placement of control and display labels so that when critical, the operator's hand will not cover up an important label that may be used in adjusting the control.

As a general principle, labels should be located in a consistent manner with respect to being above or below a control or display; i.e., avoid having some labels above a control, while others are below a control—at least on the same panel.

However, the choice of label position should also reflect consideration of the expected operator's eye reference; i.e., if the control is above the eye reference position, the label is more likely to be visible than if it were placed above the control.

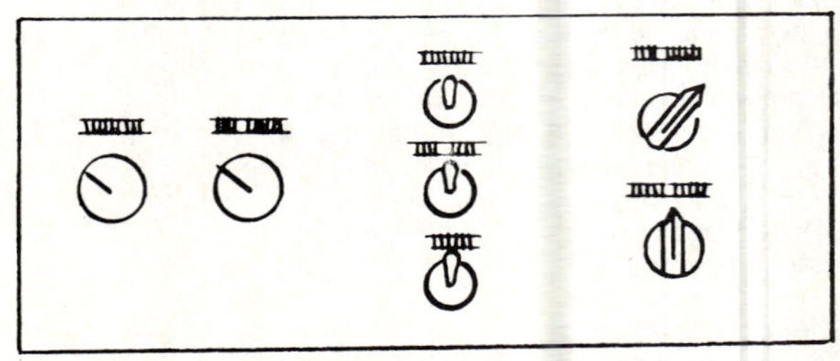

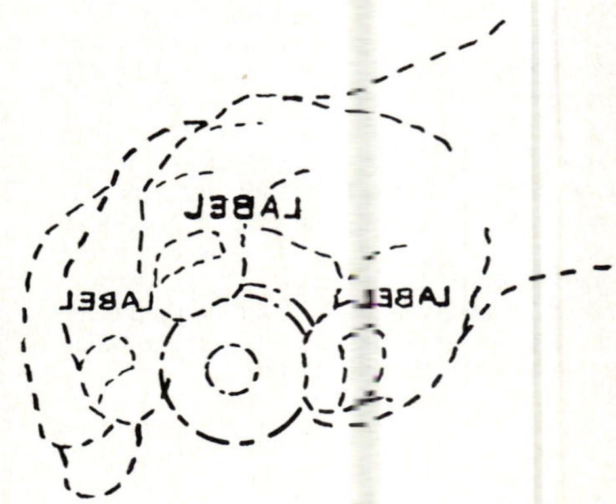

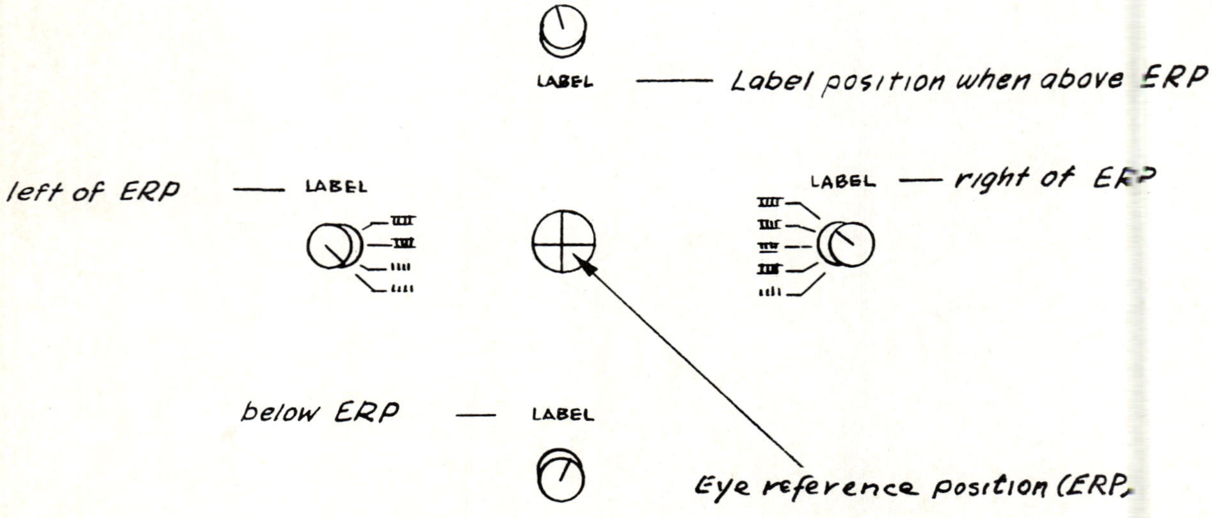

Display Arrangement

When the circumstance includes an array of displays that must be monitored on a regular basis to determine whether conditions are approximately "normal," the check-reading task is made easier for the operator if the displays are arranged so that their indications (e.g., pointer positions) generally reside in the same relative place on the instrument (see the accompanying sketch). Arrange the normal pointer reading at the twelve o'clock position when the array is vertical and at the nine o'clock or three o'clock position when the array is horizontal.

If displays are ordinarily read in sequence, arrange them in order, either horizontally or vertically. The normal reading pattern is from left to right or from top to bottom.

When a display is adjustable by means of an associated control, the control should be located as close to the display as practical. However, since there may be a directional relationship, as shown in the accompanying sketch, care must be taken to position the control relative to the display so that there is no doubt in the operator's mind as to the direction in which the control should be moved in order to cause the displayed element to move in the right direction. Controls which are placed below their respective displays are less confusing than those which are placed to the right or left of the displays.

When it is impractical to place controls close to the displays they affect, try to arrange both the displays and the controls in similar patterns (see the accompanying sketch) so that there is a similarity between the two.

When a control causes something to move in space outside the console (e.g., an antenna), control-element directions of motion should also be similar, i.e., clockwise, counterclockwise, left to right, up and down, etc.

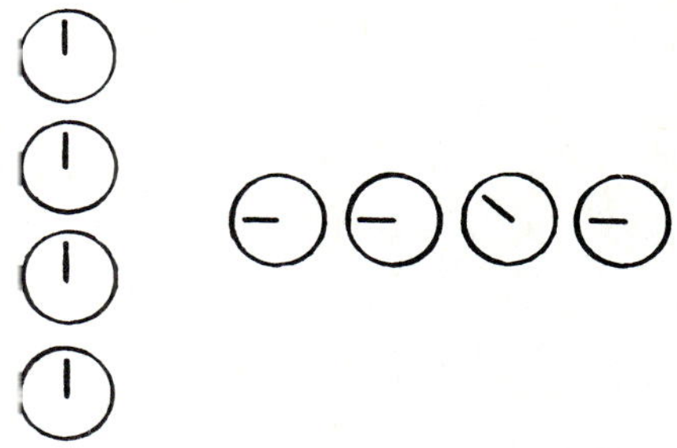

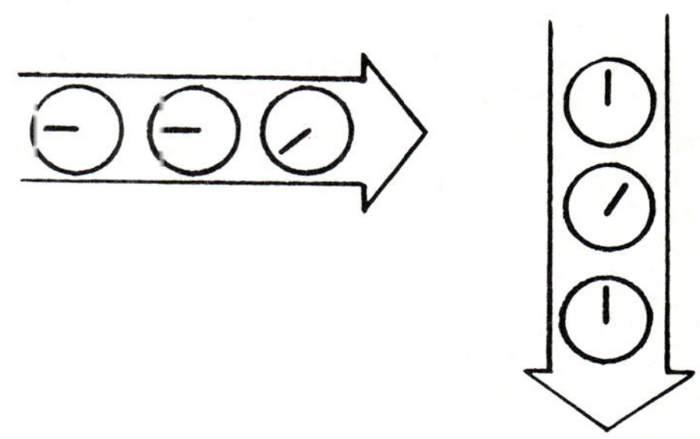

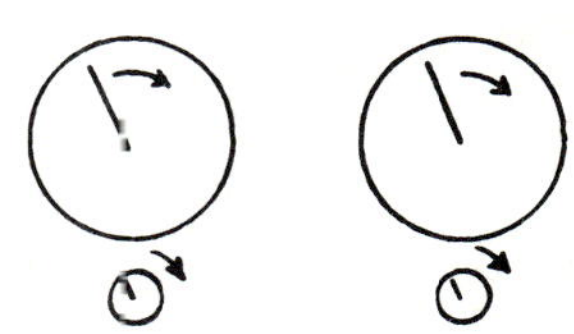

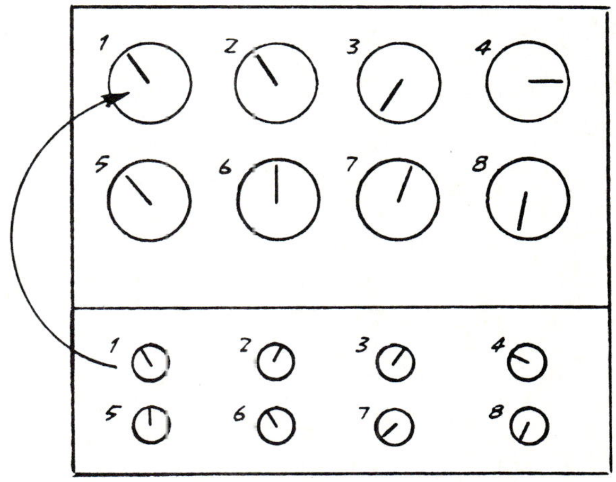

An important objective in panel layout should be to make the functional grouping of controls and displays readily apparent to the operator. There are various effective methods for doing this, some of which are illustrated in the accompanying sketches:

1. Spatial separation is effective when there is sufficient panel area. This is the preferred method.
2. When the panel is necessarily space-limited, one can use lines or borders around the various functions. This is particularly effective if the functional grouping is irregular.
3. If one functional group is especially important for the operator, it can be made more conspicuous by using a different color for the entire background for the group from that which is used for the rest of the panel.
4. Subpanels—either the protruding or the inset type, as shown by the accompanying sketches—are effective for setting off the various functional groupings.
5. In some cases, a separate package can be attached to the main package, especially if the package is small but needs a special physical orientation to make it easier to use.

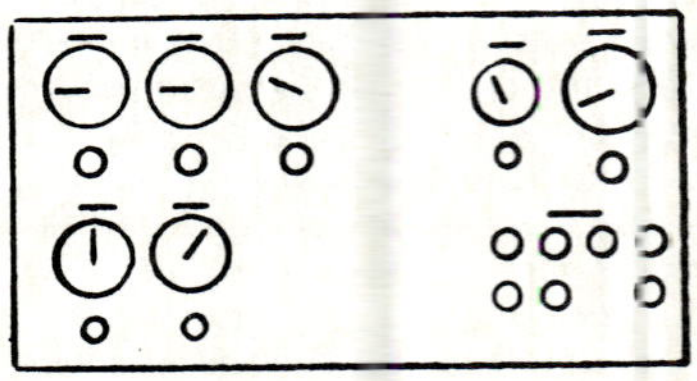

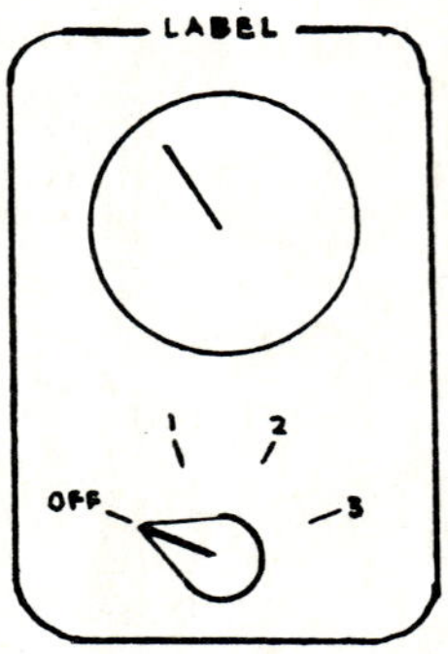

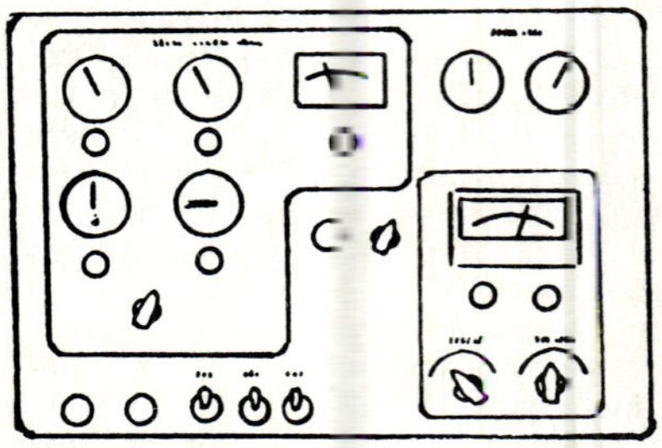

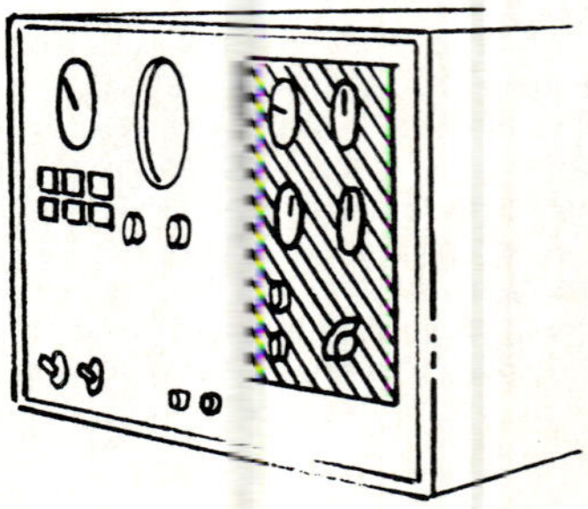

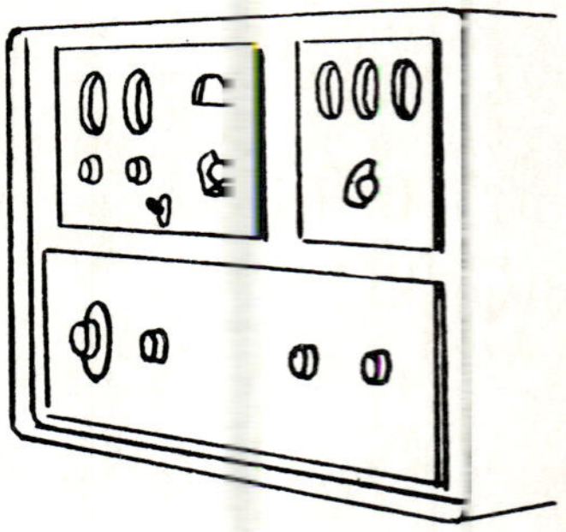

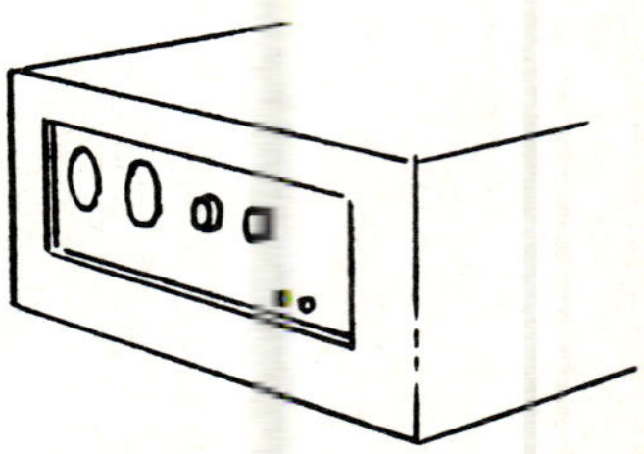

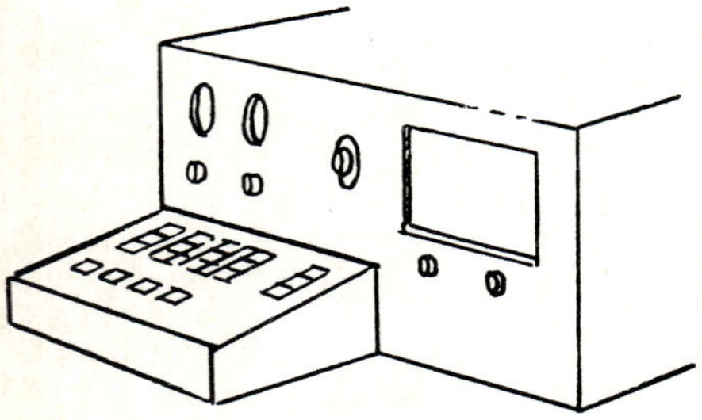

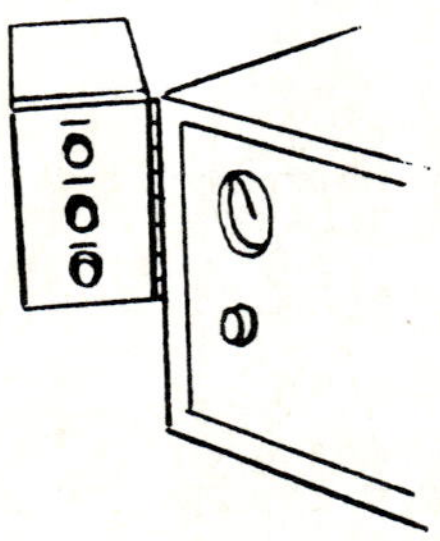

STORAGE SPACE AS A SPECIAL WORKPLACE

The location and design of a storage space should receive the same systematic appraisal and attention as other work stations, especially when the storage space is utilized frequently. Above all, avoid the temptation to treat the problem of storage space as something to be taken care of after all the more important operator work-station problems are solved. Design of the storage subsystem should consist of more than "making do" with whatever leftover space one can find in the facility or vehicle.

1. General Principles

Provide storage space where it is needed, not at some distant, out-of-the-way, or inaccessible location.

Design the space to "fit" the items to be stored; do not design just a generalized volume of space or shelf area.

Provide enough space not only for the predicted use but also for unexpected or future expansion.

2. Content Considerations

Size, shape, and weight

Special storing conditions, e.g., package position and protective and/or safety precautions to prevent deterioration and damage

Label exposure, i.e., the necessity to have identification labels immediately accessible as the storage container or storeroom door is opened and the contents are viewed

3. Special Feature Considerations

Content separation, securing, and preserving

Content adjustment as needs change

Ease of ingress and of storage and retrieval

Ease of cleaning the storage area (dirt and spills)

Content security, i.e., inaccessibility to unauthorized persons

Illumination so that the user can move about safely and/or can see to identify and manipulate stored items, make space adjustments, or clean

Environmental control, i.e., control of heating, cooling, ventilation, humidity, shock and vibration, etc.

4. Key Element Design Considerations

Location

Shelving and partitioning

Doors and covers

Handles, latches, and locks

Content securing

5. Storage Type and Location in Relation to Order and Frequency of Use

In many situations storage cabinets are built with little thought given to what type of storage is needed at what locations, e.g., a kitchen in the typical home. Although this is not always the case, kitchen cabinets may be created merely to fill in the spaces between appliances. The cabinetmaker may be told to provide a variety of drawers and cabinets to fill specified dimensions. As a result, the user often finds that things cannot be stored next to the area in which they are needed because the storage space provided does not accommodate them. Examine the use sequence and/or the frequency for each given storage requirement and make sure that the storage plan has some direct relationship to the storage and retrieval needs of the eventual user.

Storage Shelf Depth in Relation to Shelf Height and Package Weight

Avoid the temptation to provide general shelf storage space without regard to human limitations in terms of reaching and lifting. Obviously, people cannot reach very far into a shelf that is very high or very low. In addition, people cannot support or manipulate as heavy a package when a shelf is either very high or very low as they can when the shelf is somewhere between waist and shoulder height.

The accompanying sketch provides some general guidelines with respect to package weight and shelf height. Obviously, these are not absolute values, since the size, shape, and other geometric features of specific packages tend to modify the capabilities of the individual who is storing or retrieving them. Note that, because of a balancing problem, a worker in the kneeling position cannot support a very heavy or large package.

The accompanying sketches illustrate the difficulties created by extremely deep shelving. The following are general guidelines for specifying shelf depth:

1. Below waist height, the maximum shelf depth should not exceed about 18 in (46 cm).
2. Above shoulder height, the maximum shelf depth should not exceed about 12 in (31 cm).
3. At waist to shoulder height, the maximum shelf depth should be about 24 in (61 cm).

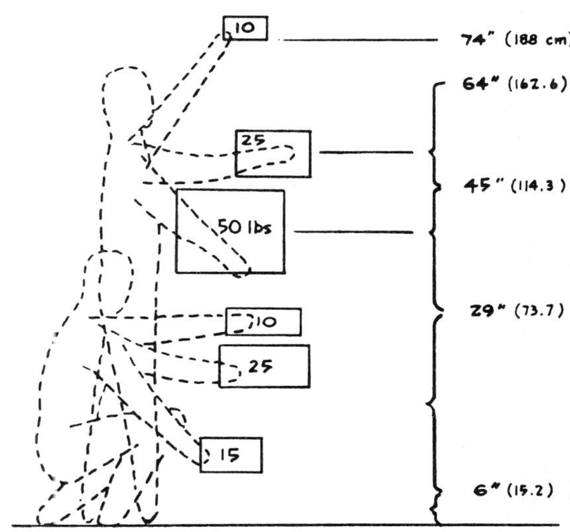

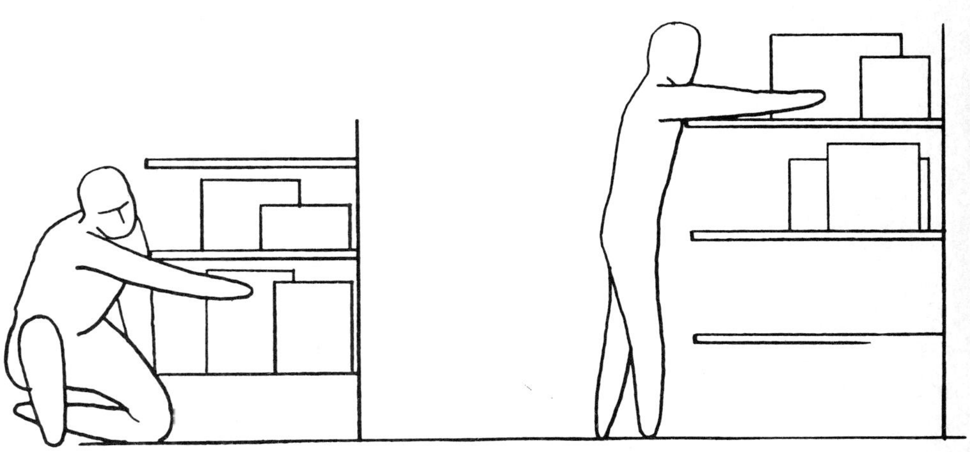

Shelving and Partitioning

In planning and/or selecting modular shelving, it is desirable to consider some of the added conveniences that could be provided for the user, as shown in the accompanying illustrations.

Special storage units designed for separation of such materials as stationery can be extremely useful. However, they should be designed so that materials cannot be shoved back so far that they are difficult to retrieve.

Using cutouts is one obvious way to make material retrieval easier when the spaces between shelves are too small for a person to get his or her entire hand into the opening.

Vertical storage spaces like the one shown in the accompanying sketch are often handy beside the worker or operator. However, they also collect dirt and other debris. Clean-out holes should be provided so that the storage bin can be cleaned out once in awhile.

Movable shelves are generally more useful because they can be adjusted to meet changing storage needs (see the accompanying illustrations). Movable partitions are equally beneficial.

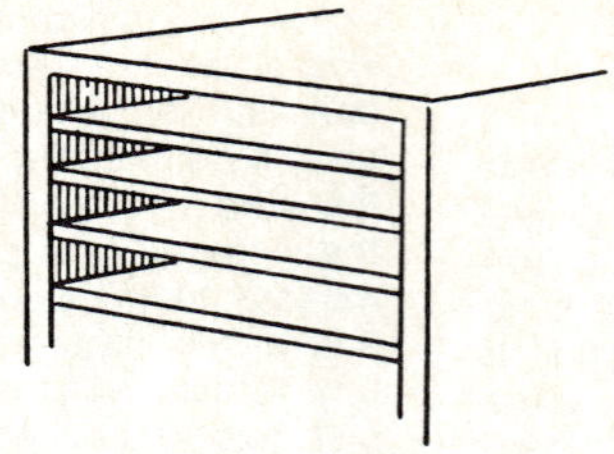

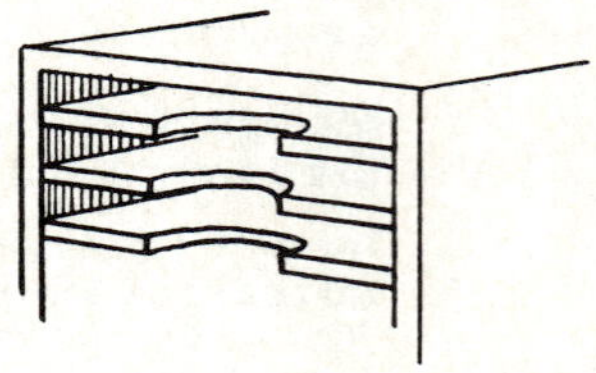

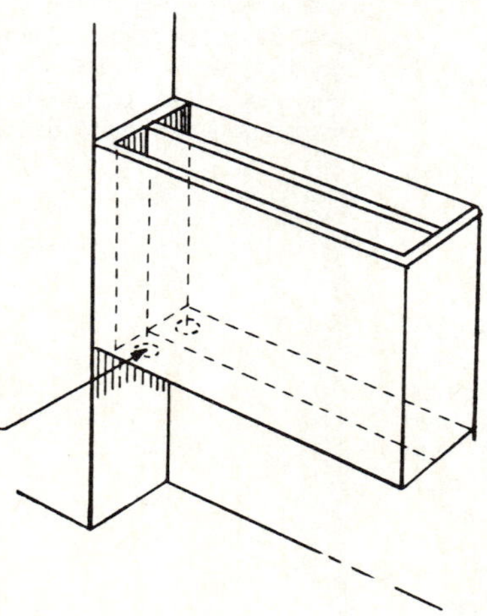

Adustable Shelves

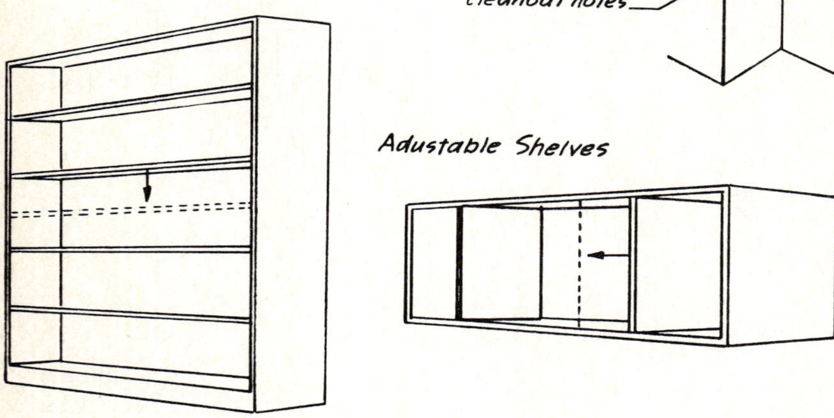

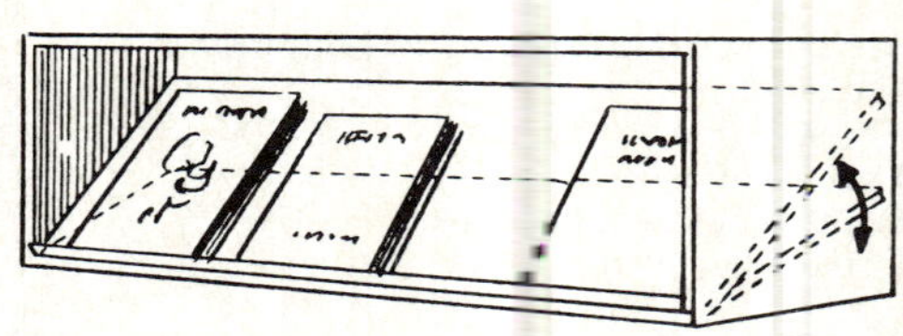

Storage shelves for articles that may be viewed for selection should be adjustable to the extent that the article position can be made compatible with the viewing angle of the observer. In the case of magazine racks, the accompanying sketch illustrates the point. Such shelves usually require a ledge at the bottom so that the magazines will not slide off when the shelves are tilted.

Similar tilting of parts bins makes it easier for workers to look into the bin to see which part they want.

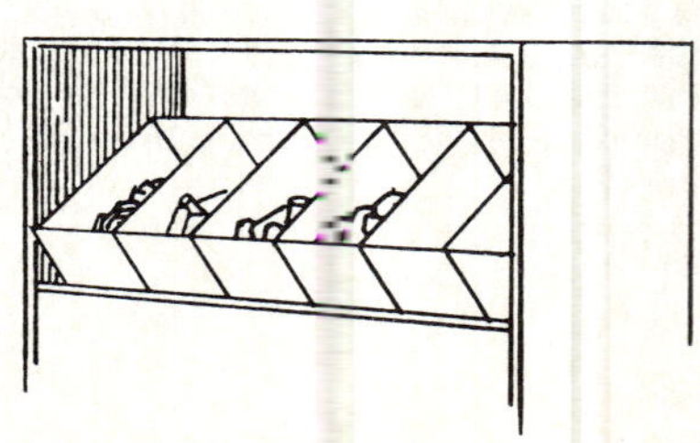

A desirable design objective should be to "bring the stored item to the user." That is, instead of designing fixed shelves, bins, etc., consider the possibility of making these elements slide out, rotate, or move into more convenient positions for the user to store or retrieve articles. The accompanying sketches illustrate some of the schemes that various designers have used successfully.

In executing such designs as shown here, it is extremely important that the mobility devices operate smoothly and easily and that there be proper means for securing shelves or drawers so that they stay where the user wants them. Equally important is ensuring that a drawer, for instance, cannot inadvertently be pulled all the way out, causing its contents to be spilled on the floor.

Always provide handles that are visible and easy to grasp on all such mobility systems.

Heavy-duty, smoothly functioning slide or motion systems should be provided, especially if the shelf or drawer supports heavy articles.

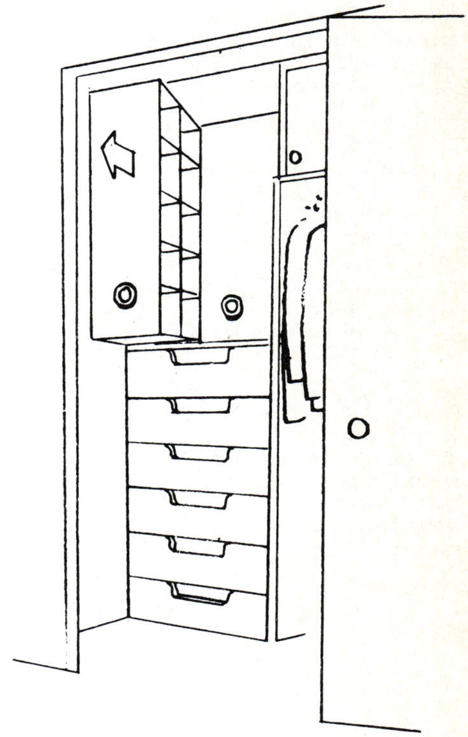

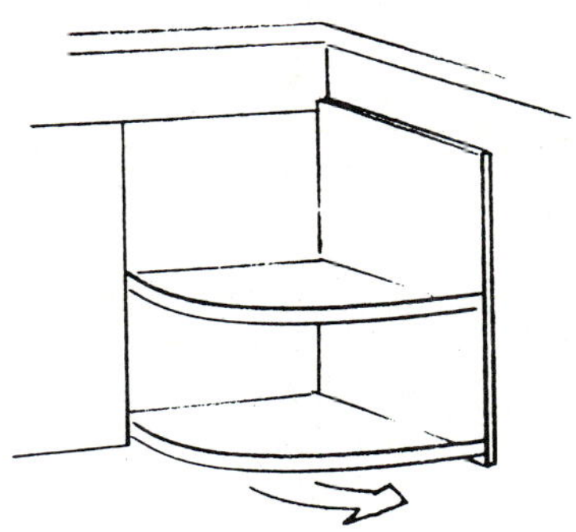

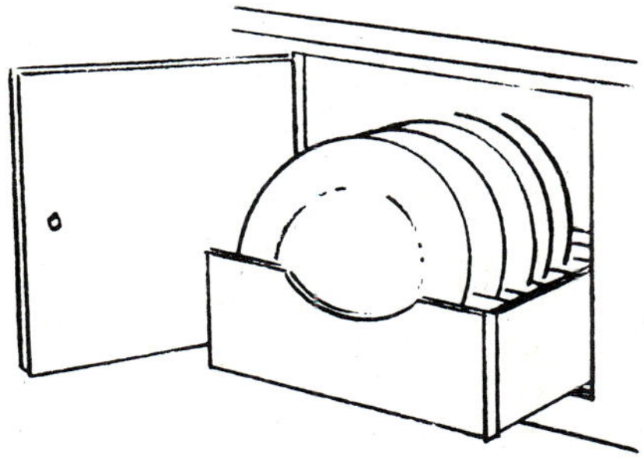

Storage Cabinet Drawer and Door Design

Although there are some aesthetic as well as constructional advantages to certain types of cabinets, these sometimes create problems for the user. The accompanying sketches provide some guidelines for designers, whether they are selecting ready-made units or designing new cabinets:

1. Lap-front drawers or doors are not preferred because they make it easier to catch and pinch one's fingers and because any liquids that are spilled down the front of the cabinet tend to enter the drawer or cabinet.
2. Handles should be on the front of drawers and doors so that the user knows where they are.
3. Corner adjacent drawers should be avoided, and similar door arrangements should be planned so that when two adjacent doors swing open, their handles do not coincide.

Magnet or spring-return types of door closures are preferred over latching types (except for those cases where dynamic forces might cause the door to fly open, as in vehicular applications).

LAP FRONT

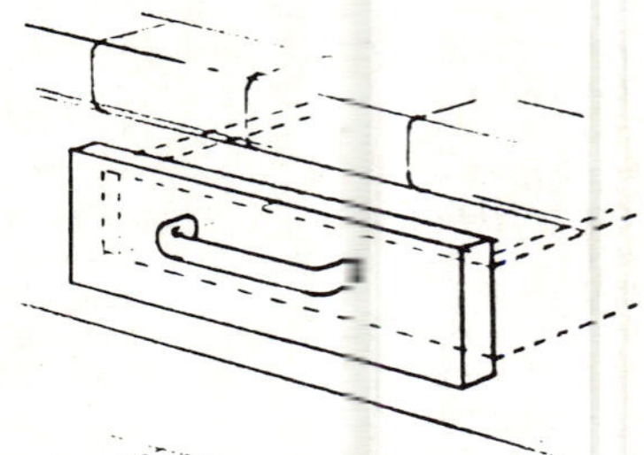

FLUSH PREFERRED

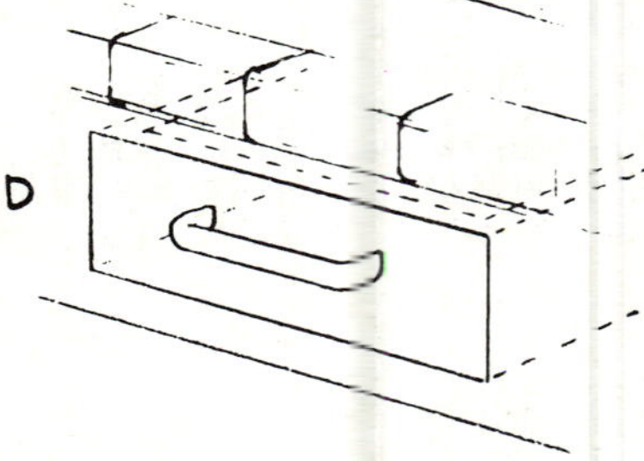

LAP FRONT

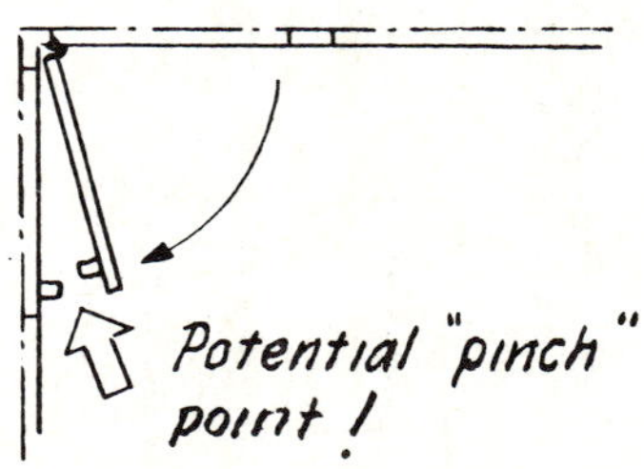

FLUSH PREFERRED

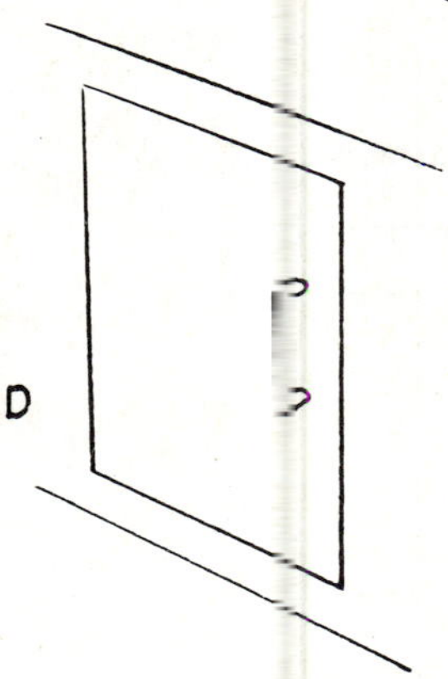

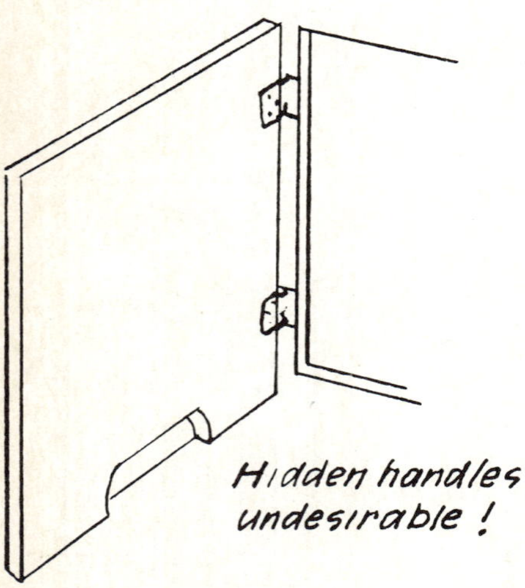

LIGHTING SYSTEM DESIGN

1. Purpose

Lighting is typically provided for two main purposes: (*a*) to enable people to see what they need to see in order to perform various tasks and (*b*) to create decorative effects. It is important to differentiate between these purposes because the objectives of each are different. In some cases decorative lighting schemes may interfere with task illumination unless there is a constant awareness that the observer, to see well, must have enough of the proper type of light, placed where it does the most good, and that there should be no extraneous light or reflections that might interfere with the observer's seeing process.

Seeing tasks are not always the same; thus the amount and type of lighting and the method for providing illumination should be tailored to fit the particular seeing task. A few examples will serve to illustrate how different tasks and lighting system requirements can be:

a. Reading a book or newspaper
b. Reading instruments or cathode ray tube displays in a darkened room
c. Performing surgery or sewing on dark material
d. Inspecting for flaws in material or metal surfaces
e. Matching colors for material selection
f. Developing film in a darkroom

2. Types of Lighting

a. Natural light: The proper manipulation and use of natural light is very desirable for many applications, especially because it tends to make things we look at seem more "natural."
b. Artificial light: Artificial light is used when there is not enough natural light available for the desired seeing task and/or when it may be necessary to control the color of the light for a special seeing task.

3. Lighting Methods

The two basic types of illumination either for illuminating seeing tasks or for creating decorative lighting effects are *(a)* general illumination and *(b)* supplementary illumination

a. General illumination: General illumination should be used when the seeing requirements are broad and varied, including seeing to move about safely, observing other people or general features within the surrounding area, performing housekeeping chores, moving materials or equipment, and performing general maintenance tasks.
b. Supplementary illumination: Supplementary illumination is required when general illumination is not available and/or when the nature of the task is such that the general illumination is insufficient, is not appropriately directed, or is of such a quality or color as to make specific seeing tasks difficult to perform. Supplementary illumination also is generally used for decorative effects since it can be precisely controlled in terms of color, direction, and amount.

4. Lighting Techniques

a. Floodlighting: As the term implies, this type of lighting is usually external to the object or objects that are to be illuminated, literally flooding them with light. Floodlighting is generally used to illuminate large areas, either indoors or outdoors, but it can also be used in more concentrated, localized situations such as for illuminating a control panel, a doorway or platform, or a parking lot. Although some control can be exercised over the flood pattern, it should be noted that the illuminated area covered by a floodlighting system will vary from the beam center to its periphery. Care must also be taken in using the floodlighting technique to consider the problems that might be associated with shadows, which are a by-product of the floodlighting technique.
b. Internal lighting: A wide variety of internally generated lighting techniques are available to make objects or visual displays "self-luminous." Typical applications for which internal lighting is appropriate include signs, panel labels, vehicle or operator console instruments and displays, shadow boards, and marker and signal lights. Internal lighting can be created in many ways, including the use of luminous phosphors, back-lighted screens, edge-lighted plastics, and end-lighted plastic "light pipes." The primary advantage of internal lighting is that more precise control can be obtained over where the light will appear.

5. Light Level

Avoid the temptation to believe that merely providing higher and higher illumination levels will result in improved seeing conditions. Although the level or intensity of light is obviously important, light can be too bright as well as too dim for the seeing task. Also remember that as light levels increase, glare problems also increase. There is usually a point below which seeing is difficult and another point above which seeing either does not improve or becomes more difficult for each type of seeing task.

6. Color

The color of the light used (i.e., for general illumination, supplementary illumination, or signal lighting) is important from several standpoints:

a. A broad-spectrum light (white light) is desirable to make the colors of objects or people seem natural. Natural light is used as the base-line reference since it is by this reference that we observe most things during everyday living. Several types of artificial light have been created to simulate natural light, but some are better than others for certain purposes. Care must be taken to select the artificial light color that is most appropriate for the task or seeing conditions being created. Incandescent luminaires typically provide more reds and yellows, which impart a warm appearance to viewed objects. Fluorescent lights, on the other hand (depending on the specific luminaire), tend to accentuate the blue. Often, a colored material selected under fluorescent light will not look the same when the material is viewed under natural sunlight.
b. Monochromatic light (i.e., red, yellow, blue, and green) may be desirable for certain types of seeing tasks and/or for decorative purposes. Such colors are not, however, desirable for general seeing conditions; they make surface colors appear abnormal and/or impossible to recognize. Monochromatic light is particularly useful for creating color-coded displays, e.g., traffic signal lights and warning displays.

7. Light Fixture Type and Usage

An extensive array of lighting fixtures, lamps, and materials is available for creating almost any type of lighting that is required. Some precautions need to be taken, however, in selecting commercially available fixtures. Some provide effective task illumination, but others are merely attractive to look at. When looking for appropriate fixtures off the shelf, designers should analyze the seeing task fully so that they can evaluate the merits (or, more often, the demerits) of the fixtures they are contemplating for a particular application.

An important factor to consider is the location of lighting fixtures, for even a good fixture's value can be lessened if it is improperly positioned relative to the seeing task. In addition, it is important to consider the light control requirements in terms of the possible need for dimming and/or selectively illuminating only certain fixtures at certain times during an operation.

General Illumination Guidelines

Task Requirements	Light Level, fc	Type of Illumination
Small detail; low contrast; prolonged viewing; fast, error-free response	100	Supplementary lighting fixture located near visual task
Small detail, fair contrast, close but short-duration work, speed not essential	50–100	Supplementary lighting and/or well-distributed and diffused general lighting
Typical desk and office work	40–60	General lighting with diffusing fixture directly overhead
Sports (e.g., tennis and basketball) or indoor recreational games (e.g., ping pong and billiards)	30–50	General lighting with sufficient number of fixtures to provide even court or table illumination
Recreational reading and letter writing	25–45	Supplementary lighting, positioned over reading so that page glare does not occur
General housekeeping, detail not required	10–25	General lighting
Visibility for moving about, avoiding people and furniture, and negotiating standard stairs	5–10	General and/or supplementary lighting (with care taken not to allow supplementary sources to project in the user's eyes)

Note: The above guidelines are only approximations. They are higher than some recommendations, not for seeing, but because these levels provide an additional psychological benefit as well. Levels relate to light levels measured at the primary seeing point, e.g., the desk or table surface on the floor or stair tread level.

Brightness ratios between the seeing task and the immediate surroundings should not exceed 5:1; between the task and the remote surroundings, 20:1; and between the immediate work area and any other remaining visual environment, 80:1.

Natural or white artificial light should be used regardless of the type of illumination; i.e., these levels do not apply to monochromatic light sources.

Special Situations Where Visual Dark Adaptation Is Required

Work Situations	Light Level, fc	Type of Illumination
Darkened control rooms and cockpits at night	0.01–0.4	Illumination on work surfaces
	0.03–0.10	Black and red instrument face detail (scale marks, numerals, and labels)
	0.10–1.00	Indicator lights (not to exceed 1 in, 2.5 cm square)
Ready rooms where crew members are required to maintain some dark adaptation prior to a mission at night	0.50–2.00	Illumination at tabletop
Movie theatre during film running	0.50–2.00	Illumination in aisle and immediate vestibule

Note: Red illumination is required in military workplaces and ready rooms. Low-level white light is satisfactory in theatres.

ILLUMINATION LEVEL GUIDELINES FOR MAINTAINING PROPER BRIGHTNESS CONTRAST BETWEEN VISUAL TASK AND BACKGROUND

Brightness Ratios

Comparison	Environmental Classification* A	B	C
Between tasks and adjacent darker surroundings	3:1	3:1	5:1
Between tasks and adjacent lighter surroundings	1:3	1:3	1:5
Between tasks and more remote darker surfaces	10:1	20:1	†
Between tasks and more remote lighter surfaces	1:10	1:20	†
Between luminaires and adjacent surfaces	20:1	†	†
Between the immediate work area and the rest of the environment	40:1	†	†

Note: Direct glare arises from a light source within the visual work field. It should be controlled by:

1. Avoiding bright light sources within 60° of the center of the visual field. Since most visual work is at or below the eye's horizontal position, placing luminaires high above the work area minimizes direct glare.
2. Using indirect lighting.
3. Using more relatively dim light sources, rather than a few very bright ones.
4. Using polarized light, shields, hoods, or visors to block the glare in confined areas.

*A—Interior areas where reflectances of entire space can be controlled for optimum visual conditions. B—Areas where reflectances of immediate work area can be controlled, but there is only limited control over remote surroundings. C—Areas (indoors and outdoors) where it is completely impractical to control reflectances and difficult to alter environmental conditions.

†Brightness-ratio control not practical.

LIGHT-SOURCE COLOR

The color of the light source should be chosen carefully to fit the application, as is indicated by the accompanying table.

Effect of Colored Light on Appearance of Object Color

Object Color	Red Light	Blue Light	Green Light	Yellow Light
White	Light pink	Pale blue	Pale green	Pale yellow
Red	Brilliant red	Dark bluish red	Yellowish red	Bright red
Light blue	Reddish blue	Bright blue	Greenish blue	Light reddish blue
Dark blue	Dark reddish purple	Brilliant blue	Dark greenish blue	Light reddish purple
Green	Olive green	Green-blue	Brilliant green	Yellow-green
Yellow	Red-orange	Light reddish brown	Light greenish yellow	Brilliant light orange
Brown	Brownish red	Bluish brown	Dark olive brown	Brownish orange

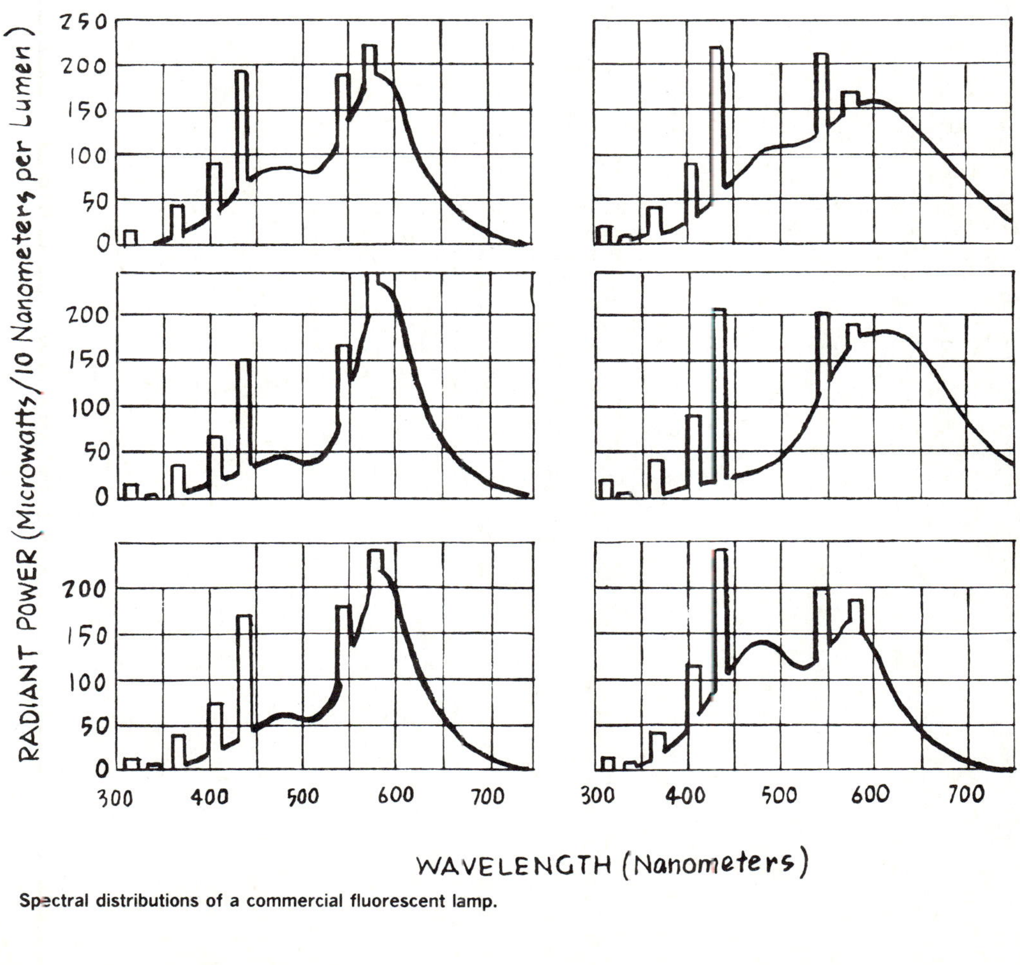

Spectral distributions of a commercial fluorescent lamp.

APPEARANCE RATINGS OF TYPICAL SURFACE COLORS WHEN VIEWED UNDER VARIOUS ARTIFICIAL LIGHT SOURCES

		Fluorescent lamps				
Color	Daylight	Standard Cool White	Deluxe Cool White	Standard Warm White	Deluxe Warm White	Incan-descent Lamps
Maroon	Dull	Dull	Dull	Dull	Fair	Good
Red	Fair	Dull	Dull	Fair	Good	Good
Pink	Fair	Fair	Fair	Fair	Good	Good
Rust	Dull	Fair	Fair	Fair	Fair	Good
Orange	Dull	Dull	Fair	Fair	Fair	Good
Brown	Dull	Fair	Good	Good	Fair	Good
Tan	Dull	Fair	Good	Good	Fair	Good
Golden yellow	Dull	Fair	Fair	Good	Fair	Good
Yellow	Dull	Fair	Good	Good	Dull	Fair
Olive	Good	Fair	Fair	Fair	Brown	Brown
Chartreuse	Good	Good	Good	Good	Yellowed	Yellowed
Dark green	Good	Good	Good	Fair	Dull	Dull
Light green	Good	Good	Good	Fair	Dull	Dull
Peacock blue	Good	Good	Dull	Dull	Dull	Dull
Turquoise	Good	Fair	Dull	Dull	Dull	Dull
Royal blue	Good	Fair	Dull	Dull	Dull	Dull
Light blue	Good	Fair	Dull	Dull	Dull	Dull
Purple	Good	Fair	Dull	Dull	Good	Dull
Lavender	Good	Good	Dull	Dull	Good	Dull
Magenta	Good	Good	Fair	Dull	Good	Dull
Gray	Good	Good	Fair	Soft	Soft	Dull

Note: Good—color appears most nearly as it would under an ideal white-light source, such as north skylight. Fair—color appears about as it would under an ideal white-light source, but is less vivid. Dull—color appears less vivid. Brown—color appears to be brown because of small amount of blue light emitted by lamp. Yellowed—color appears yellowed because of small amount of blue light emitted by lamp. Soft—surface takes on a pinkish cast because of red light emitted by lamp.

GLARE AVOIDANCE AND SHADOW PREVENTION

Because everyone has at one time or another experienced seeing problems due to shadows and/or glare, it would seem unnecessary to advise designers to exercise care in designing so that these conditions will not occur. However, shadows and glare are the most commonly encountered problems in lighting system designs. The three most serious conditions are illustrated by the accompanying sketches:

Shadows are cast by the observer because of improper luminaire positioning. Obviously, the light source should not be behind the observer. Typically, the luminaire should be located slightly to the right or left so that the light comes from over the observer's shoulder. General practice suggests that the light come from over the left shoulder because most people are right-handed. A lighting mockup is recommended for determining the best location.

The light source should not be visible to the observer; i.e., the light bulb or lamp should be screened either by some type of baffle or by a distributing screen. Once again, since layout errors have a way of slipping by, use a lighting mockup to make sure that the observer cannot see the light source directly from his or her various working positions.

Ambient glare from frontal or peripheral windows and/or similarly large and bright artificial sources not only makes seeing difficult but also produces general worker fatigue. Many methods are available for dealing with this problem, including reorienting the worker, putting shades over the windows, and adding light-filtering materials to the glazing.

Although this is not always the prerogative of the lighting system designer, try to eliminate glossy surfaces in front of the observer.

Diffusing light sources (fluorescent) are recommended to minimize glare and shadow effects.

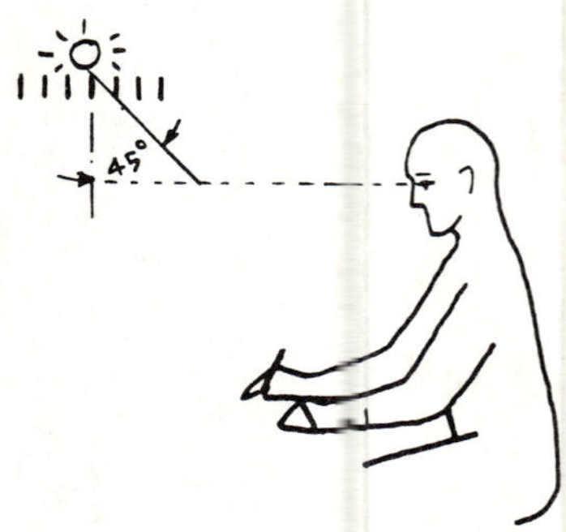

INTEGRAL LUMINOUS CEILING LIGHTING SYSTEMS

A popular concept has been the ceiling which provides a combination of overhead general lighting, acoustic treatment, and heating and ventilation, plus the basic supporting structure (as shown in the accompanying sketch). When used properly, this concept provides excellent seeing conditions for large-area installations. A common failure, however, occurs when the architect or interior designer paints the surfaces of the fixture insets a dark color, which absorbs much of the light from each fixture. This also makes the contrast between lighting fixtures and support structure extreme, which causes the lights to glare rather than create a pleasing atmosphere.

Many other overhead lighting systems fail to provide the desired seeing conditions for the same reason. Always minimize the contrast between the lighted fixture and its surrounding support structure or decorative trim.

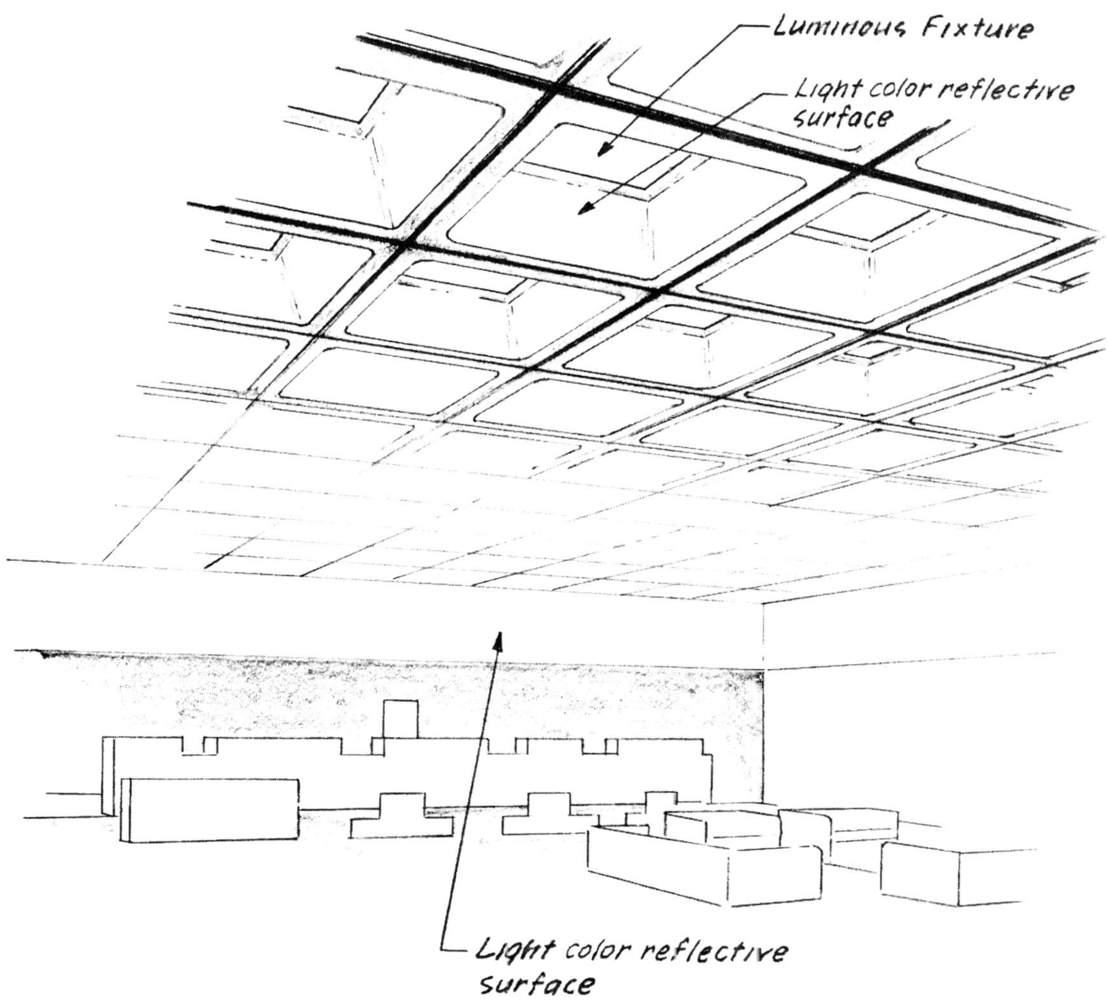

MODULARIZED, COMBINATION LIGHTING SYSTEMS

Energy conservation often can be obtained in conjunction with good lighting, as illustrated in the accompanying sketch. The objective here is to create a combination ambient and task lighting system that can "move with the workplace." Instead of trying to create a general ceiling lighting system that will be appropriate for all possible seeing situations, the designer should consider the modular approach, which illuminates only the areas that need it; in addition, since the luminaires are tailored to the modular workplace, the lighting will always be right for that workplace. This is important to note because, after a general lighting system has been installed, many seeing problems are created by the placement of furnishings in such a way that they block the intended light rays and the general light distribution.

Systems such as that shown in the accompanying illustration must be designed so that the luminaires on top of the modular cabinets are above the heads of persons passing by. In addition, to be most effective, such overhead lighting requires a normal-height (rather than a double-height) ceiling of light color.

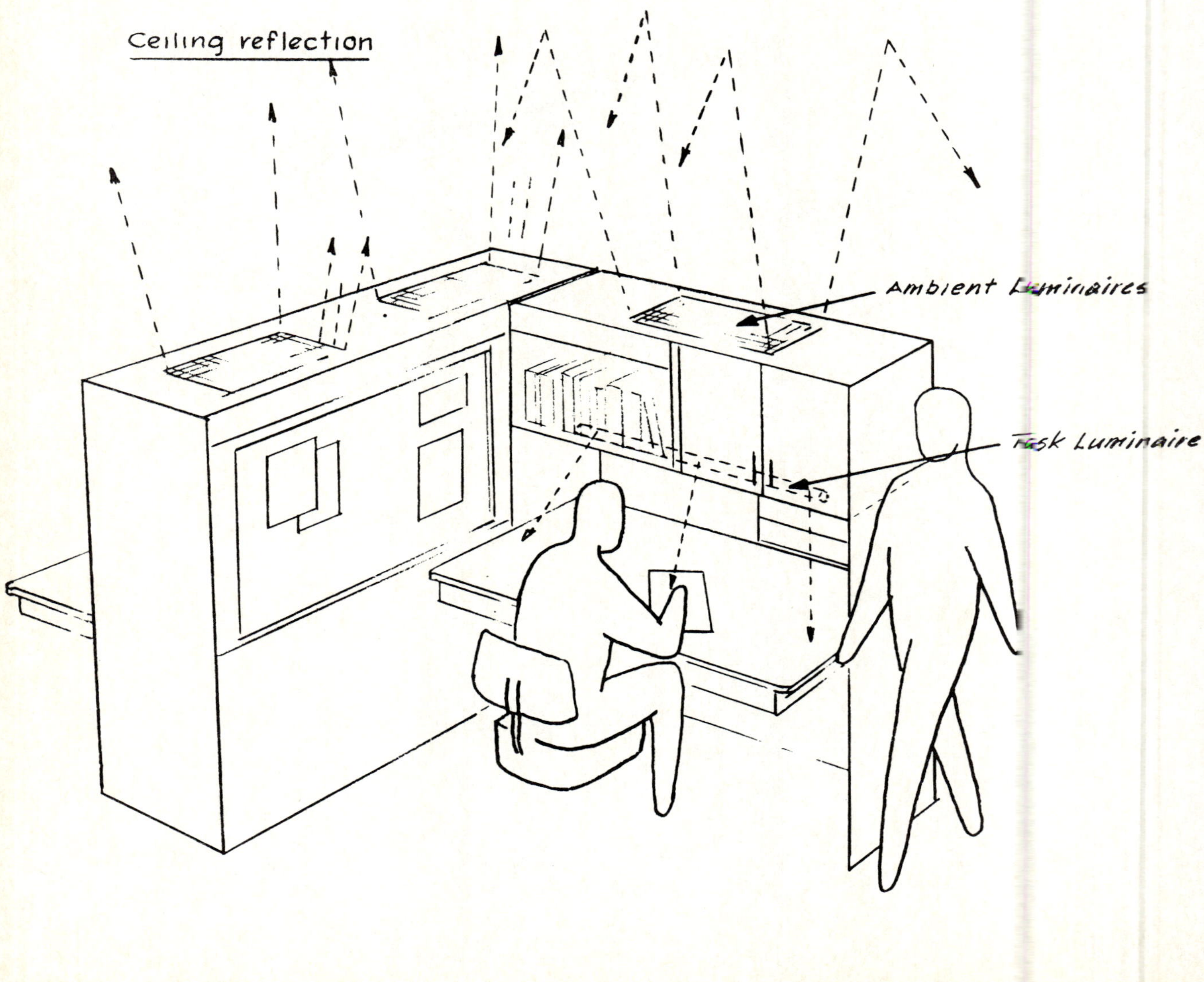

LIGHT-LEVEL CONTROL

Light-level adjustment is desirable for a wide variety of applications because it enables the operator to optimize his or her seeing conditions. The principal applications are the following:

1. Instruments and control panels: The ability of operators to adjust the light levels of a vehicle or of console instruments and/or the illumination of control panel labels is extremely important when they are confronted with changing ambient lighting conditions and/or operating needs; i.e., operators should be able to balance brightness levels among various visual displays, reduce interfering glare when they must maintain lower visual adaptation levels, and/or optimize special display viewing conditions (e.g., CRT reference scales).
2. Ambient illumination level: It is often desirable for operators to be able to adjust ambient room lighting in order to improve their ability to see special visual displays.

Note: It is generally desirable that a dimming system be designed so that the light cannot be turned completely off because, if a system has been previously turned off, it may create a problem for the succeeding operator or user, who may come into a dark room or sit at a darkened control panel and not be able to find a light switch or recognize obstacles while trying to locate the switch.

ENERGY SAVING

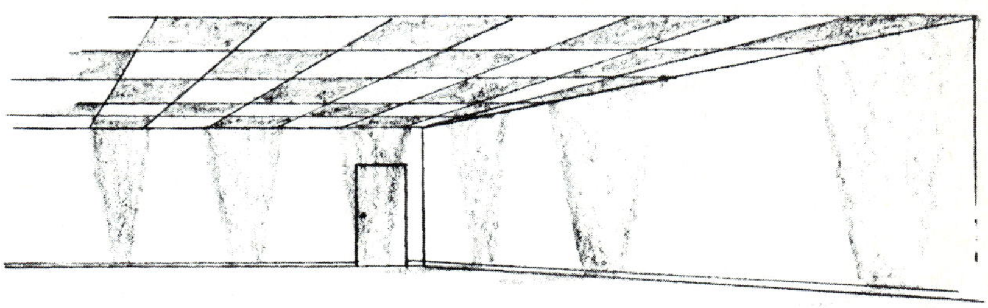

Because of the need to save energy, it often becomes necessary to use fewer lights and/or to guarantee that lights are turned off when they are not being used. Consider the following:

1. Use selective circuit control concepts so that it is possible to turn off various luminaires. Care must be taken in creating such system control capability not to destroy the general acceptability of seeing conditions; i.e., the light distribution (although less intense) should continue to be balanced for a given visual operating area.
2. Provide central control for large-area lighting systems so that the person who leaves last can turn off all remaining lights.
3. Consider the use of automatic light control (timing systems) so that certain lights are turned on or off to match operating needs. Similar automatic control of special exterior light control shutters should be considered; i.e., as long as natural light can be used, it is foolish to waste artificial light.
4. Supplementary lighting may provide greater energy saving than general lighting because the user can often obtain all the light needed (both for the immediate task and for general mobility) from the supplementary lighting system.

Using a dimmable luminous ceiling lighting system is a better energy-saving approach than selectively turning out every other luminaire because the latter approach creates light and dark spots, and often a dark spot occurs directly over a desk or other operator position.

The same consideration applies to hallway illumination; i.e., it is better to turn off one lamp within a fixture than to turn off every other luminaire along the hallway.

An automatic "user-presence" sensing system can be used effectively to turn out lights in unoccupied spaces.

LIGHTING INDUSTRY STANDARD PRACTICE

Facility	Light Level, fc
Art galleries:	
General	30
Paintings	30–60
Sculpture	30–100
Auditoriums:	
Lobbies	30
Seating area (nonperformance)	10–30
Seating area (during performance, in aisle)	1–3
Orchestra (music stands)	15–20
Banks:	
Lobby	30
Officer platform area	30–50
Teller line	50–75
Rear posting and keypunch area (work surface)	75–100
Depots and terminals:	
Waiting rooms	30
Ticket counter	50–75
Baggage room	50–75
Platform	30
Rest rooms	30
Hospitals:	
Lobbies and hallways	50–75
Patient rooms	3–50
Laboratory and examination rooms	75–100
Surgery and delivery	To 2500
Hotels:	
Lobbies	30–75
Front office	50–75
Halls and elevators	20–30
Rest rooms and guest rooms	20–50 plus reading lights to 75
Kitchen	100
Laundry	50–75
Office buildings:	
Lobbies, halls, and elevators	20–30
Offices	75–100
Residences:	
Entryway and porch	10
Hallways and stairs	10–25
Kitchen	50–75
Laundry	50
Library	50–70
Sewing room	50–100
Bathroom—general	30
sink and mirror	50
Workshop	70
Restaurants:	
Quick-service counter	50
Intimate dining area	15–30
Kitchen	70
Cashier area	50
Schools:	
Study areas and classrooms	50–100
Drafting room	100
Offices	50
Rest rooms	30
Retail stores:	
Circulation area	30
Showcase	100–150
Sports:	
Archery	10
Badminton	30
Baseball—seating	3–10
field	100–150
Basketball	50
Gymnasium, general exercise room, and	30
locker room	20
Hockey	20
Boxing	200
Football	100
Swimming pool, general overhead and	10
under water	100
Tennis	30–50
Volleyball	20

INDUSTRIAL/MANUFACTURING LIGHTING PRACTICES

Facility	Light Level, fc
Assembly:	
Rough seeing	30
Rough and difficult seeing	50
Medium-difficulty seeing	100
Fine seeing	500
Auto body:	
Frame assembly	50
Chassis assembly	100
Final inspection	200
Bakery:	
General	30–50
Hand decoration	100
Bookbinding:	
General	70
Embossing	200
Chemical laboratory	30–50
Cleaning and pressing:	
General inspection and	50
spotting	500
Hand pressing	150
Repair and alteration	200
Electrical equipment manufacturing:	
General	50
Coil winding	100
Foundries:	
General	30
Molding and pouring	50
Grinding and chipping	100
Fine inspection	500
Glassworks:	
General	30–50
Fine grinding and polishing	100
Engraving	200
Iron and steel manufacturing	20–30
Machine shops:	
General	50
Fine bench work	100
Paint manufacturing:	
General	30
Color matching	200
Printing industries:	
Presses	70
Composing	100
Proofing	150
Sheet metal works:	
General cutting and bending	50
Scribing and inspection	200
Shoe manufacturing	200–300
Textile mills:	
General	30
Carding, spooling, and spinning	50–150 (depending on color)
Inspection and weaving	100
Welding:	
General	50
Precision arc	1000
Woodworking:	
General	30
Finishing	100

**Note:* Where color matching or working on darker colors is involved, higher illumination levels are required. In order to avoid spectral glare, when the work involves close visual inspection of materials with a high-gloss finish, light levels should not be too high. Although the above standards appear to be very specific, the reader is cautioned not to assume that the specified illumination levels are either adequate or necessary. Each visual task should be examined in detail to determine seeing difficulty. A lighting mockup is desirable to test the proposed lighting scheme as well as the proposed levels.

FORMULAS APPLICABLE TO GENERAL LIGHTING REQUIREMENTS

$$\text{No. of fixtures} = \frac{\text{footcandles desired} \times \text{room area per}}{\text{CU} \times \text{MF} \times \text{lamps per fixture} \times \text{lumens per lamp}} = \frac{\text{total lumens}}{\text{lumens per fixture}}$$

$$\text{Average footcandles} = \frac{\text{lamp lumens} \times \text{CU} \times \text{MF}}{\text{room area (ft}^2\text{)}} = \frac{\text{lumens per lamp} \times \text{CU} \times \text{MF}}{\text{area per lamp (ft}^2\text{)}} = \frac{\text{lamp lumens per fixture} \times \text{CU} \times \text{MF}}{\text{area/fixture (ft}^2\text{)}}$$

$$\text{Area/lamp (ft}^2\text{)} = \frac{\text{lumens per lamp} \times \text{CU} \times \text{MF}}{\text{footcandle level desired}}$$

$$\text{Area/fixture (ft}^2\text{)} = \frac{\text{lumens per lamp} \times \text{no. of lamps per fixture} \times \text{CU} \times \text{MF}}{\text{footcandle level desired}}$$

$$\text{Total watts/ft}^2 = \frac{\text{footcandle level desired}}{\text{overall lumens per watt} \times \text{CU} \times \text{MF}}$$

where CU is the coefficient of utilization and MF is the maintenance factor.

GUIDELINES FOR LIGHTING FIXTURE INSTALLATION: GENERAL ILLUMINATION

To provide the best seeing conditions as well as the most pleasant seeing conditions (i.e., minimum annoyance), consider the *luminous ceiling*. The intensity of the light will not have to be as high, and the distribution will be even and free from extreme brightness contrasts.

The next best system consists of *translucent fixtures,* which reflect light from the ceiling and also direct light through the fixture screen. The egg-crate configuration has the advantage of not collecting dirt or insects, which eventually cut down the amount of light obtained.

Translucent globe fixtures provide excellent general light distribution and minimize the collection of dirt, which can reduce the lighting effectiveness of the fixture.

Note: Suspended fixtures require sufficient ceiling height in order to prevent direct glare from the fixture.

Indirect lighting systems rely on "bounce light"; i.e., the light is never direct and thus annoying to the observer. This type of lighting is useful when there is no requirement for a high degree of brightness on the visual task.

The following are suggested applications:

Luminous ceilings—professional offices
Translucent fixtures—work spaces with low ceilings
Translucent globe fixtures—areas with higher ceilings
Indirect lighting—areas where reading or other critical tasks are not required

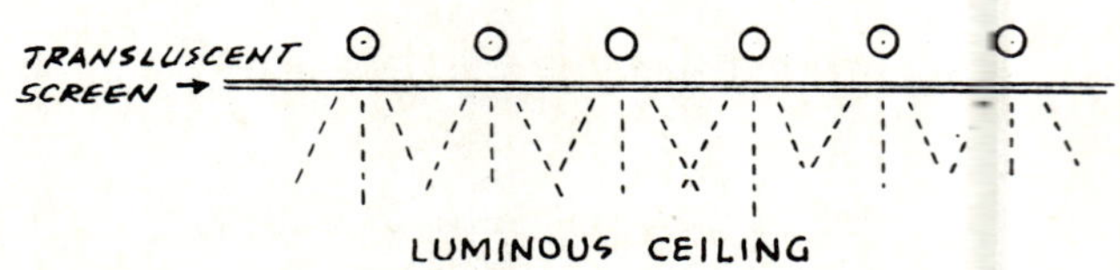

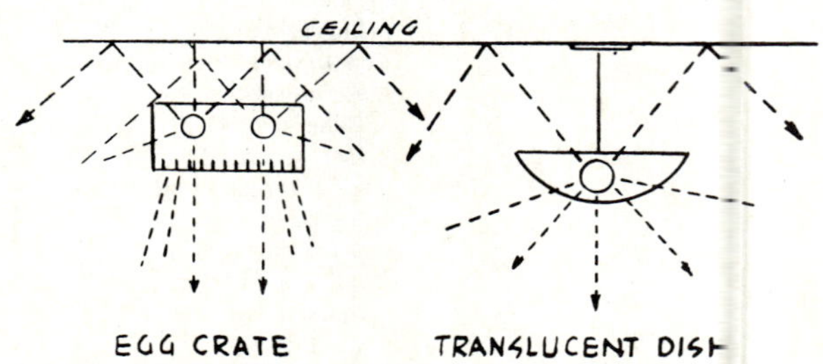

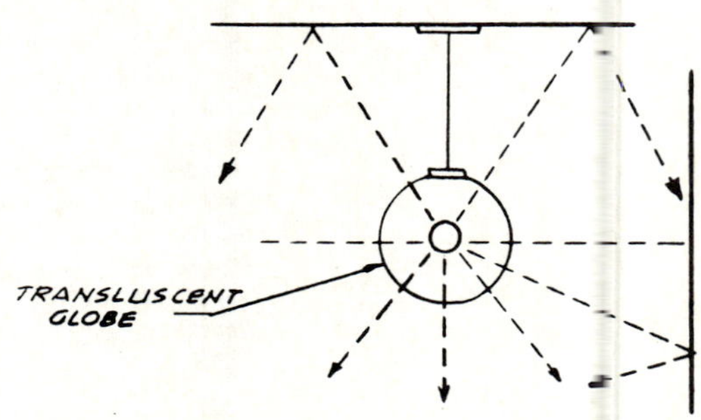

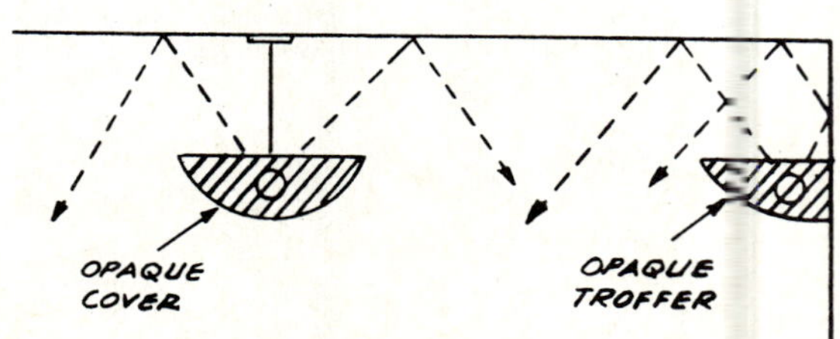

The *inset screened luminaire* is less desirable than previously discussed fixtures because it does not provide an opportunity for the light to reflect from the ceiling and thus improve general, even distribution. It can be effective, however, for "over-the-task" illumination as long as the fixture and task positional relationships remain constant.

The *hooded fixture* provides high efficiency for high-bay plant applications.

The *cannister fixture,* although popular for decorative lighting, is perhaps the least desirable luminaire system for good seeing. It tends to create a "spotlighting" condition in which extreme shadow is created whenever anyone walks beneath the light. Although it can be used as a general illumination system, an inordinately large number of fixtures are required in order to provide reasonable light distribution.

In selecting any of the above, one should pay special attention to the manner in which bulb or lamp replacement and filter removal and cleaning can be accomplished.

Suggested applications include the following:

Inset screened fixtures: These are appropriate for a small office where one or two fixtures can provide even light distribution over the entire space.

Hooded fixtures: These are recommended for high-bay areas where fairly high-intensity light is desirable, but not for close individual visual work. The fixtures need to be high enough to minimize direct view glare from high-intensity lamps.

Cannister fixtures: These are recommended for decorative purposes and/or where there is a special need for localized illumination of a doorway threshold, etc. Care should be exercised in using such systems where the casting of shadows could create seeing problems.

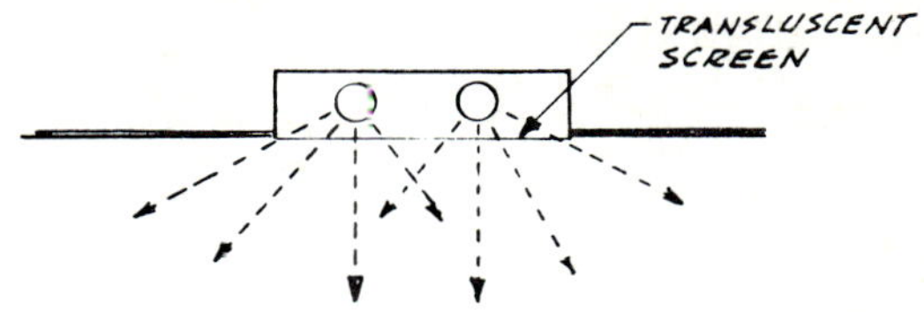

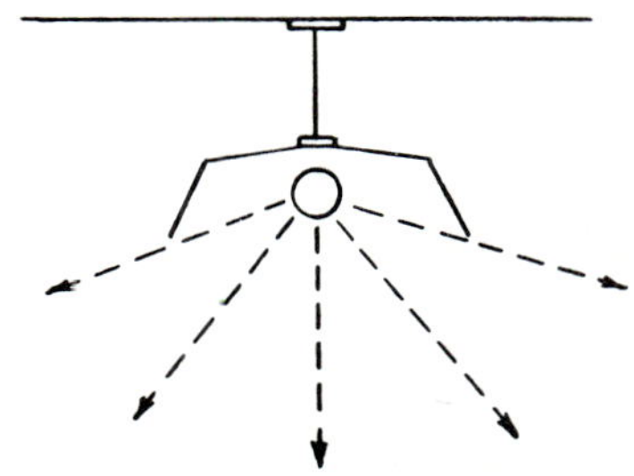

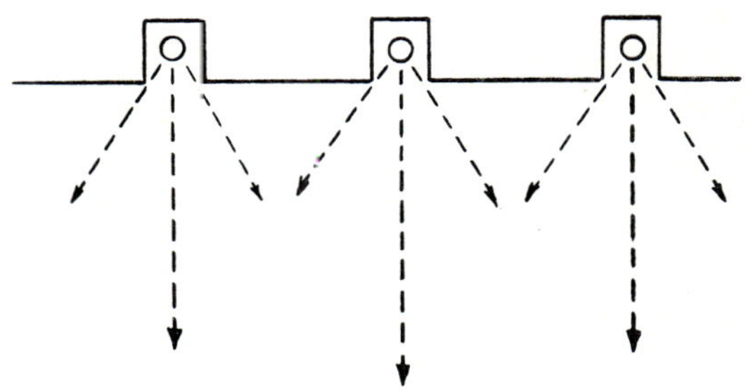

FIXTURE SERVICING

The ease with which light fixtures can be cleaned and relamped is perhaps as important as the overall lighting system design concept. It has been estimated that a typical lighting system output is reduced by 25 percent six weeks following its installation. In selecting fixtures for any lighting system, consider the following servicing features:

1. The fixture should be readily accessible using only the simplest stepladder. Avoid placing fixtures at extreme heights or in positions where it becomes necessary to employ special devices and/or where there is access only through difficult crawl spaces.
2. It should be easy to remove intervening light control panels without damaging them, and the method of fastening and releasing these devices should be obvious and not require special tools. Many current schemes that are offered by manufacturers or designed on a custom basis seem to purposely camouflage the method for removing a ceiling panel, egg-crate screen, or translucent filter panel. Removable screens or panels should not be so large that they are unwieldy to handle efficiently, and they should not be made of material that is too fragile and easily damaged.
3. Light-filtering or light-distributing panels or covers should be made of materials that do not discolor with age.
4. Fixture housings should be designed so that there is sufficient clearance to get hold of the lamps and manipulate them in and out of sockets.
5. Choose fixtures which either allow dirt or dust to drop through naturally (open egg-crate screen) or which are semisealed against dust and dirt collection.
6. Preferably, use fluorescent fixtures that contain lamps no longer than 40 in (102 cm). Longer tubes are difficult to handle safely.

Note: Do not assume that by leaving relamping to a maintenance crew, damage to fixtures will be avoided. In spite of their supposed experience, many maintenance workers are more careless than regular staff personnel in the office.

SUPPLEMENTARY LIGHTING FOR THE INDIVIDUAL WORKPLACE

Floodlighting techniques for individual workplace illumination should be designed so that the operator can select or adjust the best position for his or her seeing task; sometimes this changes.

The key consideration is to place the luminaire so that it does not cause light rays to reflect from the work surface into the operator's eyes and so that the light source (lamp) is not directly visible to the operator.

In general, the best lamp position is behind and to the side of the operator. If the operator is right-handed, the lamp should be to his or her left; if the operator is left-handed, it should be to his or her right. Thus it is necessary to provide a flexible positioning system.

The accompanying sketch illustrates a typical flexible-arm luminaire that provides a wide range of positioning possibilities.

For some special tasks, such as engraving, the light source must be placed so that the light rays actually cast shadows because these provide important guidance cues for the worker.

A low position for the light source, as shown in the accompanying sketch, is typical of the special situations in which shadow casting is advantageous.

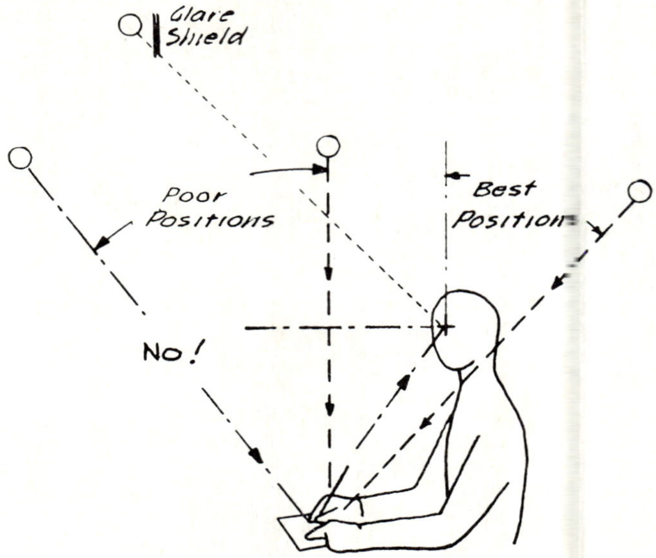

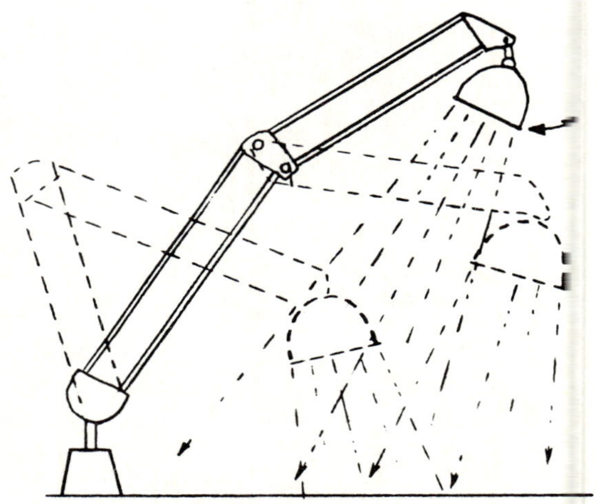

COMMUNICATIONS

GENERAL

Communications Systems

In conceptualizing a communications system, whether it is large and complex (such as a worldwide satellite system) or relatively simple and limited (such as an internal closed-circuit television system in a plant), the common denominator should always be the human need. That is, communications between people should be analyzed and planned on the basis of human task needs and thus should be designed to meet these needs in the most effective manner. This requires special consideration of human input-output characteristics, or, if you will, human channel capacities and sensory channel sensitivity. In developing a communications system concept, consider at least the following:

1. Communications mode (i.e., visual, auditory, tactile, and/or some combination)
2. Communicating environmental constraints (i.e., noise, environmental interference, and the need for security or privacy)
3. The time factor (i.e., how much information must be transmitted within some prescribed time frame)
4. Reliability requirements (i.e., whether you can afford to repeat or lose some of the information without degrading system performance objectives)
5. Convenience (i.e., whether immediate accessibility is critical)
6. Multiuser requirements (i.e., whether the same communication is important to several users simultaneously)
7. Personal innuendo (i.e., the value of face-to-face expression)
8. International language (i.e., whether it is critical that people with differing language backgrounds understand the communication)
9. Freedom from interruption (i.e., whether it is critical that a given communication not be interrupted prior to its complete transmission)
10. System cost, both the initial cost and the cost of maintenance
11. The effectiveness of the human-machine interface design (i.e., ease of use of handsets, microphones, earphones, controls, etc), considering the environmental and other constraints involved in use of the interface (temperature, noise, vibration, clothing restrictions, etc.)

General Considerations for Development of Communications System Concepts

The following discussion is limited primarily to visual and auditory communications between people within the working community, as opposed to communications for the purpose of recreation or entertainment. A primary objective should be to create a system that supports other human task activities, e.g., the exchange of information among individuals and the transmission of instructions, commands, warnings, orders, etc. Such systems should be based on the capabilities and limitations of the human as well as on the information transfer needs of the users. Consider the following:

1. Type of information (spoken messages, pictures, coded messages, digital data, etc.)
2. The time factor (i.e., how rapidly the information must be transmitted)
3. Security (i.e., whether the information is to be accessible to only one person or to anyone who might be within hearing or seeing distance of the receiving device)
4. Number of intended recipients (i.e., whether the message is for general broadcast for the benefit of a large

group of recipients)

5. The receiving environment (i.e., whether certain interferences at the receiving station may mask the incoming communication)
6. Reliability requirements (i.e., whether the situation permits repetition of the message if it is not understood the first time)
7. Quality (i.e., whether the information requires high fidelity in order to be fully understood)
8. Language (i.e., whether it is critical that people of differing language backgrounds understand the communication)
9. Personal factors (i.e., how much value there is to seeing facial expressions or hearing voice inflections)
10. Directionality (i.e., how important it is to associate the transmission with a direction or source)
11. Message capture (i.e., whether it is necessary to have a record of the transmission)
12. Condition of the addressee (i.e., whether he or she is busy, is using his or her hands or eyes, or is under special physical stress, such as high *g* forces)

Communications hardware technology has reached the state at which the planner can prescribe almost any level of fidelity or reliability and almost any mode. Therefore, the choice is limited primarily by cost. However, since the primary purpose is to allow people to communicate with one another, systems concepts should generate from human needs and from basic human sensorimotor characteristics and limitations. In defining the characteristics of any proposed system, consider the following:

For Auditory Communications

1. Loudness (intensity)
2. Pitch (frequency)
3. Timbre (signal and voice profile)
4. Signal-to-noise ratio
5. Signal pattern (duration, intermittency, and frequency shift)
6. Sound location
7. Public as opposed to private

For Visual Communications

1. Brightness
2. Color
3. Contrast
4. Duration
5. Pattern and shape
6. Alphanumeric legibility

The following are some of the important human factors objectives for communications systems:

1. Detectability: The intended receiver can sense the signal.
2. Recognizability: The intended receiver can tell what the signal is.
3. Intelligibility: The intended receiver can tell what the signal means.
4. Conspicuousness: The signal is attention-getting.

Although communications systems designers obviously are concerned with the basic transmission medium (i.e., they must select radio, radar, sonar, or another basic scheme for the system), the human factors relate primarily to the immediate interface between the system and the sender, listener, or visual observer. Although tactile communication interfaces may have some application in isolated instances, the primary questions are whether the message originator should transmit by voice, encoded aural signal, or written or graphic means; whether the message will be received by aural or visual means; and whether there is increased value (e.g., reliability) if the message is transmitted using both visual and auditory modes.

1. When to Use Auditory Communication

a. When the message is simple
b. When the message is short
c. When the message does not necessarily have to be referred to later
d. When the message deals with events in time
e. When the message calls for immediate action
f. When the visual system is already overloaded
g. When the intended receiver cannot look at a visual display or may have to move about continually
h. When the receiving environment is incompatible with seeing, (e.g., when it is too bright or when it is necessary to avoid visual signals that could interfere with the intended receiver's adaptation to the dark)
i. When the intended receiver's visual capacity is impaired as a result of a dynamic stress (e.g., high *g* forces)
j. When the intended receiver's eyes may be closed (e.g., when he or she is asleep)
k. When a visual signal strength is borderline and the addition of a coincident auditory signal would provide redundant, reinforcing confirmation

2. When to Use Visual Communication

a. When the message is very complex
b. When the message is long
c. When the message needs to be referred to later
d. When the information deals with location in space
e. When the message does not necessarily call for immediate action (e.g., when the receiver needs to finish some other task)
f. When the auditory system of the receiver is overburdened
g. When the receiving environment is too noisy to ensure reliable receipt of an aural message
h. When the receiver will remain in a position where he or she can continue to watch the visual display
i. When the information characteristics cannot be reliably described by words
j. When face-to-face communication may provide added understanding between communicators

3. When to Use a Private Rather than a Public Interface

An important decision in developing any communication system concerns whether, and under what circumstances, to use individual rather than group visual or auditory displays. Although there are no hard-and-fast rules, the following should be considered:

a. Use group display interfaces when:
(1) Several people should receive the same information simultaneously.
(2) Several people should all receive the information in the same format.
(3) Some people require general information at the same time that they are busy sending or receiving individual communications.
(4) The transmitter may not know who the persons are who should receive the information.
(5) The individual channel that the transmitter wishes to use is temporarily unavailable.
(6) The receiving individual is temporarily away from his or her individual communication interface.
(7) Input to a general status display is being generated by several individuals on a continuing basis.
(8) It is inconvenient for the required number of listeners or viewers to gather around an individual display device.

b. Use individual display interfaces when:
(1) The transmission involves only two persons.
(2) The information need not be shared with any other persons.
(3) The transmission is or should be private.
(4) The information is being modified by the individual for his or her own benefit.
(5) The individual needs to record the information for future reference.

4. When to Use Handsets, Intercoms, or Public Address Systems

Although there are obvious overlaps in the manner in which handsets, intercoms, and public address systems can or should be used, the following are guidelines for making this decision:

a. Handsets should be used primarily when privacy is desired.
b. Intercoms should be used primarily when the individual is too busy to look up a number and dial it.
c. Public address systems should be used primarily for making public announcements.

5. Other Communicating Techniques and Rationales for Their Use

a. Colored and/or flashing light signals, horns, bells, claxons, sirens, etc., are particularly useful for issuing warnings and conditional communication because they can be made conspicuous without materially interfering with other communications.
b. Digitized communications can be used to reduce the amount of paper transfer.
c. Encoded signals can be used to maintain communication security.
d. Flag, light, and hand signals can be used to minimize the cost of more expensive systems.

ALARMS

Guidelines for Selecting Auditory Alarm Systems[1]

Types of Alarms, Their Characteristics and Special Features

Alarm	Intensity	Frequency	Attention-getting ability	Noise-penetration ability	Special features
Diaphone (foghorn)	Very high	Very low	Good	Poor in low-frequency noise Good in high-frequency noise	
Horn	High	Low to high	Good	Good	Can be designed to beam sound directionally Can be rotated to get wide coverage
Whistle	High	Low to high	Good if intermittent	Good if frequency is properly chosen	Can be made directional by reflectors
Siren	High	Low to high	Very good if pitch rises and falls	Very good with rising and falling frequency	Can be coupled to horn for directional transmission
Bell	Medium	Medium to high	Good	Good in low-frequency noise	Can be provided with manual shutoff to insure alarm until action is taken
Buzzer	Low to medium	Low to medium	Good	Fair if spectrum is suited to background noise	Can be provided with manual shutoff to insure alarm until action is taken
Chimes and gong	Low to medium	Low to medium	Fair	Fair if spectrum is suited to background noise	
Oscillator	Low to high	Medium to high	Good if intermittent	Good if frequency is properly chosen	Can be presented over intercom system

Summary of Design Recommendations for Auditory Alarm and Warning Devices

Conditions	Design recommendations
1. If distance to listener is great	1. Use high intensities and avoid high frequencies
2. If sound must bend around obstacles and pass through partitions	2. Use low frequencies
3. If background noise is present	3. Select alarm frequency in region where noise masking is minimal
4. To demand attention	4. Modulate signal to give intermittent "beeps" or modulate frequency to make pitch rise and fall at rate of about 1–3 cps
5. To acknowledge warning	5. Provide signal with manual shutoff so that it sounds continuously until action is taken

[1]C. T. Morgan, J. S. Cook, A. Chapanis, and M. W. Lund (eds.), *Human Engineering Guide to Equipment Design*, McGraw-Hill Book Company (*1st Ed*), New York, 1964.

ARCHITECTURAL FACTORS

Architectural Planning for Optimum Communication

Communications systems requirements should be considered at the beginning of architectural concept development because the effectiveness of the final communication depends on the inherent characteristics of the architectural configuration; that is, avoid the tendency to try to add a communications system to an architectural plan that is frozen. Few, if any, communications systems are completely satisfactory unless the architectural and communications systems are optimized as an integrated concept.

Key factors in planning the architectural and communications systems include the following:

1. Plan specifically for the total array of expected communications, i.e., direct person-to-person speech (without benefit of intervening aids such as the telephone), private telephone conversations, public address systems, emergency alarms, special intercom connections, mobile communications (e.g., by means of walkie-talkies), closed-circuit television, signal lights, special visual displays (e.g., movies and other audio-visual aids), and directional and instructional signs.
2. Establish requirements concerning privacy and prevention of annoyance.
3. Determine both inside and outside communication interference possibilities and identify the exact nature of these in terms of loudness, frequency distribution, etc.
4. Examine and define the potential physical interactions between each type of communicating mode and the proposed architectural features that may produce undesirable effects on the communication link (reverberation, echos, sound time lag, sound shadow, visual barriers, effect on sound localization, visual glare, seeing distance, illumination level, etc.)
5. Organize the facility in terms of ease of developing a logical and meaningful signing system.

Key architectural factors that should be considered relative to auditory communications systems include the following:

1. The shape and size of interior spaces and other occupied areas (e.g., hallways and stairwells)
2. The surface texture and shape of floors, walls, and ceiling and of furnishings that are expected to be relatively permanent
3. The physical positions of speakers (e.g., people) and listeners
4. The position of communications system hardware elements (e.g., loudspeakers, intercoms, and microphones)
5. The number of people who may occupy a space and thus will tend to absorb sound

Key architectural factors that should be considered relative to visual communications systems include the following:

1. Viewing distances
2. Ambient lighting effects
3. Artificial lighting effects (e.g., decorative lighting)
4. Day versus night ambient and artificial lighting effects and their influence on internal seeing conditions
5. Individual display and light control needs relative to interference among different facility users

Maintenance of both the communications system and the related architectural features of the planned facility should also be considered early and simultaneously. Like plumbing, many communications systems become inaccessible for maintenance because of the tendency to hide wiring and hardware. Although the aesthetic objective is laudable, all visual and audio communications systems hardware that can be expected to need service or replacement should be made accessible, especially so that wear and tear on architectural features (e.g., ceiling and wall panels) is minimized.

Finally, if a proposed communications system cannot be fully implemented at the beginning or if there is any possibility of future system expansion, plan the architectural facility for ease of adding new wiring, controls, connections, etc.

GENERAL GUIDELINES FOR THE SELECTION AND DESIGN OF VISUAL DISPLAYS

General Principles

1. *Use the simplest display concept* commensurate with the information transfer needs of the operator or observer. The more complex the display, the more time it takes to read and interpret the information provided by the display, and the more apt the observer or operator is to misinterpret the information or fail to use it correctly.
2. *Use the least precise display format* that is commensurate with the readout accuracy actually required and/or the true accuracy that can be generated by the display-generating equipment. Requiring operators to be more precise than necessary only increases their response time, adds to their fatigue or mental stress, and ultimately causes them to make unnecessary errors.
3. *Use the most natural or expected display format* commensurate with the type of information or interpretive response requirements. Unfamiliar formats require additional time to become accustomed to them, and they encourage errors in reading and interpretation as a result of unfamiliarity and interference with habit patterns. When new and unusual formats seem to be needed, consider experimental tests to determine whether such formats are compatible with basic operator capabilities and limitations and/or whether the new format does in fact result in the required performance level.
4. *Use the most effective display technique* for the expected viewing environment and operator viewing conditions (lighting, acceleration, vibration, operator position, mobility restrictions, etc.). Match the display technique to the operator's constraints; do not make the operator match the display.
5. *Optimize the following display features:*
   a. Visibility: Viewing distance in relation to size, viewing angle, absence of parallax and visual occlusion, visual contrast, minimal interference from glare, and adequate illumination
   b. Conspicuousness: Ability to attract attention and distinguishability from background interference and distraction.
   c. Legibility: Pattern discrimination, color and brightness contrast, size, shape, distortion, and illusory aspects.
   d. Interpretability: Meaningfulness to the intended observer within the viewing environment; requirements for interpretation, extrapolation, special learning, and training; and general reliability in terms of retention of meaning

Note: Consideration should be given to possible visual anomalies of the expected user population, i.e., lack of normal visual acuity, color deficiencies, and nearsightedness and farsightedness. These user limitations should be considered not only in terms of selecting the proper display concept but also in terms of executing the specific features that go to make up the display.

Consideration should also be given to creating displays that are easy to read and interpret, as opposed to displays that are merely attractive and/or startling to look at. Avoid the temptation to make displays decorative at the expense of their being readable and interpretable.

When you do not have the choice of creating a display, i.e., when you are required to select ready-made components, carefully study the available products until you find the one that most nearly meets the requirements and guidelines provided in the remaining parts of this section.

Design or select dynamic displays (instruments that have moving elements, warning indicators, etc.) that let the observer know if the display is malfunctioning. If it is not obvious (e.g., a pointer that suddenly drops to zero when it should be indicating some value), consider a special indication of malfunctioning, such as a warning light or flag.

VISUAL DISPLAY SELECTION

Display Requirements	Type of Display	Rationale	Example
Go/no-go, start/stop, ON/OFF	Light	Easy to tell when the light is on or off	
Identification	Light or lights Symbol or symbols	Easy to detect; can be coded by means of location, number, color, repetition, or on-off duration	
Instructional	Labels and printed words Arrows Pictorial symbols	Single word or multiword message clarifies meaning; arrow implies direction; pictorial illustrates feature	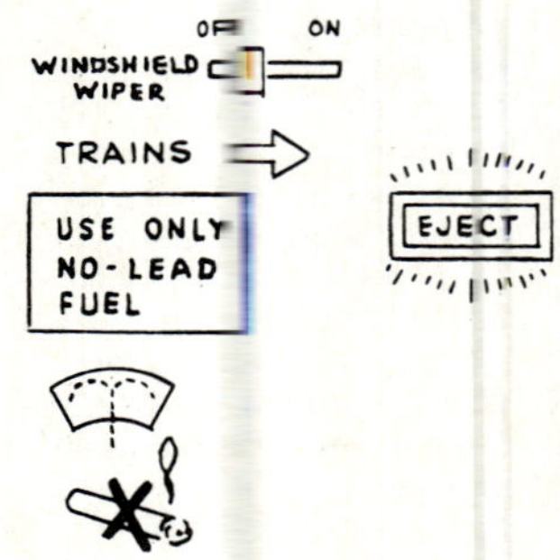
Exact quantity	Digital readout	Single number value requires no further interpolation or extrapolation	
Approximate quantity or positional relationship	Moving pointer and fixed scale with numbers; moving scale and numbers with fixed pointer	General pointer orientation gives quick cue to scale-pointer relationship; rate of change; reference to scale limits	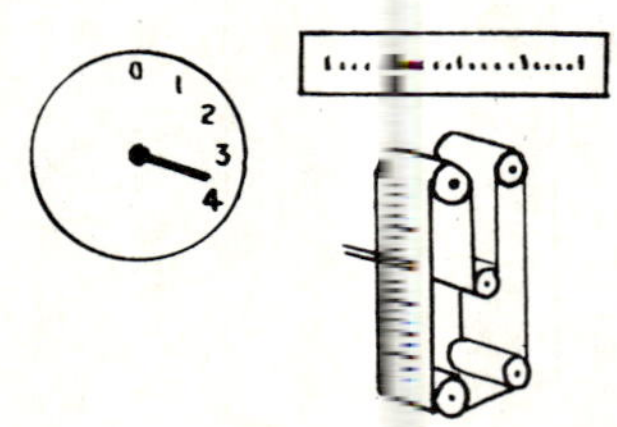
Data entry	Keyset	Alphanumeric keys provide immediate data selection reference	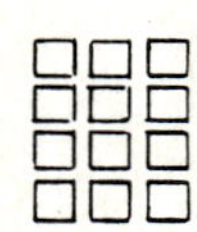
	Pointer knob	Easy to position pointer adjacent to desired scale mark or number or label	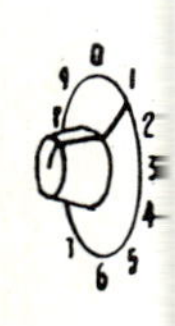
Tracking and hooking targets	CRT with associated data input and pick-off controller (joy stick, track ball, pantograph, etc.); indexing marks introduced on CRT with continuous capability for repositioning marker and adding alphanumerics or reference symbology	Provides convenient means for "pointing" to target and adding orientation reference marks or alphanumerics	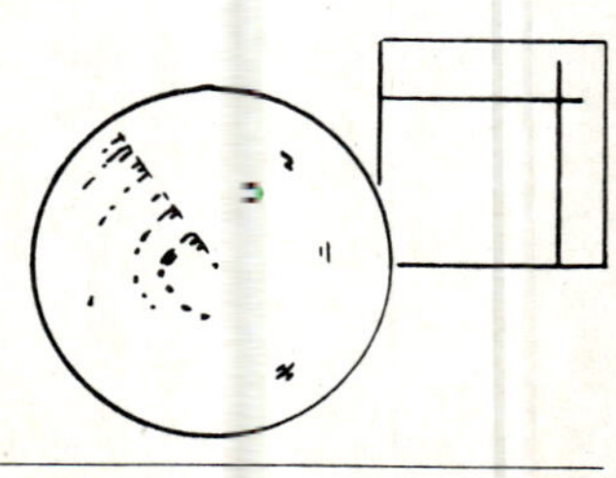

VISUAL DISPLAY SELECTION (continued)

Display Requirements	Type of Display	Rationale	Example
Geographic position	Plan position reference analog: map or electronic map and grid display	Shows direct comparison of operator's position with geographic area and features; can show superimposed targets of interest	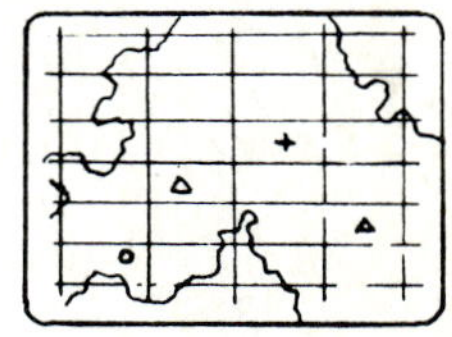 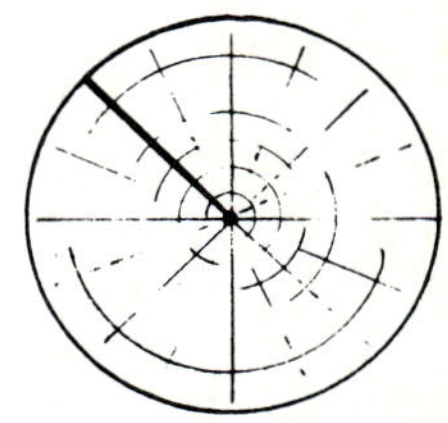
Equipment performance analysis	Meters CRT Pen recorder	Single parameter is easy to interpret Multiparameter relationships are easy to see Permanent record for later analysis	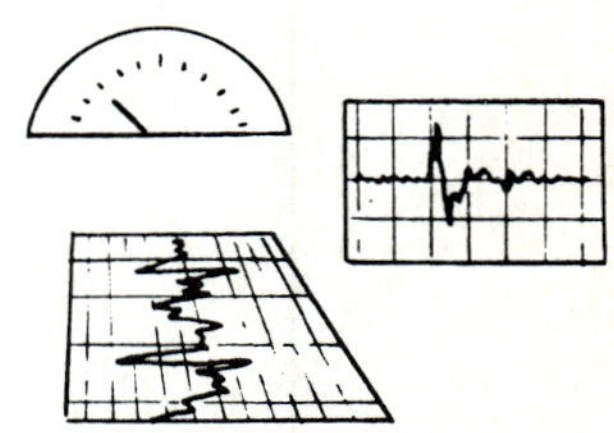
Surface traffic control	Lights, signals, signs, and markers	Multiple observer usage requires displays on or juxtapositioned with roadways, runways, and taxiways	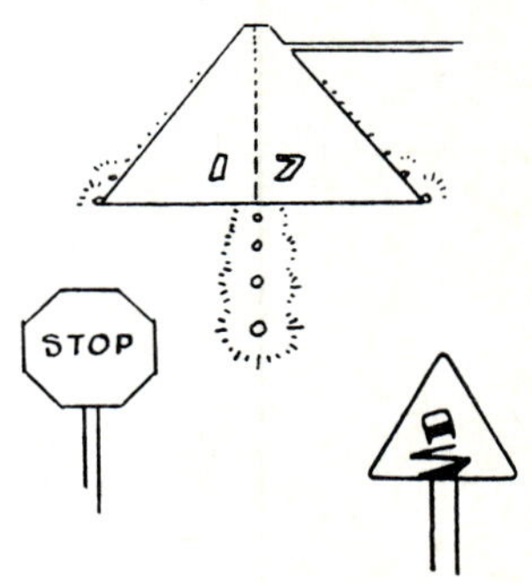

VISUAL DISPLAY SELECTION (continued)

Display Requirements	Type of Display	Rationale	Example
Large-group monitoring and signboard and scoreboard displays	Static, painted, back-lighted, or multilight signs (lights effective only at night)	Least expensive to install and maintain; permanent	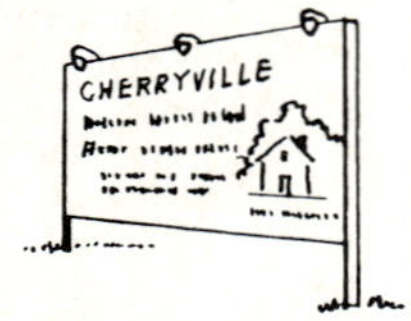 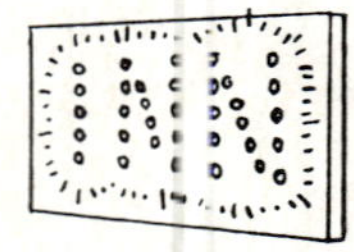
	Slide projection (front and rear; effective only in darkened room)	Static information; temporary exposure of successive pictorial or data items	
	Matrix type, moving message, or pictorial sign (multilight)	Effective at night for advertising or sports scoreboards	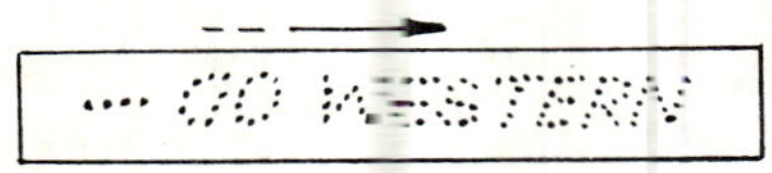
	Mechanical character-changing message board	Useful for message status boards in transportation terminals	
	CRT and TV	Use only indoors for short viewing distance	
	Motion pictures	Primary use for movies, although has potential for training simulation	
	Edge-lighted plotting displays (darkened rooms only)	Command and control operation centers—plotting by hand on rear side	

LIGHTS TO DISPLAY INFORMATION

The following are examples of how a simple light or set of lights can be used to convey information by the fact that the light is on or off, by the color of the light, by the position and/or shape of the light, and/or by whether the light is steady or flashing.

1. Power ON Lights
These are used to indicate that electrical power is available to the equipment. Such indicators should be limited to one per piece of equipment or work station since several lights will confuse the operator unless each pilot light is labeled.

Pilot lights should not be brighter than necessary to determine that they are either lit or not lit (depending on the prevailing ambient illumination). Otherwise, they may interfere with other visual tasks since they tend to cause glare if the general ambient light conditions are fairly dim.

Entire control panel lit — signifies power ON

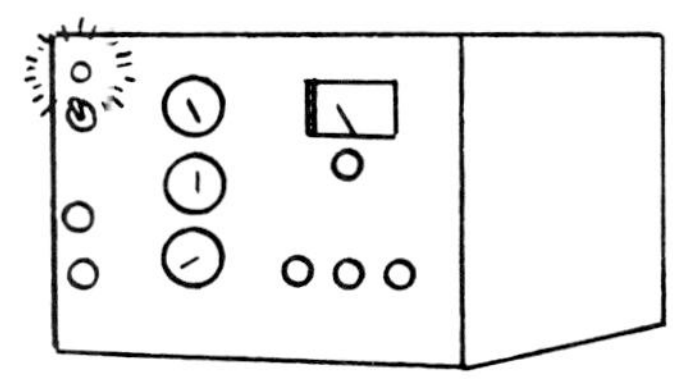

Power ON pilot light

2. Function Active Lights
These are used to indicate that a particular function is occurring, as in the case of the hand-held recorder shown in the accompanying sketch; i.e., the pilot light advises the operator that the recorder is in the "dictate mode" and that the tape reel is running.

Record mode pilot light

3. Special Hazard Condition Lights
Some equipment, products, or tools present potential hazards that the operator may not be aware of unless a warning light is provided. Two examples are illustrated. The soldering iron will be hot when power is fed to the soldering element and the operator forgets and becomes careless about touching the soldering element and/or lays the tool down before it has cooled. The pilot light is helpful in this case to remind the operator to be careful.

In the case of the stove with several burners, a pilot light arranged with each burner and/or control is helpful in reminding the user not to touch a live or hot burner. This hazard has become especially critical with the advent of the newer ceramic-top stoves, which provide no visual cue that a burner is hot; with an older stove, the change in color tells the user that a burner is hot, but no such visual indicator is provided on the ceramic-top stove.

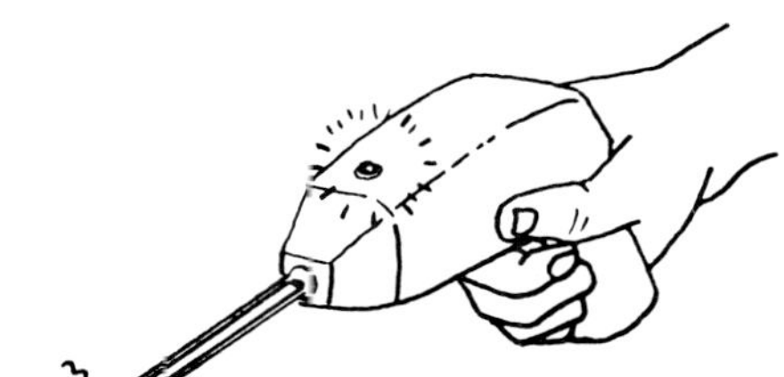

Element hot light

4. Operating Mode Lights
A single push-button control switch is often used on some equipment in order to reduce the number of switches. This is practical because the push button can be back-illuminated, not only with a single light, but also with both a white and a red back-lighting source; the red light indicates that the machine is in a warm-up mode, and the white light indicates that the machine is ready for operation.

Many different approaches can be taken in using a simple light to indicate an operating mode. However, it is important to recognize that such advisory lights must be understood by the particular user; i.e., the mere fact that a light comes on does not necessarily mean that information is being conveyed.

In some instances, however, almost everyone associates a certain type of light signaling with a particular function. One example is the simple light on the automobile driving panel that indicates when the headlights are switched to the high-beam condition. The recent use of blue as opposed to red has confused many drivers. Such arbitrary changes should be avoided.

A similar panel light is the turn signal indicator. Although some designers have used a single indicator light for both left- and right-turn indication, this is poor practice. In addition, since a primary concern of the driver is to know whether the control is signaling right or left, the position of the two indicators, as well as their shape, should reinforce the needed identification. These lights, by tradition, should be green.

Burner No. 4 ON

LIGHT DISPLAY VARIABLES

Viewing Angle

In most situations it is desirable that the light display be visible from the widest viewing angle possible. An extended lens is suggested to increase the effective viewing angle. Except for special cases where it may be desirable to limit the viewing angle so that only certain observers can see the light, avoid burying the light below the surface of an instrument panel or other structural surface.

As shown in the accompanying illustrations, when it is desirable to make the light visible only to certain observers, use louvers and/or turn the display toward the intended observer. The limited view is appropriate for such devices as a stop light which is intended only for one group of motorists and which would be confusing for other motorists.

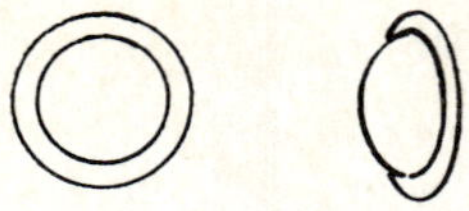

Wide-angle view

Limited angle of view

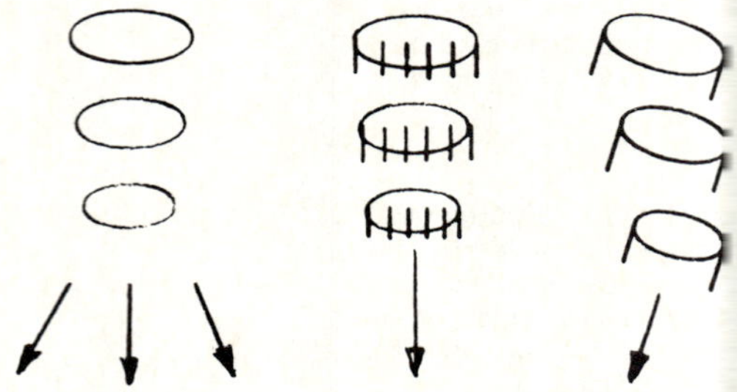

WIDE VIEW

VIEW LIMITED

Colored Lights

Certain color-coding conventions are fairly well established and should be followed; i.e., unless it can be demonstrated that changing the color code would enhance performance, do not change colors that already have significant meaning to the majority of observers. The following are some of the conventions:

Red lights: Use for danger, warning, fire, taillights for highway vehicles, and portside navigation lights.

Yellow or amber lights: Use for caution, slow, power on, front turn signals for highway vehicles, and marinecraft identification.

Green lights: Use for go, ready, functioning correctly, and interior turn signal indicator for highway vehicles.

Blue lights: Use for high-beam headlight indicator for highway vehicles and taxiway lights for airfields.

White lights: Use for airfield runway lights.

SIZE AND COLOR CODING

When a number of light indicators are required within a single operator station or several related operator stations and when the level of importance varies for different functions, a combination of encoding techniques can be effectively applied.

The accompanying table illustrates the use of size and color codes to help differentiate among the several functions and urgency categories. It must be recognized, of course, that the value of such encoding lies in the fact that the devices are repeated in several places, thus allowing operators to learn to recognize the size differences.

Color as an Indicator of Operational Condition

Red: Red should be used to alert an operator to the fact that some situation makes the system inoperative, e.g., an error, a failure, or a no-go or life-endangering condition.

Yellow: Yellow should alert an operator to a situation in which caution, recheck, or delay is necessary, i.e., a condition that, if not attended to, could lead to a dangerous situation.

Green: Green should indicate that equipment is operating satisfactorily or that one can proceed. It can be used to indicate the successful completion of steps within a process, thereby establishing a basis for continuing with the next step.

White: White should indicate such items as status, alternative functions and selection modes, a test in progress, or any similar items that imply neither success nor failure of system conditions.

Blue: Preferential use of blue should generally be avoided because it has no standard meaning, except when it has been assigned a special significance within a given operating system; e.g., a blue reference light has been used by some automobile designers to indicate that headlights are in the high-beam condition.

A flashing red light should be used when an extreme danger is present; i.e., the flashing light indicates a situation of more importance than a continuous red light. A flash rate of between three and five flashes per second, with approximately equal on and off times, is recommended.

The flash coding technique is also suggested for other colored lights when it is desirable to increase the conspicuousness of the signal. Typical examples include a flashing amber traffic signal in advance of a regular stop-lighted intersection and an alerting signal on a communications status board.

For additional suggestions regarding the use of color coding for light indicators, refer to U.S. Army MIL-HDBK-759.

Coding of Simple Indicator Lights

Size and Type	Color			
	Red	Yellow or Amber	Green	White
½-in (12.7-mm) diameter or smaller; steady	Malfunction; failure, stop action, no-go	Marginal condition; possible problem arising	Condition acceptable; OK to go ahead; in tolerance; ready to perform function	Indicates status of function or physical position; some action in progress
1-in (25.4-mm) diameter or larger; steady	Master summation (system or subsystem)	Extreme caution (impending danger)	Master summation (system or subsystem)	
1-in (25.4-mm) diameter or larger; flashing (3 to 5 per second)	Emergency condition (impending personnel or equipment disaster)			

FLASHING LIGHTS

Flashing lights are recommended wherever it can be anticipated that an observer needs reminding that a situation exists or is about to occur; in certain cases this need arises when there is a high probability that the observer may be visually occupied and needs the benefit of a more conspicuous visual signal.

The main benefit of the flashing light is that it is more conspicuous. On the other hand, too many flashing lights easily destroy the conspicuousness of any one signal. The same is true of color. Red was at one time the most effective color to use for taillights because they were easily identified among the typical white street lights. On a crowded freeway, however, there are so many red taillights that the motorist often has difficulty knowing what is going on ahead.

BRIGHTNESS

The desired brightness level of any signal light depends on the ambient lighting conditions, the brightness of other lights in the field of view, and the need to identify a change in situation (i.e., a given signal may be made to glow brighter when a functional change has occurred).

Often it is helpful to increase the conspicuousness of a conditional change by combining a brighter light with flashing. Care should always be exercised, however, not to blind an observer at a critical moment by increasing the brightness of a light. A typical case in point is the change in brightness of the taillight of a car when the brakes are applied. Although the objective is to make the change in brightness sufficient to let the observer in the rear know for sure that the brake has been applied, the observer may be blinded by an extremely bright light and make precarious evasive maneuvers that are not warranted.

POSITION OF LIGHT SIGNALS

To be effective, of course, a signal or pilot light has to be in a spot where the observer is most likely to be looking. Other positional considerations are also important, such as the following:

A power pilot light should be located in the place where the observer first looks at the equipment (typically on the upper part of the equipment).

Warning lights (and/or lighted indicator lights) should be located within about 15° of the principal line of sight of an operator in order to make sure that the operator will notice a warning while concentrating on something else (e.g., watching a CRT display).

COMBINATIONS OF PILOT LIGHTS, ANNUNCIATORS, AND LIGHTED PUSH BUTTONS

It is important to an operator to be able to tell quickly which lighted indicator is a pilot light, an annunciator, or a telelight and which is a lighted push button. Avoid the temptation to select hardware components that all look alike; e.g., use a round fixture for a pilot light, a rectangular fixture for an annunciator or telelight, and a square fixture for a push button.

The accompanying illustration shows how much easier it is to tell the difference between components when each has a different shape.

It is also desirable to select push-button hardware in which the button extends about ⅛ in (0.32 cm) out from the panel and annunciator hardware that is more or less flush with the panel. The best pilot light is one that also extends from the panel so that an operator can tell from an angle when the light is on.

Some of the new flush control concepts that require the operator merely to touch an outlined section of a panel do not comply with the above suggestion. There are no studies to indicate whether operators are confused by such concepts. However, one can be certain of no confusion if the above suggestions are followed.

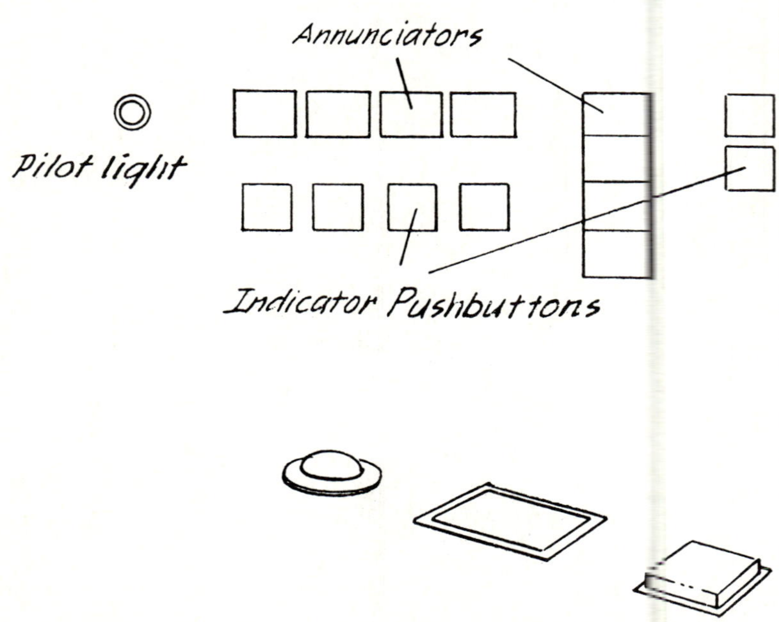

WORD DISPLAYS: SIGNS AND LABELS

The use of words as a display device adds another dimension for increasing information transfer. However, it may also add complexity and increase the time an observer requires to recognize and understand the meaning of the information one is trying to convey. The most common word display is perhaps the sign, which can be used for identification, for warning or caution, for guidance, and for instruction.

Graphics specialists often are given the task of producing signs. In some instances, they are concerned primarily with the attractiveness of the sign rather than with its readability. Refer your graphics specialist to the guidelines presented in this handbook if you are not sure that he or she is familiar with what is required to make a sign readable.

Sign Standards

Certain types of signing have been standardized in order that the size, shape, color, and wording of particular signs will be the same wherever they may occur. For example, the U.S. Department of Commerce, Bureau of Public Roads, issues a *Manual on Uniform Traffic Control Devices for Streets and Highways*. Similar standards have been devised for various industrial safety signing. It is important to acquire and use these standards rather than to invent new signs.

Traffic signs fall into several distinct categories: (1) regulatory signs, (2) warning signs, (3) guide signs, and (4) pavement and curb markings. Although the manual also covers street-name signs, the specifications are not necessarily acceptable from a human engineering standpoint because they are not large enough in most cases and because they cannot be used by the blind.

Key issues for signing are the following:

Legibility
Size in relation to viewing distance
Illumination
Clarity of meaning
Location relative to how and when the sign is to be used by the observer

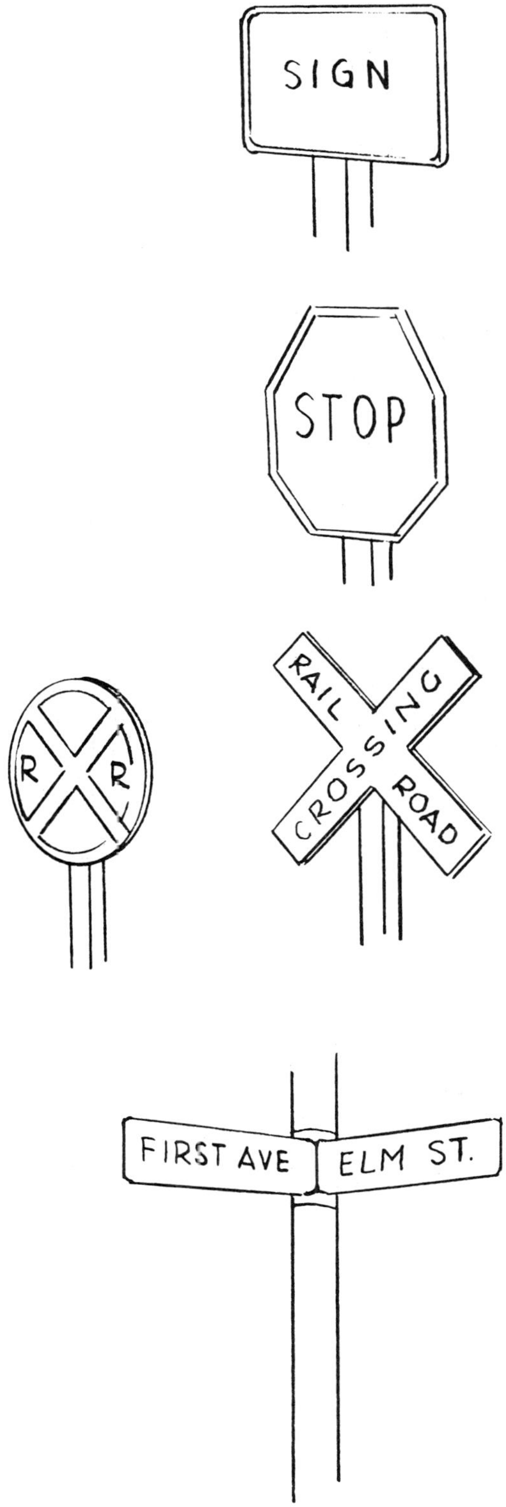

Sign Colors

Color can be used to encode and/or categorize various signs. For example, highway signs are typically red, yellow, black, or white, with other colors such as green and blue interspersed for less critical information.

The important factors to consider in deciding to use special color codes are the following:

Adhere to certain color codes that are already standardized: red—danger or stop; yellow—caution or be prepared; green—OK or go ahead.

Select colors or color combinations that enhance the legibility of the sign lettering or symbols.

Limit the number of colors selected to those which the observer can differentiate reliably, i.e., those which the observer can verbalize and recognize independently of a comparison with another color close by. The average individual can differentiate only a few colors in terms of faithful verbalization: red, yellow, orange, blue, green, brown, purple, white, and black.

In general, select saturated colors, as opposed to pale colors when you want the colors to have special significance.

Use luminous or fluorescing color to enhance color discrimination under marginal lighting or viewing conditions.

Illuminate colored signs with essentially white light in order to ensure reliability of color identification.

Use aviation red, yellow, or blue lights to ensure that color-deficient observers also can recognize these particular colors.

Do not alter critical colors to make them match an aesthetic color scheme.

Orientation Signs

Designers often fail to recognize the importance of providing signs to help people identify where they are and where they are going. Typical of these are signs that identify an entrance, a parking area, a bus stop, and an elevator. Once again, such signs should be legible and intelligible, and they should be located so that the observer can easily associate the sign with the physical features pertinent to the situation.

Emergency Signs

Identification of features that may be needed in an emergency is especially important. The typical exit sign is generally prescribed by local ordinance. Other, similar signs include those indicating an emergency escape route, a fire extinguisher, and a first-aid station.

Company Logos

Although one might argue with a specific logo used by a particular company, it is important to the company (and probably to most of its customers) that the logo be retained in any signing created for that company.

Caution Signs

In addition to the caution signs one generally associates with large outdoor installations (e.g., traffic control), similar signing is applicable to other situations, such as the sign posted on an automobile gas filler cover indicating that unleaded fuel is to be used.

Operating Control Labels

The push buttons for elevator call are typical of this type of signing. Usually, the button that is pushed also illuminates to indicate that the request has been made.

Mechanical Flags

The mechanical flag has been used successfully for indicating the functional condition of a parking meter and/or a control panel instrument. In a sense, this provides a go-no-go signal using words rather than a light.

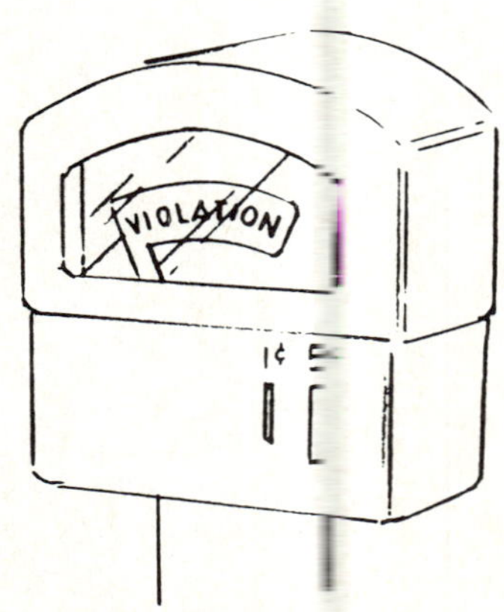

WARNING SIGNS

Regardless of whether the proposed sign falls within the specific classification of a warning or caution sign or signal or a hazard advisory or instruction, the primary objective should always be to make the device noticeable, recognizable, and understandable. Some of the key factors to consider are the following.

1. Conspicuousness
The sign should stand out among adjacent visual distractions because of its size, shape, color, or symbology or because it is moving or flashing. The sign or signal must be located where it is expected that most people will be looking; it should not be hidden behind a post or an operator control or below the heads of other observers or the tops of equipment, parked cars, etc. Above all, do not camouflage the sign or signal to keep it from spoiling the decor planned for the site.

2. Emphasis
The most prominent feature of a warning sign should be words or symbols that imply danger. Include such words as "danger," "hazard," "caution," and "warning." Use standardized symbology that by convention immediately indicates the nature of the hazard, e.g., a skull and crossbones for poison, the "fan" symbol for radiation, and the slow vehicle triangle for a roadway hazard. Add unique borders around the sign, such as yellow and black stripes or a checkerboard design.

3. Legibility
When words and messages are used, make sure that they are legible by selecting the proper letter style and size and by making the contrast between the letters and the background sufficient for whatever lighting condition may exist. Provide a sufficient border around the lettering so that other background influences, such as other signs, lights, or patterns, will not cause confusion.

4. Simplicity
Use as few words as possible; i.e., avoid long, complex advisory information or instructions that take too much time to read. Tell the observer exactly what to do or not to do. Avoid acronyms and abbreviations.

5. Intelligibility
Make it clear what the hazard is and what may happen if the warning is ignored. Especially avoid contradictions or implications that imply some doubt as to the actuality or seriousness of the hazard.

6. Visibility
Make sure the sign is visible under *all* expected viewing conditions, i.e., during the day and at night, if the artificial light fails, under marginal atmospheric conditions, in bright sunlight, etc.

7. Permanence
Choose devices and/or sign materials that are resistant against aging, wear, soil, and deterioration due to sunlight or cleaning methods, and use techniques that are resistant to vandalism.

8. Standardization
Use standard signing words, symbols, etc., wherever they already exist, such as those recommended by OSHA, the FHA, the Department of Commerce, and the military. Although some of these standards may not be in complete accord with the recommendations suggested here (i.e., they may not represent the best human engineering), they are usually so well established that changes to comply with human engineering principles might confuse the observer. When no such standard exists, however, apply the above principles and guidelines in creating any new signing.

STREET SIGNING AND MARKING

The practice of placing signs directly on street surfaces has proved useful for instructing and guiding motorists. Special care must be taken in designing such signs so that the letters (which are not viewed from the same angle from which most signs are seen) are readable at the expected approach speeds and visual angles of the motorist.

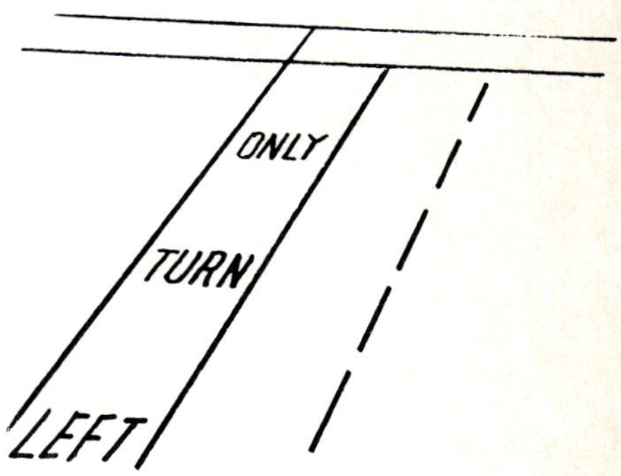

Combining Words and Lights

As noted earlier, providing both a light and a word expands the possibilities for signing visual information displays. The addition of the words increases the probability that the meaning of the sign will always be clear, and the addition of the light provides a means for emphasizing the particular sign or signal mode and/or allows one to use the same display unit to present alternate information. One of the most common devices is the lighted pedestrian sign that switches from a "walk" to a "wait" instruction.

Combination light-word signal devices have long been successfully used for vehicle and equipment control monitoring. These are variously referred to as "telelights," "annunciator lights," etc.

One has the choice of using dark printing with a lighted background or the reverse, i.e., illuminated printing with a dark background. The general rule is to reserve the lighted background for emergency or warning displays, since this makes the display much more conspicuous and attention-demanding.

The question is invariably raised as to whether the lettering of an inactive display should be visible. As a rule, it should be visible because it is important for the operator to have previous knowledge of what display comes up, and where.

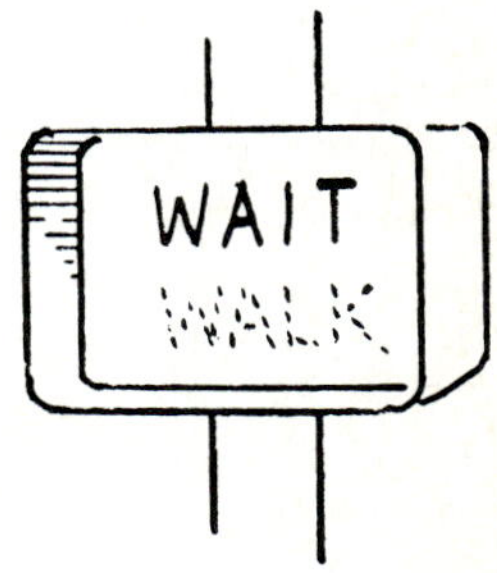

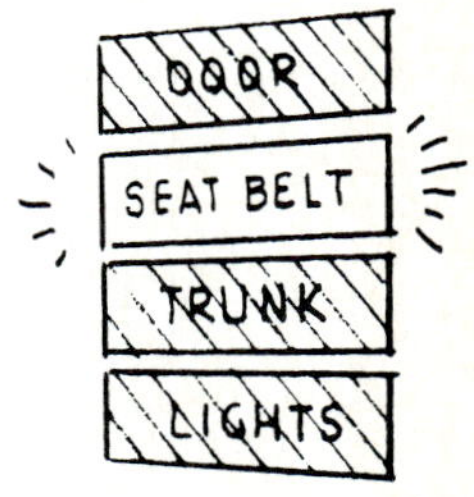

SIGN SHAPE

Sign shape can be used as a method for identifying categories of information, as in the case of highway signs and industrial safety signs. The most recognizable shapes under all conditions (i.e., size, distance, ambient light, and atmospheric interference) are shown in the accompanying sketches. Other shapes can be used, but with caution. Some of the factors to consider in using shape are the following:

Will the shape mean anything to the average observer? For example, street and highway sign shapes are categorized for various regulatory, guidance, or information purposes, but very few motorists remember these categories.

Will the shape be compatible with the spacing of words or symbol labels; i.e., will the shape crowd the lettering so that it will not be legible?

Will the shape continue to maintain its recognizability under all expected size, distance, or other viewing conditions, or will the critical features tend to blend and disappear?

Are there too many other signs of the same shape (and possibly color) that would tend to diminish the significance of the intended sign?

Note: In the examples shown, it should be noted that the basic categories of shapes are circles, triangles, rectangles, and crosses. They are varied in terms of orientation or expansion and contraction. Pentagons, hexagons, and octagons do not retain their recognizability; i.e., they fade into circles when viewed from great distances.

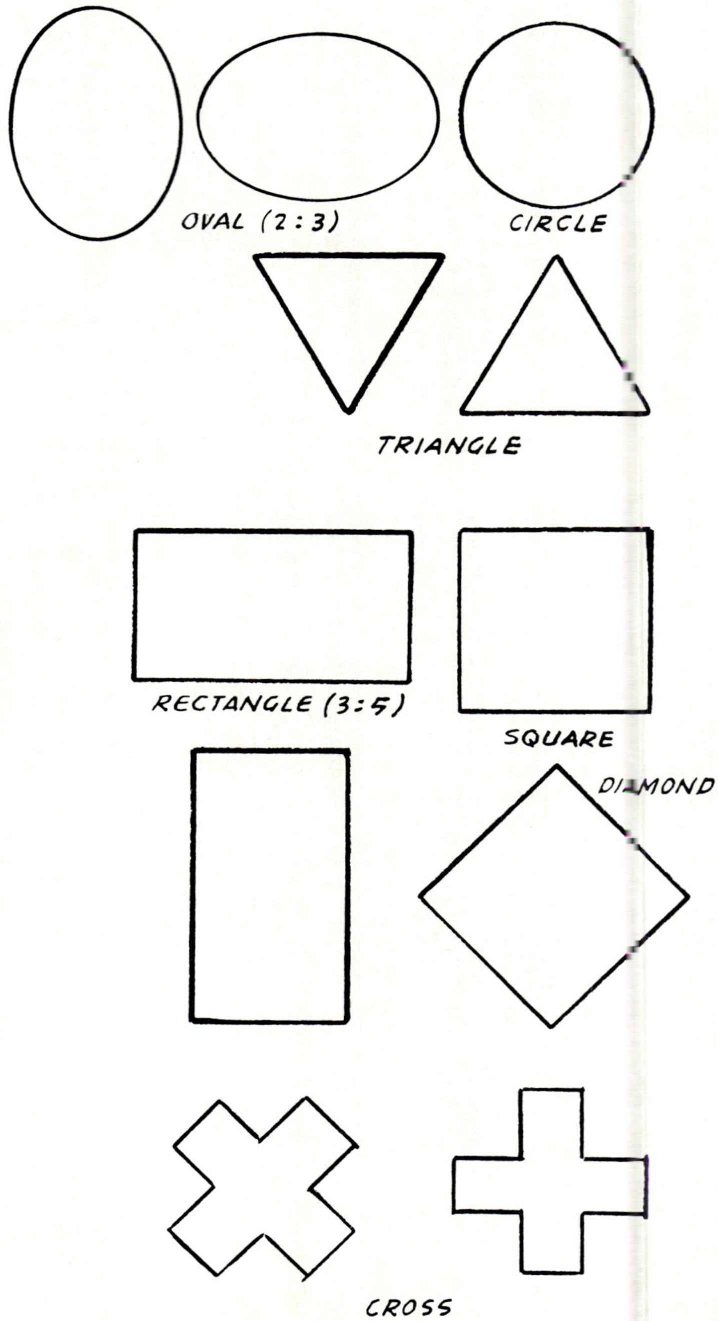

ARROW AND POINTER GRAPHICS

Directional pointers, unlike those for instrument displays, do not have to mate with other markings.

Design or select directional pointers which are recognizable as pointers and which are sufficiently conspicuous to stand out among an assortment of other geometric designs and backgrounds.

The designs shown in the accompanying illustrations have various deficiencies, although they are used frequently. The first two lack boldness and are often not recognizable under certain illumination conditions; the third is not recognizable under poor illumination conditions or from considerable distances; and the fourth and fifth are easily confused with other kinds of symbols because they lack sufficiently conspicuous geometric cues.

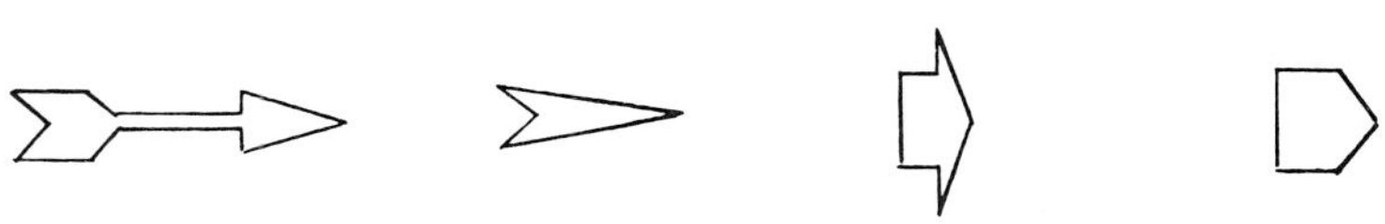

The idealized pointer proportions shown in the accompanying illustration provide proper geometric emphasis to maximize recognition.

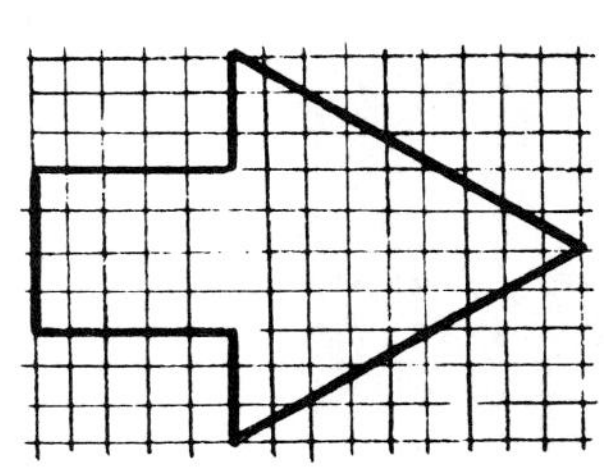

Pointer proportions for directional signs

The Arrow as a Directional Aid

Although many people are aware of the value of the arrow symbol as a directional aid for highway signs and other architectural guidance signs, not many designers have taken advantage of this special graphic device to help the person in other situations where guidance is necessary. The following guidelines may be helpful in deciding when to use the arrow symbol and how to use it effectively:

The arrow tends to attract attention since it implies "Follow this."

The use of several arrows properly aligned above separate aisles or highway lanes provides immediate identification of guidance information related to the specific aisles or lanes.

An arrow with a tail that obviously changes direction provides immediate recognition of a directional change. In a series of arrows, it is immediately recognized when one or more of the arrows point to another direction, as opposed to continuing in the direction the observer might be going.

A single arrow accompanied by simple place names immediately implies to the observer that these places may be found by following in the direction in which the arrow points.

Beware of the use of an arrow pointing upward. Unless observers can plainly associate a staircase or inclining ramp, this type of arrow orientation merely confuses them.

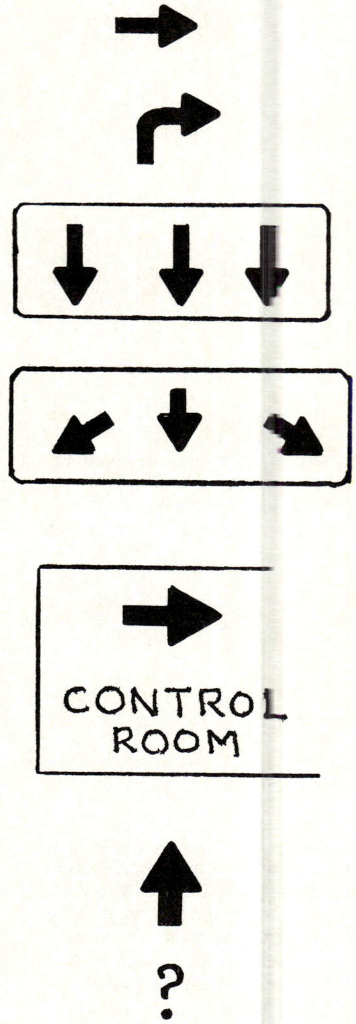

ALPHANUMERIC CHARACTERS

Each alphanumeric character's recognizability depends on the proper emphasis of specific features such as diagonals, flat tops, and easily discernible loops. Some type fonts tend to obscure these features. Some examples of how certain characters can be distorted to the point of confusion are shown in the accompanying illustration. The following points are important:

The letter A must have a clearly delineated space above the horizontal stroke.
The letter B should have approximately equal loops at the top and bottom.
C's and G's are easily confused with each other or with O's if the break is not clearly discernible, if the horizontal stroke of the G is not long enough, or if the O tends to appear too square.
The E must not have its horizontal strokes too close together.
The M and the W must have a sufficiently long center section.
The P must have a large enough loop.
The S and the 5 are easily confused if the S is too square or if the horizontal top of the 5 is not long enough.
The 1 and the I are easily confused with each other unless they are made to look different. In this instance the serif is perhaps most helpful.
The U and the V are easily confused unless the vertical sides of the U are maintained, as opposed to the slanting sides of the V.
The Y needs a long tail to be discernible from the V. However, the V-shaped top also must be definite to keep the Y from being mistaken for a T.
When the 6 and the 9 have loops that are too large and tails that are too curved, they are not easily recognized. The loops must be readily apparent, but the tails need to be straight.

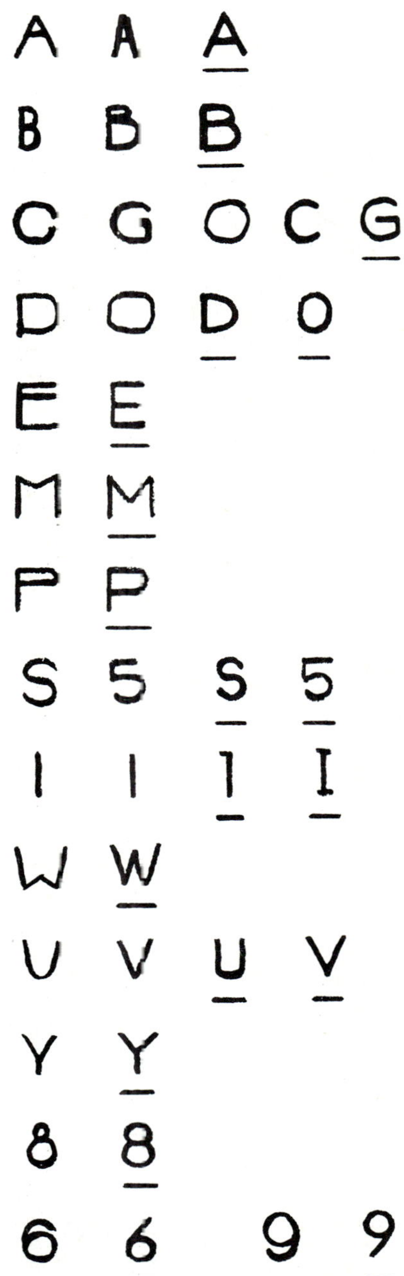

— Underlined are preferred

GUIDELINES FOR OPTIMIZING LETTERS AND NUMBERS TO BE USED ON SIGNS, CONTROL PANELS, AND INSTRUMENT FACES

In spite of some controversy over the best style of character for use in signing, labeling, and instrument marking, at least the following parameters are agreed upon as important to consider in selecting a given character style for a specific application:

1. Proper height/width ratio, e.g., character proportions
2. Proper stroke width, e.g., width of the lines making up the character
3. Maximum figure-ground contrast
4. Adequate overall size for the viewing distance expected
5. A simple rather than an elaborate style, i.e., generally a block style without serifs
6. Upright rather than slanted (italic) characters
7. Adequate intercharacter spacing
8. All capital letters for short labels and signs (as opposed to capital and lowercase letters for extended, instructional materials)

Special Type Fonts

Special type fonts such as those shown in the illustrations on this page and on page 488 have been devised for use on military control panels and visual displays. Individually, these characters are considered maximally legible. They are available in Le Roy Lettering Guide templates and are generally preferred over the standard Le Roy characters.

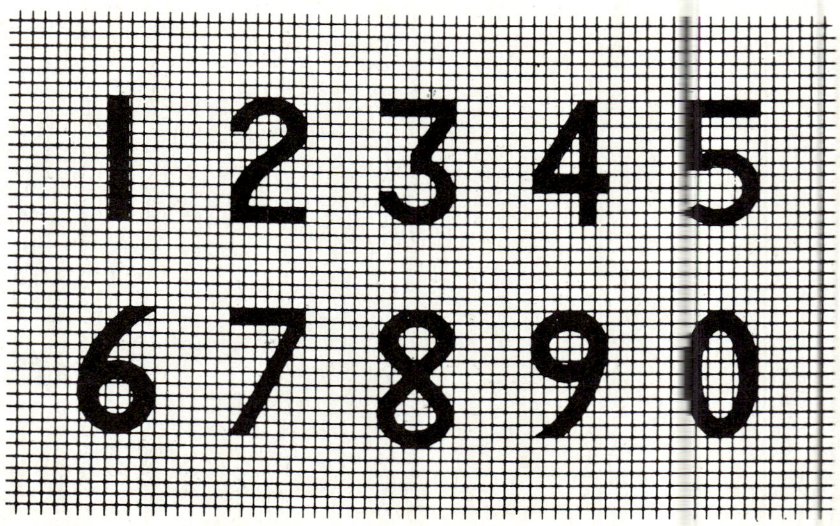

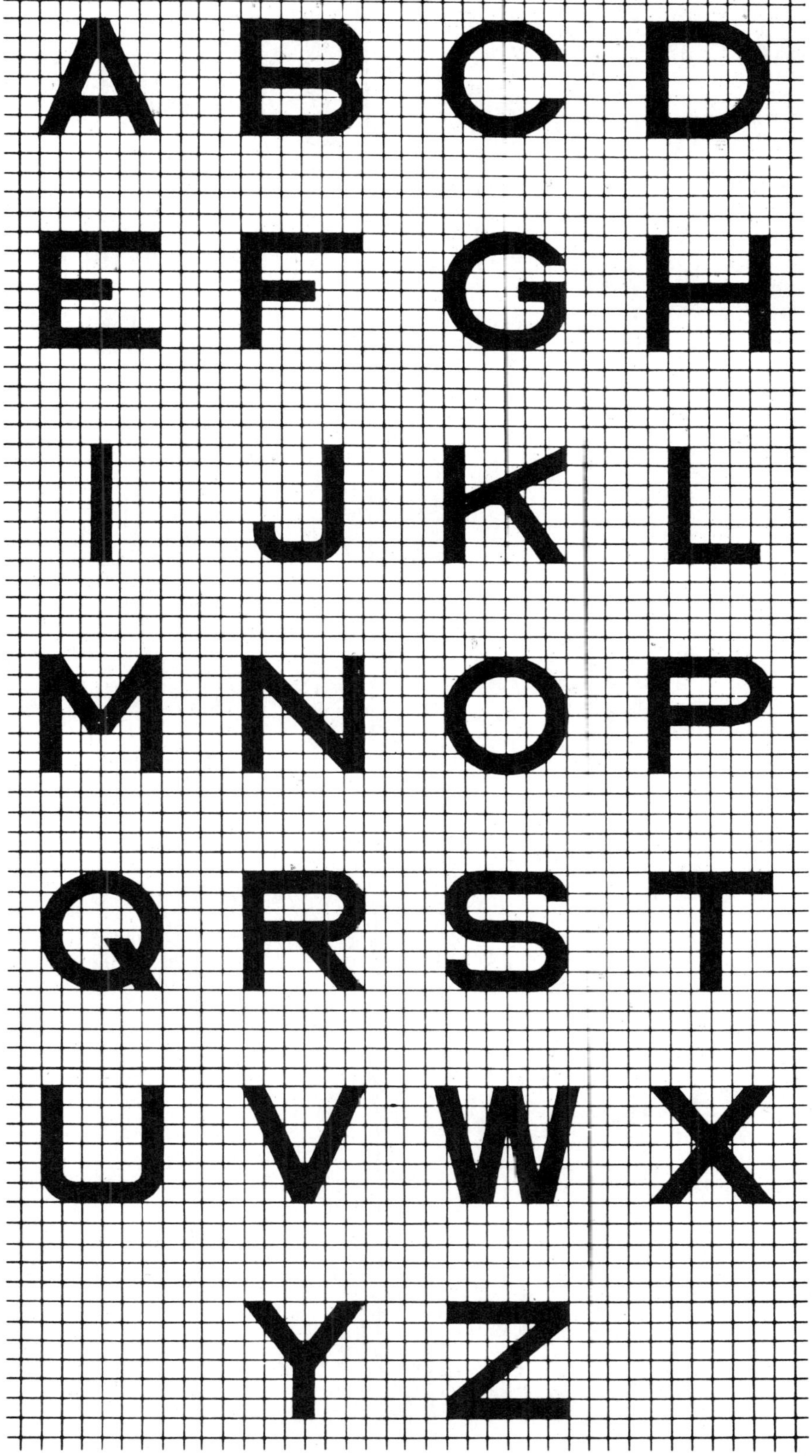
A B C D
E F G H
I J K L
M N O P
Q R S T
U V W X
Y Z

Choosing Correct Type Styles from Standard Type-Font Catalogs

Typically one must select type styles from the engraver's or printer's available type. In doing so, one should look for certain key letters or numbers that are the most difficult to read. The several comparison characters shown in the accompanying illustration are especially important to examine because they are most easily confused.

Simple block styles should be used for control panel and instrument labeling applications. Some of the standard type styles that are acceptable include Folio Book, News Gothic, Trade Gothic, Futura Medium, and Spartan Medium.

However, since the particular printer or engraver may not have these styles, one should learn to examine the available styles and select the best one for a particular application.

Type faces to be avoided are those which:

Have uneven stroke widths
Have extended serifs
Include internal patterns or stripes
Are italicized
Are stenciled
Appear like handwritten script
Are shadowed or made to look three-dimensional
Look like Old English script
Are distorted to look tall and thin or wide and fat

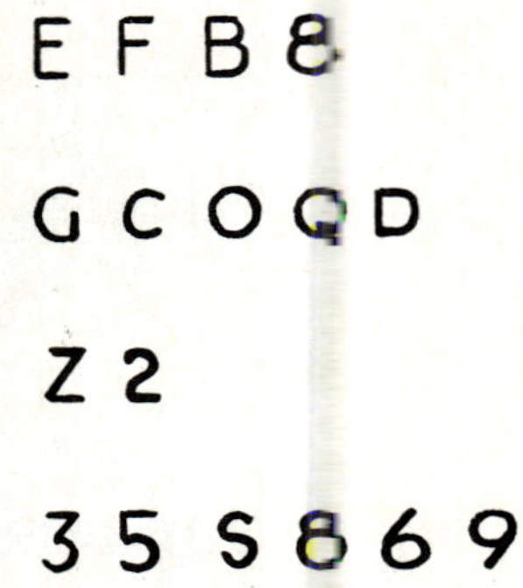

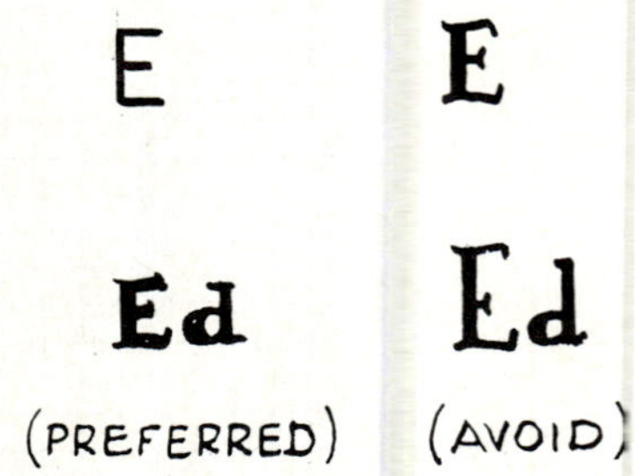

Examples of typically poor type styles that can be found in standard type catalogs are shown in the accompanying illustrations.

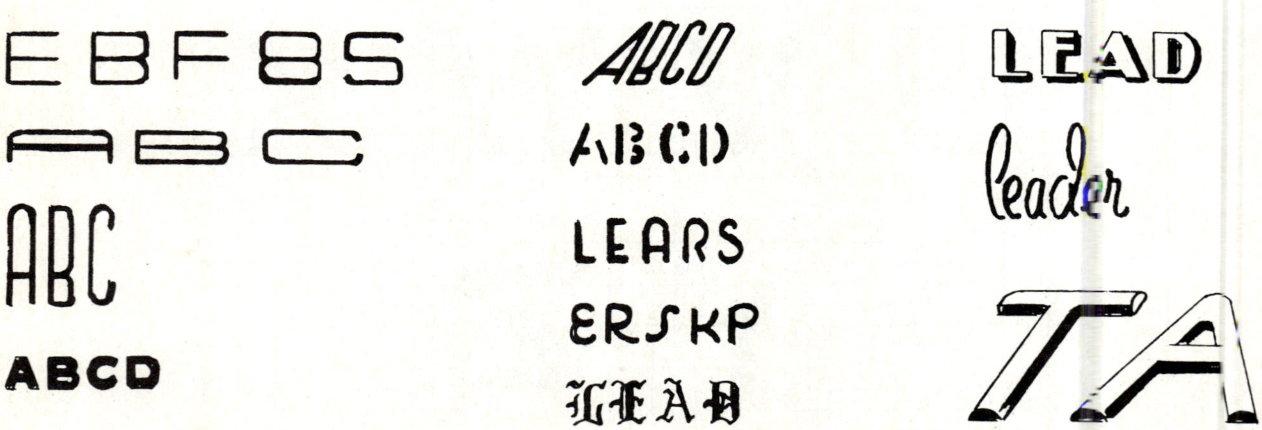

STROKE WIDTH The stroke width of letters and numerals is important in determining how legible the characters will be. If the stroke width is too narrow, there will not be enough definition to make the character clearly recognizable. If the stroke width is too thick, some of the more critical features of the character may be obscured so that it is unrecognizable. For example, the closed portions of a letter such as a B or an R or of a number such as a 6 or 9 may be obscured by a thick stroke width. Similarly, the open space between the horizontal strokes of an F or an E may not be discernible when the stroke width is too thick.

The preferred stroke width for *dark characters* on a light background is 1:6, i.e., the stroke width versus the character height.

The preferred stroke width for *light characters* on a dark background is 1:7.

The rationale for two different stroke widths is based on the fact that light falling on the lighter surface tends to "spill over" into the darker area. In the case of a lighter character, this tends to make the stroke width broaden out; in the opposite case, the stroke width appears narrower. The above recommendations compensate for this apparent artifact of light reflection.

Above all, avoid selecting extremes, either stroke widths that are extremely narrow or ones that are extremely bold.

Because letters and numbers are most often grouped, it is important to recognize that the criteria for a single character's maximum legibility do not always apply. For example, since grouped numbers or letters become too spread out for the observer to grasp the significance of the total pattern, characters should be selected that are taller than they are wide.

Depending upon the specific typeface available, the most readable height/width ratio should be somewhere between 5:3 and 3:2.

Letters or numbers within a group should be about one stroke width apart (± ½ stroke width).

The space between groups of words or number groups should be about three stroke widths (± ½ stroke width).

The space between lines of number groups or words should be at least one-third the height of the tallest character, but in no case should the space between the tallest character of a lower line be less than one stroke width from a character above it that projects below the line.

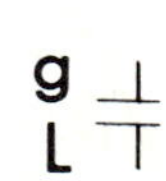

The preferred height/width ratio for typical block-type letters and numerals ranges from 5:3 to 3:2, depending on the typeface available.

For emphasis, it is satisfactory to use wider characters with ratios up to 1:1. Avoid selecting characters that are wider than they are tall, however, since such distortion tends to increase recognition time.

LEADERSHIP

LEADERSHIP

When horizontal space is limited, it is permissible to use narrower characters. In some cases, this is even more desirable than trying to crowd the so-called optimum character into a limited space or reducing the space between words.

LEADERSHIP

Avoid selecting extremely narrow characters since they tend to appear blurred and to increase both observation time and errors.

LEADERSHIP

The name of a manufacturer on a product usually is not a critical operator use consideration. A unique typeface here may be an advantage since it clearly separates the name from other operationally critical information.

LEADER

FIGURE-GROUND CONTRAST

Brightness and Color

To ensure adequate legibility, there must be adequate visual contrast between alphanumeric characters and the background against which they are viewed. The following should be considered:

Under normal illumination conditions (i.e., where the observer does not have to be dark-adapted), use dark characters against a light background.

Where the observer must maintain a dark-adapted condition, use light characters on a dark background.

The character should be at least twice as light (or dark) as the background. Greater contrast is preferred and is especially important if the expected observer population includes persons with impaired vision, if the viewing conditions will be less than optimal, or if physiological stress factors are anticipated, such as oscillation vibration, motion, or high *g* forces.

Avoid the use of glossy or highly reflective metallic materials for either the lettering or the background (except where the lettering must be illuminated remotely, e.g., by automobile headlights). Spectral unevenness usually causes portions of the letter or numeral to become illegible as a result of the variation in light reflection.

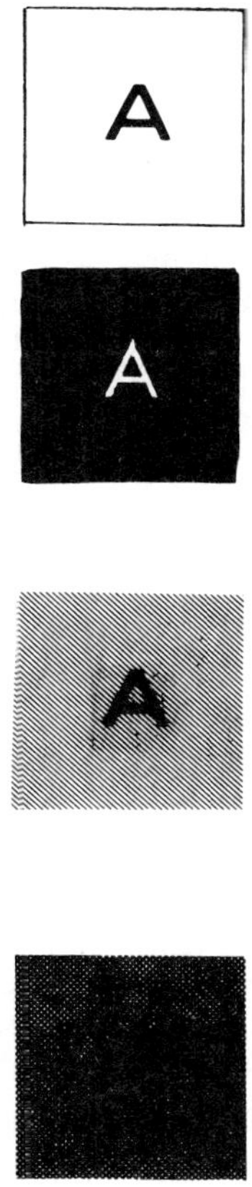

Suggested Color Contrast Selections in Order of Expected Visual Efficiency

Conditions	Characters	Background
Average or higher levels and quality of illumination	Black	White
	Black	Yellow
	White	Black
	Dark blue	White
	White	Dark red, green, and brown
	Black	Orange
	Dark green and red	White
	White	Dark gray
	Black	Light gray
Poor level and quality of illumination	Black	White
	White	Black
	Black	Yellow
	Dark blue	White
	Black	Orange
	Dark red and green	White
Dark adaptation required	White	Black
	Yellow	Black
	Orange	Black
	Red*	Black
	Blue and green	Black

*Low-level red light is required to maintain the lowest dark adaptation level.

Note: When lettering is internally illuminated so that the letter is always bright against an essentially black background, almost any color is satisfactory as long as the brightness is above the minimal threshold, i.e., as long as the light is visible. Opaque letters must be illuminated with an essentially white light; otherwise, the colored illumination may decrease the contrast between the character and its background.

CHARACTER SIZE

All things being generally equal (legibility factors, illumination, atmospheric interference, etc.), the larger letters and numbers are, the better they are seen and recognized—up to the point where the observer would be unable to see an entire character at a single glance.

Many people make a common error, however, in applying the above generalization; i.e., they believe that they should also make the character bolder as it is made larger for more distant viewing. As noted in the discussion of stroke width, to maintain legibility while increasing the size of a character, one must also maintain the proper stroke-width relationship.

The accompanying charts provide guidelines for determining how high a letter or numeral has to be for given viewing distances.

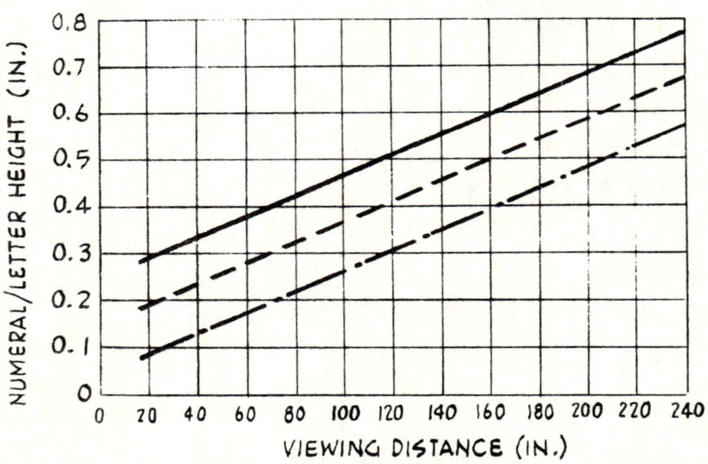

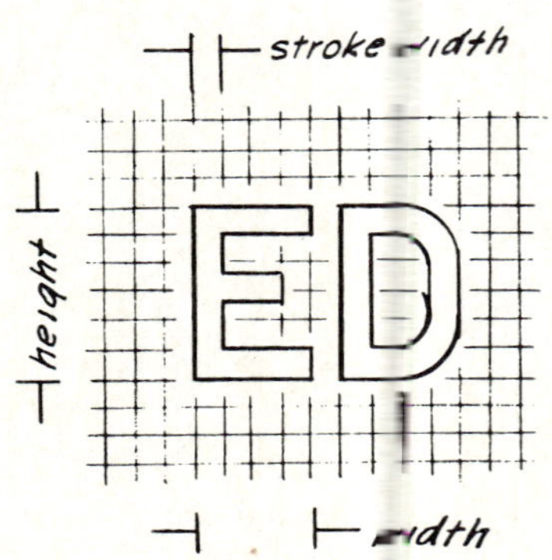

Letter height versus viewing distance and illumination level (minimum space between characters, one stroke width; between words, six stroke widths). (— For instruments where the position of the numerals may vary and the illumination is between 0.03 and 1.0 fL. --- For instruments where the position of the numerals is fixed and the illumination is 0.3–1.0 fL, or where position of the numerals may vary and the illumination exceeds 1.0 fL. –·– For instruments where the position of the numerals is fixed and the illumination is above 1.0 fL.)

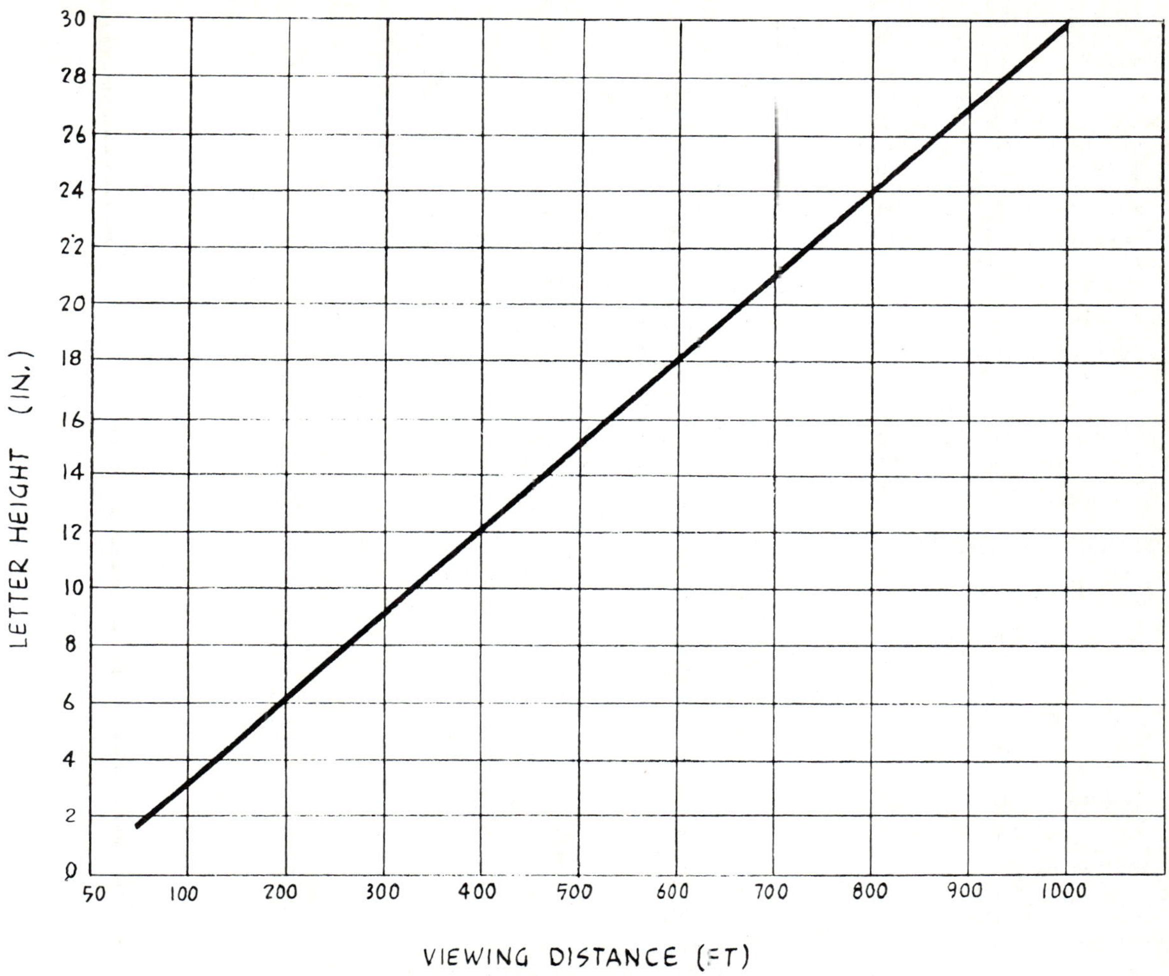

Minimum letter size recommended for highway signs. To calculate character height for greater viewing distances, letter height = $\frac{\text{viewing distance (ft)} \times 0.3\ \text{in}}{10}$

Note: The above is based on the assumption that the observer is relatively still. If a driver is moving rapidly toward or past a sign, the letter-height values should be increased approximately 10 percent, since the observer must be able to read the sign far enough in advance of a decision point to be prepared for the selected maneuver.

MESSAGE INTELLIGIBILITY

To be effective, words have to be selected carefully so that their meaning is perfectly clear to the observer. One should first try to determine the background of the expected user population. The lay person, for example, probably will not understand some words that may be clear to an engineer. Wherever possible, use common words and abbreviations so that everyone will understand the meaning intended.

Word Labels and Instructional Materials

1. Use common terms that originate from typical language usage and/or from standard lists of terms for special fields or groups of users (plant operators, etc.).

Examples of General versus Special Usage Terminology

General (Preferred)	Special
Speed	Velocity
Rear window	Back light
Rest room	Head
Brightness	Intensity
Adjust	Calibrate
Water analyzer	Ionic chromatograph

2. Use whole words rather than abbreviations wherever space permits. If abbreviations are required, refer to the list of standard abbreviations created by the appropriate "user" group. However, check these lists for duplications; it is undesirable to use the same abbreviation for two different functions.

Preferred	Avoid
Range	RNG
Bearing	BNG
Ticket	TKT
Air conditioner	A/C
Transmitter	TM
Power amplifier	PA

Acronyms may be used sparingly if they have been well established over a long period. They are least desirable for consumer products because of the wide variation in user experience.

3. To identify instruments, use terms that indicate what the instrument measures rather than the name of the device.
   OHMS not OHMETER
   RPM not SPEEDOMETER
   HRS/MIN not TIMER
4. Avoid the use of words that may be interpreted one time as a noun and another time as a verb whenever both may occur at the same workplace.
   FIRE = a fire has occurred (warning)
   FIRE = to fire a furnace (command)
5. Always use capital letters for labels and short instructions because they can be read at a greater distance than capital and lowercase letters. Capital and lowercase letters should be used only for extended sentence messages where it is necessary to provide punctuation.
6. Generally, instruction labels should be as brief as possible within the limits of clarity. If an instruction implies several sequential steps, arrange each step on a separate line, preferably with a dot at the beginning of each line.

The original instruction was:

WALK UP ONE FLOOR
WALK DOWN TWO FLOORS
FOR IMPROVED ELEVATOR
SERVICE

What the instruction really meant was: If you intend to go up only one floor or down two floors, walk instead of taking the elevator because this will relieve the use load and therefore provide better service for everyone.

Following are two ways to simplify the message:

SAVE TIME BY WALKING
IF YOU ARE GOING UP ONE OR
DOWN TWO FLOORS

TO GO UP ONE FLOOR
OR
DOWN TWO FLOORS
PLEASE WALK

7. Avoid the use of words that may have a negative psychological effect; e.g., the word "speed" itself implies traveling at a fast rate and therefore may encourage a motorist to drive fast.

THIS	*NOT THIS*
SLOW TO 10 MPH	SPEED ZONE AHEAD 10 MPH
SLOW DOWN WORKMEN IN ROADWAY	SLOW MEN WORKING
SLOW DOWN ANIMALS IN ROADWAY	CAUTION DEER CROSSING
DANGER LETHAL VOLTAGES	DANGER HIGH VOLTAGE
SLOW DOWN ROCKS ON ROAD	WATCH FOR FALLING ROCKS

8. When special precautionary words are required, select ones that provide a sense of urgency, hazard, or danger. Avoid words that seem to leave it up to the observer to decide whether there is a danger involved, since he or she may not have the necessary experience to make such a judgment.

Preferred Nomenclature	Poor
Warning	Note
Hazard	Precaution
Caution	Slow men working
Danger	Slippery
Watch out for workmen	Slide area
Watch for children	

9. The lettering for all labels should be oriented so that they read from left to right, not around corners, on the side, or up and down.
10. Control and display labels should be located consistently, either below or above the panel component, especially on the same panel. The decision to locate the label above or below components should be based on the eye position of the observer; i.e., if the panel labels are above the eye reference level, the controls may extend out so far that they obscure a label placed above the control, or the opposite might be true if the control panel is considerably below the eye reference level. Random placement of labels on the same panels makes them difficult to find and associate with the pertinent function.

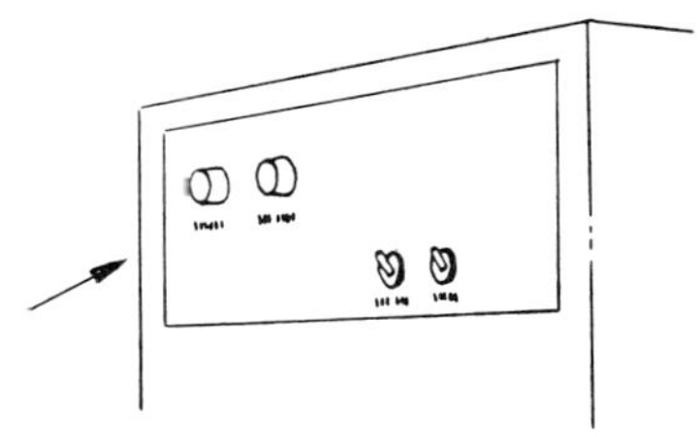

Put labels below controls when viewed from below,

above when viewed from above.

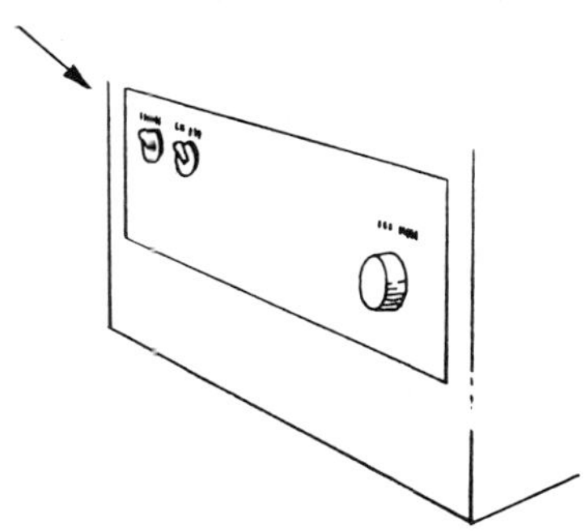

AVOID!

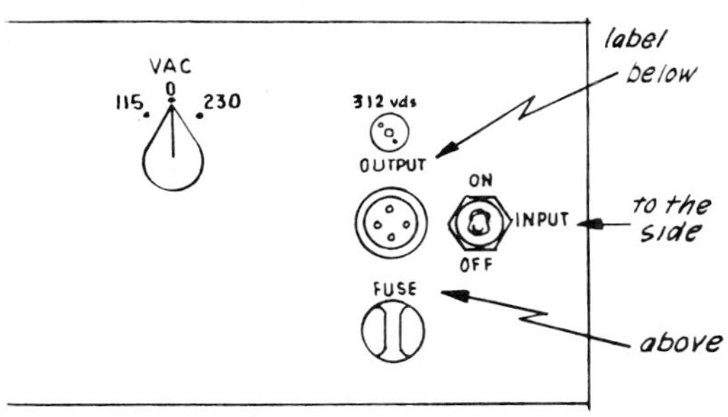

11. Avoid running adjacent labels together. This often happens because control elements were spotted on the panel without considering where the labels would go or how much space they might require.

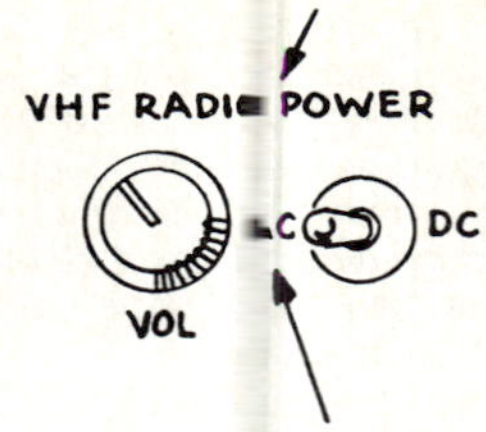

crowding = confusion

12. Utilize size and/or color coding to help the observer differentiate between levels of importance on your panel; i.e., the *system* label should be slightly larger than subsystem labels, and these should be larger than component labels. The examples shown in the accompanying sketches should be used as a general guideline.[1]

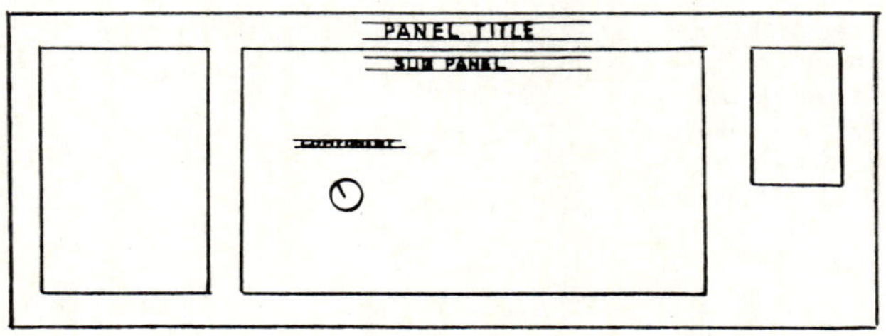

Primary title about half again larger than intermediate title – this half again larger than smallest title.

13. As a general rule, all labels should be made as permanent as possible; i.e., they should be wear- and damage-resistant. If labels are to be read under low light levels, they must be illuminated by means of ambient room lighting, special supplementary floodlighting, back or edge lighting, and/or luminescent materials or substances. If raised letters are provided (i.e., cast into a panel or package housing), there must be additional coloring to make the letters contrast with the background. Do not assume that a raised letter that is the same color as the basic background will be readable just because it is raised. Engraved or depressed lettering should also be of a color that contrasts sharply with the background. In addition, the engraving should always be filled level with the panel surface so that it will not collect dirt and eventually become unreadable. Avoid glossy inks or paints and plastic or glass dust covers, unless these are specially treated to minimize spectral reflectivity.

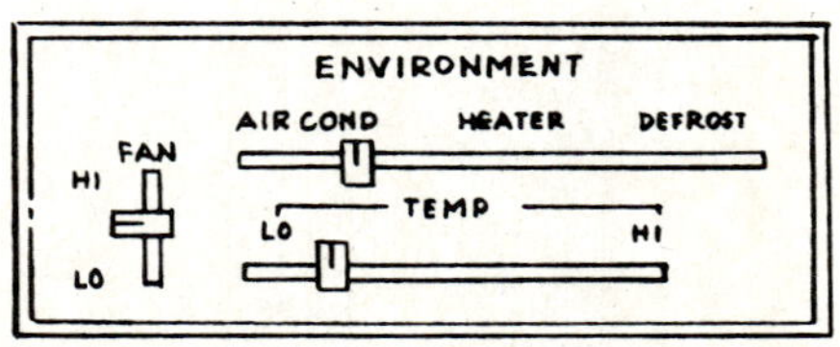

Typical good practice

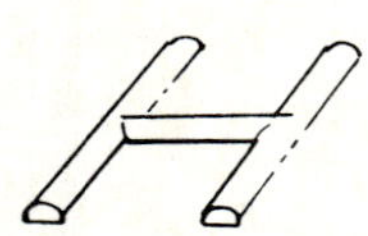

RAISED

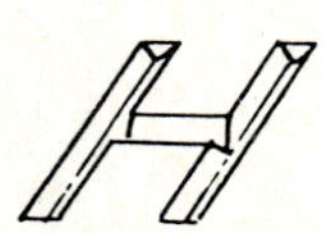

ENGRAVED

[1]The character height of the smallest label must be sufficient for the maximum viewing conditions expected. Avoid the tendency to settle for some arbitrary policy set by your engineering department for all equipment turned out by the company; this may be completely unsatisfactory.

LABEL SPECIFICATIONS

Starting with the smallest lettering size that will be compatible with the typical average viewing distance, establish specifications for each of the labeling categories illustrated by the accompanying sketch.

Typical Panel Labeling Standards

Label Designation	Letter Size	Location
1. Panel title	18 pt (0.187 in)	Centered; ¼ in from top edge of panel
2. Panel subsection	14 pt (0.156 in)	Centered at top of subsection; ¾ in from top edge of panel
3. Subtitle	12 pt (0.125 in)	¼ in above components or ⅛ in above labels of individual components
4. Toggle switch	10 pt (0.093 in)	¼ in above and below standard switch
5. Single component	12 pt (0.125 in)	¼ in above component
6. Rotary switch positions	10 pt (0.093 in)	¼ in from apex of pointer, line from pointer to label

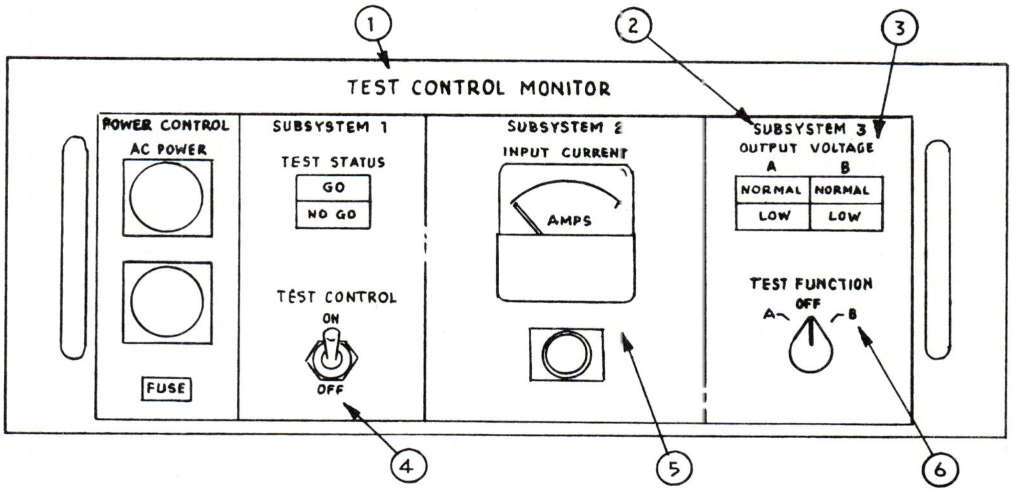

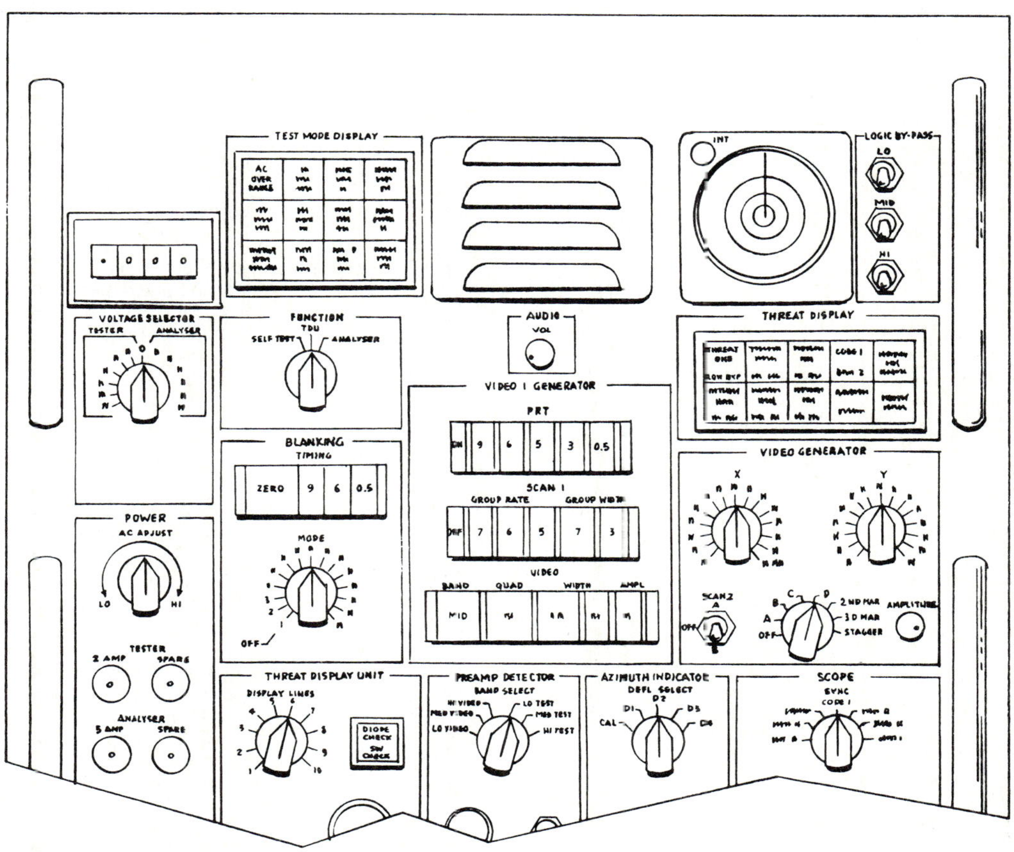

Example of a relatively well-executed labeling program for a fairly complex panel.

LABELS PLACED ON CONTROLS

Word Displays in Conjunction with Controls

Although perhaps more appropriately discussed under the general topic of "labeling," words used directly on certain types of control devices increase the utility and efficiency of control identification and operation by combining the identifying variables of control position, light, and words. The examples shown in the accompanying illustration are representative of only a few of the applications that would be appropriate. See the section on controls for a more complete discussion of this application of word and light combinations.

Several precautions are in order relative to placing word labels directly on the control device:

The control must not rotate so that the label may become unreadable.

If the control surface area is too small for a fully legible label, place the label adjacent to the control; i.e., do not feel that you have taken care of your obligation just because a label is there—it has to be readable.

Push buttons should extend out from a control panel, as compared with a telelight (i.e., not a control); otherwise, an operator will be unable to tell which similar-looking word-light displays are really controls and which are merely advisory displays.

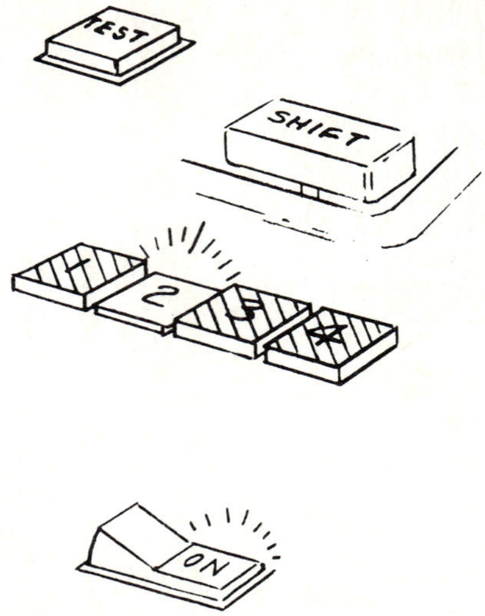

PICTORIAL SYMBOLS

Pictorial symbols make it possible for persons with different language backgrounds to recognize and understand a single symbol. Unfortunately, it is not always possible to pictorialize all informational conditions since some things would not be recognized by the observer because he or she has never seen the actual object. An example is the choke system used on some vehicles. A symbol that is significant to engineers because they have seen the actual system or device not only is meaningless to average drivers but also may encourage them to believe the symbol implies something entirely different; e.g., the typical hourglass pictorial suggested by the automotive designer could be construed to mean "time" to the lay person. Some principles to consider in deciding whether a pictorial signing method should be used include the following.

1. Pictorial Comprehensibility
If a particular situation includes items that cannot be completely pictorialized, i.e., if some items are easy to pictorialize, but others are not, it is better to stick to a word labeling system; observers will only be confused if they find some pictorials and some word labels on an operator panel.

2. Combined Pictorial and Word Labeling
Whenever there is any doubt as to whether some observers will be able to understand the meaning of a pictorial, use both pictorial and word labels.

3. Size-Distance Compatibility
Some pictorial patterns may be effective only when the viewing distance and lighting conditions are optimum; be sure that a particular pictorial pattern does not lose its identity when it becomes smaller or more distant and/or when the ambient lighting and/or atmospheric conditions are not good.

4. Orientation
Do not place pictorials on components that may be reoriented, i.e., turned or moved so that the pictorial may not appear right side up. Although the observer might figure out what the pictorial is after considerable study, a prime purpose of the single pictorial is to elicit a quick response with a single symbol, as opposed to several words in a label.

5. Sufficient Detail
In creating a pictorial to represent some actual visual element, provide just enough detail to make the symbol recognizable, and no more. Fine detail often cannot be seen under certain distance and lighting conditions and may serve only to distort the impression the symbol creates. On the other hand, be extremely careful not to overstylize pictorial symbols just to create an artistic rendition. Such stylization often makes all symbols begin to look alike.

6. Isolation
Some type of border should always be used around a pictorial; otherwise, it may blend with background images.

SAFETY-RELATED SIGNS

A primary objective for safety-related signs is to make them conspicuous. This requires the use of the following special features:

Unusual border patterns such as the black and yellow stripes shown in the accompanying illustration.
Large lettering for the critical words "caution," "warning," "danger," etc.
A bold red border (the color red is most often associated with danger).
Self-luminous paint for the sign background, which tends to make the sign appear to be lighted.
Standardized symbols (see the examples presented later in this section).

The position of safety-related signs is also critical in terms of making sure that they will be seen. Avoid placing such signs at right angles to the expected normal viewing direction; i.e., the intended observer should be exposed to a safety sign as far in advance of arriving at the actual hazard site as possible.

The sign should always be located where it will not become inadvertently covered if someone places an object in front of the hazard site.

If the hazard is especially critical and it is important to warn a person not to pass close by, it is often desirable to add a flashing red light that can be seen well in advance.

Be careful about accepting so-called warning-sign size standards; manufacturers of such signs often stock a given warning sign in several sizes, but it may be that none of the signs are large enough for the sighting or reading distance involved.

Safety-related signs should be made from materials and/or painted with paints that are extremely resistant to wear and deterioration. Since signs often are not frequently replaced, the critical aspect of conspicuousness may be lost after a time, and thus the sign will fail to draw the attention of passing observers.

Examples of Caution and Danger Signs

The Occupational Safety and Health Administration (OSHA) has attempted to standardize the general format and design of caution and danger signs for use in industrial establishments so that workers will recognize them quickly and understand what they mean. Manufacturers of safety equipment (including signing) generally can supply these signs, and thus designers should not have to create special signs of their own.

As shown in the accompanying sketches, the upper panel of the caution sign should consist of yellow letters on a black background; the lower section should have a yellow background with black letters.

Certain special pictorial graphics have been standardized, as shown in the accompanying illustration of a biological hazard sign. It is particularly important not to create an alternative pictorial graphic to replace these standardized symbols because this would tend to confuse the observer.

In the case of a danger or warning sign, as illustrated by the accompanying sketches, the top panel should consist of the word "danger" in white letters on a red oval background. The lower section should consist of black letters on a white background.

Signs are usually available in various sizes. Select the size that is compatible with the greatest required viewing distance.

Additional examples of various types of safety signing are shown in the accompanying illustrations.

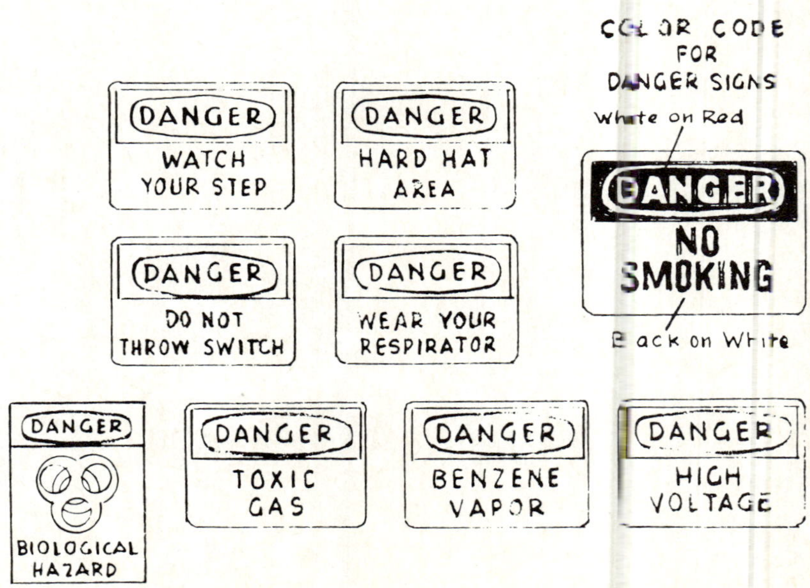

Radioactive warning signs generally should contain the standardized pictorial symbol shown in these examples.

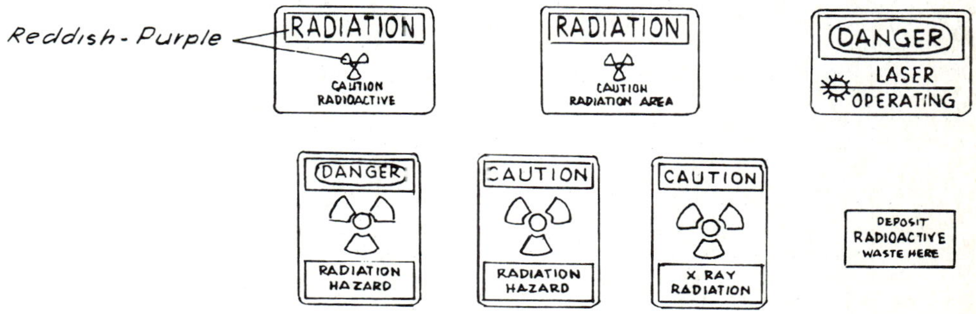

Other special pictorials have been developed for various hazardous environments, as shown in the accompanying illustration. Note, however, that all pictorials should be accompanied by a word set to make sure that the observer gets the message, since he or she may not have learned the meaning of the pictorial.

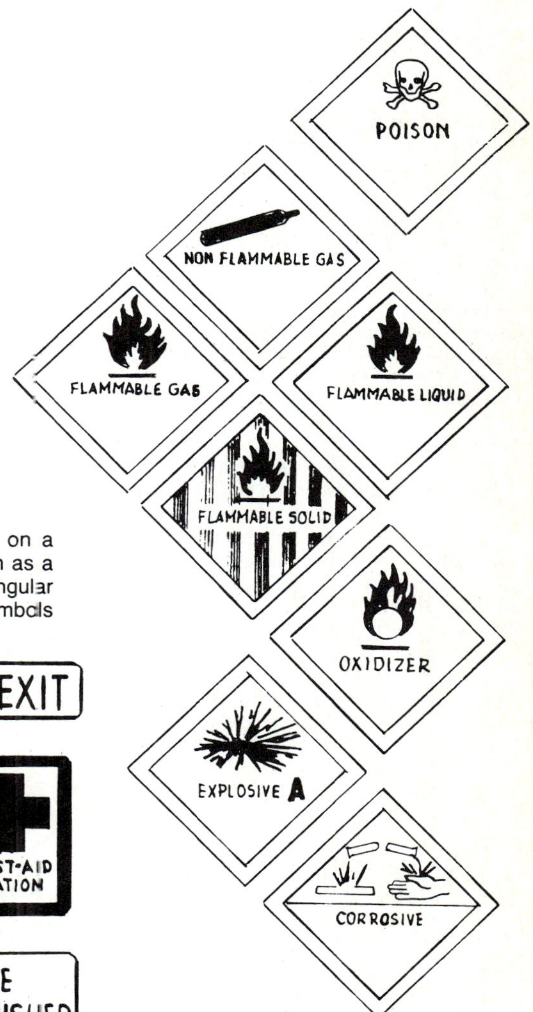

Safety-Related Advisory Signs

The examples shown in the accompanying sketches represent typical advisory signs which are safety-related but which do not necessarily fall into the warning or caution categories.

Such signs are also available from the typical safety equipment manufacturer and are designed to meet OSHA standards. Do not try to create a new type of advisory sign before checking to see whether a similar accepted sign is already available.

When it has been determined that a desired sign has not been standardized by some governmental agency, then consider the following in developing a new advisory sign:

1. Use as few words as possible to convey the message clearly.
2. Use a pictorial whenever it can be made clearly obvious to the expected observers, considering their background and experience. Do not invent vague symbols that have no relationship to something real, i.e., something that people have seen and will recognize.
3. Use maximum figure-ground contrasts and avoid the use of "pretty" colors just to be different. The best approach is to utilize the same color combinations that are used for standardized signs:
   Black on white or yellow
   White on black or red

When in doubt about colors to use on a facility that is already color-coded, such as a fire box or circuit box, provide a rectangular white background on which words or symbols are painted in black.

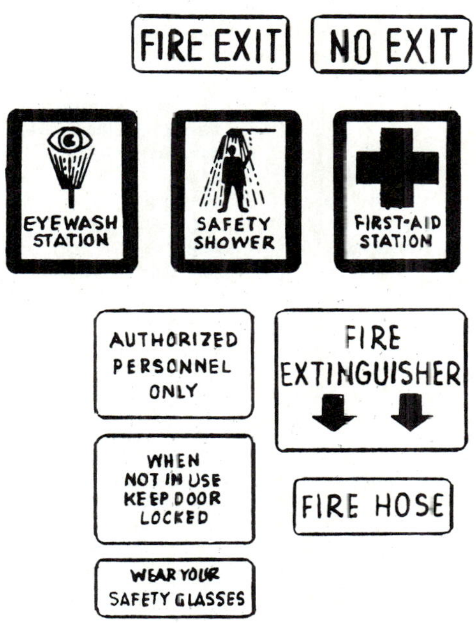

USE OF COLORS FOR SIGNING AND MARKING SAFETY HAZARDS[2]

Red should be used to identify:
- Fire protection equipment, fire apparatus, and fire alarm boxes
- Fire blanket boxes
- Fire buckets or pails
- Fire exit signs
- Fire extinguishers
- Industrial fire hydrants
- Fire pumps
- Fire sirens
- Post indicator valves for sprinkler systems
- Cans containing flammable liquids
- Stop bars on hazardous machines

Orange should be used to identify:
- Dangerous parts of machinery
- Energized equipment that could cut, crush, shock, or otherwise injure people
- The inside of a cabinet door to indicate when the door is open, when guards around moving parts are removed, or an electrical circuit box is open, exposing high voltage potential

Yellow should be used to identify:
- Possible physical hazards such as being struck, falling, stumbling, tripping, or being caught between moving and stationary devices. Yellow and black striping around the hazard is recommended

Green should be used to identify:
- Safety equipment
- First-aid equipment

Blue should be used to identify:
- Situations in which equipment should not be started or moved
- Situations in which equipment is being repaired or worked on
- Electrical circuit boxes

Purple should be used to identify:
- Radiation hazards including x-rays, alpha rays, beta rays, gamma rays, neutrons, protons, deuterons, and mesons. Yellow should be used in combination with purple for markers such as tags, labels, signs, and floor marking

Black and white should be used as basic colors for designating traffic and housekeeping markings. Black and white striping and/or checkerboard borders and solid areas may be used for added conspicuousness.

[2]For additional information, refer to *Safety Color Code for Marking Physical Hazards,* ANSI Z53.1.

Recommended Munsell Designations for Color Coding Visual Displays

Color-Code Preferences for Four Different Color-Code Schemes (Munsell Color Designations)

Eight-Color Code		Seven-Color Code		Six-Color Code		Five-Color Code	
n	p	n	p	n	p	n	p
1R	999	5R	1008	1R	999	1R	999
9R	892	3YR	890	3YR	890	7YR	884
1Y	946	5Y	1128	9Y	1131	7GY	960
7GY	960	1G	1103	5G	1101	1B	1093
9G	1099	7BG	1095	5B	1087	5P	1007
5B	1087	7PB	1133	9P	1005		
1P	1135	3RP	1003				
3RP	1003						

Note: n = book notation in Munsell Color System. p = Munsell Production Number. R = red. Y = yellow. G = green. B = blue. P = purple. RP = reddish purple, YR = yellowish green, etc.
Source: Adapated from Conover and Craft, *The Use of Color in Coding Displays,* WADC, WPAFB, WADC-TR-55-471, October 1958.

EQUIPMENT COLOR-CODING GUIDELINES

Equipment Color Codes

Color	Coding
Red:	
Meaning	Fire protection equipment and apparatus
Typical use	Fire alarm boxes; fire exit signs; fire extinguishers, buckets, and pails; fire hose locations; fire blanket boxes; fire hydrants, pumps, and sirens; fire sprinkler piping
Meaning	Danger
Typical use	Danger signs; safety cans or other portable containers of flammable liquids having a flash point at or below 80° F (26.9° C) (excluding shipping containers); red lights on temporary obstructions or construction
Meaning	Stop
Typical use	Stop button used for emergency stopping of machinery; emergency stop bars on hazardous machinery
Orange:	
Meaning	Dangerous parts of machines or energized equipment which may cut, crush, shock, or otherwise injure; purpose is to emphasize danger when enclosure doors or guards are open and a safety hazard exists
Typical use	Safety starting buttons; exposed edges of pulleys, gears, cutting devices, rollers, power jaws, etc.
Yellow:	
Meaning	Physical hazards, such as stumbling, tripping, falling, or striking against an object
Typical use	Handrails, guardrails, or top and bottom treads of stairways where caution is needed; lower pulley blocks and cranes; pillars, posts, and columns which might be struck; materials-handling equipment, such as industrial tractors, trucks, trailers, forklifts, conveyors, or gantry cranes; piping systems containing dangerous materials
Meaning	Caution
Typical use	Caution signs
Green:	
Meaning	Safety
Typical use	Safety bulletin boards
Meaning	Location of first-aid equipment
Typical use	Gas masks; first-aid kits; stretchers; safety deluge showers; safety signs
Blue:	
Meaning	Caution against starting, using, or moving equipment under repair or in use
Typical use	Scaffolding and ladders; electrical controls; valves
Purple:	
Meaning	Radiation hazards (used in combination with yellow for tags, labels, signs, and floor markers)
Typical use	Containers of radioactive materials; disposal cans for contaminated materials; signal lights to indicate when radiation-producing machines are in operation
Black and/or white:	
Meaning	Traffic and housekeeping markings
Typical use	Directional signs; dead ends of aisles or passageways

Electrical Conductor Color Coding

Instructions	Cable Coding Standards*		
	No. of Conductors	Basic Color	Tracer
1. Enter table at number of conductors desired to be coded.	1	Black	None
	2	White	None
	3	Red	None
2. Colors appearing to the right of the entry number are the appropriate combination for that particular number of conductors.	4	Green	None
	5	Orange	None
	6	Blue	None
	7	White	Black
	8	Red	Black
3. Example: If a cable consists of 12 conductors, the twelfth color combination would be black and white, the eighth color combination would be red and black, and so on.	9	Green	Black
	10	Orange	Black
	11	Blue	Black
	12	Black	White
	13	Red	White
	14	Green	White
	15	Blue	White
	16	Black	Red
	17	White	Red
	18	Orange	Red
	19	Blue	Red
	20	Red	Green
	21	Orange	Green

Note: Only saturated colors should be used. Special care should be taken to select hues which are not easily confused (i.e., red and orange) and which are bright enough to be seen under typically poor lighting conditions (blue may be confused with black if the blue is too dark).

*If the cable has concentrically laid conductors, the first combination or color applies to the center conductor; if the cable contains various sizes of conductors, the first color applies to the largest conductor, and the sequence continues in order of conductor size.

Source: Human Factors Engineering Design for Army Material, MIL-HDBK-759, 1975.

Hydraulic Conductor Color Coding

Function	Color	Function Definition
Intensified pressure	Black	Pressure in excess of supply pressure induced by a booster or intensifier
Supply pressure	Red	Pressure of power-actuating fluid
Charging pressure	Intermittent red	Pump inlet pressure higher than atmospheric pressure
Reduced pressure	Intermittent red	Auxiliary pressure lower than supply pressure
Metered flow	Yellow	Fluid at a controlled flowrate (other than pump delivery)
Exhaust	Blue	Return of power-actuating fluid to reservoir
Intake	Green	Subatmospheric pressure, usually on intake side of pump
Drain	Green	Return of leakage of control-actuating fluid to reservoir
Inactive	Blank	Fluid within the circuit but not serving a functional purpose during phase being represented

Source: Human Factors Engineering Design for Army Material, MIL-HDBK-759, 1975.

Gas Cylinder Color Coding

Type of Gas	Area to Be Coded A (Top)	B (First Band)	C (Second Band)	Remainder of Cylinder
Fuel:				
Petroleum, liquefied	Yellow	Orange	Yellow	Yellow
Petroleum, nonliquefied	Yellow	White	Yellow	Yellow
Hydrogen	Yellow	Black	Yellow	Yellow
Manufactured gas	Brown	Yellow	Yellow	Yellow
Acetylene	Yellow	Yellow	Yellow	Yellow
Refrigerant:				
Ammonia	Brown	Yellow	Orange	Orange
Freon	Orange	Orange	Orange	Orange
Methyl chloride	Yellow	Brown	Orange	Orange
Oxidizing gas:				
Oxygen	Green	Green	Green	Green
Oxygen, aviator's	Green	White	Green	Green
Air, oil-pumped	Black	Green	Green	Black
Air, water-pumped	Black	Green	Black	Black
Helium, oxygen	White	Green	Black	Black
Oxygen, carbon dioxide	Gray	Green	Black	Black
Inert gas:				
Carbon dioxide	Gray	Gray	Gray	Gray
Helium, oil-pumped	Gray	Orange	Gray	Gray
Helium, oil-free	Gray	Orange	Orange	Gray
Nitrogen, oil-pumped	Gray	Black	Gray	Gray
Nitrogen, water-pumped	Gray	Black	Black	Gray
Fire extinguisher:				
Carbon dioxide	Red	Red	Red	Red
Methyl bromide	Red	Brown	Red	Red

Gas Cylinder Color Code Meanings

Color	Fed. Std. 595*	Meaning
Orange-yellow	13538	Flammable materials; materials commonly known to be flammable
Brown	10080	Toxic and poisonous materials; extremely hazardous to personnel
Light blue	15102	Anesthetics and harmful materials
Green	14110	Oxidizing materials; materials which readily furnish oxygen for combustion
Aircraft gray	16473	Physically dangerous because of state of temperature, pressure, etc.
Insignia red	11136	Fire-extinguishing materials
Jet	17038	No significant meaning
Insignia white	17875	For general use where specified
Orange	12246	No significant meaning
Middlestone	30266	For use on selected groups for segregation purposes

*Federal Standard 595, *Colors*.

Gas Cylinder Color Coding Requirements

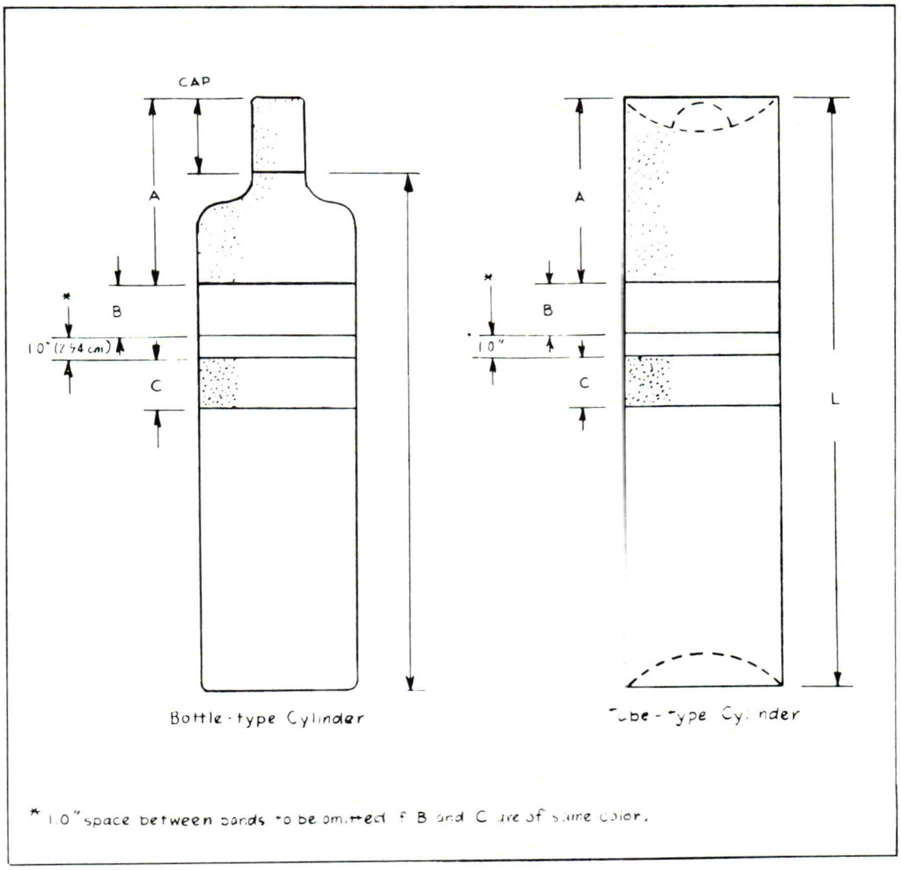

Dimensions of Color Coded Areas (in)

Cylinder Length *L*	Cap and Shoulder *A*	Color Bands *B* and *C*
30.0 (76.2 cm) and under	Bottle-type: one-fourth of *L*-cap Tube-type: one-fourth of *L*	2.0 (5.08 cm)
Over 30.0	(independent of cylinder length)	3.0 (7.62 cm)

QUANTITATIVE READOUT DIGITAL DISPLAYS

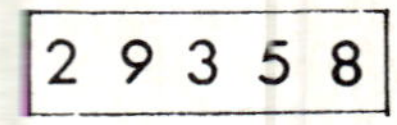

The digital readout generally provides the most rapid and accurate method for presenting purely quantitative information. In selecting or designing such displays, consider the following:

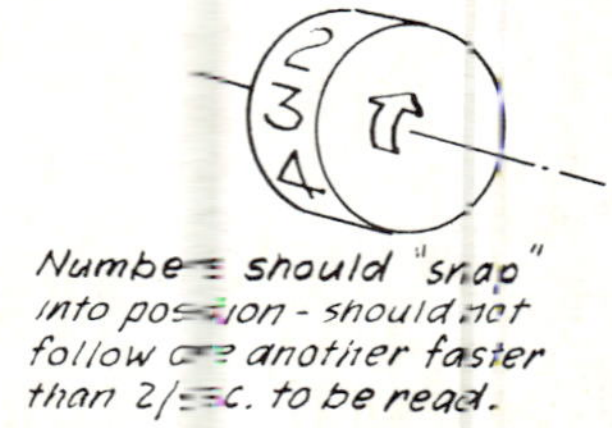

1. The numeral characters should be as legible as possible in terms of character style, height/width ratio, stroke width, and figure contrast.
2. Mechanical counter readouts (drum type) should increase in value by an upward movement of the drum. The face of each drum (and its numerals) should not be buried deeply in the counter window. Individual drums that separate numerals are not desirable because they tend to obscure the numbers and make the total set of numbers harder to read. Wide separators between drums are similarly undesirable. However, such separators are useful for imprinting decimal points and/or commas when the numerical value contains many characters, e.g., 1,000,370.

3. Numbers should always read left to right; vertical arrays are undesirable.

4. Whenever possible, select display systems in which you can use optimized characters (including the drum type, folding character types, etc.). However, when it is necessary to use other types because of a requirement for more rapid response, dot and line matrices are permissible. The accompanying sketches illustrate line or bar matrices that can be used to create effective numeric characters. Although sloping matrix characters have been used extensively, these are less desirable. The slope should never exceed about 11°. Electrically pulsed counter rates should not exceed 50 per second.
5. Although LED (light-emitting diode) displays are effective for use in well-controlled ambient light conditions and/or for nighttime applications, LCDs (liquid crystal displays) are more effective for most general applications because they are easier to see under all ambient light conditions.

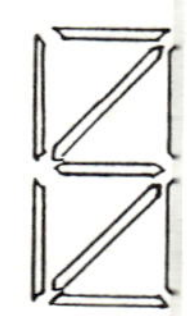

6. Dot matrix characters are generally more readable than other matrix-type displays if there is a sufficient number of dots to create the critical features of the harder-to-read characters, i.e., 2, 3, 4, 5, 6, 8, and 9. The 9 $\times$ 7 matrix shown in the accompanying illustration is the best format. Note that, although only full steps are required for the vertical dimension, half steps are required horizontally in order to provide proper diagonal strokes for numbers like the 4 that is illustrated.[3]
7. Electronically generated digital characters should be as free of brightness variations, jitter, and flicker as possible.
8. Although various colors are acceptable when they have special meaning, the preferred colors are black on white and white on black for general use (and/or a bluish color if the display medium is a CRT). In the latter case, avoid long-persistence phosphors, which may produce ghost images.

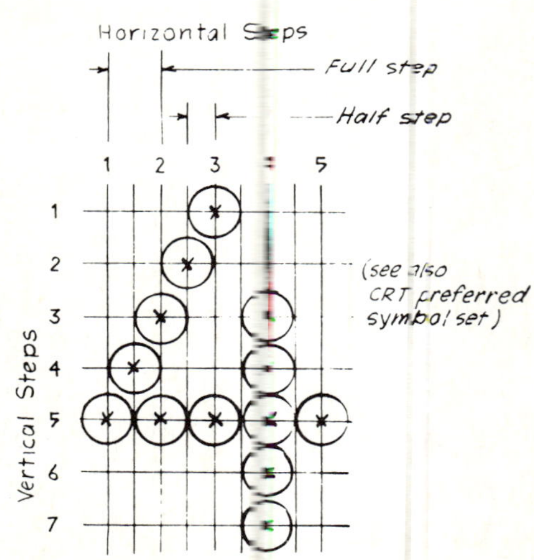

[3]A number of recent studies have indicated that, although general recommendations for the design of alphanumeric characters suggest that properly designed solid or continuous-stroke characters provide more error-free, rapid readout, matrix-type numerical characters are just as good, if not better, in some cases. Examination of such studies indicates that these conclusions are influenced by the fact that the displays studied typically do not control all the critical variables to draw a general conclusion that would contradict the previous study results. From a practical point of view, therefore, almost any type of digital character generation technique can be made to produce an adequately legible readout if the design of the characters is such that a sharp, evenly lighted character is produced that contains the critical pattern features that make it discernible from another character that tends to look like it.

DIALS AND GAUGES

Dials and gauges are used primarily to provide quantitative information. However, as opposed to a digital display (which provides extremely precise quantitative information), the dial or gauge also provides some additional information in the form of advance warning, rate of change, and/or opportunity to make "cross-dial" extrapolation; this is due to the fact that the pointer position and motion act as an additional qualitative cue as to what is happening.

First, and perhaps foremost, whenever possible avoid the use of dials and/or gauges in which double pointers or scales are used. A single scale and pointer format is best because it provides the least reading error.

As shown in the accompanying sketches, both round (or half-round) and linear or rectangular formats are possible. It has not been shown that either format results in any difference in readout effectiveness.

However, the accompanying illustrations comparing the round and rectangular gauge formats provide some interesting insights into the possible value of the rectangular format over the round format:

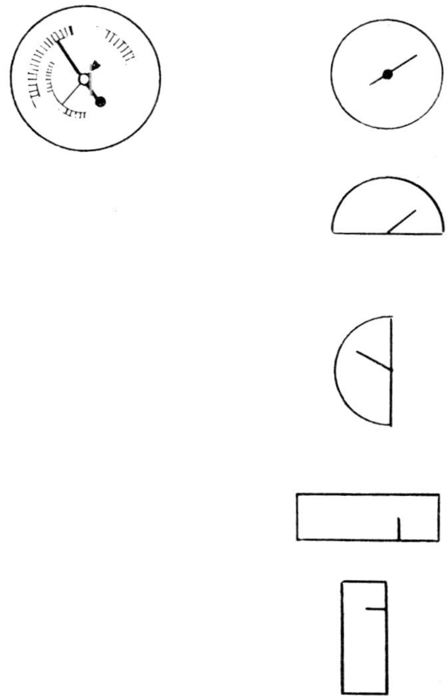

When several gauges are placed side by side, it is obvious that the relative positions of the pointers are more readily assessed.

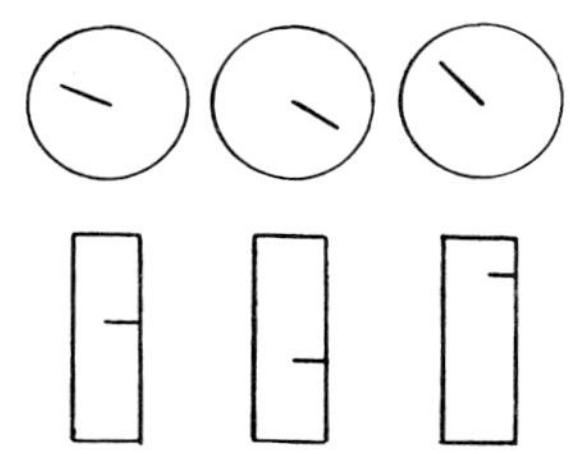

The direction of pointer increase and/or the value level is always clear in the rectangular format, whereas in the round dial, it is more difficult to tell by the position of the pointer just where it is, depending on which quadrant of the dial the pointer is in; i.e., when the pointer is on the left or top half of the dial, the direction of pointer movement for increase and decrease is "natural," but when the pointer is in the right or bottom half of the dial, the increase and decrease direction of motion is reversed.

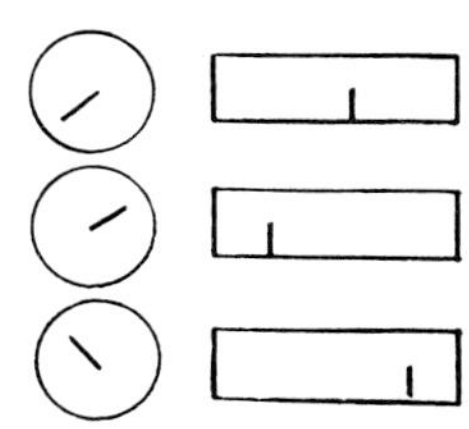

In the case of an automobile speedometer, for example, the horizontal scale can be placed higher on the instrument panel (that is, the entire scale can be placed higher); this makes it easier for the driver to glance at the instrument without taking his or her eyes too far from the road.

Often, several rectangular gauges can be placed closer together than round dials. This, of course, depends on the particular instrument package shape behind the panel.

When several round dials have to be clustered for ease in check readings (i.e., when an operator must glance at them quickly to ascertain that things are functioning properly), the instruments can be oriented so that all the pointers are aligned alike when the instruments are indicating normal operation. This makes a single pointer misalignment easy to see.

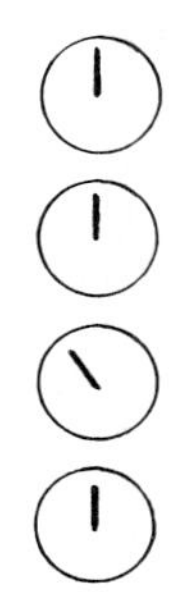

Vertical arrays should be oriented so that normal pointer positions are vertical; horizontal arrays should be oriented so that normal pointer positions are horizontal (see the accompanying sketches).

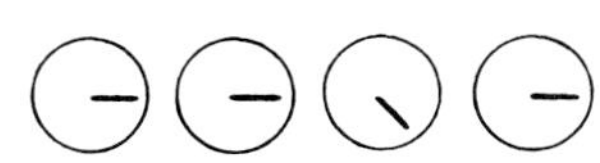

Dial Size

The size of a dial or rectangular gauge normally should be determined by the number of scale graduation marks required (which is a matter of the inherent precision required). Instrument manufacturers have tended to standardize their instrument package sizes for the sake of production convenience and therefore have sometimes put more markings on the instrument face than can be accurately read. The spacing of scale graduation marks must be great enough so that the observer can discriminate between one mark and another, and also see the relation between the marks and the pointer, without taking an inordinate amount of time peering at the instrument.

In addition, the size of the dial depends on how deeply inset the face is below the bezel. When the manufacturer's case design includes a deeply inset dial face, the bezel tends to limit the viewing angle, with the result that (depending on the viewing angle) some of the numbers and/or scale marks may not be visible. Some designers have created poor formats to overcome this problem, such as placing the numbers "inside" the scale to make sure that they are not obscured. As will be noted later, this format is counter to good pointer-scale relationships and results in a pointer that partially covers up the scale numbers.

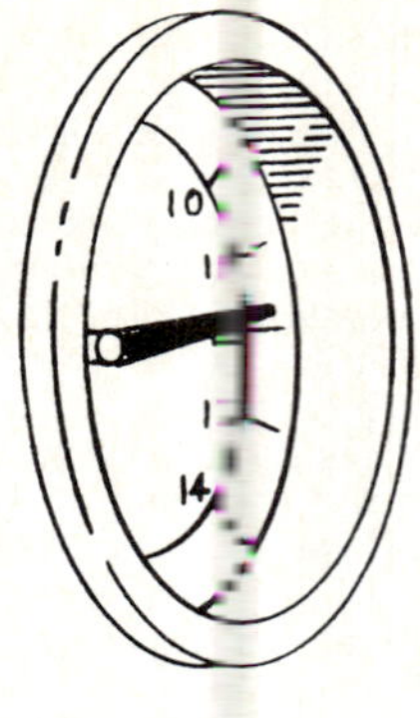

Dial Size in Relation to Required Number of Scale Marks

Too many scale marks, too close together, make dial reading difficult and error-prone. In fact, it is impractical to have a dial that is so small that the inner annulus of the scale is less than 1 in (2.5 cm) in diameter. Use the accompanying table as a guide in determining the size of dial necessary to accommodate the required number of scale marks. For example, if the viewing distance is approximately 12 ft (3.7m) and if 50 scale graduation marks are required, the inner annulus of the scale must be at least 5 in (13 cm) in diameter.

Minimum Diameter of Inner Annulus at Various Viewing Distances*

	Viewing Distance				
No. of Scale Marks	20 in	3 ft	6 ft	12 ft	20 ft
50		1.3 in	2.6 in	5.0 in	9.0 in
100	1.4 in	2.6 in	5.0 in	10.0 in	17.0 in
150	2.0 in	3.9 in	8.0 in	15.0 in	26.0 in
200	2.9 in	5.0 in	1.0 in	21.0 in	34.0 in
250	3.5 in	6.4 in	13.0 in	26.0 in	43.0 in
300	4.0 in	7.7 in	15.0 in	31.0 in	51.0 in
350	5.0 in	9.0 in	18.0 in	36.0 in	60.0 in

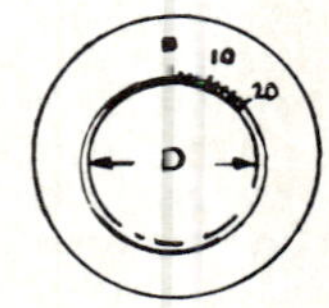

A common design error occurs when the designer believes that a specified level of accuracy must be provided and therefore crowds the scale marking in order to get a certain number of marks within the constraints of the dial-face diameter, as shown in the accompanying illustration. The result is that the observer cannot differentiate between the marks and not only makes errors in reading but also spends more time reading; thus the basic objective is lost.

The designer should either reconsider the need for the level of accuracy thought to be required (i.e., the instrument itself may not provide the level of accuracy it was assumed to) or use fewer marks, since the observer can probably interpolate the pointer position between fewer, properly spaced marks more accurately and reliably than he or she would be able to do with the poorly marked dial.

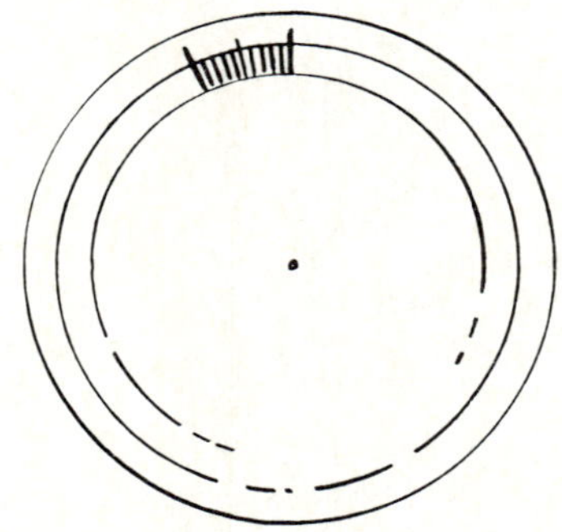

POOR

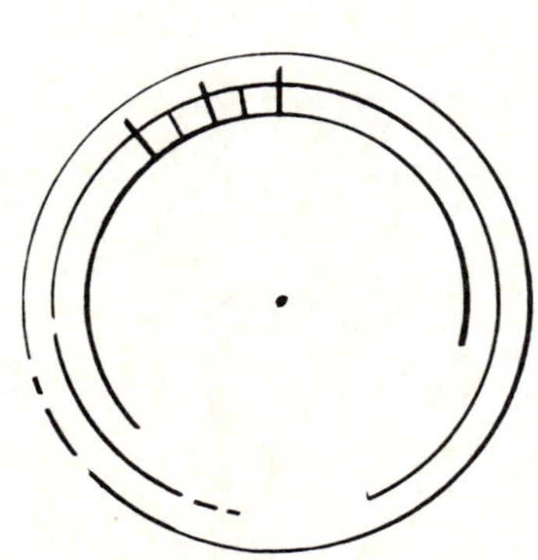

PREFERRED

The following are some typical scale-marking design errors to be avoided:

1. Dots instead of lines
2. Thick marks
3. Marks joined by a heavy base line
4. Long marks spaced close together
5. Irregularly spaced marks about an arc, which cause the observer to shift his or her spatial reference
6. Uneven spacing, which leaves the observer in doubt as to the value of each mark
7. Significant spacing variation between the ends of the marks, which, because of the tight radius of the dial, makes it hard for the eye to follow along the scale-pointer axis

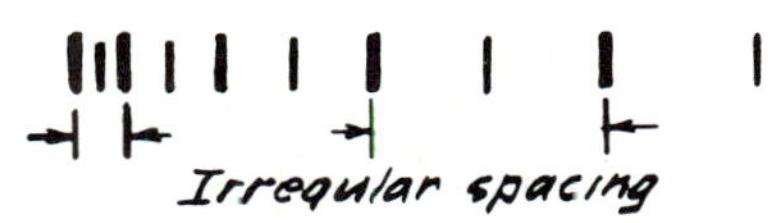

Pointer-Scale Relationships

For ease of reading, the scale-pointer relationship should be as shown in the accompanying sketch. The point tip should be the same width as the scale mark, and it should fall just short of the mark.

Placing the pointer just inside the scale annulus ensures that it will never be confused with a mark. However, the plane of the scale and the pointer also should be the same; otherwise, the observer may misread the scale-point association because he or she was looking at the instrument from an angle (parallax). The accompanying sketches illustrate the proper arrangement for pointer and scale planes.

Finally, the numbers associated with the scale should be outside the scale ring so that the pointer will not obscure the number.

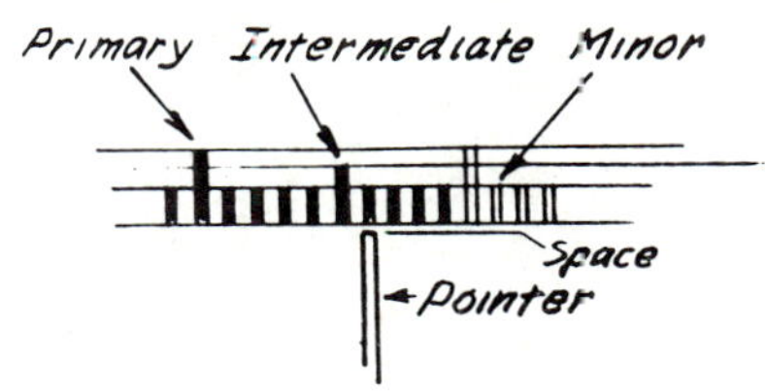

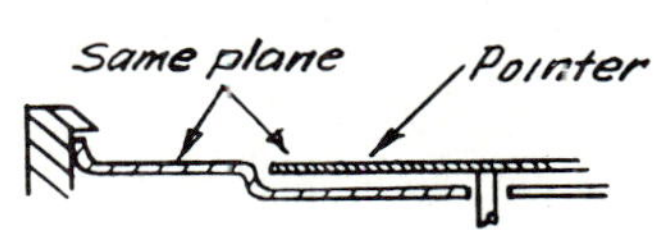

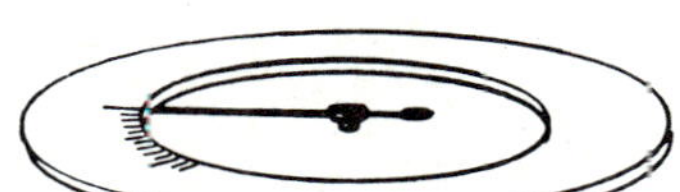

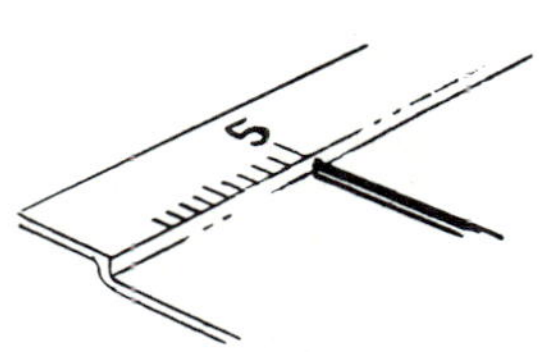

Instrument Scale-Marking Guidelines

Experiments by military psychologists produced the scale-marking specifications shown below; scales were read with fewer errors and faster response when these dimensional characteristics were present. The 28-in (71-cm) viewing distance is typical of the aircraft cockpit situation. For other viewing distances, the proportions should be maintained, even though values are increased or decreased; i.e., multiply each dimension times the viewing distance, or

$$\text{Dimension at 28 in} \times \frac{x \text{ in}}{28}$$

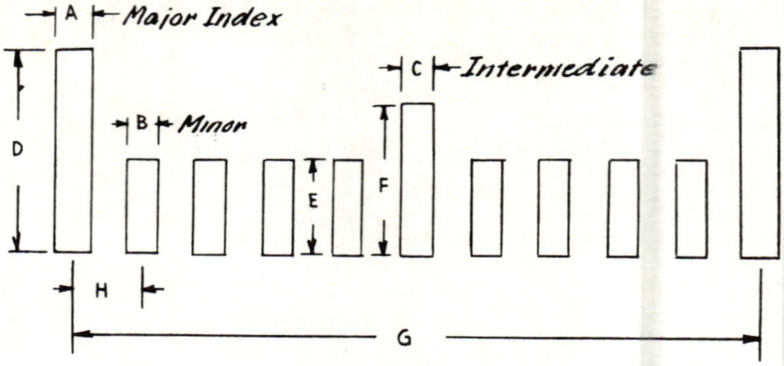

Scale Mark Dimensions for 28-in (71-cm) Viewing Distance

Dimension	Black on White	White on Black
A width	0.035 in (0.069mm)	0.125 in (0.032 mm)
B width	0.025 in (0.054mm)	0.125 in (0.032 mm)
C width	0.030 in (0.076mm)	0.125 in (0.032 mm)
D length	0.22 in (0.56mm)	0.22 in (0.56 mm)
E length	0.10 in (0.25mm)	0.10 in (0.25mm)
F length	0.16 in (0.41mm)	0.16 in (0.41 mm)

Note: Although the major, intermediate, and minor index mark widths are shown to vary for black on white markings, it is suggested that all mark widths be equal for most applications so that a pointer tip can be used that is the same width as all the marks. Use the minor index width to define the pointer tip.

Direction of Pointer Movement

Scale numbering should reflect the expected direction of pointer movement; i.e., pointers should move to the right, up, or clockwise to indicate a value increase.

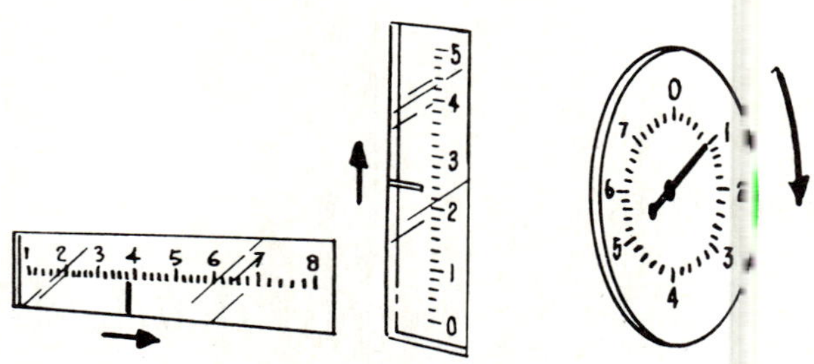

Scale Interval

For most people, scale intervals of 1 through 10 (or multiples thereof) are easiest to interpret without error. The next easiest are intervals of 2, 4, 6, 8, and 10. However, intervals such as 3, 6, and 9 (or 4, 8, and 12) can cause confusion.

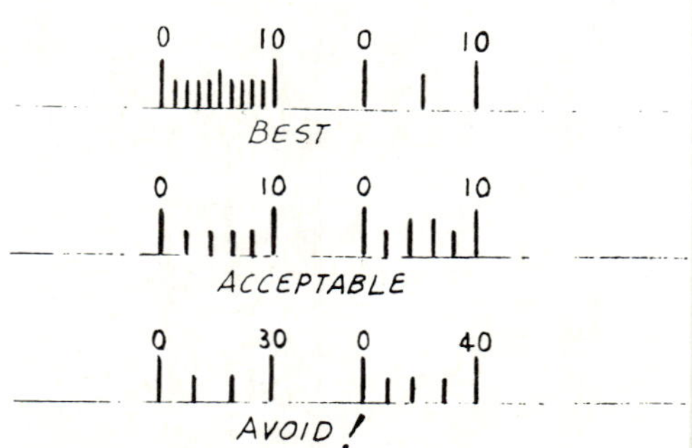

Other Design Considerations

Make sure that numerals are oriented upright rather than slanted and turned on their side. The exception to this rule is the situation in which the dial may rotate, in which case it is desirable to orient each numeral so that its base is toward the center of the dial, which makes the numeral appear upright when it rotates to an indexing position, preferably at the top of the dial. Alternative orientations may be necessary if the display exposes only one-half of the dial and the display is oriented to emphasize an up-and-down direction of motion.

If a scale covers only part of a circumference, attempt to balance the number positions so that they present a symmetrical appearance.

Consider the use of patterned or colored range strips when the primary readout is to determine when a system is performing within specified ranges, i.e., when it is less important to read specific values.

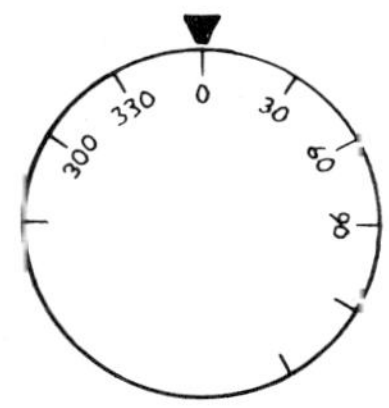

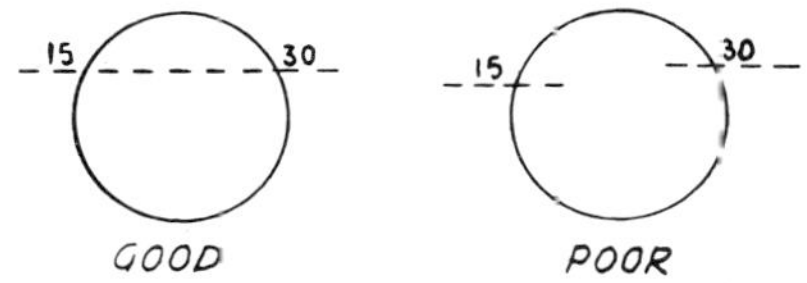

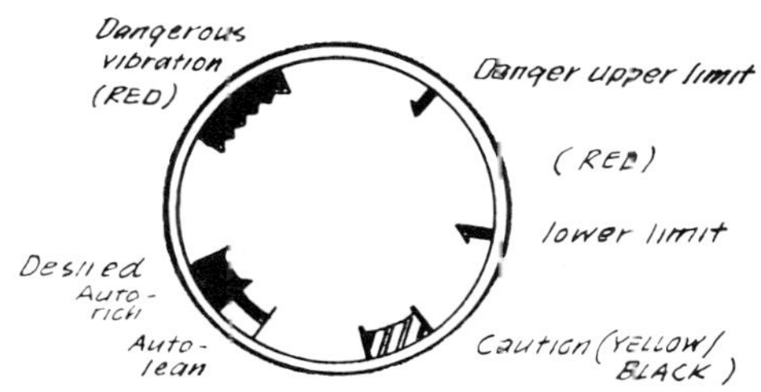

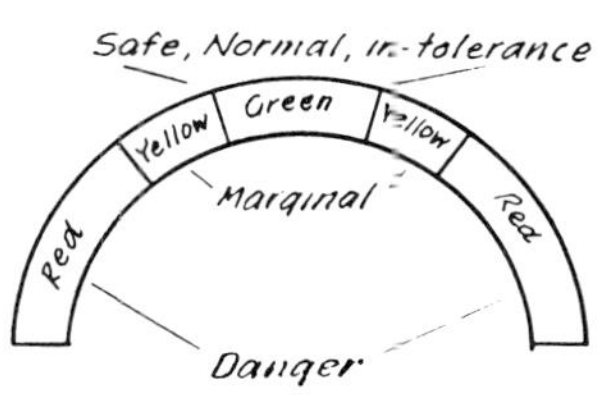

Although it is always desirable to optimize the size and shape of numerals used on dial displays, circumstances may dictate a smaller instrument face. It is better to use smaller numerals and thus avoid the clutter and illegibility that result when characters of a so-called optimum size are crowded together.

Whenever the dial scale is of finite length, be sure to leave a sufficient break between the beginning and end of the scale so that observers do not mistakenly assume that the scale continues and/or that they should continue reading around the scale, but with some special extrapolation.

When several dial or rectangular, scaled instruments or gauges are mounted within a single work station or on the same instrument panel, make sure that the scale breakdown is alike on all the instruments; i.e., do not mix two different scale breakdowns, as shown in the accompanying illustration.

Instrument faces for use when night or dark adaptation is not critical are more easily seen if the markings are black on a light-colored face. Even though the instrument may have to be illuminated at night (e.g., an automobile speedometer), the lighter background can still be dimly lit and provide a more readable display. In the case of the automobile speedometer, the display is also easier to see when sunlight reflects on the cover.

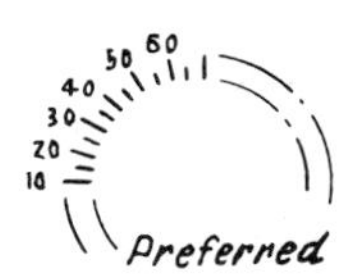

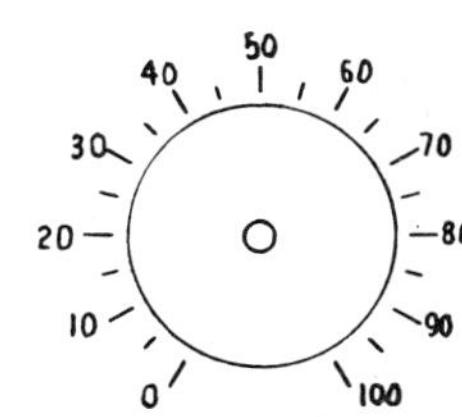

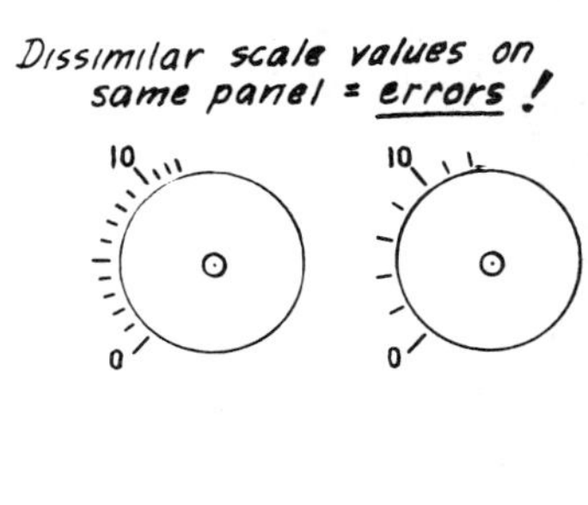

Design of Pointers for Dial and Gauge Instruments

For maximum reading accuracy, a pointer should be of equal width throughout so that the parallel sides visually project to match the parallel-sided scale marks to which the pointer must be associated.

The width of the pointer should be exactly the same width as the scale marks. That is why, in a previous discussion of scale marking, it was noted that all scale marks should be the same width.

In many applications, however, the thin pointer is not conspicuous enough, especially when the user has to depend on a brief, quick glance to see how a particular system is performing. Thus, it is necessary to increase the width of the pointer, i.e., to make it bolder and more obvious during a quick scanning look. In such cases, the pointer should be designed as shown in the accompanying sketch. Note, however, that although the main stem of the pointer is enlarged, it is still tapered at the tip so that the width of the tip still matches the scale-mark width.

Generally speaking, if a pointer has a "tail" on it, the observer can judge slight movements of the pointer better than if the pointer is tailless. The tail should not be more than one-third the length of the pointing segment, however; otherwise, there may be confusion as to which end is the pointing end.

Above all, avoid artistic pointer designs.

Pointer illumination is often difficult. A translucent dial face provides a good background for silhouetting a black pointer. Another useful technique, shown in the accompanying sketch, consists of a plastic pointer ring inside a scale ring. Both can be either back-or edge-lighted with considerable success.

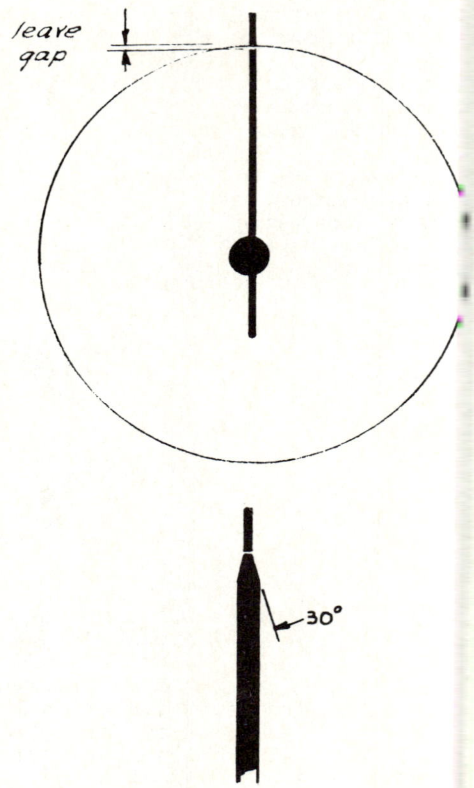

Correct

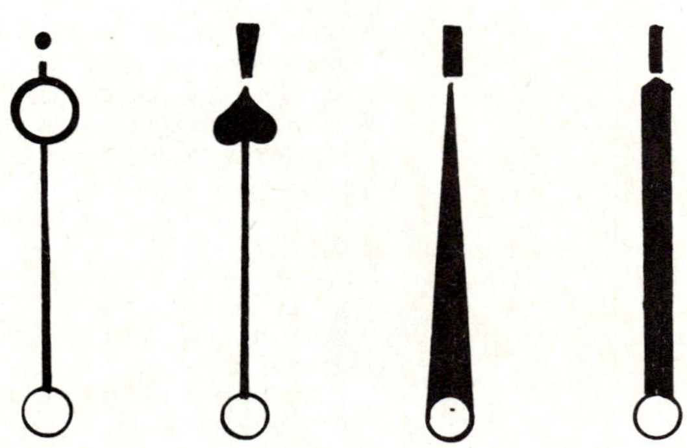

Poor

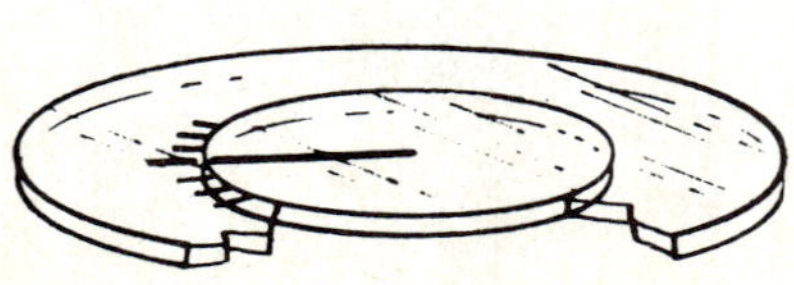

Multiple Pointer Display Considerations

First, avoid multipointer instruments if at all possible. When absolutely necessary, consider the following:

A problem of pointer visibility and discrimination can occur when two or more pointers are necessary. (Three pointers are to be avoided if at all possible.) Several typical design errors are shown in the upper sketch. The two pointers are the same color, the shorter pointer tip is too far from the line of marking, and the tails of both pointers are the same length.

The second sketch shows several ways to improve the visibility and discriminability of the dual pointer system. The color contrast and the variation in the length of pointer tails help clarify pointer identification. The length of the shorter pointer has been increased so that the tip of the pointing end is not so far from the line of marking, and a point has replaced the blunt tip, illustrated in the first example. Other ideas for aiding discrimination between pointers are shown in these illustrations:

When two pointers are required on the same instrument dial, the upper one is generally too far from the surface on which its corresponding scale marking is located. One way to solve this problem, as illustrated by the accompanying sketch, is to bend the upper point in order to bring the pointer tip closer to the dial surface. Another method is to place the scale marks on a raised plane so that the marks are about midway between the two pointers.

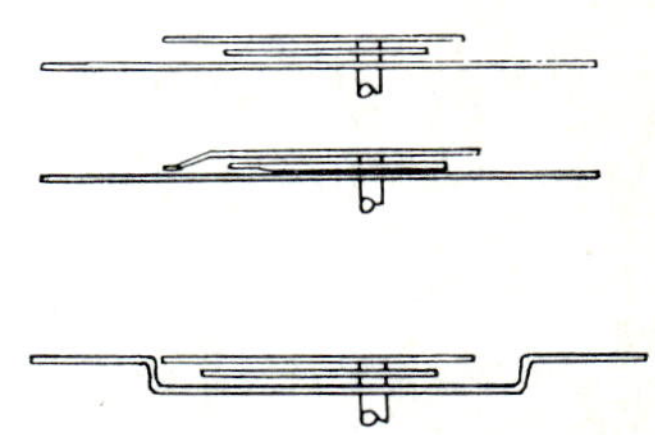

When the particular instrument requires use of a "moving scale" which must be referenced against an indexing mark or pointer, it is usually necessary to make the index marker or pointer more conspicuous than an ordinary pointer because of the lack of length and pointer or index motion. The accompanying sketches illustrate acceptable design practice. Special attention is called to the double scaled instrument shown in the accompanying illustration. Here, making the central pointer pictorial has made it more conspicuous and identifiable; i.e., it represents the ship on which the operator is working. In other words, the operator knows immediately that the ship symbol represents the ship's heading, while the diamond pointer represents the "true" heading.

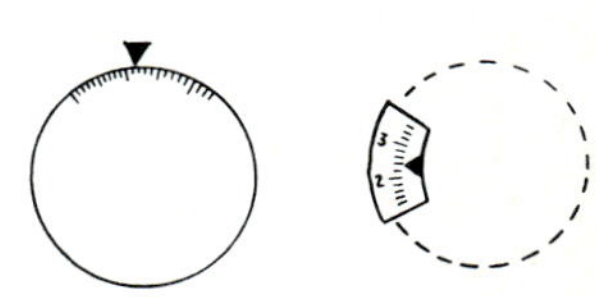

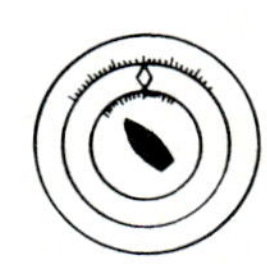

When a moving-tape type of gauge is used in order to provide an expanded scale, the indexing pointer can be designed either as a pointer positioned at the side or as an indexing or "lubber" line across the tape. In some cases two pointers may be required: one to indicate where the observer's position is now, and the other to tell the observer which way to go.

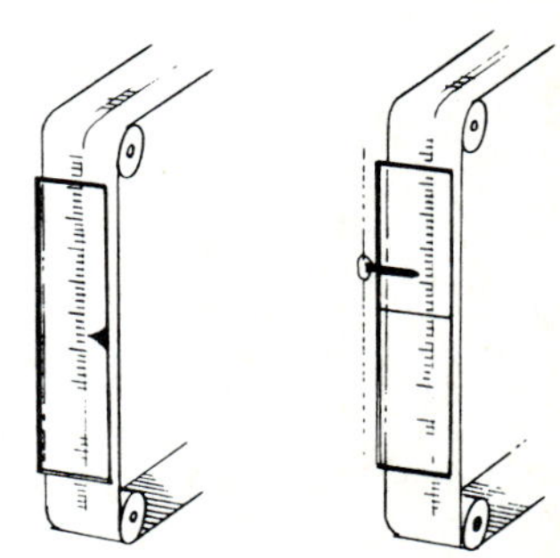

Example of a Thermostat Display and Control

The accompanying sketch shows a popular thermostat control unit that combines the display of temperature, the displayed setting, and the controller all in a seemingly compact package. In an informal survey, over 50 percent of the persons asked reported the wrong temperature reading. They had read the scale at the top, which is the scale for setting the control. Unfortunately, the values of setting did not coincide with the temperature readout value.

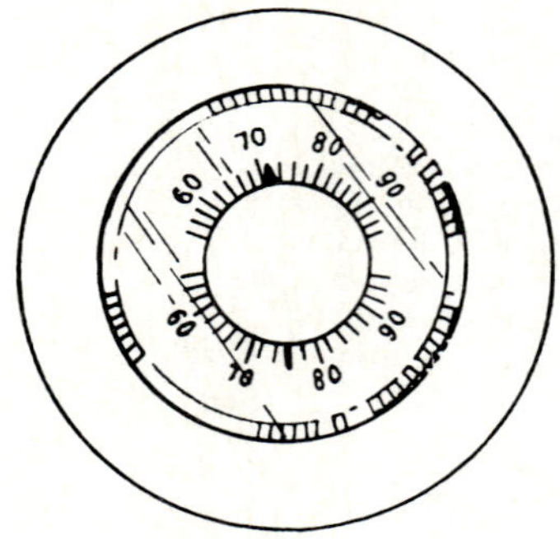

This illustrates several poor human engineering features. In addition to the obvious problem cited above, the index marks are gold on a white background, which makes them extremely difficult to see through the plastic (knob) cover.

The pointers cover the index marks; thus at certain angles, it is difficult to tell exactly where the pointer is pointing.

The accompanying sketch illustrates a preferred display approach and also the application of a number of the good human engineering principles discussed earlier in this section.

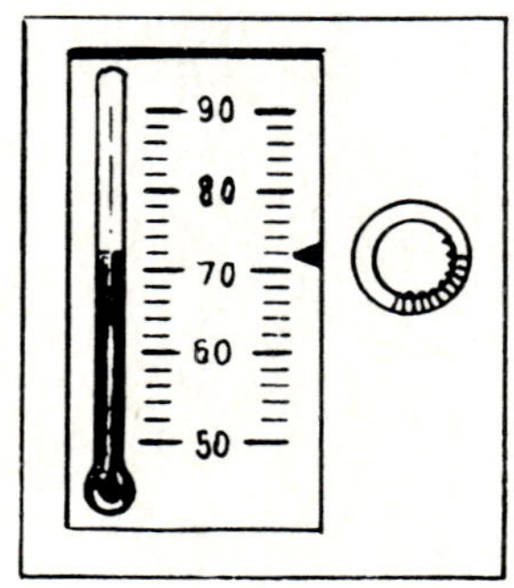

The scales are vertically oriented, which is more typical of people's expectation of colder or warmer. The pointer does not cover the scale on the control side, and the adjacency of this scale, the pointer, and the control knob ensures proper association. To further ensure that pointers are not confused, an actual thermometer is provided, although this could just as well be instrumented by means of another pointer.

Examples of Functional-Group Display Optimization

The accompanying sketches illustrate the optimization of automobile driving instrument displays:

The accompanying sketch illustrates several salient features of conventional instruments and back-lit indicators. The advisory indicators to the left advise the driver of conditions that should be checked. The combined advisory lights and the fuel, oil, temperature, and ampere indicators advise the driver when a condition is not optimum. The speedometer instrument has an evenly spaced scale. The high-beam advisory light is close to the nominal line of sight as the driver watches the road ahead. Although round dials also could be used in the same relative positions, the rectangular scalar instruments take up less space, thus allowing the entire display grouping to be raised to a position near the driver's nominal line of sight.

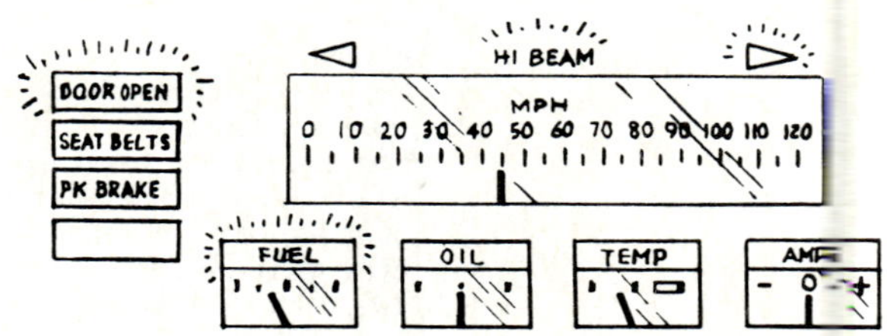

With more exotic solid-state display techniques, considerably more information can be displayed in essentially the same or less space as illustrated in the accompanying sketch. Everything on the display can be displayed by purely electrical means, including pointer position. In this case, only speed, fuel quantity, time, and mileage would appear on a continuous basis. Other information would appear only as required, thus reducing the operator's visual load and the number of distractions from his or her external viewing task.

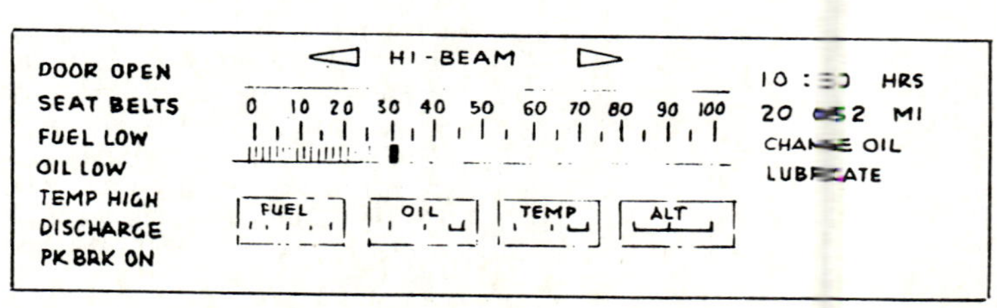

LUMINOUS IMAGE DISPLAYS

Electronically generated image displays such as the CRT (cathode-ray tube) offer maximum flexibility for presenting a wide variety of images and have the advantage of almost instantaneous image modification (as evidenced by the television systems widely used in homes, business, and industry). The following guidelines are offered to assist the designer in making specific choices regarding types of tubes, size, and display mode.

General

CRT-type displays should be viewed from a position perpendicular to the screen; thus the display should be packaged and/or positioned so that the intended observer normally assumes the proper viewing position.

The overall size of the CRT screen should be determined on the basis of the smallest significant detail size that must be visually resolved by the eye at its expected viewing distance.

The screen should be protected from direct high ambient light because it tends to reflect and obscure images and/or excites the phosphor, reducing the inherent contrast of the display.

CRT displays should be packaged so that all adjustment controls for the tube are on the front of the package i.e., so that they are accessible to both an operator and a maintenance technican. Purely service controls can be placed under a hinged cover to minimize tampering by inexperienced operators.

A transparent safety screen should always be provided over the CRT to prevent radiation or implosion injury.

Except for critical detection, CRT displays should not be viewed under complete blackout ambient light conditions.

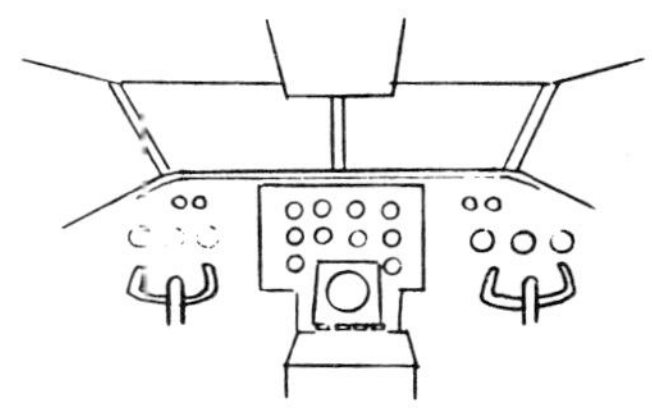

Desirable CRT Performance Characteristics

TUBE SIZE The diameter of the tube should be a minimum of 12 to 19 in (30 to 48 cm) for typical console operator viewing distances (all signal types). Larger tube sizes are acceptable for TV-type viewing and status-board displays that may be viewed by several operators. Sizes larger than 48 in (122 cm) are not recommended because of the loss in brightness and contrast as the viewers are forced to stand further away from the screen.

DISPLAY RESOLUTION The maximum number of lines possible within the state of the art is always desirable for complex motion displays. Typical TV scan lines and symbolic character height is 10 for adequate character legibility. For digitally generated images that may include handwritten characters, the scan lines and symbolic character height should be about 125 lines per inch.

TV DISPLAY ASPECT RATIO The standard width/height ratio is 4:3, but ratios of 5:7 or 2:3 provide the greatest legibility.

ACCEPTABLE BANDWIDTH The acceptable bandwidth is 4.0 to 10 MHz.

GRAY LEVEL There should be a minimum of five levels for TV; a single level (black and white) is acceptable for most digitally generated images.

DISPLAY BRIGHTNESS A line brightness of 50 fL (± 40) is required under normal ambient light levels; there should be lower adjustable brightness for lower ambient light conditions.

BRIGHTNESS CONTRAST The contrast ratio should be as near 90 percent as is practicable.

FLICKER Display pulse rates should be compatible with critical flicker frequency response of the eye (CFF); i.e., the particular phosphor-driver combination should not generate pulses in the 30- to 55-Hz range.

GEOMETRIC DISTORTION Displacement of any image element should not exceed 2 percent of the image height.

COLOR MISREGISTRATION For additive color systems, the maximum acceptable misregistration is ± 65 percent.

PPI DISPLAY SWEEP RATE Any rate between 1 and 70 rpm is acceptable.

ALPHANUMERIC CHARACTER WIDTH/HEIGHT RATIO A ratio of 2:3 to 3:5 is best for maximum legibility and fastest recognition.

ALPHANUMERIC CHARACTER STROKE WIDTH Ratios of 2:6 to 1:10 are recommended.

ALPHANUMERIC CHARACTER STYLE The closer the character style is to the character legibility guidelines noted elsewhere for printed matter, the more readable the character is. However, when rapid readout is not an absolute requirement, modified, matrix-type characters are acceptable. Although sloping (italic) characters are frequently used and are acceptable for single values (i.e., when there is at least one character-height separation between character lines), vertical characters are preferred.

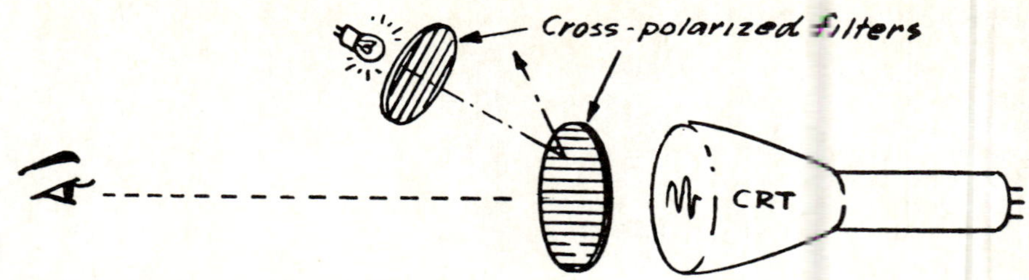

Cross-polarized lighting system for CRTs.

DISPLAY FORMATS Display formats should be compatible with the manner in which the display image is generated; i.e., a plan position indicator sweeps an arc, and therefore the shape of the display should provide a round or circular face, and a rectangular face should be used for horizontal or vertically swept display systems, used typically for TV motion pictures and/or alphanumeric status or message boards. For image previewing, a scrolling technique is suggested so that an operator can key in information from one image while viewing the next.

VIEWING DISTANCE The optimum for the typical console situation is 18 to 20 in (45.7 to 50.8 cm), e.g., a 12- to 19-in (30.5- to 48.3-cm) screen. A maximum of about 20 ft (6.0 m) should be planned as the limit, depending on the size of the display.

VIEWING ANGLE This should not be less than 30° from the perpendicular axis. For seated operator console operation, the operator's sight line should not exceed about 30° vertically or horizontally, i.e., about 15° on either side of the operator's normal center-of-the-display viewing angle.

ADJUSTABILITY PROVISIONS As a minimum, capability should be provided for the operator to adjust CRT brightness and contrast. In addition, however, it is also desirable to provide capability for operator tilt adjustment of the CRT or CRT cover face so that the operator can remove disturbing reflections and glare from the display.

POSITIVE VS NEGATIVE IMAGE A positive image (i.e., a dark image on a light background) is preferred when the display is used under normal or bright ambient lighting conditions. A negative image (i.e., a light image against a dark background) is preferred under dimout or blackout ambient lighting conditions, since it is desirable to reduce the amount of light that could affect the observer's dark adaptation level.

COLOR For normal ambient lighting conditions, use a phosphor color that is near white in color (bluish white, blue, and green are also acceptable, however). For dimout or blackout ambient lighting conditions, use an orange or reddish phosphor to minimize reduction in the observer's dark adaptation level.

When a filter cover is used over the display tube, use a neutral density filter to minimize distortion of the color from the CRT.

Under dimout ambient lighting conditions where reflections may be a serious problem, use a circularized Polaroid filter or a cross-polarized filter-light system to prevent reflections and/or tube phosphor excitement.

By placing the light filter 90° to the CRT filter, it is possible to provide illumination for panel controls without reducing the brightness contrast of the CRT signal.

Multicolored CRT displays are suggested when it is necessary to differentiate between scales, curves, call-outs, etc. Red, green, blue, and yellow are the most reliably differentiated colors.

General Checklist for CRT Displays[4]

- Whenever possible, mount the scope face perpendicular to the operator's normal line of sight. If the operator is standing, his or her line of sight is about 3° downward; seated, it is 15 to 20° downward. Ideally, the operator's line of sight should be perpendicular to the tube center.
- Recommended viewing distance is 16 to 20 in (40.6 to 50.8 cm).
- Tube size should be proportional to the required resolution. Scopes as small as 2 to 5 in (5.1 to 12.7 cm) in diameter may be used for special cases, e.g., infrequent calibration or tuning.
- Scopes with a 5- to 7-in (12.7 to 17.8-cm) diameter may be used when they adequately display the necessary information and when no plotting is required.
- For plotting, the scope diameter should be 10 in (25.4 cm) or larger.
- The shape of the CRT should be compatible with the information format, i.e., round for plan position displays (PPI) and rectangular for A-scan presentations, document images, etc.

[4] *Human Factors Engineering Design for Army Material*, MIL-HDBK-769, 1975.

For a CRT displaying synthetic video, as opposed to raw or unprocessed video, the system should be capable of presenting a symbol of the required size (at least 20 min of arc) for any return resolvable by the system.

For synthetic video, in which signals are clipped for digital presentation to the discrete points resulting from digital conversion, the number of points per inch is

$$\frac{\text{Number of points per tube diameter}}{\text{Tube diameter in inches}}$$

For a 14-in (36-cm) screen at 1024 points per tube diameter,

$$\frac{1024}{14} = 73 \text{ points per inch}$$

Symbols should be drawn as point-to-point representations. Symmetrical symbols require an even number of points. For the screen described above (with a viewing distance of 16 in, 41 cm), a 20-min visual angle would subtend 0.093 in (0.236 cm), or 6.77 points. A symmetrical square would have sides extending over 8 CRT points, or ⅑ in (0.28 cm). The area in space which is masked by a symbol equals

$$\frac{\text{Number of points on symbol diameter}}{\text{Number of points on screen diameter}} \times \text{range of coverage}$$

If the screen described above covers 100 km (62 mi) the area masked by the symbol is

$$\frac{8}{1024} \times 100 \text{ km} = 0.782 \text{ km}$$

If the symbol is accompanied by identifying alphanumeric information, each character covers an area representing an area 0.782 × 0.782 km² (0.289 mi²) in the real world. If the scale size is changed to magnify the area covered by the CRT, the area masked by the symbol decreases. For 4× scale (25-km, 16-mi coverage), the area masked by a single symbol is

$$\frac{0.750}{4} \times 0.188 \text{ km}^2$$

Control devices used in conjunction with CRTs (to permit an operator to request information or direct action) should provide a centering accuracy of approximately ±4 CRT points. All such devices (e.g., joy sticks, tracking balls, "stiff sticks," and light pens) require a switch action to indicate that the operator's action is directed to the area currently designated.

A light pen (or pointing device) recognizes only bright line structures specifically called by the computer program and falling within a field of approximately one symbol diameter. To avoid problems, the light-pennable character should not be blink-coded, two individually pennable characters must be separated by at least one symbol, and the cord connecting the pen to the equipment should be attached at a point where it will cause minimal interference with the display viewing and will not accidentally catch on console features.

Summary of Coding Method for Symbols

Code	Number of Steps in Code	Evaluation	
Alphanumerics	Unlimited	Excellent	High information-handling rate; unlimited number of coding steps
Geometrics	20 or more	Excellent	Certain shapes easily recognized; many coding steps
Color	4	Excellent	Difficulty in techniques of reproducing for CRT; objects easily and quickly identified
Blink	2	Poor	Distracting and fatiguing; interacts poorly with other codes; best for attention getting; few steps in code
Brightness	2	Poor	Limited number of steps; fatiguing; detrimental to decoding performance
Line lengths	4	Fair	Limited number of steps; clutters displays
Angular orientation	12	Fair	95 percent of estimates correct within 15 percent
Inclination	24 or more	Fair	Many coding steps, especially with combinations
Visual number (dots)	6	Fair	Few steps
Combinations	Unlimited	Good	Avoid overloading symbols with too much information; complex combination can degrade decoding speed and accuracy

Point resolution is conventionally a fixed percentage of display size. On a 12-in (30-cm) CRT, resolution is about 85 points per inch (1023 × 1023 per display surface). A 4-ft (1.2-m) display of the same matrix would have a resolution of about 21 points per inch. This has little effect on alphanumeric data but puts a definite limitation on graphic displays. On such a 4-ft (1.2-m) screen, lines or points must be separated by 0 05 in (0.13 cm). In both CRT and large-screen displays, line thickness is also proportional to display area.

A 12-in (30-cm) CRT is contained within a visual angle of 41° for a 16-in (41-cm) viewing distance. An operator is able to scan 8.6 in (22 cm) (+15°) comfortably without head movement. The same viewing angle is available at 64 in (163 cm) on a 4-ft (1.2-m) screen. If the larger screen is viewed at a greater distance, the total angular view is decreased; however, element sizes must be increased. With a 16-ft (4.9-m) viewing distance, a 4-ft (1.2-m) screen can be viewed comfortably by a small working group; however, for detail resolution comparable to that of a 12-in (30-cm) CRT, the larger screen can represent only 4 in (10.2 cm) of the CRT display. A choice must be made between the area which can be shown and the amount of detail that can be presented.

Ambient illuminance should not contribute more than 25 percent of screen brightness through diffuse reflection and/or phosphor excitation. The ambient illuminance in the CRT area should have appropriate intensity and color with respect to other visual tasks, e.g., setting controls, reading instruments, inspecting maps, and performing various maintenance and housekeeping tasks, but it should not interfere with the visibility of signals on the CRT display.

The luminance range of surfaces immediately adjacent to scopes should be between 10 and 100 percent of the screen background luminance. With the exception of emergency indicators, light sources in the immediately surrounding area should not be brighter than scope signals.

Viewing hoods or glare-reduction devices should be provided when very faint signals must be detected, and/or a suitable light control filter system should be used to maintain maximum ratios of signal to CRT background luminance.

Electrically or optically generated displays should conform to MIL-STD-884. Radar CRTs display information more effectively when a symbolic shape code is used in association with, or in place of, raw video pips. The following encoding principles are suggested: The number of discrete symbols should be held to a minimum. Symbol meanings should be standardized.

Symbols should be recognizable without reference to some comparison standard.

If combination codes are used, the respective meanings should be decodable separately without confusion. Two codes maximum combination.

Symbol shapes should reflect a natural relationship to the event or type of object the symbol is supposed to represent.

Codes should be easy to learn and retain.

Symbols should differ widely in shape; avoid variations of a single form, e.g., a circle and an ellipse.

Redundant cues (i.e., those which associate a symbol with a particular size) increase coding efficiency.

External modifiers surrounding a symbol should be avoided.

Reserve blinking coding for emergency situations. There should be no more than two blink rates.

Two brightnesses are the maximum for brightness coding.

Four line lengths can be identified with minimum error.

No more than three blip sizes should be used.

Video Signal	Vertical Resolution Center (Lines)	Vertical Resolution Corner (Lines)	Horizontal Resolution Center (Lines)	Horizontal Resolution Corner (Lines)
Monochrome	400	400	800	700
Red, green, or blue	400	400	800	700

Typical TV Equipment Standards[5]

Television signal characteristics should be based on a 525-line scanning standard, interlaced 2:1, with 60 fields and 30 frames per second. A video bandwidth of 4.5 MHz should be used for color picture signal transmission. A video bandwidth of 10 MHz should be used for monochrome picture signals.

Color Picture Monitors

SCAN SIZE The normal scan should provide a display in which all four corners of the raster are visible. Width and height controls should have sufficient range to vary the raster size from −10 to +20 percent while maintaining specified linearity.
RESOLUTION The limiting horizontal and vertical resolution of the monitor should be as specified in the above table for both composite and noncomposite operation.
ASPECT RATIO The width/height ratio of the raster should be 4:3.

Monochrome Large-Screen TV Projector

SCAN SIZE Width and height controls should have sufficient range to vary raster size from −10 to +20 percent of the nominal dimensions.
ASPECT RATIO The width/height aspect ratio of the normal picture should be 4:3.
RESOLUTION The limiting horizontal resolution with the specified brightness level should be at least 800 lines at the picture center and 700 lines at the corners. Vertical resolution should be at least 400 lines, attained when a composite or noncomposite monochrome video signal is applied to the projector input.
BRIGHTNESS The projector should provide brightness levels as shown in the table below. Variation in any area of the screen plane should not exceed ±30 percent of the center of the plane.

Monochrome Television Project

Image Size	Screen Brightness (fL)
6 × 8 ft	62
9 × 12 ft	28
12 × 16 ft	16
15 × 20 ft	10
24 × 32 ft	4

Note: 1.0 ft = 0.3 m.

[5]Defense Communications Agency, *Television Technical Characteristics*, vol. I, *Picture Generation and Display Equipment*, AD 805-174, November 1965.

GRAY-SCALE REPRODUCTION Nine shades of gray and the white background should be distinguishable.
GEOMETRIC DISTORTION The combined effects of all distortion should not displace any point on the projected display from its correct position more than 1.0 percent of the picture height.
TRAPEZOIDAL DISTORTION The projector should be capable of correcting "keystone" or trapezoidal distortion resulting from vertical tilt of the screen, i.e., screen tilt from the perpendicular to the optical axis of the projector within ± 15°.
INTERLACE Displacement of any scanning line from a center position between lines of the alternate field should not exceed 10 percent of the distance between the lines of the alternate field.
WHITE BALANCE The TV monitor should be capable of producing, from a monochrome input signal, a white that corresponds to CIE illuminant "C" ($x = 0.310$, $y = 0.316$).

TV Quality Variations

At resolutions of 8, 10, and 12 lines, the quality of TV equipment has little significant effect on the accuracy or speed with which standard alphanumerics can be read. At 6 lines, readability is good with high-quality TV (i.e., a minimum of 945 lines).

For group viewing of TV, a minimum vertical resolution of 15 lines/character height is recommended when small visual angles are involved. At 15 lines/character height to total display height is 1/33, and 16 rows of characters can be put on the screen (i.e., as long as the screen visual angle is within 8 min of arc).

For a symbol resolution of 15 lines, the recommended maximum viewing distances for various TV monitor sizes are as follows:

27-in (69-cm) monitor—18 ft (5.5 m)
24-in (61-cm) monitor—15 ft (4.6 m)
21-in (53-cm) monitor—13 ft (3.9 m)
17-in (43-cm) monitor—11 ft (3.4 m)

For pictorial TV viewing, the recommended maximum viewing distances are as follows:

21- to 30-in (53- to 76-cm) monitor—20 to 30 ft (6.1 to 9.2 m)
19- to 23-in (48- to 58-cm) monitor—10 to 20 ft (3.1 to 6.1 m)
17- to 19-in (43- to 48-cm) monitor—6 to 10 ft (1.8 to 3.1 m)
15- to 17-in (38- to 43-cm) monitor—30 in to 6 ft (76 cm to 1.8 m)
9-in (23-cm) monitor—18 to 30 in (46 to 76 cm)

Effect of TV Surround Brightness

The mean value of surround brightness preferred by viewers of broadcast TV is shown in the accompanying graph (plotted for three surround areas at each of five values of peak screen luminance).

Ambient illumination glare should be minimized; i.e., sources should not be located within 60° of the observer's central visual field.

The optimum brightness distribution on the surface of the TV display is approximately 17 fL measured from the central axis and not less than 13 fL measured at the largest angle of view off center. Assuming an ambient light level of 1 fc on the display, this permits viewing in about 10:1 contrast for symbols with regard to the background.

Brightness ratios required for comfortable viewing of large-screen displays are determined by locating two values: (1) the minimum ratio required for adequate viewing and (2) the maximum measure of brightness without annoying aftereffects. The maximum brightness for group displays should not be more than 35 fL. Higher brightness may produce afterimages if the display is viewed for extended periods. An increase in brightness over the 15- to 35-fL maximum contributes no significant improvement in visual acuity.

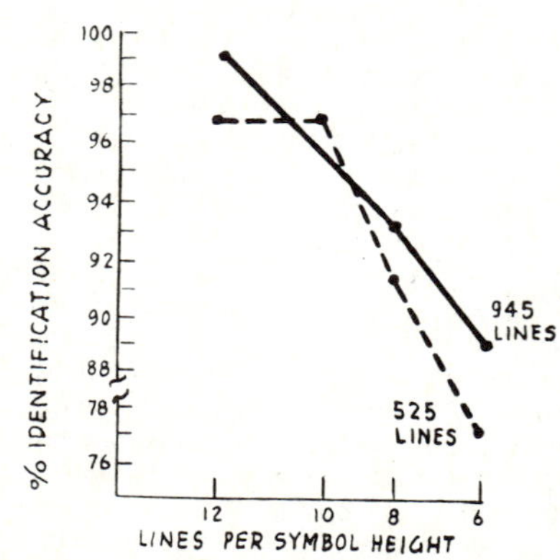

Accuracy of symbol identification for good-quality versus low-cost TV.

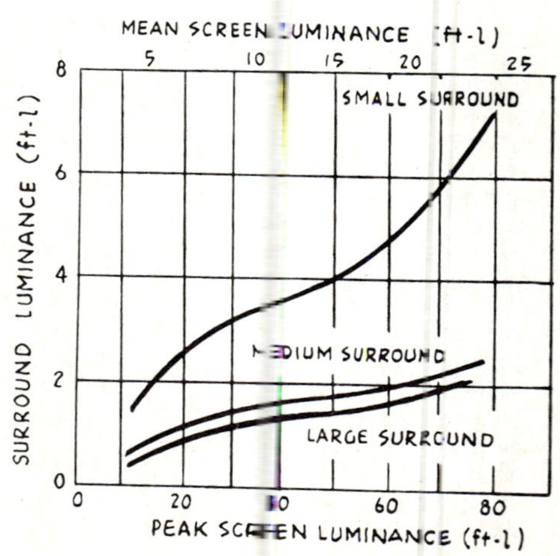

CRT Symbol Characteristics

DOT MOSAIC The coarsest acceptable mosaic is 5 × 7, as shown in the accompanying illustration. Even these may be marginal compared with larger matrices. Only 35 decoded lines are required. If only numbers and a limited number of symbols are required, such as horizontal or vertical lines or crosses, the number of dots may be further reduced to 35.

STROKE MOSAIC Bar, stroke, or segment patterned symbols can be created as shown in the accompanying illustration. Such matrices are suitable only for numerical characters, not letters. Readability is not as good as with dot matrices if speed and accuracy are required. The seven-segment format is the minimum configuration that is acceptable. Additional segments can be added to improve the legibility of characters. The character format can be slanted, but legibility suffers if the slope is more than 5–7°.

Dot matrices are more readable than segmented types because of the capability to produce curved portions of characters.

SYMBOL SPACING At low brightness levels (1 fL), spacing should be 25 percent of character dimension; at high brightness levels, spacing should approach 200 percent for maximum readability of a single character. For typical applications (e.g., clear text messages or grouped numbers), use 50 percent of average character width for between-character spacing within a word or number group and 75 to 100 percent spacing between words or groups. Spacing between lines of characters should be at least 50 percent of character height.

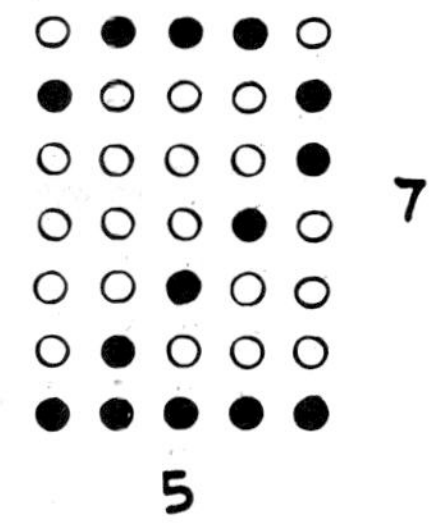

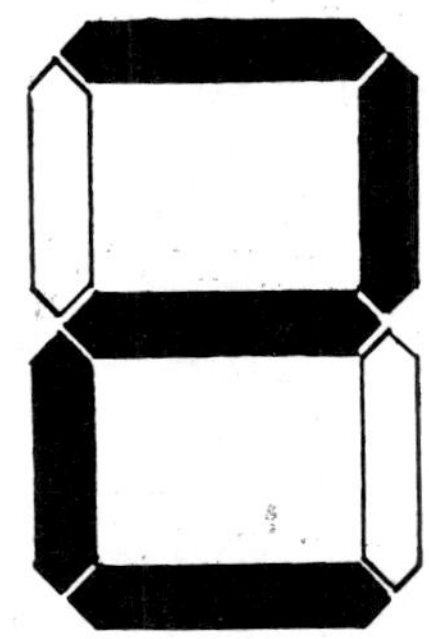

Dot matrices are more readable than segmented types because of the capability to produce apparent curved portions of characters.

A 5 × 7 matrix of dot characters is minimally acceptable; a 7 × 9 matrix, as shown in the illustration on page 264, is preferred.

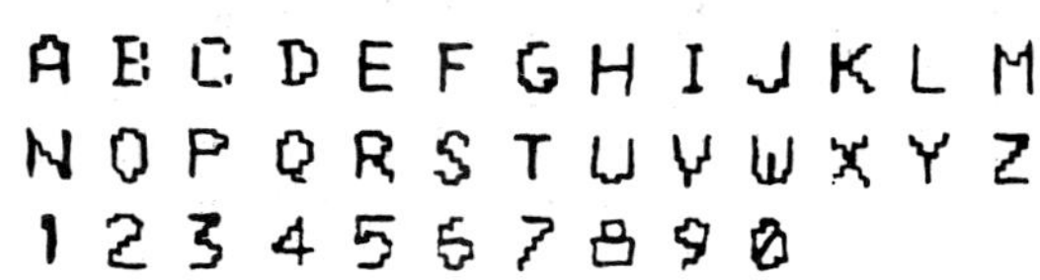

Minimum character height	3.1–4.2 mm
Maximum character height for 5 × 7 dot matrix	4.5 mm
Width/height ratio	3:4–4:5
Stroke/height ratio	1:8–1:6
Minimum number of raster lines	10 lines

Suggested CRT Character Optimized Set Using a 7 × 9 Matrix

The suggested character formation shown in the accompanying illustration contains most of the alphanumerics and supporting symbols required for data processing. They are designed to minimize recognition errors, or at least possible confusion between one character and another. Although lowercase letters have been designed, they are not recommended because it is difficult to create lowercase features that are not easily confused, especially if the observer has to read information rapidly.

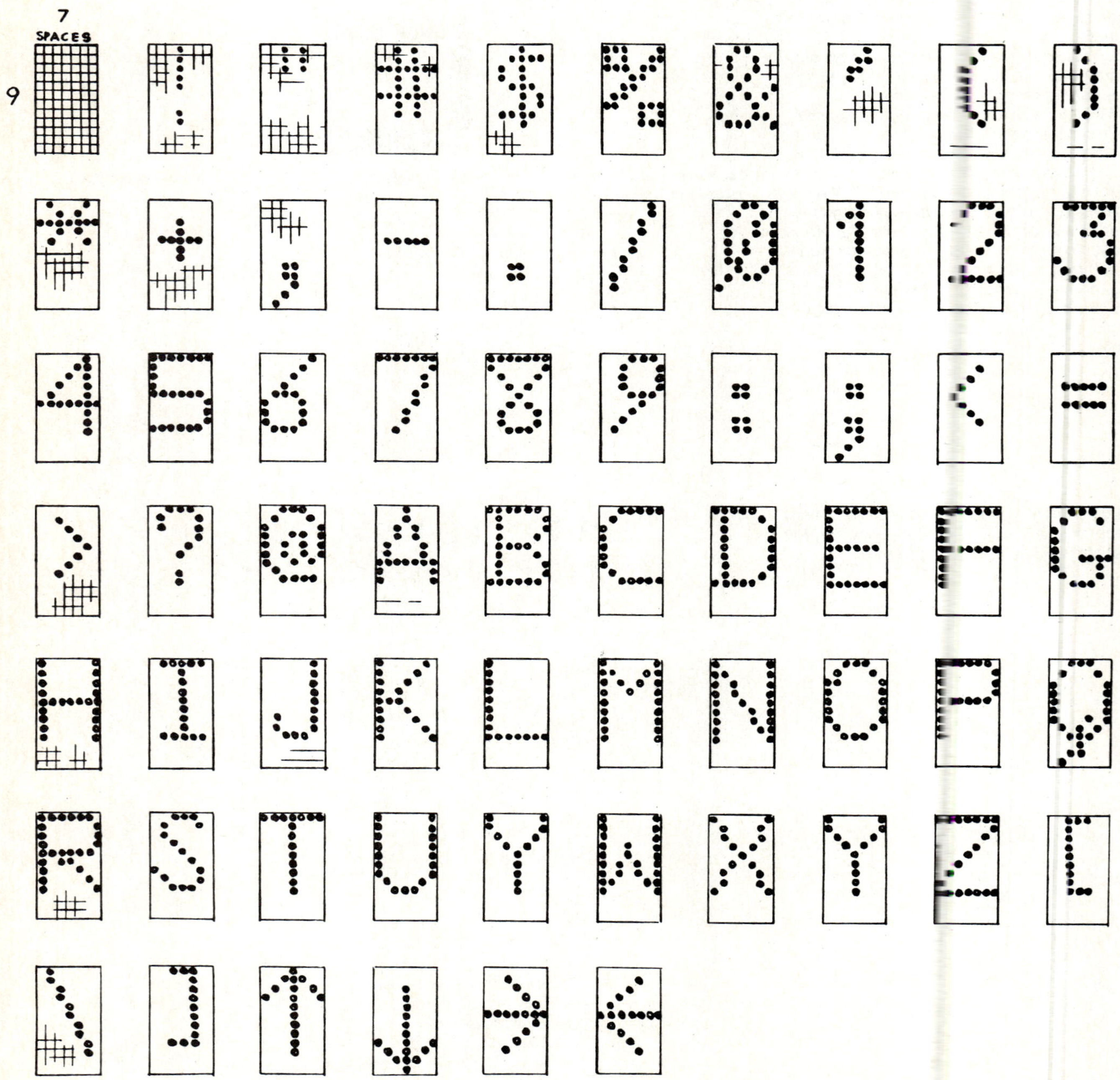

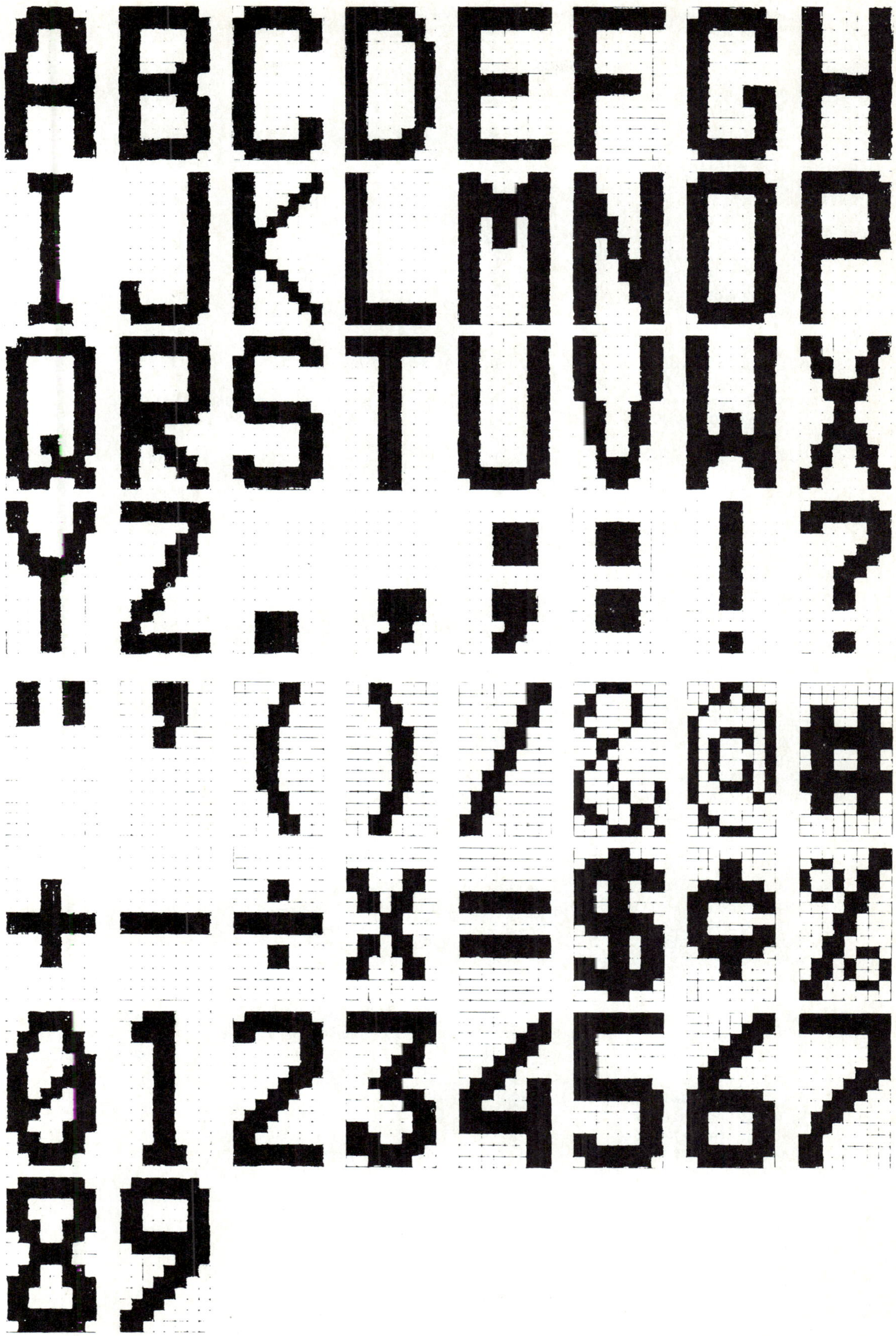

Example of an 8 × 14 matrix character set.

General CRT Phosphor Characteristics and Applications

Identification	CIE Coordinates x	CIE Coordinates y	Persistence	Fluoresence	Phosphoresence	Decay Time (m/s)	Typical Applications
P-1	0.218	0.712	M	YG	YG	24	Radar and test equipment oscilloscopes
P-2	0.279	0.534	M	YG	YG	35–100	Oscilloscopes
P-3	0.523	0.469	M	YG	YO		
P-4	0.270	0.300	M/S	W	W	25	Monochrome TV
P-5	0.169	0.132	M/S	B	B	25	Photography
P-6	0.338	0.374	S		W		
P-7	0.357	0.537	L	Y	YG		Radar, sonar, and oscilloscopes
			M/S	B	B		
	0.151	0.032	M/S	W	YG	40–60	
P-11	0.139	0.148	L	B	B	25–80	Photography
P-12	0.605	0.394	M/S	O	O		Radar
P-14	0.504	0.443	M	YO	O		Displays where repetition rate is 2 to 4 s after excitation is removed
	0.150	0.093	M/S	B	O		
			M/S	PB	YO		
P-15	0.246	0.439	VS	G	G		TV pickup of photographs by flying spot scanning
P-16	0.175	0.003	VS	BP	BP	0.12	TV pickup of photographs by flying spot scanning
P-17	0.302	0.390	L	Y	BP		Military displays
			VS	G	BP		
			S	YW, BW	Y		
P-18	0.333	0.347	M	W	W		Low-frame-rate TV
P-19	0.572	0.422	L	O	O		Radar
P-20	0.444	0.536	M/S	YG	YG	0.01–2	High-visibility displays
P-21	0.539	0.373	M	RO	RO		Radar
P-22	0.155	0.060	M	B	B	25	Color TV
	0.285	0.600	M	YG	YG	60	
	0.675	0.325	M	OR	OR	0.9	
P-23	0.375	0.390					Interchangeable with P-4
P-24	0.245	0.441	S	G	G		Flying spot scanner tubes
P-25	0.557	0.430	M	O	O		Long-persistence displays up to 10 s
P-26	0.582	0.416	VL	O	O		Radar
P-27	0.674	0.326	M	RO	RO	27	Color TV monitors
P-28	0.370	0.540	L	YG	YG	0.5 s	Radar and sonar
P-29	(P_2 +	P_{25})	M	W	YW		Aircraft indicators and radar
P-31	0.193	0.420	M/S	G	G	4	Oscilloscopes
P-32			L	PB	YG		Radar
P-33	0.559	0.440	VL	O	O		Radar
P-34	0.235	0.364	VL	BG	YG	40 s	Oscilloscopes and radar
P-35	0.286	0.420	MS	G	B	1	Oscilloscopes
P-36	0.400	0.543	VS	YG	YG	0.25	Flying spot scanning tubes
P-37	0.143	0.208	VS	B	B	0.155	Flying spot scanning tubes
P-38	0.561	0.437	VL	O	O	1040	Integrating phosphor for low-repetition-rate displays and radar
P-39	0.223	0.698	L	YG	YG	150	Integrating phosphor for low-repetition-rate displays and radar
P-40	0.276	0.3117	M	W	YG		Integrating phosphor for low-repetition-rate displays and radar

B = blue, G = green, O = orange, P = purple, W = white, Y = yellow.
VL = very long, 1 s or over.
L = long, 100 ms to 1.0 s.
M = medium, 1 ms to 100 ms.
M/S = medium short, 10 μs to 1.0 ms.
S = short, 1 μs to 10 μs.
VS = very short, less than 1 μs.

Special Notes:

P-1—High efficiency, resolution, and resistance to burn; lacks low-level or short persistence.
P-2—Decrease in decay with increase in beam current.
P-4—Sulfide version.
P-7—High efficiency and resistance to burn; amber filter required to obtain long persistence.
P-19—Slow refresh rate for flickerless display; low light output; low burn resistance.
P-22—Sulfide blue and green, vanadate for red.
P-25—Desired low-level persistence; high resistance to burn; low light output.
P-26—Slow refresh rate for flickerless display; low light output and burn resistance.

(continued)

P-31—Curve has blue peak at 450 nm; high efficiency, resolution, and resistance to burn; lacks low-level or short persistence.
P-33—Decay decreases with beam current decreases; burns rapidly when used with stationary or slow-moving beam.
P-34—IR stimulatable; Y-phosphor.
P-35—Resists burning compared with P-11.
P-39—Similar to P-1 but with longer decay.
P-40—Similar to P-7 but with longer decay.
P-37—Similar to P-11 but with shorter decay.
P-36—Similar to P-20 but with shorter decay.

Source: After H. R. Luxenberg and R. L. Kuehn *Display System Engineering,* McGraw-Hill Book Company, New York, 1968.

CRT Flicker Thresholds of the Average Observer

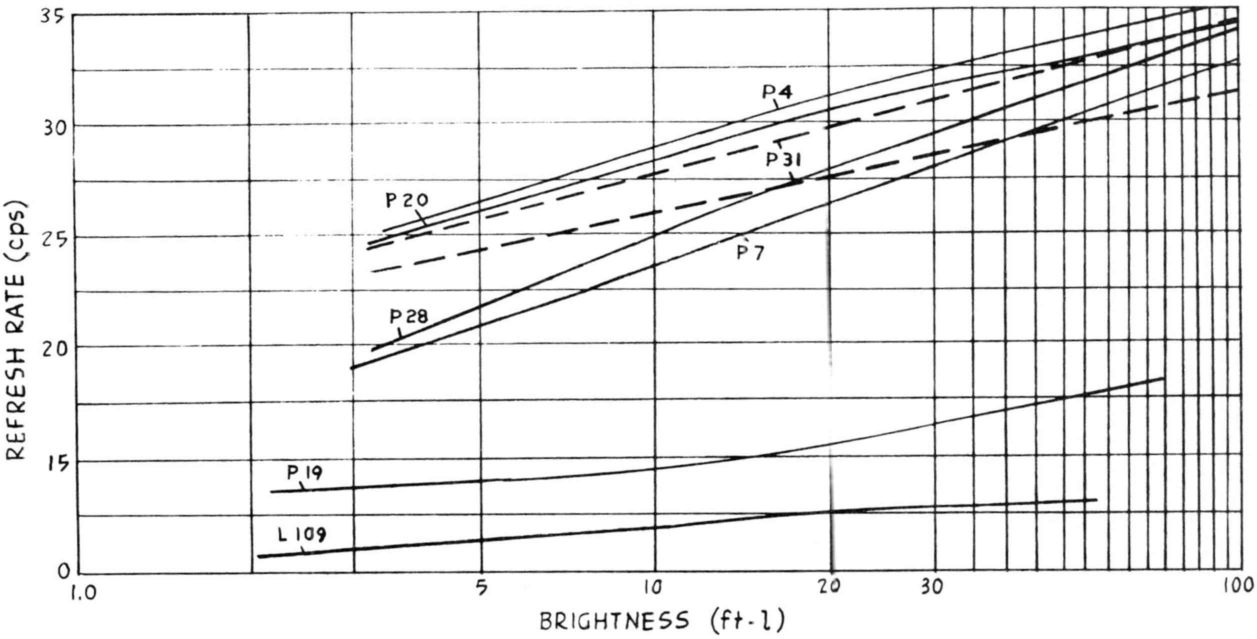

For character displays the pulse rate should be greater than about 30 to 40 Hz so that the characters will not appear to blink. Some flicker may be noticed with average display brightness values unless the repetition rate is maintained at least at 50 Hz.

Flicker will not be noticed in TV at 60 fields per second unless the display brightness exceeds about 180 fL; 50 fields per second is usually acceptable if the display brightness is reduced to 30 fL.

GUIDELINES FOR SELECTING A TONAL VERSUS A SPEECH MODE

1. Use a Tonal Signal:

When immediate action is required on the part of the listeners, i.e., when vocal explanations or directions are not necessary for the listeners to know what the signal means and what they should do

When a specific point in time (that has no absolute value) is to be indicated, e.g., when the sound of a gong tells the listeners that something has happened or is about to happen, that they should be prepared for a message, etc.

When a spoken message would compromise the security of a situation, i.e., when a coded tonal signal would be unrecognizable to persons not privy to the code

When noise conditions are unfavorable for receiving spoken messages

When speech channels are overloaded

When a spoken message could annoy listeners for whom it is not intended or when the spoken message could mask other messages

When the intended listeners are familiar with the tonal signal implication or the tonal signal code

When it is desired to use the simplest audio signal

2. Use a Spoken Message:

When more message flexibility is needed than a tonal signal can convey

When it is necessary to identify the source of the information

When listeners have not had training in a special tonal signal code

When there is a need for rapid two-way exchanges of information

When the intended information deals with a future time and when preparation is required, e.g., during a countdown preparatory to initiating some operation

When use of a tonal signal countdown could result in a miscount

When operational stress surrounding the intended listeners could cause them to forget the meaning of a tonal signal code

AUDITORY DISPLAY SELECTION GUIDE

Display Configuration	Use Criteria	Human Engineering Considerations
Steady continuous tone 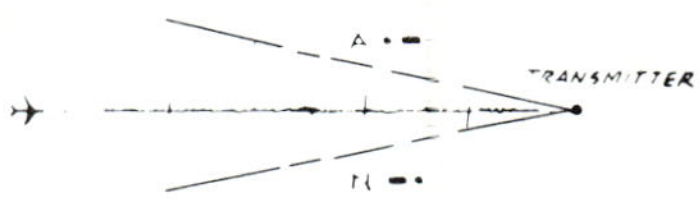	Use in both ears for tracking, i.e., both ears equally loud for "on target," shifting left or right when "off target."	Select the frequency which is least subject to other nonverbal auditory signals or typical vehicle system noises.
Prerecorded spoken message *Now hear this!* *"this is your captain speaking"* *calling Doctor Brown . . . come to emergency!*	Use when communication flexibility is necessary. Use when it is necessary to identify the source of a message. Use when a stressful situation might cause the listener to forget the meaning of a coded signal. Use when a simple coded signal cannot adequately give directions or instructions to the listener. Use when ambient masking noise characteristics obviate the use of simple tonal signals. Use when other complex tonal signal possibilities have already been exhausted, i.e., have been assigned and cannot be duplicated.	Utilize the most concise message possible that still retains maximum intelligibility and minimum ambiguity.
Intermittent tone 	Use for "start and stop" timing. Use for continuous information where rate of change of input is low.	Avoid a "train" of signals which might be confused with similar electrical system noise.
	Use for irregularly occurring signals, i.e., alarms.	
	Use as indicator of speed, i.e., a change in rate of signal occurrence. (Faster signal repetition indicates an increase in vehicle speed.)	
	Use for continuous-wave message transmission, i.e., Morse code.	
Warble and undulating tones 	Use when masking noise characteristics are unknown or cover a broad frequency spectrum in random pattern.	Avoid a signal that sounds like any characteristic navigation signal or radio signal such as might occur when two carriers present a "beat frequency" effect. Where distraction is a critical factor, abrupt onset of the audio signal should be avoided, as well as apparent movement of the signal in space.
	Use pitch differences to represent "up and down" relationships.	
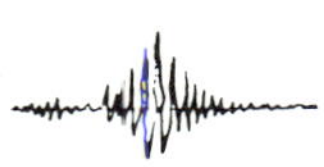	Use for locating a null position such as in orienting a radio beam.	
Bell, buzzer, horn, siren, and whistle	Use bell for fire alarm. Use horn for emergency warning, equipment malfunction, pressure leak, hatch open, etc. Use whistle (typical of "Bosun's whistle") for "All Hands Alert" (followed by further communication). Use buzzer for individual operator alert to receive further communication. Use Claxon for "Report to Duty Stations."	The signal should be "distinctive" and unlikely to be obscured by other noises. Where employees' attention may be highly concentrated on a task, use signals with relatively high alerting capacity. Signals should be at least 10 dB above the ambient noise level. They should be repeated so as not to be missed. Caution signals should be provided with reset and volume controls. Consider a multistage warning or alarm system, i.e., an initial tone signal followed by the critical signal, followed by speech where appropriate, to indicate the specific nature of the hazard. Alarm signals should focus attention on the first 0.5 s, with all essential information within the first 2.5 s. Use "sudden" onset. Concentrate signal energy between 250 and 2500 Hz, with signal readily identifiable on the basis of components below 2000 Hz.
Microphones 	Provide microphones for each operator position for shirt-sleeve operation. Microphones should be head-mounted for all situations where the operator will have both hands occupied.	Provide maximum convenience and minimum discomfort.

AUDITORY DISPLAY SELECTION GUIDE *(continued)*

Display Configuration	Use Criteria	Human Engineering Considerations
Telephone handsets 	Use at operator stations where the operator does not perform communications as a primary activity and can pick up the handset with essentially no interference with his or her regular duties. Use in situations where it is desirable to reduce the fatigue caused by wearing a headset.	
Earphones 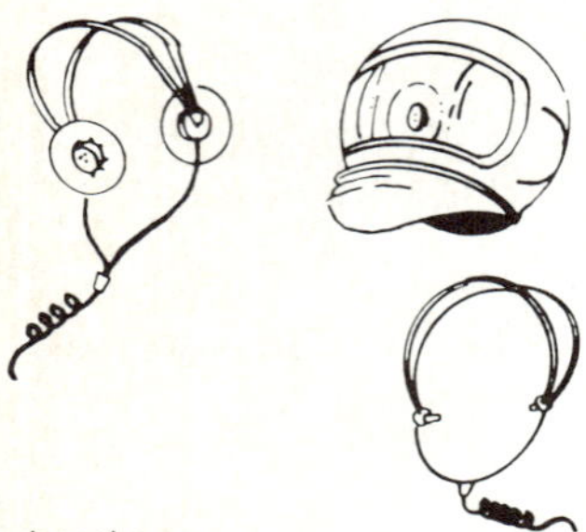	Provide earphones in all pressure helmets. Provide two portable head-mounted earphones for each operator position for shirt-sleeve operation.	The comfort and convenience of the operator as well as audio quality must be considered so that the operator will use the equipment as designed.
Loudspeakers 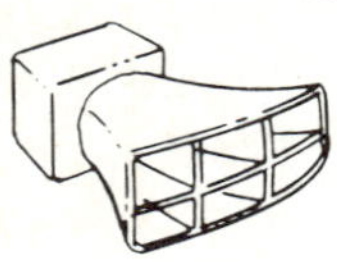	Use loudspeakers only when it is desirable to communicate with more than one person at the same time and when it is anticipated that one or more of the parties concerned will not be near a phone or will be wearing headsets (i.e., a public address system).	
Audio recorders and playbacks 	Use wherever it is desirable to preserve an audio communication for: 1. A record of normal operating messages 2. A record of emergency message transmission for evaluation of a failure or catastrophe Use for presenting a prerecorded message (verbal alarm, procedural instruction, etc.). Use as an on-site test, evaluation, or research analysis aid where the operator's activity precludes making a manual record and/or to capture critical messages that may be used to analyze events prior to an accident.	All units should be packaged for convenience of use and servicing. Units should be packaged and located for maximum protection from destruction during the accident.

TYPE OF SYSTEM IN RELATION TO APPLICATION

Although the choice of system may be obvious in most cases, there are specific operational reasons for selecting one type of communications system over another. The following guidelines are intended to help the planner or designer decide which type of system to use for a particular operational environment:

Public address system: Use for nonspecific emergency announcements for all diversely located listeners. Do not use for routine messages intended for one individual if the announcement may interfere with localized communication.

Intercom system: Use for small group-specific, two-way communication and/or for individual communication between people who are physically isolated from one another, where telephone dialing would create a considerable waste of time (e.g., for communications between a shop supervisor and personnel at factory work stations). Do not use where conversational privacy is necessary.

Telephone: Use for situations requiring individual conversational privacy.

Radio: Use for situations where there are long distances between the conversants and/or when one or more of the conversants are moving from one location to another, (e.g., a security guard on foot or a police officer in a patrol car).

Closed-circuit TV: Use where face-to-face visual reference and/or reference to the object under discussion adds considerably to communication reliability.

Type of Communicating Interface

Handset (telephone): Use when at least one hand can always be free to hold the handset.

Headset (earphone or earphones and microphone): Use when both hands are often occupied.

Wireless microphone: Use when a fixed microphone or hand-held microphone with cable would restrict the user's freedom to move about.

GENERAL PRINCIPLES FOR AUDITORY DISPLAY DESIGN

SITUATION COMPATIBILITY The effectiveness of any auditory display depends on proper consideration of the sound environment within which the display must operate; i.e., a spoken message is easily obscured by other spoken messages, a particular tonal signal is easily masked by other similar tonal signals, and any auditory signal or message may be masked by frequent, extremely loud bursts of noise.

COMMON SIGNAL CONVENTIONS Certain types of signals (sirens, bells, claxons, etc.) are recognized and associated with certain activities, such as those of fire fighters and the police. Such characteristic signals should not be used for other purposes when the situation is such that the more common convention is in use. Other natural relationships, such as high frequencies being associated with "up" and low frequencies being associated with "down," should be taken into account when considering the use of pitch as an auditory code.

STANDARDIZATION Once a particular auditory signal code is established for a given operating situation, the same signal should not be designated for some other display.

LISTENER SENSITIVITY The signal frequency of auditory displays should be compatible with the midrange of the ear's response curve for both pitch and loudness; i.e., avoid the use of signals at the extreme ends of the sensitivity curves, where response reliability is more easily masked.

DUAL MODE When the noise environment is unknown or suspected of being difficult to penetrate, consider one or both of the following: (1) using a shifting frequency signal that passes through the entire noise spectrum and (2) combining the auditory signal with a visual signal.

QUANTIFIABLE SIGNALS When the tonal signal information is to be quantifiable, provide a reference tone, e.g., a base-line loudness or pitch against which the primary signal can be compared.

PRIVACY If a signal must occur in an area in which only certain personnel should be privy to its purpose and others are not to be unduly annoyed, select a simple bell tone that can be recognized among ambient speech sounds without being loud.

SELECT LISTENERS Consider a simple repetition code signal to indicate who is to respond, e.g., to answer a phone or to respond to a previously established set of task events.

ANNOYANCE Avoid the use of extremely loud signals that may startle listeners, and add to the overall noise levels, or interfere with local speech activity. When the ambient noise level gets too high, use earphones.

AUDITORY DISPLAY FOR WARNING PURPOSES: GENERAL OBJECTIVES FOR USE AND DESIGN

Auditory warning displays are particularly good because they do not require the intended listener to be facing the signal source, and, with the exception of impossibly loud noise environments, they can be designed to cope with most acoustic interferences. In addition, because the audio channel of the operator is typically last to be affected by such stresses as dynamic oscillation of the listener and vibration, audio signals can be used under these stress conditions more successfully than visual warning devices.

Operational Objective	Guidelines
Audibility in high-noise: level environment	The signal energy should be concentrated within a narrow frequency band that is lowest in the background noise spectrum. A signal-to-noise ratio of 5:1 should be maintained if at all possible. The audio signal power at the listener's ear should be approximately 300 mW. If the ambient noise is such that the warning signal would have to reach intensity levels that could be painful or injurious, provide the listener with a headset or a combination headset within an acoustic helmet. If the nature of the ambient noise is such that it is loud across most of the audible frequency range, consider the use of a repeating signal (with shutoff control) and provide a noise-operated gain control to maintain a compatible signal-to-noise ratio as the background noise changes. Use a noise-canceling microphone so that background noise will not be picked up along with the warning signal or the speaker's voice. Consider delaying the signal to one of the listener's ears (by 0.01 s) or reversing the phase of one earphone of the headset.
Fast operator response	Make the signal brief and concentrate the signal energy at the beginning. If a canned voice signal is used, provide a short alerting signal (0.5 s) just before the speech message to prepare the listener.
Operator distraction	Provide the operator with a temporary "silencing control" so that he or she can prevent the warning signal from interfering with a particularly critical ongoing task. The shutoff control must, however, have an automatic reset feature so that the warning system is not permanently disabled. Do not provide the operator with a volume control to quiet an interfering warning signal; the operator will generally turn it down to the point where he or she may never be alerted to an emergency.
Interference of warning signal with other important signals	Deliver the warning signal to one earphone via a headset and the other signals to the other earphone. Alternate the warning and general signal input positions every 2–3 s. Avoid frequency bands normally used by common communication or navigation signals.
Reduction of signal confusion	Avoid selecting warning signals whose particular characteristics are easily confused with common signals such as typical electrical interference.
Transmission mode	Where practicable, send warning signals via a different system from those being used for speech communications.

NONVERBAL AND ALARM SIGNALS

1. Two-Phase Signals

Attention should be focused on the first 0.5 s of the alerting signal and then upon the first 2.0 s of the identifying or action signal. Alerting signals should precede action signals by a time interval of 2.0 to 8.0 s for isolated signals and of 0.3 to 2.0 s for action signals occurring in sequence. When the alerting signal consists of a "glissando," the audible signal should spend at least 0.1 s in each octave band from lowest to highest.

Where masking is a critical factor, warning signals should be concentrated in frequency bands that are unused or little used by primary communication (command, navigation, advisory, etc.). Major concentration of energy should be between 250 and 2500 Hz/s and the signal should be readily identifiable on the basis of sound components below 2000 Hz. Where possible, avoid the following:

a. Trains of impulses that resemble electrical interference
b. Simple glissandi which might be confused with sounds made by carriers, i.e., beat frequency and oscillator effects
c. Periodic impulses similar to radar signals
d. Random noise similar to that generated by such equipment as an air conditioner
e. Musical sounds
f. Noise similar to static or sporadic radio signals
g. Modulated or interrupted tones resembling navigation signals or coded radio transmissions

Where warning signals delivered to headsets might mask other essential auditory signals, separate channels may be provided to direct the warning signal to one ear and the background signal to the other ear. Presentation of warning signals may also be dichotic and alternating with the dichotic presentation of communication signals.

Where earphones cover both ears of the operator during normal system operation, the warning signal should be directed to the operator's headset as well as externally.

Audio warning signals should be at least 10 dB above the loudest expected ambient noise level. Such signals should not exceed a maximum intensity of 110 dB in areas (near the sound source) which must occasionally be occupied by unprotected personnel for as long as 10 min duration. The signal energy should be concentrated in the frequency band in which background noise is lowest.

2. Single-Phase Signals

All essential information should be contained in the first 0.5 s of the signal. Considerations regarding masking and signal strength are the same as for the two-phase signal.

3. False Alarms

Audio warning systems should be designed in such a way that any possibility of false alarms is minimized.

4. System Independence

The circuitry of audio warning systems should be independent of the system in which they are to provide warning information so as to preclude warning system failure in the event of primary system failure.

5. Operator Control

All audio warning systems or devices should be provided with a test button, switch, or other means of verifying the audio aspects of the signal at any time. Whether audio warning signals are designed to terminate automatically, by manual control, or both, an automatic "reset function" should be provided. The reset function should be controlled by a sensing mechanism which should recycle the signal system to a specified condition as a function of time or the state of the signaling system. The signal duration should be at least 0.5 s and should continue until an appropriate response is made by the operator. The completion of the proper corrective action should automatically terminate the warning signal. A shutoff switch should be provided that is controllable by the operator, the sensing mechanism, or both, depending on careful consideration of the operational situation and personnel safety factors. (Caution signals should be provided with manual reset and volume controls.) Shutoff switches should not prevent the presentation of other warning signals and should reset automatically after an interval of time.

6. Steady, Continuous Tonal Signals

Signal frequency should be confined between 400 and 1500 Hz.

The signal should exceed its masked threshold by at least 15 dB. Do not exceed a sound power level of 135 dB for intermittent operation (120 dB for extended periods). Set the signal level 60 dB or more above the absolute threshold.

For listening in noise, select a signal midway between the masked threshold and about 110 dB.

Provide automatic volume control when the dynamic range is 20 to 30 dB or less.

7. Intermittent Tonal Signals

Signal frequency should be confined between 400 and 1500 Hz.

The signal should exceed its masked threshold by at least 15 dB. Do not exceed a sound power level of 135 dB.

Set the signal level 60 dB or more above the absolute threshold. Select a signal midway between the masked threshold (noise) and about 110 dB.

Provide automatic volume control when the dynamic range is 20 to 30 dB or less.

8. Warble or Undulating Tonal Signals

The signal frequency should be confined between 500 and 1000 Hz. Use a pitch rise and fall rate of about 1 to 3 Hz. Where an alerting signal consists of a glissando, the audible signal should remain at least 0.1 s in each octave band from the lowest to the highest frequency.

Where the operator's task is to detect frequency change, set the signal at least 30 dB above the absolute threshold.

VARIOUS SIGNAL DEVICES AND CHARACTERISTICS

Buzzers: Usually low intensity ranges, low frequency (150 to 400 Hz). Generally suitable for relatively quiet situations where the distinctive sound commands attention but does not cause "alarm."

Bells: Normally higher intensity and frequency than buzzers. Suitable for situations where the ambient noise is higher (although they can also be used at lower intensity in fairly quiet environments).

Horns: Usually high intensity (90 to 100 dB) and middle to lower frequency. However, horns can be used with higher intensity and higher frequency up to 3000 to 4000 Hz. Because they generally are of longer duration, they are not suitable for quiet environments, but rather for high-noise-level situations where urgent emergency warning is desired.

Chimes: Similar to quiet bells, typically in the frequency range of 500 to 1000 Hz. Suitable primarily for situations where the resident personnel require signaling without disturbing visitors.

Sirens: Can be made to penetrate almost any noise environment. The combination of high intensity and constantly changing frequency is not easily masked by spurious or continuous ambient noise conditions. Primarily for emergency signaling by police, fire fighters, etc.

Characteristic Intensity Ranges and Predominant Frequencies for Specific Signaling Devices

Signal	Average Intensity At 10 ft	At 3 ft	Predominant Audible Frequency
	For Large Areas (High Intensity)		
4-in bell	65–77	75–83	1000
6-in bell	74–83	84–94	600
10-in bell	85–90	95–100	300
Horn	90–100	100–110	5000
Siren	100–110	110–121	7000
	For Small Areas (Low Intensity)		
Heavy-duty buzzer	50–60	70	200
Light-duty buzzer	60–70	70–80	400–1000
1-in bell	60	70	1100
2-in bell	62	72	1000
3-in bell	63	73	650
Chime	69	78	500–1000

VERBAL (SPEECH) SIGNALS

1. Word Selection

In selecting words for use in auditory warning signals, priority should be given to aptness, conciseness, and intelligibility (in that order). When a set of standard words are chosen, i.e., a phonetic alphabet (Able = A, Baker = B, Charlie = C, etc.), select words of two or more syllables rather than one-syllable words and test to make sure that no two words in the set sound similar.

2. Talker Characteristics

Voice communication systems should be designed to make maximum use of features that will preserve and enhance the intelligibility of speech produced by the "average talker" with a generalized "American accent" (for the United States), average intensity level (approximately 70 to 75 dB sound pressure level [SPL]), and average speech rate. When a message is to be conveyed in another language, select a talker whose basic language training and experience are in the particular language; e.g., extreme care should be taken in selecting an English-speaking translator to provide messages in German, Spanish, or French to make sure that his or her pronunciations will not be misconstrued by the listeners because of improper emphasis, colloquial discrepancies, etc.

3. Presentation

Verbal signals that occur frequently (e.g., status and caution) should be presented in a fairly formal manner, as opposed to an impersonal one. Although emphasis is important in circumstances where a sense of urgency must be conveyed, care should be taken to train the talker not to convey emotional stress or anxiety.

Less frequent and more urgent signals may be less formal and delivered in a more personal manner to instill confidence.

4. Signal Characteristics

Speech signals should fall within the range of 200 to 6100 Hz.

The audio signal power should be approximately 300 mW at the listener's ear.

The signal-to-noise ratio should be at least 5:1.

Provide distortion-free signals and minimize masking effects that may occur as a result of nonuniform frequency response or a lack of uniform amplification.

Maintain intelligibility nearly equal to that available from a given system at sea level by using a pressure-sensitive device to adjust the gain of the amplifier (as pressure changes, e.g., at higher altitudes).

Provide microphones and earphones that have been designed to give uniform frequency response at all altitudes.

Verbal warning signals should be "processed" only when necessary to increase or preserve intelligibility, i.e., by increasing the strength of consonant sounds relative to vowel strength. Where speech signals must be relatively intense because of high ambient noise, use "peak clipping" to protect the listener against auditory overload.

Where speech communication signals are presented dichotically via earphones, one of the following methods should be used for the recorded warning signals:

a. Dichotic presentation (individual input to each ear)

b. Distinctive, nonmasking, nonspeech component to identify the speech as being part of the warning signal, i.e., superimposed on a speech component

c. A very distinctive voice for recording warning signals, e.g., a female voice rather than the typical male voice.

Where confusion might exist, avoid scrambled speech effects that might be confused with "monkey chatter" from adjacent channels.

5. Verbal Warning Devices

a. Initial alerting signal (nonspeech) to attract the listener's attention and to designate the general problem

b. A brief, standardized speech signal (message) which identifies the specific condition and suggests the appropriate action that should follow

Verbal alarms for critical functions should be at least 20 dB above the speech interference level (SIL) at the operating position of the intended listener. SIL describes the effectiveness of noise in masking speech. It is the average (in decibels) of the sound levels of masking noise in the 600- to 1200-, 1200- to 2400-, and 2400- to 4800-octave bands.

For most applications where "killer" warning messages are required, use a mature male voice; although female voices often provide a unique variation to attract attention, there is some evidence that certain listener groups, such as pilots, sometimes lack confidence in the training and experience of the female operator.

The warning message should always be presented in a calm manner, with the intent of eliciting a rational reaction from the listener.

TELEPHONE HANDSET PACKAGING AND DESIGN

Packaging

The key factors in packaging the telephone handset are the following:

1. The earphone and mouthpiece should be accommodated to the user's head so that regardless of the size of the head or its unique shape, the user can hold the earphone close to the ear, and the mouthpiece close to the lips (without breaking contact with the ear). Ideally, the axis of the speech projection should coincide with the receiver axis, as shown in the accompanying sketch. Because of differences in head size, this is not possible without an adjustable handle. An acceptable compromise is shown in the accompanying specification.
2. The handle should fit the hand comfortably. The accompanying sketch indicates the maximum acceptable dimensions. Avoid square cross-sectional shapes and edges.
3. When a "talk switch" is provided on the handset, place it as shown in the accompanying sketch so that the handset can be used with either hand.
4. Use coiled or retractable cords with handsets to minimize cord tangles. Handset cords that are used in fixed operator stations are easily knocked on the floor. Unless there is a need for a longer cord, shorten the cord so that it prevents the handset from ever striking the floor and causing damage to the handset.
5. The total weight of the handset should not exceed 10 to 11 oz (0.28 to 0.31 kg).
6. Where normal communicating requirements necessitate the frequent use of more than one handset, locate the various handsets for convenience according to use priority. Consider the following potential user problems:
   a. Cord interference; i.e., the cord of the phone in use should not drag across another phone and knock it off its hook or base.
   b. Nominally locate the phone for left-hand use so that the user can use his or her right hand for writing (although this inconveniences left-handed operators, they are in the minority). Alternatively, provide a flexible configuration that allows users to arrange their phones for their own convenience.
   c. Phones used in vehicles should be provided with "securing cradles" and/or other devices that will prevent the movement of the vehicle from unseating the handset.
7. Handset bases should be designed so that they are not easily displaced, especially if the base has a dialing device. Consider the use of rubber feet, suction cups, and/or other special holding devices.
8. In-handle channel and line selection concepts may be helpful for some applications as long as the following do not occur:
   a. The dial or push buttons cause the handle to be awkward to hold onto.
   b. The dial or push buttons are too small for accurate finger manipulation, or the push buttons are placed too close together.
   c. The handset becomes too heavy.
9. Except in special cases, any color should be acceptable for the handset package. A given color can be used to identify a special channel, i.e., a "hot line."

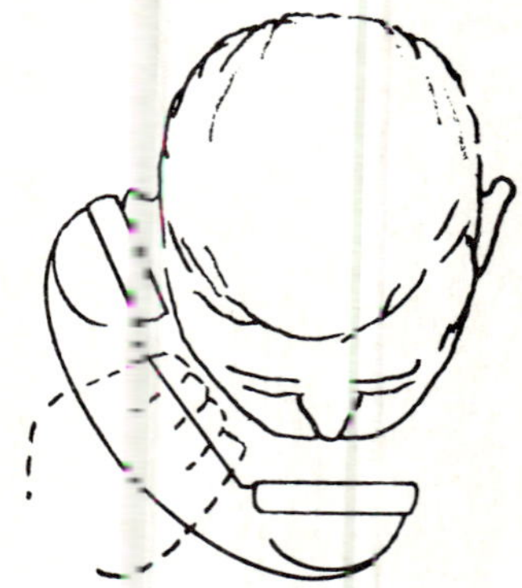

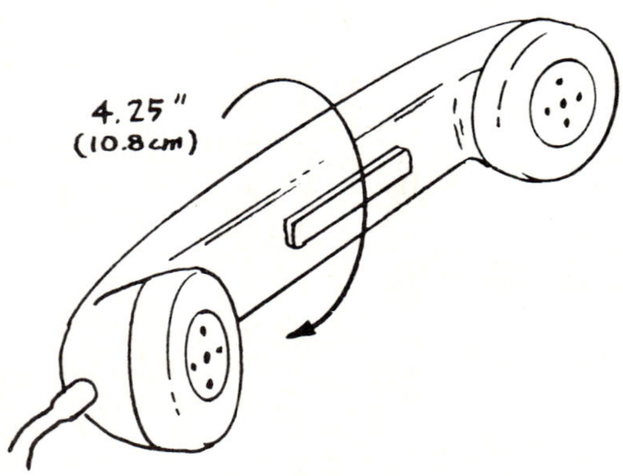

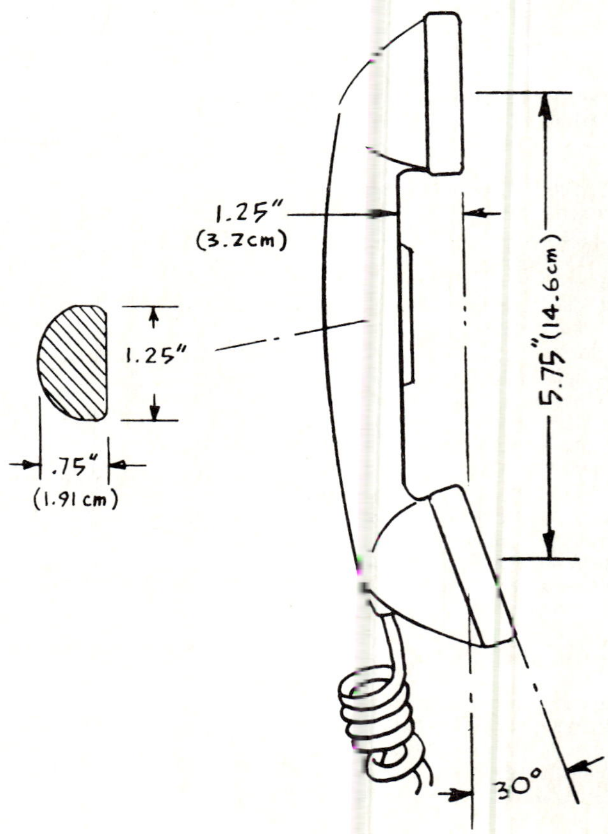

Special Considerations

A console-mounted phone to keep the desk clear (or because the console has no horizontal surface) and a phone that is inset to keep it from being accidentally knocked from its cradle

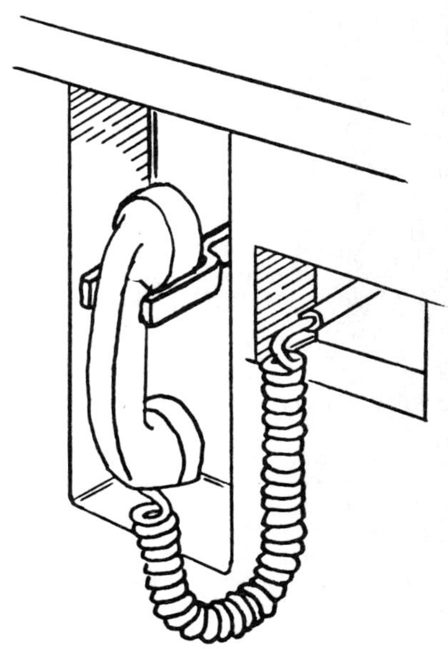

A conference phone supplement so that several individuals can participate in the same telephone conversation

Operating Characteristics

The microphone (mouthpiece) should be:

1. Highly sensitive to acoustic speech signals.
2. Of sufficiently high quality to provide faithful transduction of the speech signal into an electrical signal.
3. Capable of rejecting other acoustic signals and noise.
4. Designed for "close talking." Do not use large pressure-gradient velocity-type microphones; instead, use noise-canceling types in the low-frequency sound range (100 dB overall SPL). Avoid "Ribbon Mikes" for close talking.
5. Designed so that the talker has to talk into the axis of the microphone, so that the talker's lips naturally fall close to the microphone, and so that the operator's hand never covers the microphone holes inadvertently.
6. Designed to have a smooth frequency response between 200 and 6100 Hz.
7. Designed to provide a dynamic range (when working with the selected amplifier) great enough to admit a minimum of a 50-dB variation in signal input. A close-talking microphone should not overload with signals as high as 130 dB.
8. Equipped with a noise shield for applications where the ambient noise level is extremely high (above 100 dB). The shield volume should be at least 250 cm³, but should not be so large as to be unwieldy for the operator to handle and use. The shield should fit tightly against the operator's face with normal hand pressure or head-strap tension; i.e., the operator should not have to press the shield with considerable force in order to get it to fit. Provide exhalation holes in the shield, but place these as far from the actual microphone as practicable. Sound-absorbing material should be enclosed in acoustically transparent, waterproof material. The shield (if worn) should not impede the talker's mouth or jaw motion, thus causing speech distortion.

PACKAGING AND DESIGN OF HEAD-MOUNTED MICROPHONES AND EARPHONES

Packaging

Provide complete adjustability of headset components.

There should be lightweight, minimum tension to hold the earphones snugly against the head.

Provide a soft cushion which is large enough so that it does not press on the user's ear but which seals around the ear to minimize external noise interference.

Operating Characteristics

1. Provide smooth frequency response characteristics which are as broad as those of the remainder of the system.
2. Provide a dynamic range, without appreciable distortion (at least 40 dB), capable of power-handling capacity for peak amplifier output.
3. Provide a combination earphone and socket (or cushion) for which the earphone sensitivity and the earcup attenuation together will provide an adequate signal-to-noise ratio of 5:1.
4. Provide an earphone cushion which is comfortable for long-duration use and which is not prone to developing fungus or collecting moisture.
5. Connect the earphones to operate out of phase.
6. Delay the signal reaching one ear about 500 μs.
7. Use binaural rather than monaural headsets if listeners will be in an intense noise environment.
8. Provide a sidetone (delay of 0.05 s) to make the talker increase his or her voice level of effort.
9. Provide sufficient electrical power to drive the peak sound pressure level to 131 dB when using two earphones.
10. Provide a gain control with a dynamic range sufficient to make the signal at least 15 dB more intense than the ambient noise.
11. Provide a reduction in frequency range below 500 Hz and above 4000 Hz if it results in an increase in the average power of the audio signal.
12. Provide a uniform frequency response of receiver and headset between 300 and 4000 Hz to avoid unpredictable distortions.
13. Provide a pressure-operated gain control switch to compensate for the effects of altitude in aircraft cabins.

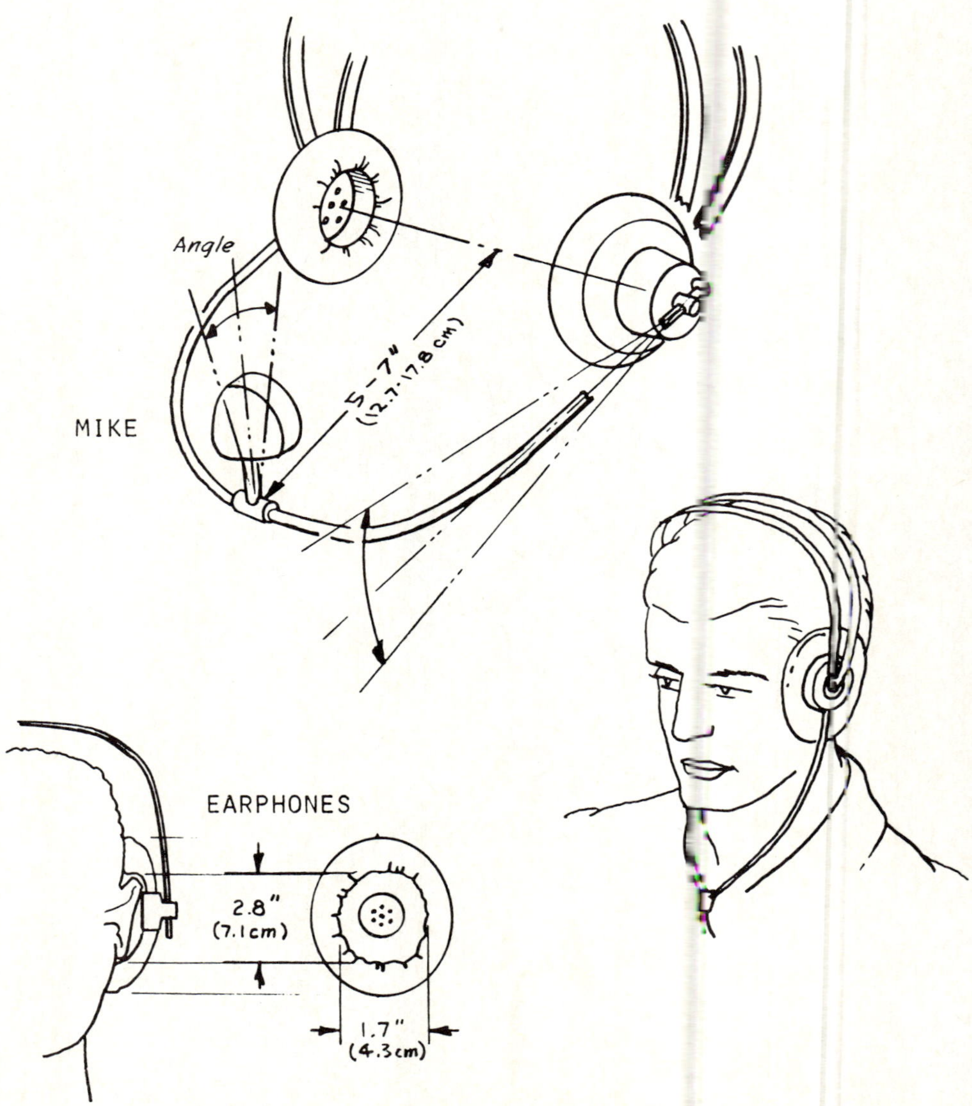

LOUDSPEAKERS

Loudspeakers should be subject to the same frequency response characteristics as other elements of the communication system, except that loudspeakers for use in multispeaker installations (i.e., where several speech channels are to be monitored simultaneously) should respond uniformly over the range of 100 to 8000 Hz.

Use Considerations

1. Where an individual loudspeaker is provided a single operator, locate the speaker as close as practicable to the operator (within 3 ft, 0.9 m) so that the operator can reduce the signal of his or her particular speaker and thus refrain from adding to the overall noise level of an operating area.
2. Locate speakers and microphones so as to minimize speaker feedback into the microphone.
3. Where several channels are to be monitored simultaneously by means of loudspeakers, separate the speakers by at least 10° with respect to the principal listener so that he or she can localize the separate speaker sounds. If additional channel differentiation becomes necessary (i.e., if the 10° does not provide clear definition), utilize low-pass filtering (f_c = 1800 Hz) to signals fed to loudspeakers on one side of the operator, high-pass filtering (f_c = 2000 Hz) to signals fed to loudspeakers on the opposite side of the operator, and unfiltered signals to loudspeakers in front of, or behind, the operator.
4. Position speakers approximately at the head (or ear) height of listeners, except in auditoriums, where the speaker should be at podium height or higher, and in multilevel work areas, where it may be important to recognize that a sound emanates from a position below the operator.

CONTROL SELECTION GUIDE

Control Configuration	Use Criteria	Human Engineering Considerations
Hand-operated push buttons: Square push button 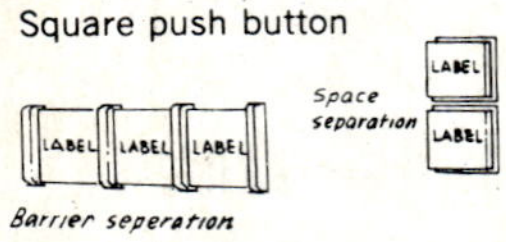	Use for noncritical operations which start an action, select a channel, or turn on a piece of equipment. Use for "event stack" combination lighted push buttons.	Locate within easy reach of the operator, considering the constraints of special garments, acceleration restraint systems, etc.
Round (guarded) push button 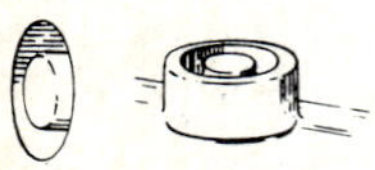	Use for critical operations (such as for a "panic" button) to stop a "runaway" equipment operation, fire and emergency escape sequence, etc.	Locate so that inadvertent operation is impossible (i.e., so that the switch cannot be actuated by bumping it accidentally).
Round (extended) push button 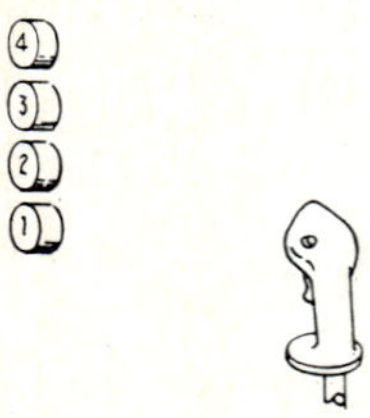	Use in conjunction with a primary hand controller for changing control mode, initiating communication, etc. Use for multibutton keyboard arrays and circuit-breaker panels.	Locate within easy reach of the operator's thumb, considering the constraints of a glove. Locate so that the switch will not be activated accidentally during normal manipulation of the parent controller.
Bar-type push button 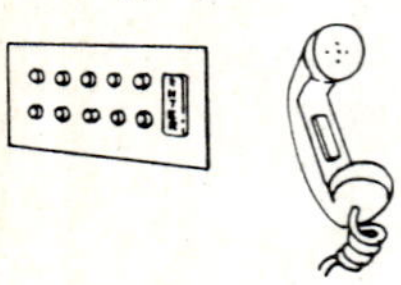	Use for either finger or thumb operation of a communication handset. Use as an "entry" tab for inserting preset numbers into a computer from a keyboard.	Consider location within keyset configuration for optimum operational sequence. Consider position on the handset for convenient use by fingers or thumb for both left- and right-hand operation.
Toggle switch 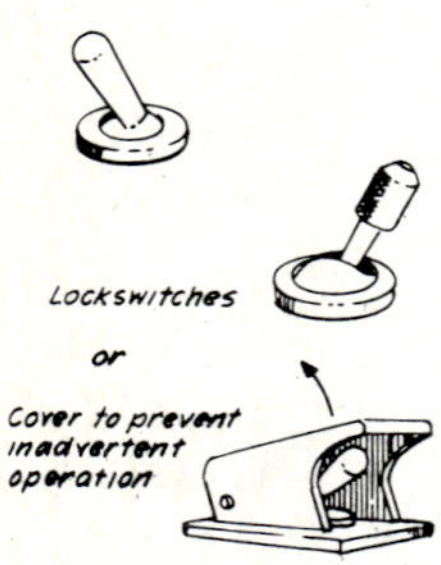	Use a standard two-position toggle switch for START/STOP, ON/OFF, etc. Use a "momentary" (spring-return) toggle switch for check-reading an instrument or circuit or use a spring-return toggle switch for a "slewing" operation. Use a three-position toggle switch for combining alternatives of a single function such as AUTOMATIC, MANUAL, or OFF. *Note:* Do not use four-position switches unless absolutely necessary because of space limitations.	Consider particularly the hazards associated with accidental contact with the toggle switch, and also the difficulties of operating in a glove.
Rocker Switch 	Use in lieu of a toggle switch when the protruding toggle handle would be a considerable hazard, i.e., when a crew member could be injured by bumping or striking the handle or handle cover accidentally. Use an illuminated rocker switch in lieu of a toggle switch where illumination conditions make it impossible to see the toggle position, assuming this requirement to be critical. Use in lieu of a toggle switch when panel space is inadequate.	Special precautions should be taken in placement of rocker switches to obviate accidental operation. Minimize intermixing of toggle switches and rocker switches in the same panel area; i.e., use one or the other.
Rotary selector control (mechanical detents) 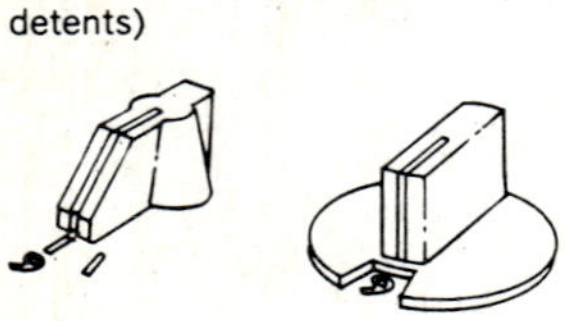	Use (where it is desirable to select two or more—but no more than 24—channels, circuits, etc., with a single control) to save operator time and panel space which would be required by separate control switches. *Note:* Do not use for continuous frequency selection, audio intensity control, panel illumination control, etc.	Make the "pointing" end of the switch knob obviously different from the nonpointing end. Provide mechanical detents so that the operator can "feel" and "hear" when the control is properly positioned. The knob should be shaped like a bar or pointer to differentiate this type of control from continuous-function controls, which have round knobs. Provide an index mark on both the top and the end of the pointer or provide a cutout in the knob skirt that exposes the numerical setting.

CONTROL SELECTION GUIDE

Control Configuration	Use Criteria	Human Engineering Considerations
Rotary adjustment control (continuous motion)  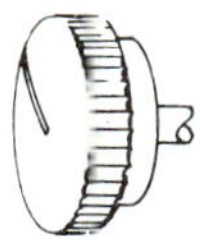	Use where smooth, continuous adjustment of audio intensity, light intensity, random radio frequency selection, etc., is required. *Note:* Do not use on detent-type switches.	Use a round knob, which allows ease of finger positioning. Use suitable knob surfaces which provide necessary finger security but which will not interfere with manipulation dexterity. Use a larger-diameter knob when "fine" adjustment is required.
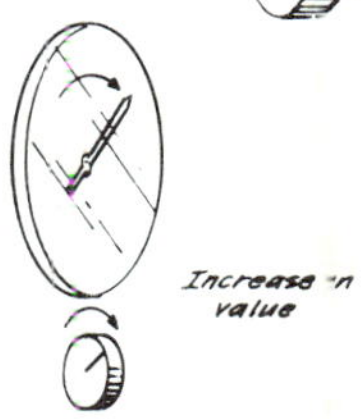	Use to provide continuous control of a related display pointer.	Maintain proper direction-of-motion relationships. Place the control where the operator's hand will not cover up the display when he or she is adjusting the display. Use the proper CD-ratio to ensure ease and accuracy of display adjustment.
Rotary valve control 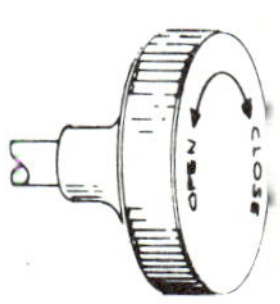	Use to control flow of materials, fuel, air, liquid oxygen, etc.	Use a round handle configuration suitable for gripping by the hand as a whole, but with sufficient finger clearance for pressurized glove operation. Whenever possible, locate valve handles far enough from other control knobs (which are similar in shape) to avoid direction-of-motion confusion (i.e., accepted usage for valves is for a clockwise turn to provide a decrease in flow, whereas the same motion in other types of controls provides an increase in functional value). In addition to placement separation, emphasize clear, unambiguous marking and other appropriate code techniques to minimize the chance of direction-of-motion of operator error.
Thumb-wheel controls (continuous smooth motion) 	Use for limited-range, continuous adjustment functions (panel light dimmer, audio volume, etc.). *Note:* Use only as a second choice to a normally mounted finger knob in the event some advantages may accrue in terms of saving panel space.	Coarse serrations should be provided on the edge of the knob. Sufficient exposure of the knob rim must be provided for convenient manipulation with pressurized gloves, but guarded against accidental activation.
Thumb-wheel controls (discrete stepping motion) 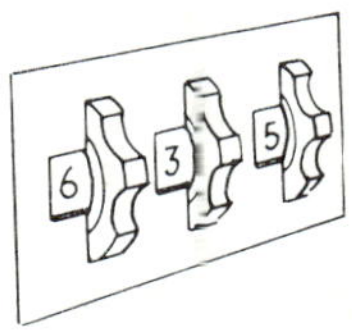	Use in conjunction with detented counter-type selector control and display combinations.	The edge of the thumb wheel should provide protrusions, or "teats," which coincide with the position of the numbers on the appropriate counter drum. These devices should be placed and/or guarded so that they cannot be accidentally displaced, or bumped, causing injury to personnel or damage to garments.
Push-pull controls: Stirrup handle 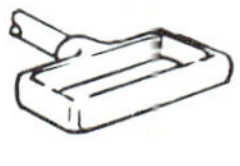	Use for an ON/OFF control which (because of mechanical reasons) requires more force than can be applied by a finger-operated control.	Locate controls so that they cannot be accidentally struck by personnel and cause injury or garment damage or interfere with normal or emergency crew movement. Make the stirrup opening large enough to accommodate four fingers (including gloves when worn).
T-shaped handle 	Use in lieu of a stirrup handle when space is not available for whole-hand operation.	Make handle extensions on each side of the central shaft long enough to accommodate the index and middle fingers on one side and the third and fourth fingers on the other.
Ball handle 	Use in place of the above when the orientation of a stirrup or T-shaped handle would cause serious inconvenience to the operator and possibly delay his or her total reaction time in the event of an emergency.	Select a ball diameter that will comfortably accept a "fist grip" if the force requirement is high and the thumb and at least two fingers if the force requirement is low.

CONTROL SELECTION GUIDE *(continued)*

Control Configuration	Use Criteria	Human Engineering Considerations
Levers (console type) 	Use for main power setting to provide forward vehicle thrust or machine acceleration.	The handle length should be long enough to accommodate a four-finger grip.
	Use for vehicle attitude control.	All lever-type controls must be designed and located so that the full control motion (excursion) is within the limits of reach of the restrained operator wearing a special garment.
	Use for discrete (bang-bang type) power application (e.g., translation).	
Levers (extended arm operation) 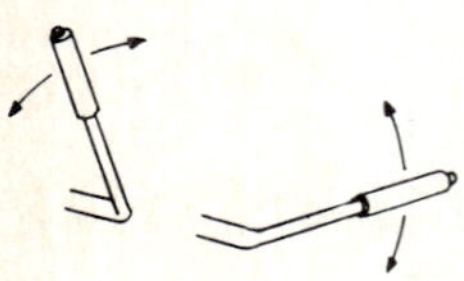	Use where medium to heavy manual force application is required.	
Hand wheel 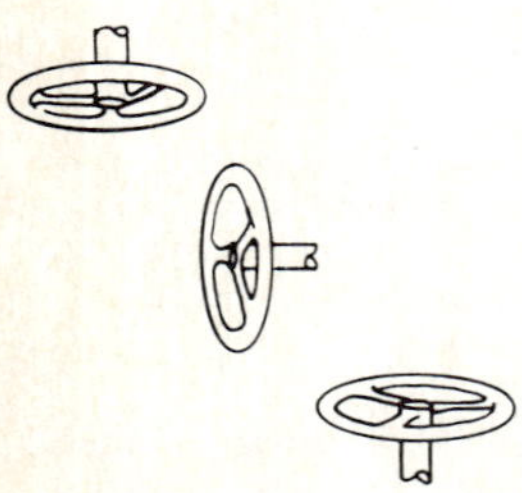	Use for opening and closing pressure-tight hatches where extremely heavy force application may be required, for high-torque valves, etc.*	The diameter of the wheel and the diameter of the wheel rim must be compatible with operator dimensions, considering the constraints of special garments.
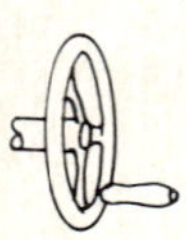	Use for low-torque, multirevolution cranking operations where many turns are made in rapid fashion.*	Select an optimum diameter for the expected force, rate, and operator position variables involved.
Steering column levers† 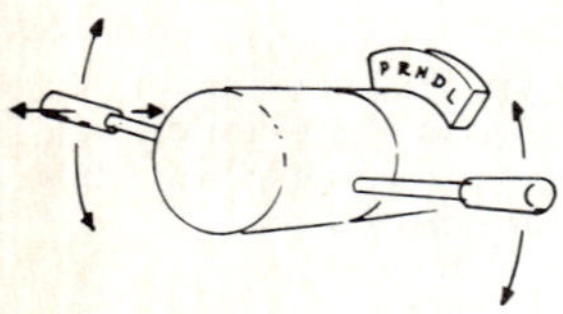 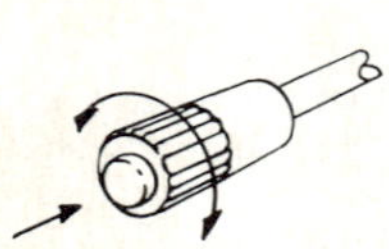	Use for trucks, etc. Use when it is desirable to bring critical control functions within "fingertip" reach of the driver (i.e., for easy reach from the normal hands-on-steering-wheel position). Typical functions: Automatic gear selection Turn signal operation	Consider typical driver control placement and motion expectancy, i.e., certain functions for right-hand operation and others for left-hand operation. Avoid too many different functions or motions on a single lever—especially if inadvertent "cross-coupling" might occur. Use only one lever per hand. Use rotation action sparingly. If used, provide knurling to indicate to the driver that this motion is possible.

Rotary controls for discrete positioning should use the well-established principle of "a moving pointer versus a fixed scale," as shown in the first illustration. The second illustration shows that when values are imprinted on the knob skirt, they progress in the opposite direction to that in which one normally expects to read them (i.e., the principle of "a fixed index versus a moving scale"). The latter configuration leads to frequent positioning errors.

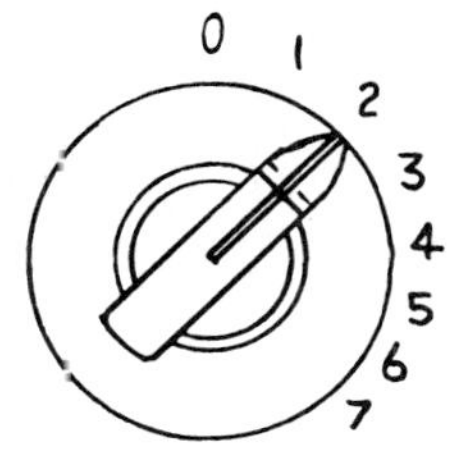

PREFERRED

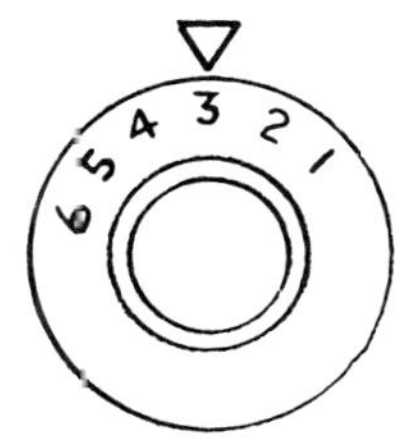

AVOID

CONTROL AND DISPLAY DIRECTION-OF-MOTION RELATIONSHIPS

The most reliable expectancy occurs when the right-hand arrangements illustrated by the two upper sketches are used.

A weaker expectancy occurs when the arrangements for left-hand operation shown in the lower sketches are used.

These expectancies occur because of the apparent mechanical relationships that the operator observes; i.e., the operator, purposely or not, observes the directional movement of the control perimeter and the scale pointer as moving together. In the upper vertical display, the pointer must be adjacent to the side-mounted control. In the lower vertical scale, the pointer is moved to the other side so that it, too, is adjacent to the left side-mounted control.

The lower vertical scale-control arrangements are less desirable because the visual expectancy conflicts with the normal "blind" control movement expectancy noted earlier. Thus, if by chance the operator did not actually look at the scale (i.e., adjusted the controls blind), he or she would probably turn the knob in a clockwise direction for "increase."

Note: Because of the circular geometry of both scale and control knob, left-hand expectancies are the same as right-hand arrangement.

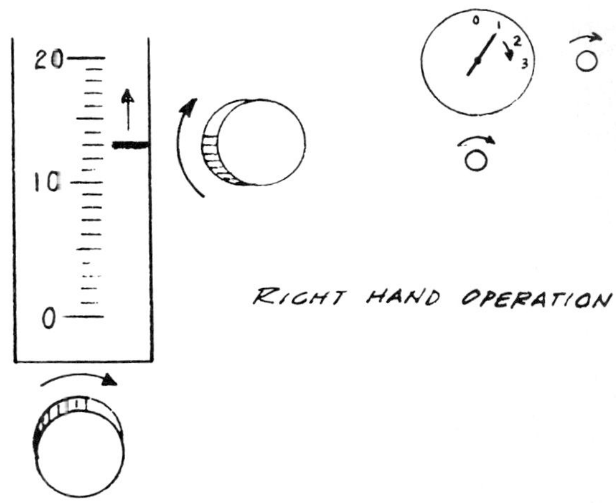

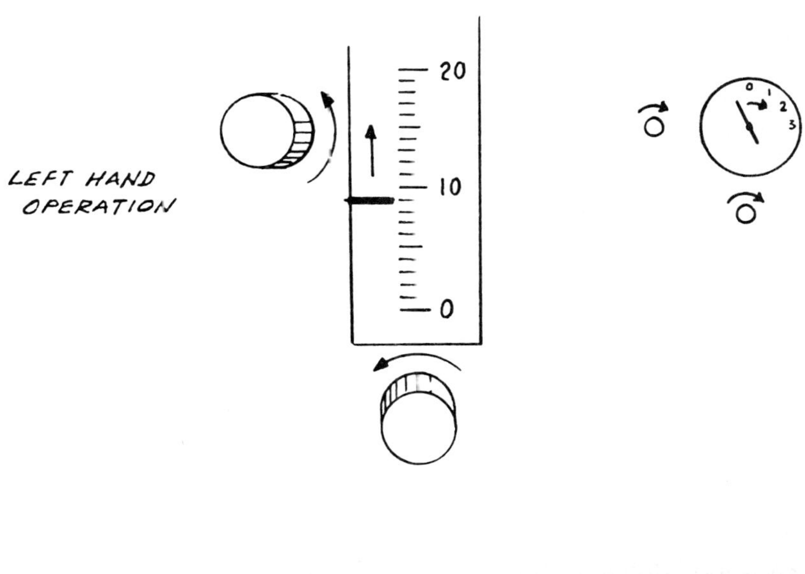

Valve operation is a special case of predetermined convention creating expectancies opposite to those established by human behavioral research. The single valve direction-of-motion stereotype convention is mechanically related; i.e., the valve is turned to "unscrew" a valve seat and thus allow liquid, gas, or other substances to "flow." In other words, to increase the functional factor, the valve handle is turned counterclockwise—as opposed to the behaviorally oriented principle of a clockwise motion to effect an increase in functional value or factor.

This stereotype is so well established now that any attempt to reverse the standard should be avoided. This applies to single valves only, as will be explained below.

When dual valve handles are involved, i.e., when one is controlled with the right hand and the other is controlled with the left hand, the above convention ceases to remain constant. The average user becomes somewhat confused and tends to revert to another direction-of-motion referencing philosophy. Although this stereotype is not as well established as others, it involves the following interpretive factors:

1. Motion toward the user to obtain flow increase
2. Mirror-image operation, i.e., both hands performing similar motions

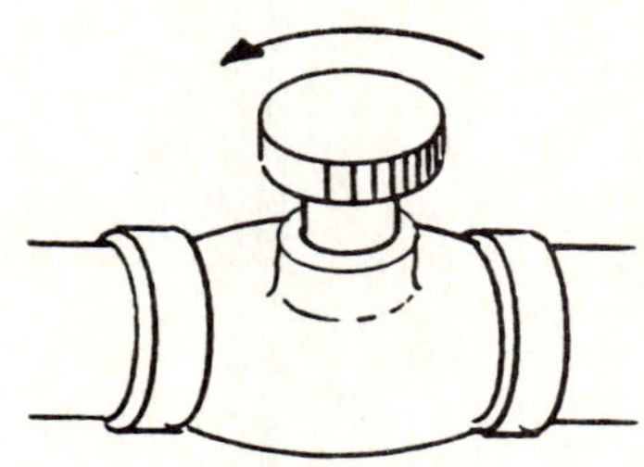

INCREASE

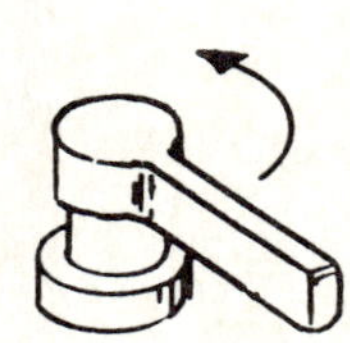

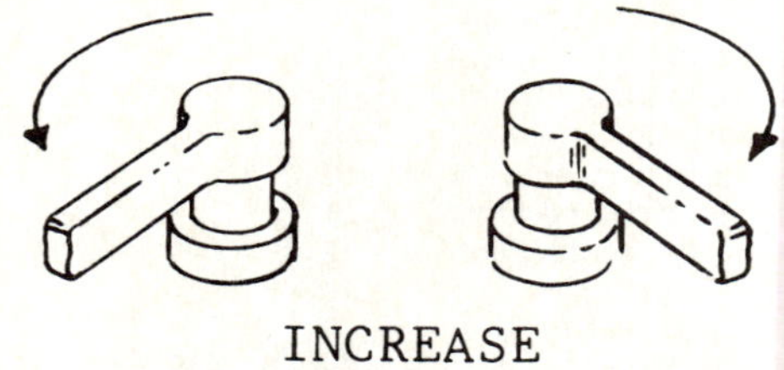

INCREASE

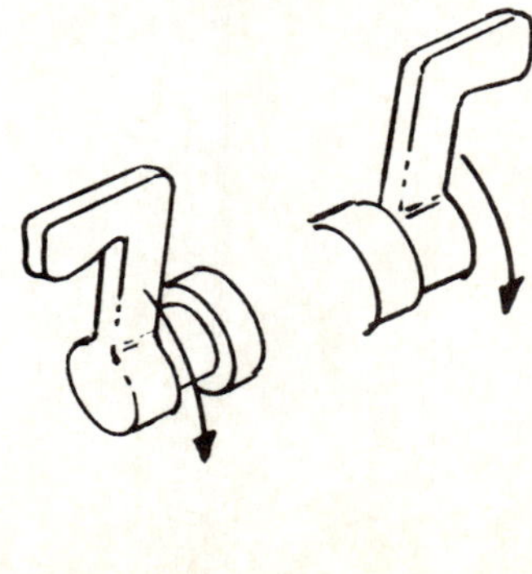

Automotive, foot-operated control conventions are well established, and even though they appear to violate some of the previously noted direction-of-motion stereotypes derived from laboratory research, these conventions should not be reversed.

The accompanying sketches show the typical foot-operated controls for automobiles, trucks, and buses. As indicated, the motion required to accomplish each function is forward—regardless of the vehicle's motion relationship.

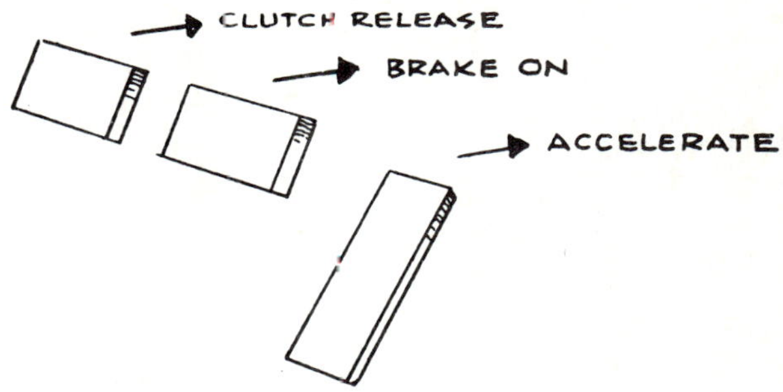

The automotive gear-shifting functional motion position convention is also well established and is in fact standardized by U.S. governmental regulations.

It should be noted, however, that if it were not for such previous standardization, a more operator-compatible configuration could be created that probably would reduce the confusion that makes the training of new drivers more difficult. A so-called optimized configuration is illustrated in the accompanying sketches. These are presented primarily as ideas that could possibly be transferred to some as-yet-unproposed control system.

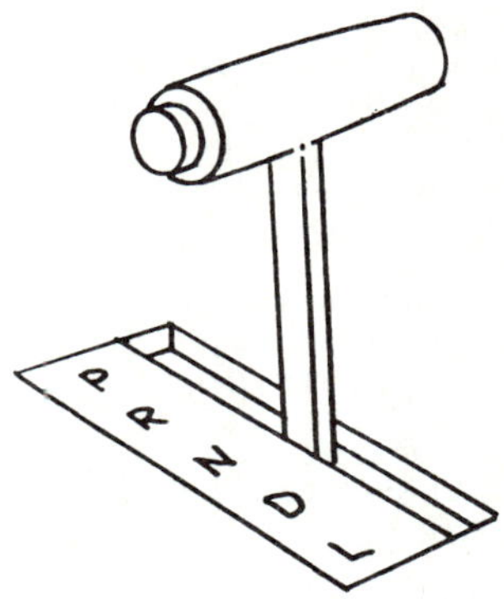

Automatic transmission

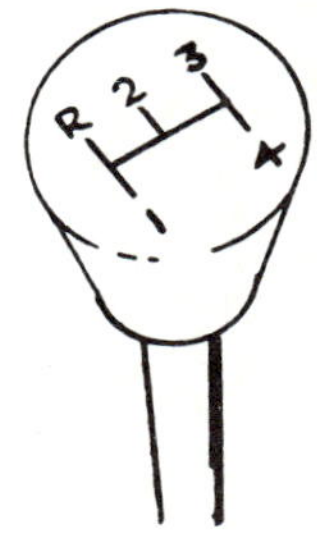

Manual transmission

Connector-fastener devices are often required to secure cables, hoses, and other, similar elements to each other or to some fixed element. Thus these connector-fasteners are in fact controls.

For most applications, the well-established direction-of-motion stereotype is illustrated in the accompanying sketches; i.e., a clockwise motion secures and tightens the connection, and a counterclockwise motion loosens the connection.

Note: Exceptions to this rule include nuts on wheels, where the stereotype must be reversed to prevent the movement of the wheel from loosening the nut.

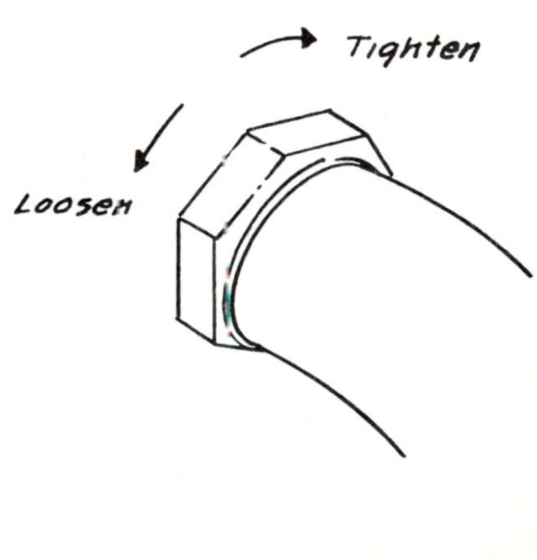

CONTROL OPERATION EXPECTATIONS VERSUS CONTROLLER APPEARANCE

One should not use a controller configuration that is confusing to the potential operator. The following typical expectations should be observed:

1. A push button should extend out from the panel and thus imply that it is to be pressed or pushed in. Some push button and indicator displays are made to look alike; i.e., they both are flush with the panel, making it difficult to tell which one is a control and which is merely an indicator.
2. A pointer-shaped knob implies that the device is a positioning control that can be positioned in discrete steps.
3. A circular-shaped knob implies that the knob turns continuously. It also makes it easier for the operator to reposition his or her fingers for multiturn manipulation.
4. The waferlike knob indicates that it is to be pulled out. If it also is to be turned, it should have serrations around the edges, implying rotational possibilities. (Do not use serrations if the knob is not intended to be turned.)
5. The rocker-shaped control implies that either end of the switch cap can be pushed. However, the angular cues may not be apparent if the control is positioned at a poor visual angle. Make sure that the operational cues are apparent in the use position before selecting this type of control.
6. The toggle-switch handle provides a readily apparent cue that the handle can be alternately positioned from one side to the other—especially when several of the switches are arranged in a horizontal or vertical pattern.

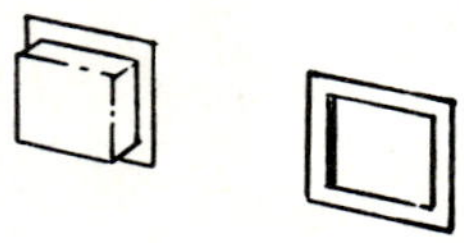

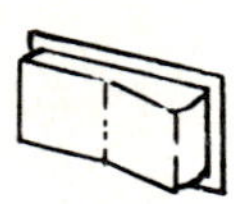

Deliberate Confusion

Sometimes deliberate confusion is created by the appearance of a control device, as illustrated by the sketches at the right.

Lack of an obvious "pointing" feature makes it possible for the operator to turn a selector knob to the wrong position. Symmetrical bar handles on rotary knobs should be avoided.

Confusion with other panel-mounted features often leads the operator to believe the control is something other than a controller; i.e., it may look like a lighting fixture or a panel fastener.

An otherwise effective pointer-selector knob may be so distorted that when the operator manipulates the control in the dark or without actually looking at it, he or she tends to point the handle at the desired setting because the handle feels like a pointer.

A sleeve handle that is designed to slide inward is seldom recognized as performing that type of movement.

An extremely thin, waferlike rotating knob is often mistaken for a push button.

Push buttons deeply inset in the end of a control handle are usually not observed when the operator views the handle from the side. Thus the operator fails to find the button switch.

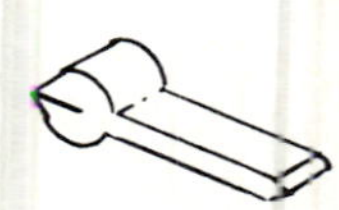
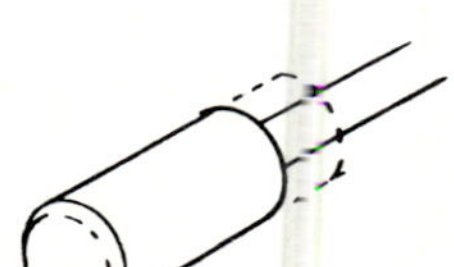
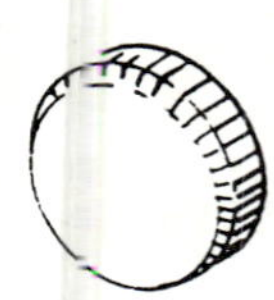
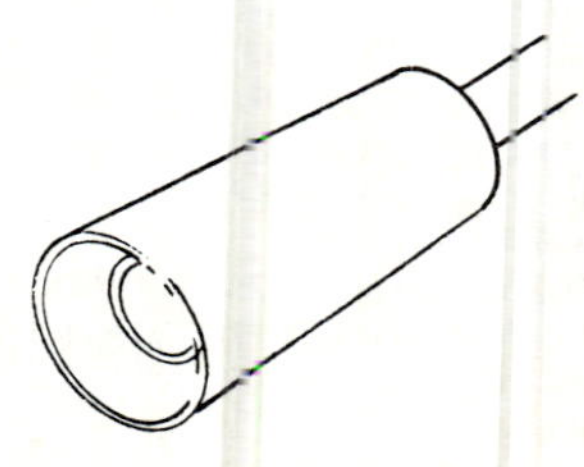

PUSH BUTTONS

A variety of push-button sizes and shapes are acceptable as long as certain conditions are maintained. The smallest diameter should not be less than about ¼ in (0.64 cm) for bare-hand operation or ½ in (1.3 cm) when gloves are worn.

Normally, the button should extend far enough above a panel so that when the button is in the depressed position, a reasonable amount of it is still exposed (otherwise, the operator may not be able to press the switch far enough to make contact).

When buttons are approximately ¾ to 1 in (1.9 to 2.5 cm) in diameter, it sometimes helps to provide a concave top to the button (finger centering). An additional "location aid" may be provided in the form of a tapered, recessing bezel (such as is found in the typical doorbell).

Illuminating the push button is useful when the button needs to be located in the dark.

The push button is an effective emergency control device for shutting off machines in a hurry. In such applications, the button should be larger so that the operator does not have to be so accurate, i.e., so that the button can be operated either with the end of a finger or with the heel of the hand. As the button is made larger, consider a concave surface.

When it becomes important to preclude inadvertent push-button actuation, recess the button so that the button is actually below the adjacent panel surface plane. The "well" for such configurations needs to be at least 1 in (2.5 cm) in diameter for bare-hand operation and at least 2 in (5.1 cm) for gloved-hand use. Tapering the sides of the well helps the operator find the button without having to be so accurate.

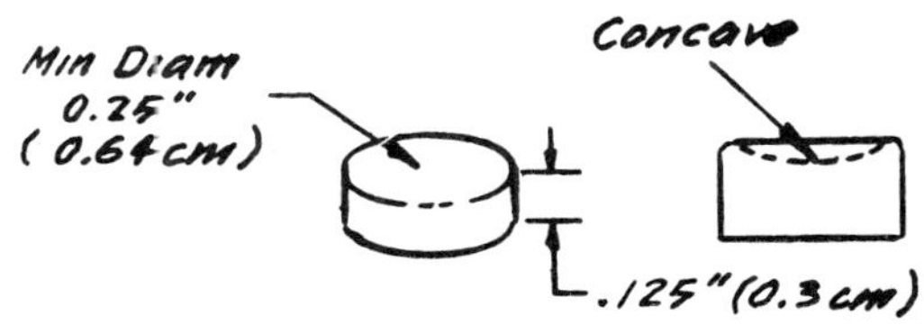

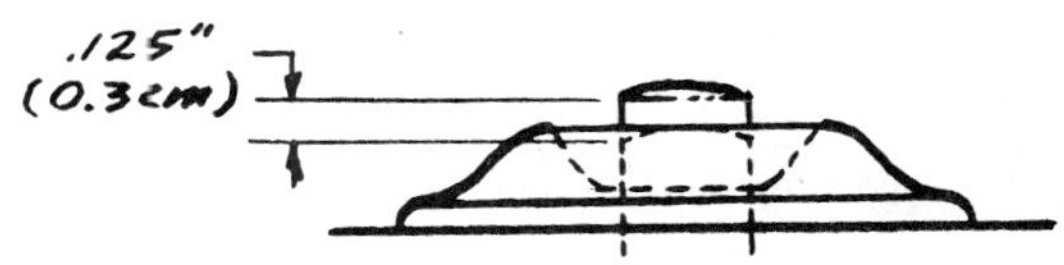

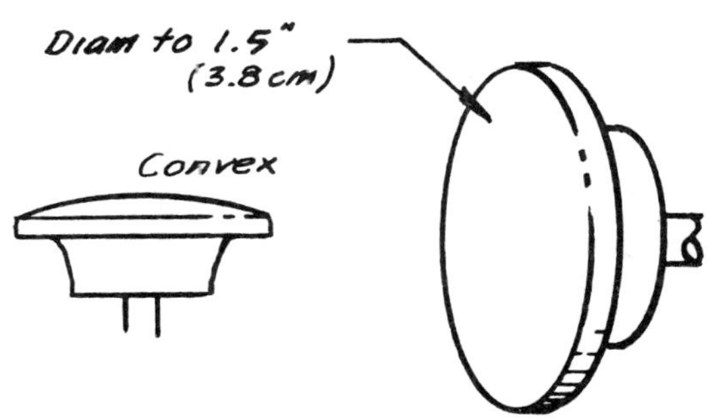

Grouping Push Buttons

As indicated by these guidelines, adequate button separation is required to avoid inadvertent operation of adjacent switches.

Minimum Button Separation

Arrangement	A	B	C	D	E	F
Vertical plane:						
No gloves	0.75 in (1.9 cm)	0.625 in (1.6 cm)	0.1875 in (0.476 cm)	0.25 in (0.64 cm)	1.25 in (3.17 cm)	0.75 in (1.9 cm)
With gloves	1.75 in (9.45 cm)	1.625 in (4.12 cm)	0.375 in (0.95 cm)	1.25 in (3.17 cm)	2.25 in (5.72 cm)	1.75 in (4.45 cm)
Horizontal plane:						
No gloves	1.0 in (2.54 cm)	0.875 in (2.22 cm)	0.4375 in (1.11 cm)	0.50 in (1.27 cm)	1.50 in (3.81 cm)	1.0 in (2.54 cm)
With gloves	2.0 in (5.1 cm)	1.875 in (4.76 cm)	0.375 in (0.95 cm)	1.25 in (3.17 cm)	2.25 in (5.72 cm)	2.0 in (5.1 cm)
Under severe vibration or oscillation	3.0 in (7.6 cm)			3.0 in (7.6 cm)		3.0 in (7.6 cm)
For blind selection	6.0 in (15 cm) apart in front of operator; 12 in (31 cm) apart when buttons are located in the peripheral areas.					

Note: The above guidelines apply to any shape of button.

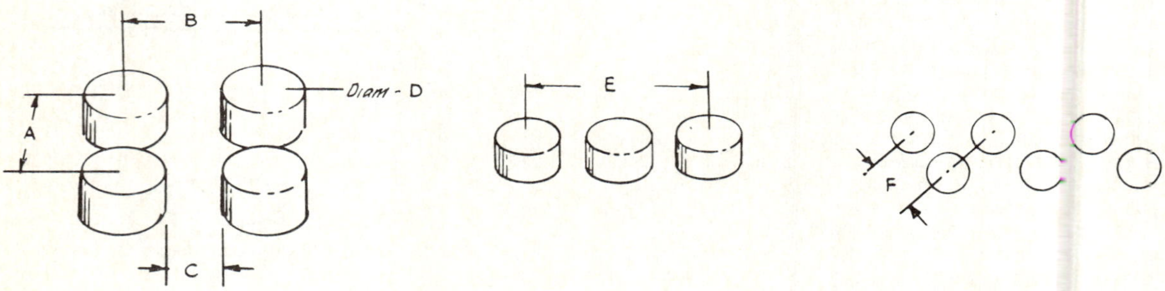

Push-Button Shape

From a biomechanical point of view, it makes no difference whether one selects a round, square, or rectangular push button. However, the square or rectangular button provides more useful area for labeling, and obviously a rectangular button laid on its side provides more label space than one that is standing on end.

Shape can also be used as a cue for determining the functional significance of a particular push button. This is especially important when advisory indicators are used in combination with push buttons. A round push button is seldom mistaken for an advisory indicator.

Off-the-shelf push-button configurations such as those shown in the accompany illustration are popular for many applications. The separator bar between buttons is also helpful in preventing the operator from accidentally pressing an adjacent button. These devices are generally available for either horizontal or vertical button arrays. The dimensions shown are representative.

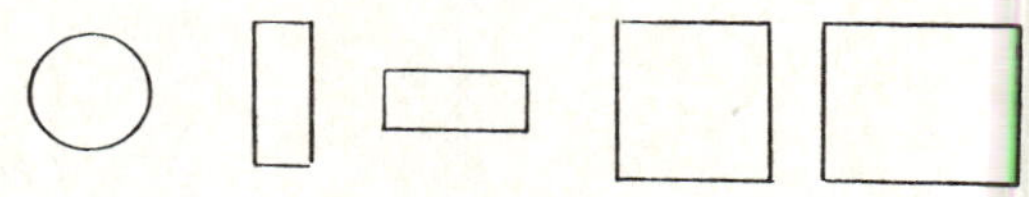

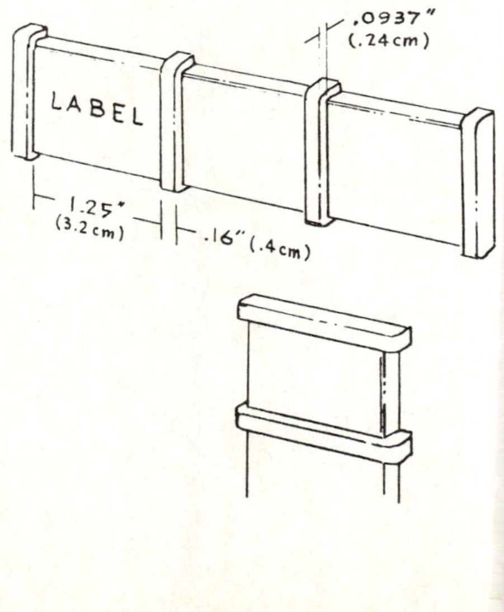

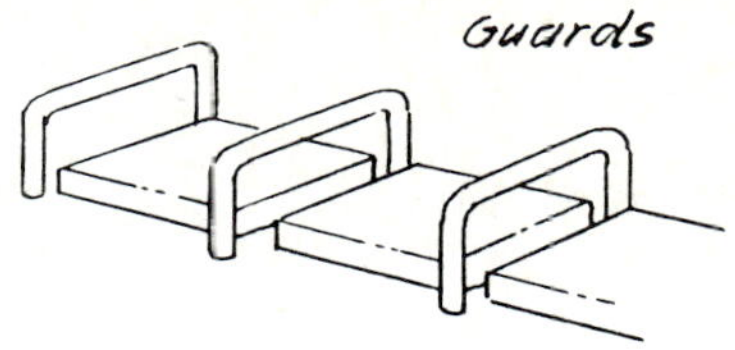

When inadvertent button operation could be serious, other, more accentuated guard systems should be considered. The one illustrated is effective because the height of the guard prevents the operator's finger from "slopping over" into the adjacent button area, and it also allows the labels to be seen with minimum interference.

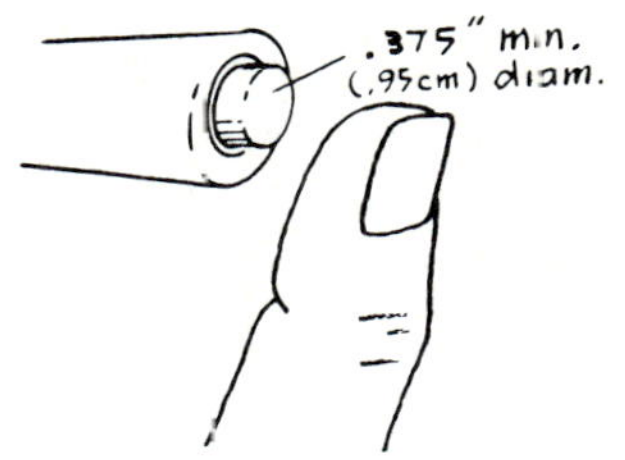

A push button located in the end of a handle should be visible from the operator's normal eye reference position.

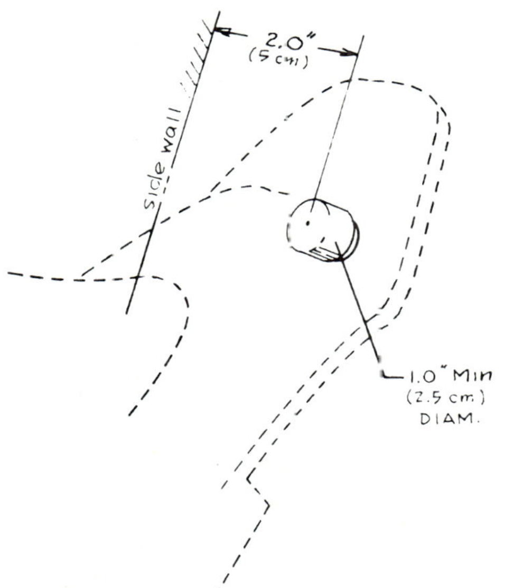

Foot-operated push buttons are often useful to alleviate the work load placed on an operator's hands. More than one button per foot is not recommended, however. Make sure that there is space between any adjacent structure and the button so that there is plenty of room for the foot. However, too much space may not be desirable since an operator may use the adjacent structure as a guide in finding the button.

Push-button forces should be within the range of 10 to 20 oz (283 to 567 g) (although they can be as high as 40 oz, 1134 g) in order to reduce the possibility of inadvertent actuation.

Push-button action should be positive; i.e., there should be elastic resistance aided by slight sliding friction, starting slowly, building rapidly, and with a final sudden drop to indicate activation. An audible snap action is helpful under high-noise-level conditions, although the snap action must not be too heavy if the button is to be operated frequently over extended periods of time.

Contact Switches

Although not a true push button, the contemporary "contact switch" is often used to replace the mechanical push button. Since this type of switch provides no three-dimensional cue for the operator, it is important to provide extremely clear graphic delineation of each switch boundary.

It must also be made clear, by means of appropriate labels, which are switch functions and which are purely visual readouts, since both the switches and the readouts are flush with the panel surface.

Note: Contact switches should not be used where continuous, long-duration operations are expected, since it has been noted that operators tend to strike the push-button surfaces harder because of a lack of tactile or auditory feedback. Some complain of sore fingers after a long operating session.

TOGGLE SWITCHES

Toggle switches come in a variety of shapes and sizes, most of which are satisfactory. One should, however, choose a toggle switch in which the "throw" or displacement is at least 30° so that the operator can tell at a glance which position the switch is in.

A toggle-switch handle should be at least 0.5 in (1.3 cm) long, and preferably no longer than 1.0 in (2.5 cm).

For smaller, shorter switch handles, the resistance of the switch should not exceed about 8 oz (227 g); for larger switches, it should not be more than 15 oz (425 g).[1]

The sketches at the right illustrate typical handle shapes that are generally available. The first is the most typical. The second is often representative of very small toggle switches. Such a shape is not desirable when the handle is long, however, because the user can be injured if he or she bumps against the handle. The third handle shown provides added tactile identification when a switch is used in the dark. The fourth handle is especially good, not only because it makes it easier to identify the switch position, but also because the special handle can usually be obtained in various colors, making it possible to color-code certain switches.

Switch separation is especially important when several switches are arranged side by side or vertically in columns. Note the different spacing requirements for vertical and horizontal arrangements.

Although three-position toggle-switch hardware is available, it is not recommended. It is not possible to maintain the 30° "throw" when more than two positions are provided. It is suggested that, when three or more switch positions are required, some other type of switch be selected.

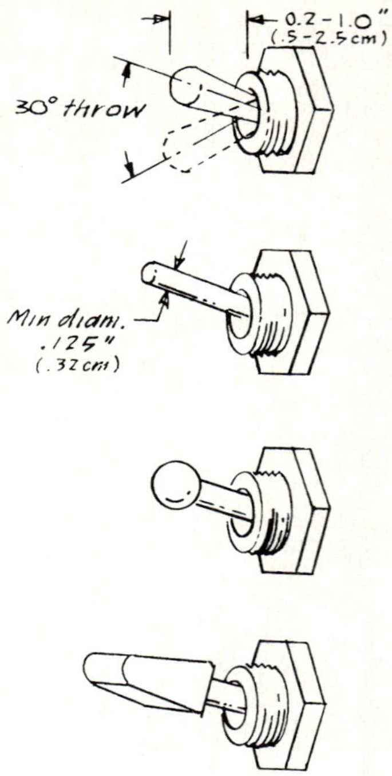

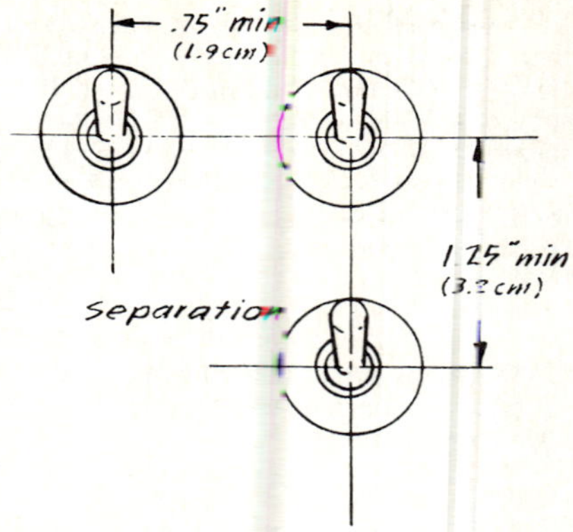

[1]Toggle switches should snap into place with an audible click.

ROCKER SWITCHES

Rocker switches are a good substitute for the more common toggle switch to provide good physical indication of switch position. Like toggle switches, rocker switches, to be effective with regard to visual identification of position, should be designed so that the slope of the handle from the nominal plane is about 30°, and the nominal plane should be at least 0.125 in (0.32 cm) above the plane of the panel on which the switch is mounted.

Rocker switches may be broad and thus provide space for labeling, or they may be made narrower, as illustrated. In either case, however, the switches should be mounted so that the switch centerlines are no closer than 0.75 in (1.9 cm) when the switches are arranged side by side.

Switch resistance should be between about 8 and 12 oz (227 and 340 g) so that the switch is easy to operate and yet is not easily disturbed inadvertently.

Rocker switches typically can be obtained in different colors and thus can be color-coded for ease in differentiation. In fact, the two faces of a single switch usually can be obtained in two different colors, which makes it even easier to tell which switches are activated and which are not.

Rocker switches (like toggle switches) should be designed to snap into position with an audible click.

The rocker switch is preferred over the toggle switch when switches are laid out on a horizontal panel (see the accompanying sketch) because the toggle switch tends to protrude in this arrangement and is easy to become snagged on if an operator has to reach across it.

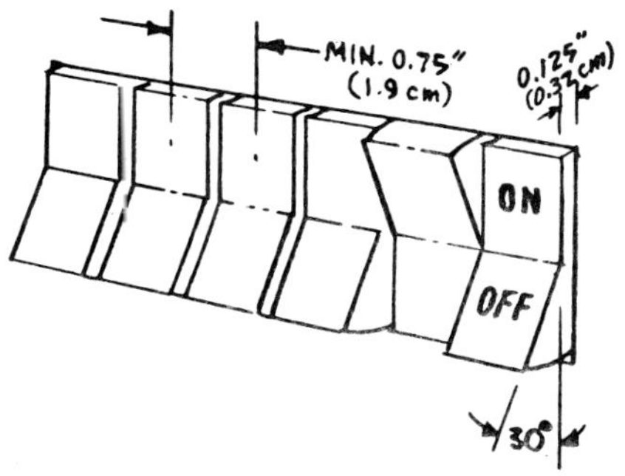

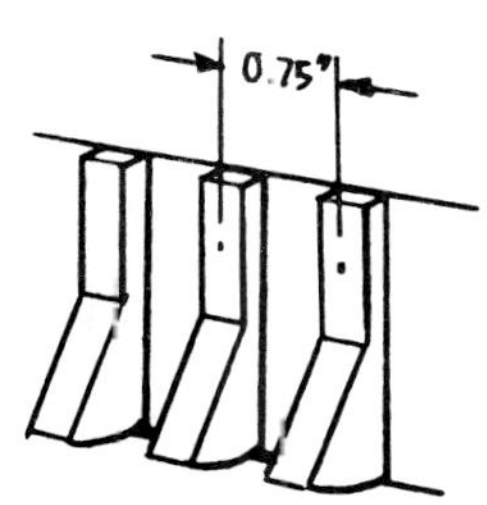

ROUND CONTROL KNOBS

Round knobs should be used only for applications in which the rotation is continuous and smooth and/or when multirevolutions are required.

The smallest useful knob should have minimum dimensions of 0.25 in (0.6 cm) in diameter × 0.75 in (1.9 cm) in depth, and the knob should be either serrated or knurled. A maximum resistance of 2 to 4 oz (57 to 113 g) should be the most force expected by an operator. Such configurations should be used only sparingly, such as for a clock setting or for gross adjustment of certain noncritical displays.

The preferred minimum size of the knob for convenient operation of volume or illumination potentiometer controls is 0.50 × 0.50 in (1.3 × 1.3 cm), with appropriate serrations to ensure comfortable and effective gripping. The upper limit to ensure ability to make fine adjustments is approximately 4 oz (113 g) of resistance. The edge of the knob can be tapered slightly, although this adds nothing to the effectiveness of the knob. Avoid severe tapering, however, for the operator's fingers may tend to slip from the knob.

For most normal panel layouts, select a knob whose dimensions are approximately as shown in the accompanying sketch. These are easily accommodated even by an operator who is wearing gloves. Resistance can vary from 4 to 10 oz (113 to 283 g) as the diameter of the knob is increased.

Be particularly careful in accepting some of the decorative knobs produced by some manufacturers, for they sometimes possess features that make the knob difficult to handle. An acceptable alternative to the previous shape is shown in the accompanying sketch.

Note: Choose serration patterns that provide good gripping characteristics, i.e., sharper peaks or points rather than "humps."

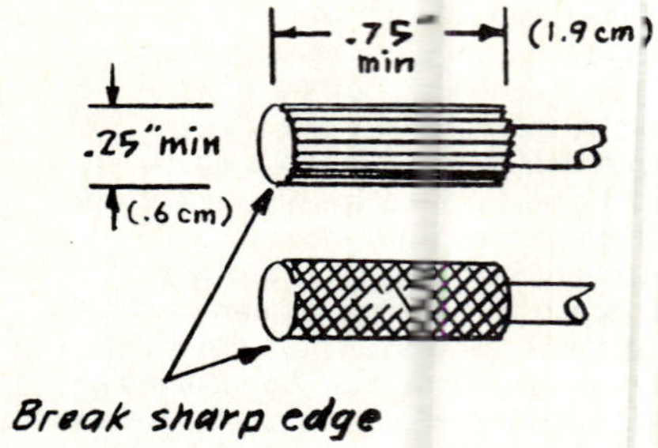

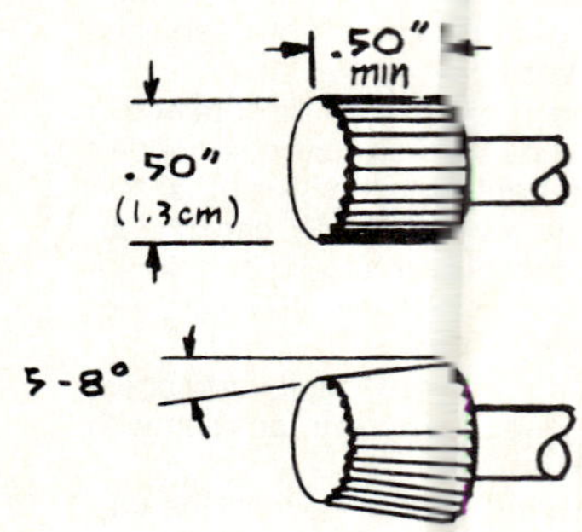

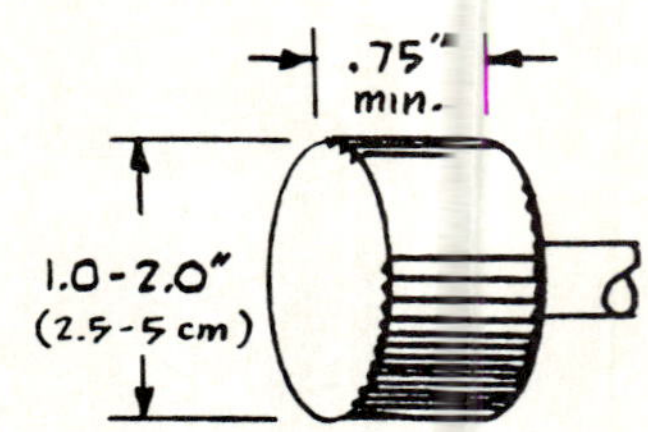

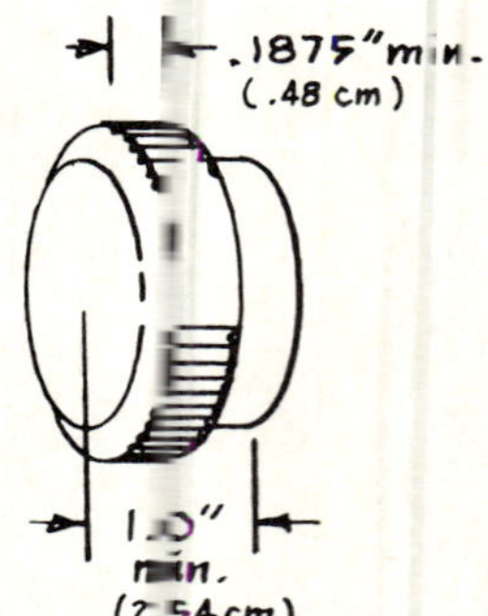

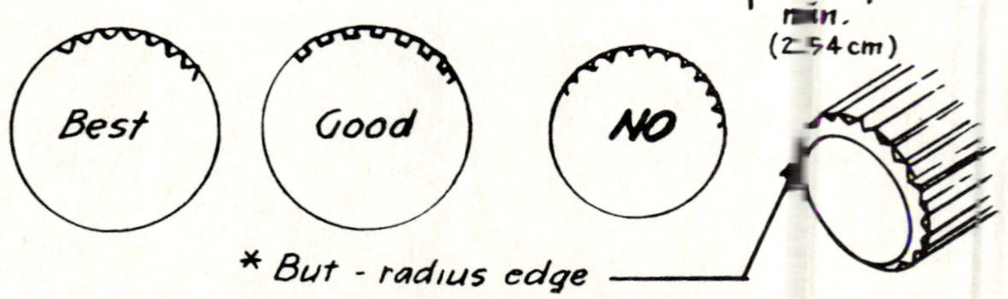

Control knob or switch or potentiometer operating characteristics should include viscous damping as opposed to static resistance. A certain amount of resistance is desirable (approximately 2.0 in/oz in order to minimize the possibility of inadvertent knob movement).

Generally, more precise knob movement is possible with larger knob diameters (1.5 to 3.0 in, 3.8 to 7.6 cm) because more fingers can be brought in contact with the knob surface. The knob torque should not exceed about 4 in/oz.

For high-torque applications, select knob diameters between 2 and 3 in (5.1 and 7.6 cm) or 3.25 in (8.2 cm) maximum. Maximum torque should not exceed 15 ft/lb (for husky male operators). The table below provides approximate values for knobs of different diameters and depths. Instead of serrations, use knurling on the edges of high-torque knobs for maximum gripping. The edges should be slightly beveled so as not to injure the operator's hand.

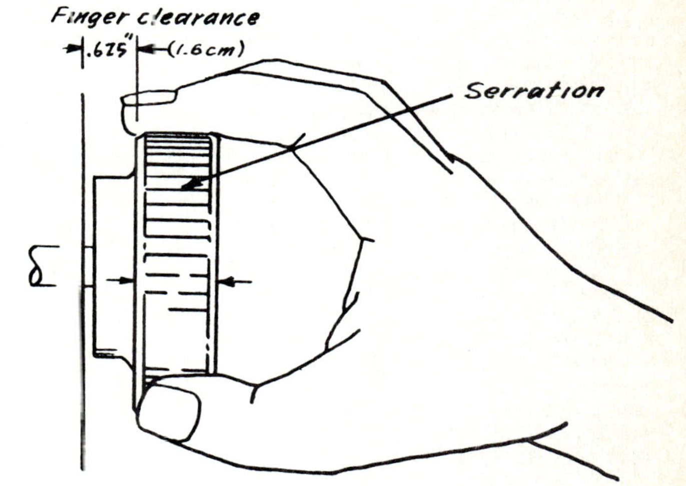

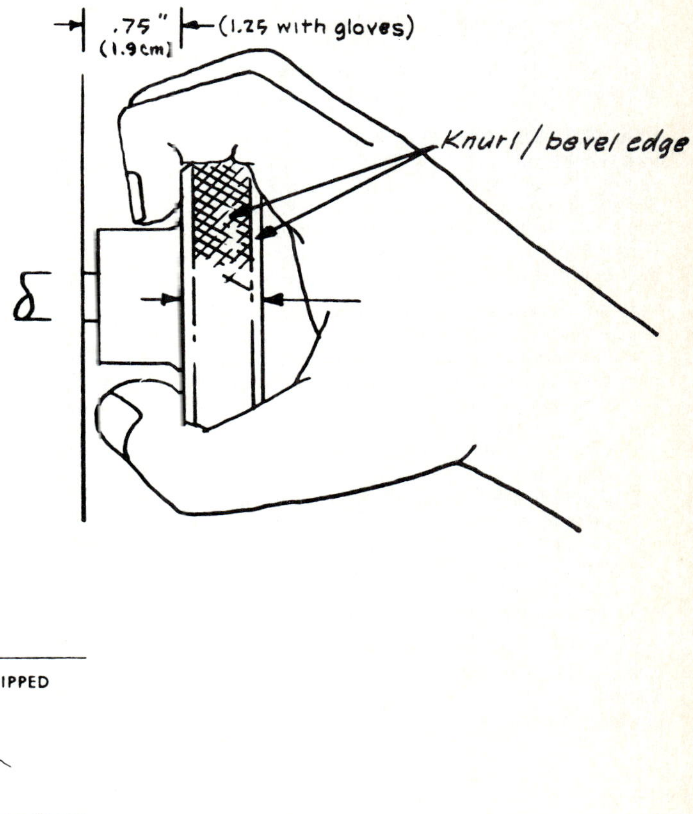

KNOB DIAMETER (INCHES)	KNOB DEPTH (INCHES)			
	FINGER GRIPPED 		FULL-HAND GRIPPED 	
	0.50	1.0	0.50	1.0
0.50	5 in. lb	6 in. lb	11 in. lb	16 in. lb
0.75	6 in. lb	8 in. lb	20 in. lb	29 in. lb
1.00	8 in. lb	10 in. lb	5 ft lb	6 ft lb
1.50	13 in. lb	15 in. lb	7 ft lb	10 ft lb
2.00	20 in. lb	24 in. lb	11 ft lb	13 ft lb
3.00	6 ft lb	6 ft lb	14 ft lb	16 ft lb

Control and Display Movement

The amount of control movement required to effect a display movement or change is referred to as the "control/display ratio" or "CD ratio." Although it should be optimized for each situation whenever possible, some general guidelines are shown in the accompanying illustration. A complete rotation of a control knob should cause about 1 to 2 in (2.5 to 5.1 cm) of movement of the displayed element, e.g., the pointer on the scalar instrument.

In the case of an auditory signal used to find a station, about 30 to 60° of rotation of the knob should be used to "null out" the station position with minimum necessity for unnecessary searching.

A dual-mode sensitivity capability may be required for some kinds of control-display relationships. For example, if a pencil joy stick is being used to control an electronic hook on a CRT display, a single CD ratio would be unsatisfactory since the operator typically needs to "get into the vicinity of a target" as rapidly as possible, after which he or she requires a more sensitive CD ratio in order to finally pinpoint the target location. As a general rule, the gross CD ratio should be approximately 1:1; i.e., a full sweep of the joy stick should translate the hook all the way across the screen. However, for the fine adjustment around the target site, the control lever should move approximately 1½ to 2 times as far (relatively speaking) as the hook moves on the scope.

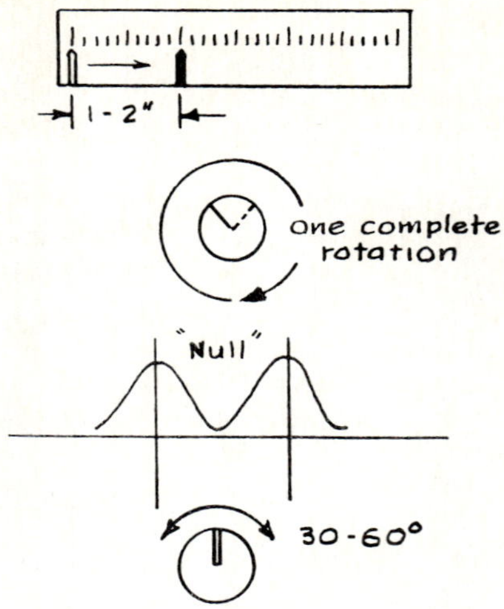

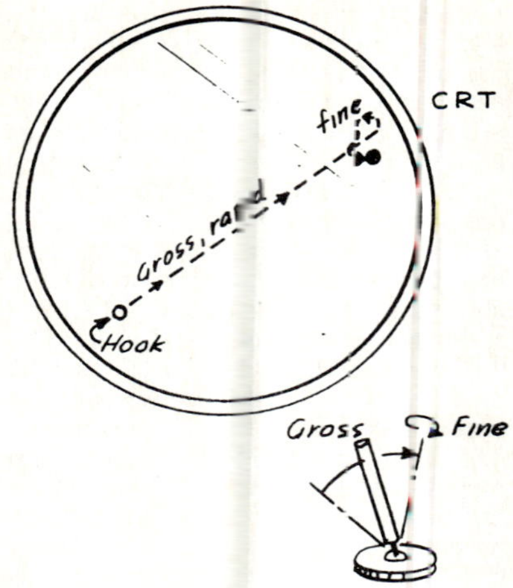

STACKED (OR GANGED) KNOBS Up to three knobs may be ganged on the same shaft center to save space. This should not be done, however, unless space is really at a premium. When this concept seems necessary, consider the following in order to effect the best operation and the least operator error:

1. The largest knob should be used for operation 1, as shown in the accompanying illustration, so that the operator works outward and thus is less apt to disturb a previous setting.
2. Gross tuning should be assigned to the smaller knob, and fine tuning to the larger one.
3. If the two knobs work sometimes together and sometimes independently, design them so that the operator depresses the top knob to engage the bottom one, with spring loading to return the upper knob to its independent status.

The "ideal dimensioning" for ganged controls has been determined experimentally, as illustrated. The accompanying sketch shows unintentional contact conditions to look out for.

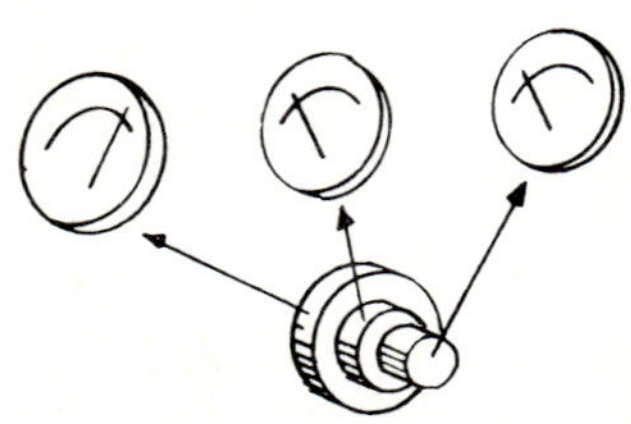

Largest knob associated with left hand dial, smallest with right hand dial.

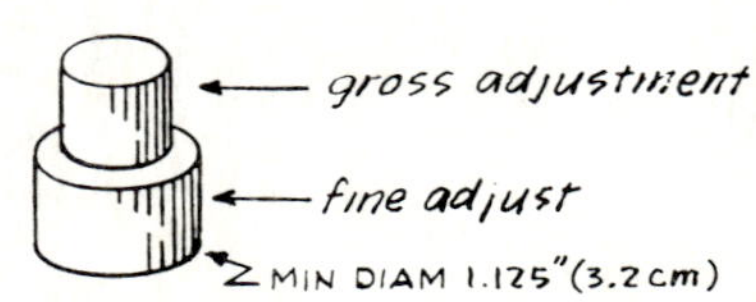

Coarse vs fine adjustment designation

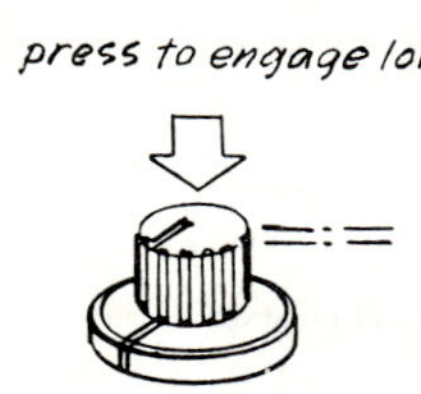

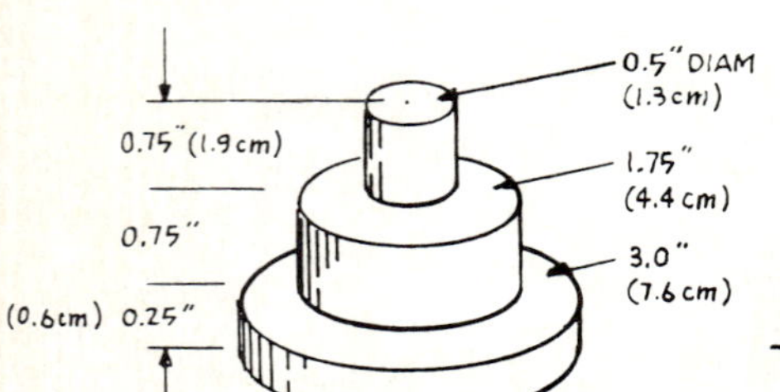

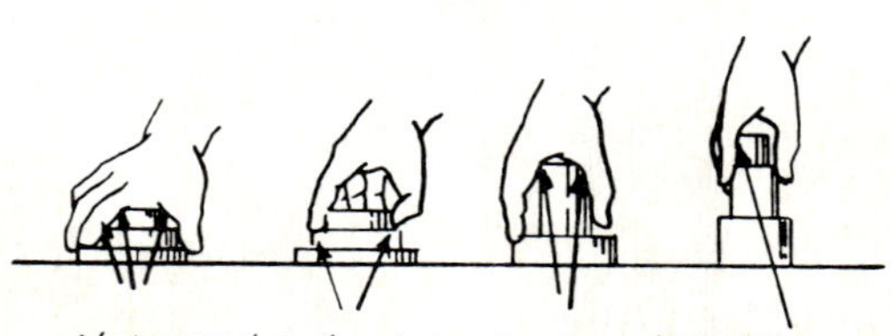

Note inadvertent contact probabilities

SKIRTED KNOBS Knob skirts are useful for at least two reasons: They provide a convenient place for a more visible pointer and/or for numbers or labels, and they help prevent the operator's fingers from scrubbing the panel surface. A properly designed skirt can prevent the operator's fingers from covering up the pointer or numbers. The design shown in the accompanying sketch provides relatively optimum relationships to accomplish the above objectives.

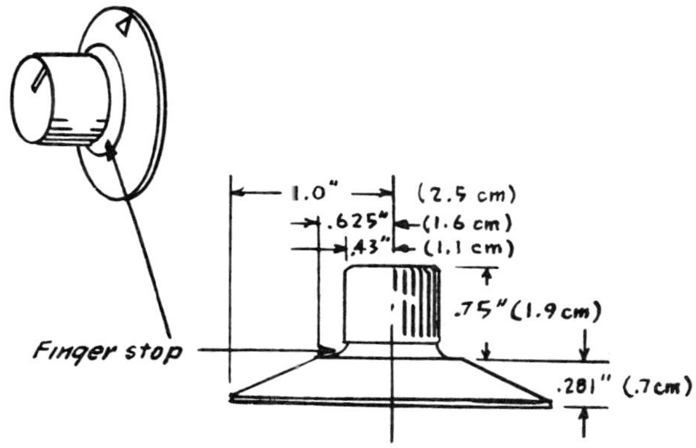

COMBINATION KNOBS AND CRANKS When it is desired to provide for rapid multiturning for initial gross setting, followed by a need for more accurate setting, the combination knob and crank design is useful. Two versions are shown. The folding crank handle is recommended for applications where it is probable that an operator might inadvertently bump the control or catch his or her clothing on the extended handle. Suggested dimensions for such control knobs are as follows:

Minimum radius—0.625 in (1.6 cm)
Maximum resistance—2 lb (0.9 kg)

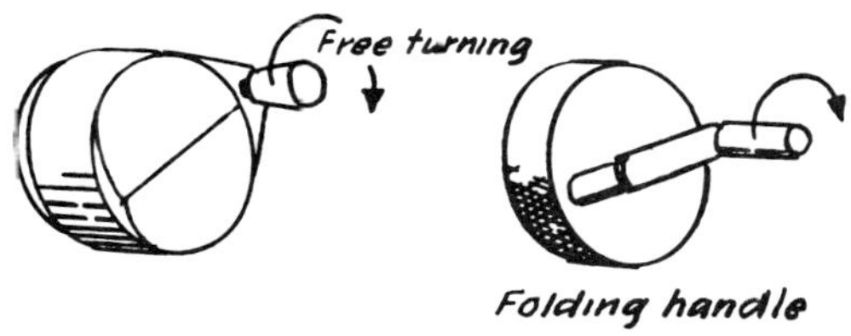

Cranks should not be mounted directly in front of an operator, particularly on the operator's centerline, but rather to one side or the other. If considerable torque is required, place the crank control as shown. If, however, very rapid cranking is required, place the crank either in front but to one side of the operator, or preferably, on an angle to one side of the operator, as illustrated.

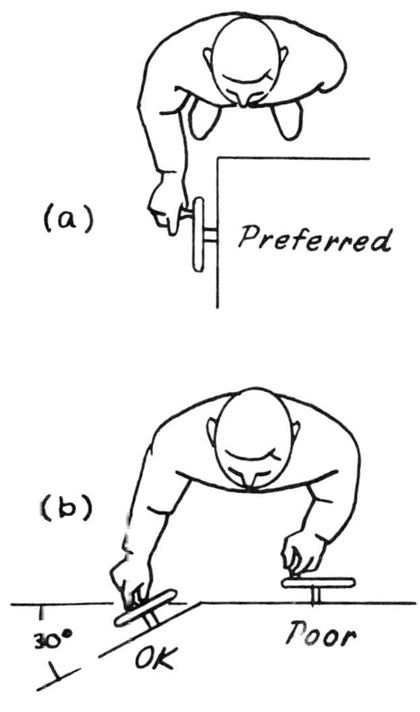

WHEEL AND CRANK CONTROLS

Wheel and crank controls frequently are useful for machine applications as well as for electronic console operations. The guidelines shown in the accompanying illustration and table provide approximate design goals when any of the separate parameters may already be fixed.

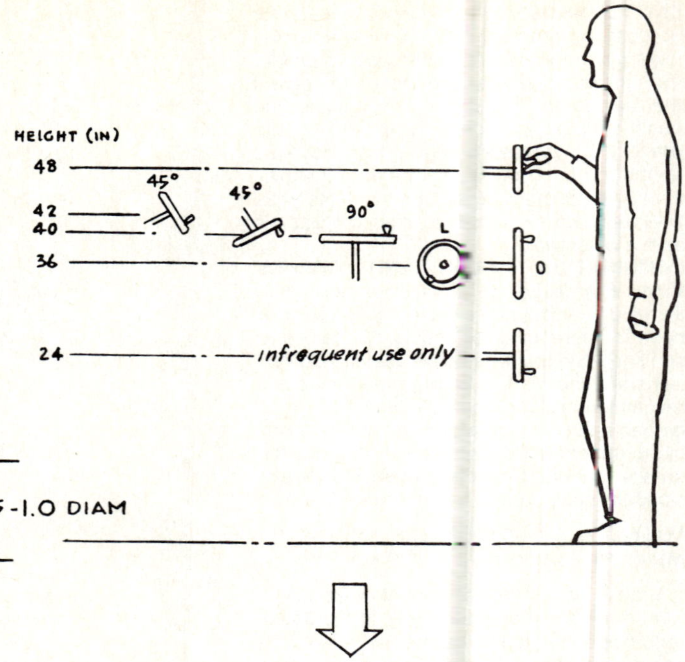

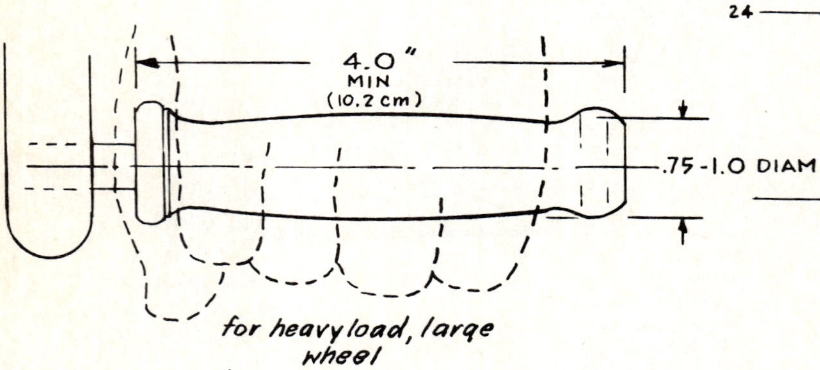

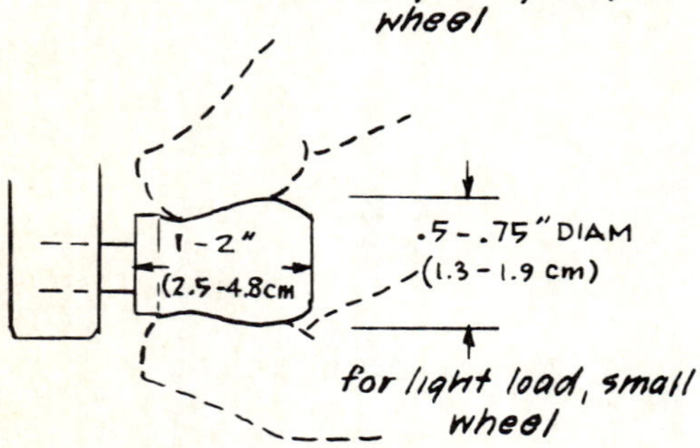

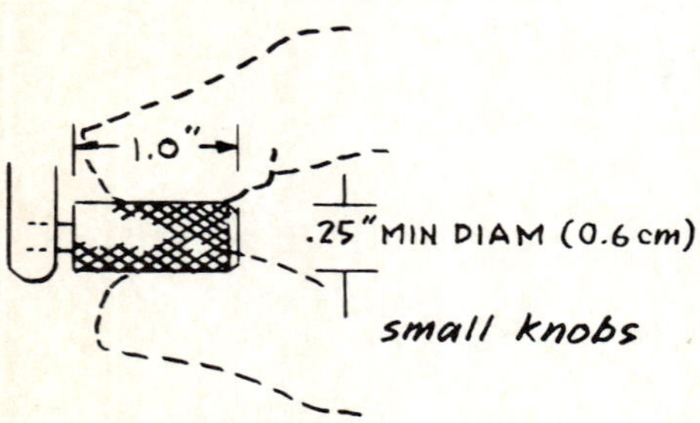

Note: Handles must rotate freely.

Optimum Force Limits versus Control Position

			Size Handwheel W, Diameter in, Crank C, Radius, in At Torque of		
Height, in	Position, deg	Type	0 in-lb	0 in-lb	90 in-lb
24	O	W	3–6	10	16
36	C	W	3–8	10–16	16
	L	W	3–6	10	10
	O	C	1½–4½	4½–7½	4½–7½
39	90	W	3–10	10–16	16
	90	C	2½–4½	4½–7½	4½–7½
40	45	W	3–6	6–16	10–16
	−45	C	2½–7½	4½–7½	4½–7½
42	45	W	3–6	10	10–16
	45	C	2½–4½	2½–4½	4½
48	O	W	3–6	8–16	10–16
	O	C	2½–4½	4½	4½–7½

Note: C = crank; W = wheel. 1 in = 2.54 cm; 1 lb = 0.4536 kg.

THUMB WHEELS

Although operable with either the thumb or the forefinger, the control configuration illustrated by the accompanying sketches is commonly referred to as a "thumb wheel." From a human engineering point of view, thumb wheels are seldom considered the most desirable type of control. In cases where they are used, they are applicable only to smooth, continuous functions (audio volume, light intensity, etc.) where nonprecise adjustment is required.

The principal guidelines for selecting or designing these devices are illustrated in the accompanying sketches:

A minimum of 1.0 in (2.5 cm) of the knob edge must be exposed, and/or about 90° of the circumference of the wheel.

The width of the wheel can be as narrow as 0.125 in (0.318 cm), but a minimum of 0.15 in (0.38 cm) is preferred. More than 0.50 in (1.3 cm) is undesirable.

Resistance should not be more than about 6 oz or less than 2 oz (170 to 579 g).

Acceptable multiple arrangements and spacing minimums are shown in the accompanying sketches. Note that horizontally arrayed wheels should not be arranged one above the other or one in front of another.

Thumb wheels should not be used if the operator has to wear gloves.

Wheel motions upward or to the right should provide "increasing function."

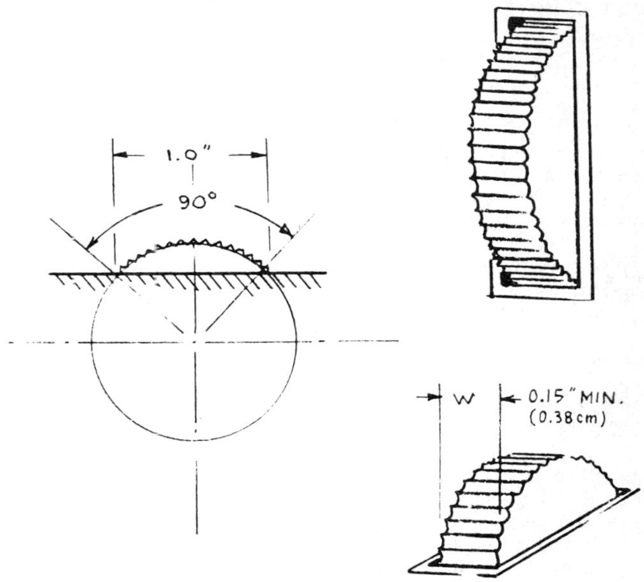

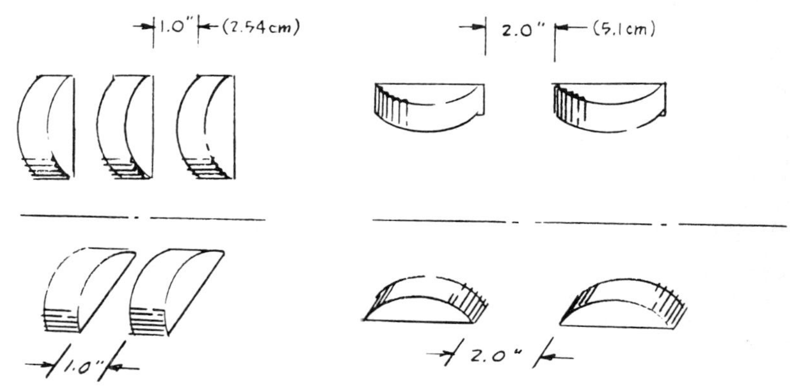

Discrete-Positioning Thumb Wheels

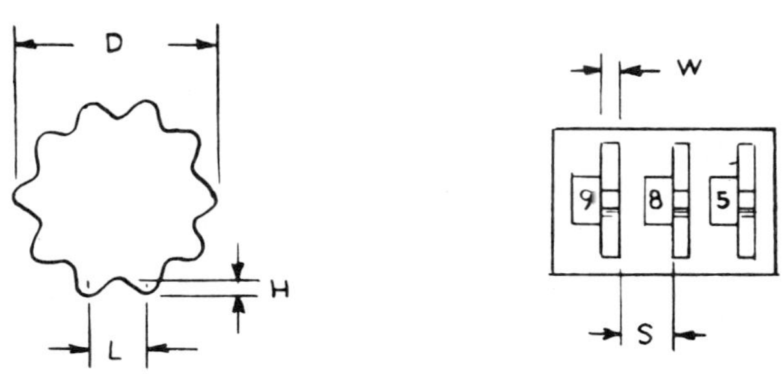

	D (Diameter)	L (Trough Distance)	W (Width)	D (Depth)	S (Separation)	Resistance
Minimum	1.5 in (3.8 cm)	0.45 in (1.14 cm)	0.1 in (2.54 cm)	1/8 in (0.3 cm)	0.4 in (1.0 cm)	1 in-lb (0.12 m · N)
Maximum	2.5 in (6.4 cm)			1/2 in (1.3 cm)		3 in-lb (0.34 m · N)

POINTER KNOBS

In contrast to the smooth, round knob recommended for continuous, multiturn controls, a "pointer-shaped" knob should be used for controls when the operator may wish to switch from one discrete position to another. The switch should have mechanical detents so that the operator can "feel" and perhaps hear when the knob is properly seated or positioned.

A prime objective of the pointer knob is to provide (1) a sufficiently discriminable pointing shape so that the operator knows which end of the knob is the pointing end and (2) adequate depth so that the operator can get a firm enough grip to turn the knob.

Various knob shapes are generally acceptable as long as certain details are not compromised. For example, the accompanying sketches show typical "bar-shaped" controls that provide adequate pointing characteristics. The *d* dimension should always be at least 0.50 in (1.3 cm); the *l* dimension should be at least 0.50 in (1.3 cm); and the *w* dimension should not be less than 0.25 in (0.6 cm).

The third knob shape is generally a better configuration than the first two because of the sloping pointer end, which makes it easier for the operator's eye to follow the pointer directly to the point at which it becomes adjacent to an index mark on the panel.

The fourth pointer shape provides additional length to assist the operator when more force is required to position the knob.

The minimum torque requirement for pointer knobs should be about 10 in/oz (283 g); the maximum should be around 45 in/oz (1275 g). The switches should have elastic resistance which builds up and then decreases as the switch "seats."

When panel space dictates use of a very small switch knob, slight variations are possible, as shown in the accompanying illustration. The left-hand knob is not especially desirable, but it can be used under extreme conditions. The others are preferred because they are more comfortable to use.

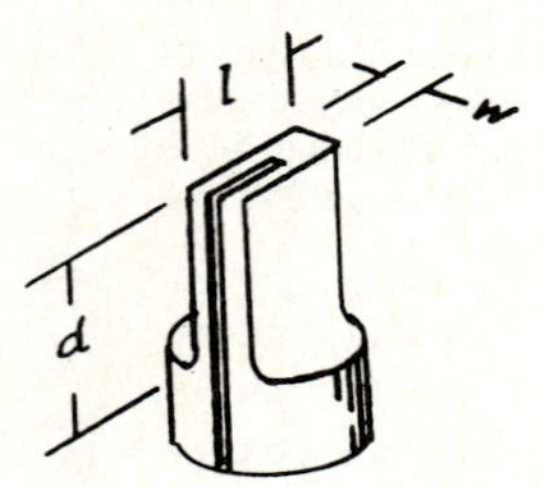

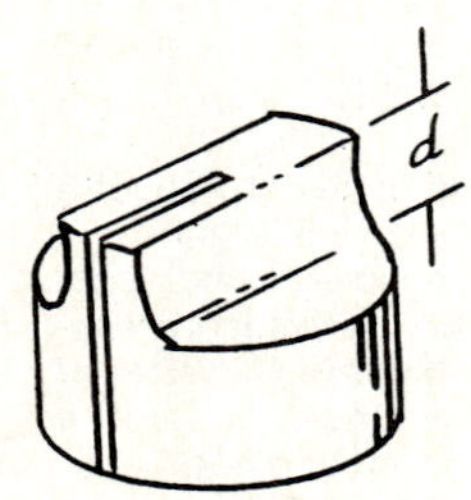

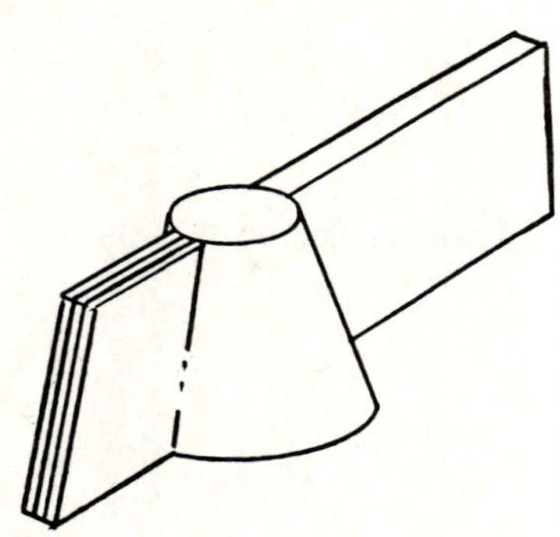

When it is known that a pointer knob will be used at night or under darkened ambient lighting conditions, the knob pointer should be illuminated. An effective method is to use a transparent plastic knob, cover it with a coat of transluminescent white paint or material, and then place an opaque coating over that. The pointer should then be inscribed through the outer opaque coating so that the white inner coating shows through and so that either edge lighting or back lighting from the control panel can shine through the pointer areas (see the accompanying sketch).

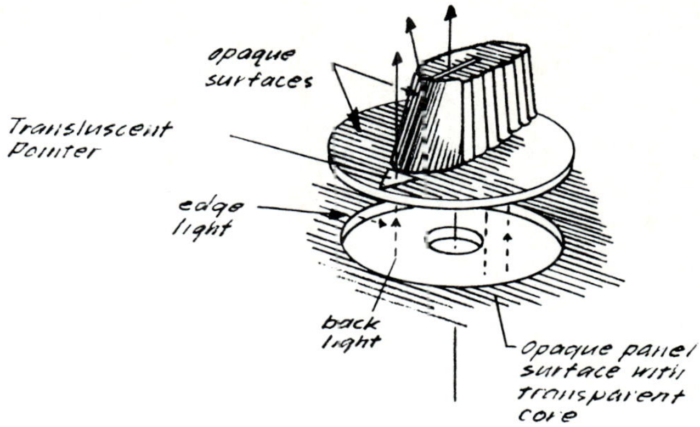

Lighted pointer knob.

Idealized Pointer Knob

The idealized pointer-knob shape shown in the accompanying sketch illustrates several desirable features to look for when selecting a knob:

A 45° pointer slope ensures a better view of the pointer tip from most viewing positions.
The long, tapering handle provides good leverage for rotating the knob, without having to cover the pointer tip with the fingers.
All sharp corners are eliminated.
The skirt provides a finger rest so that the fingers do not scrub the panel.
Not only is the configuration attractive, but also it will never be confused with round knobs, nor will the operator be in doubt as to which is the pointing end of the knob, whether he or she is looking at the knob or not.

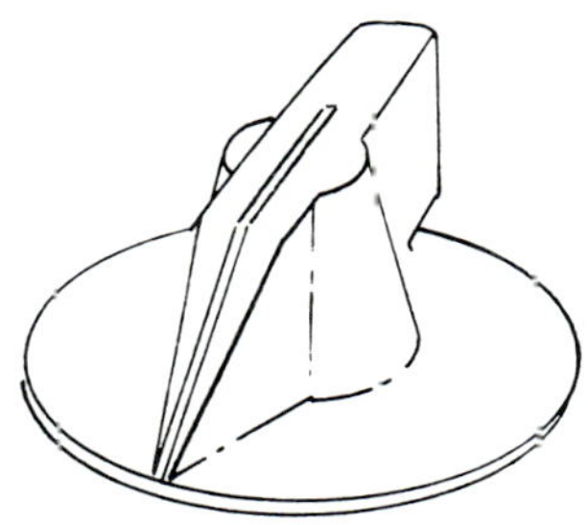

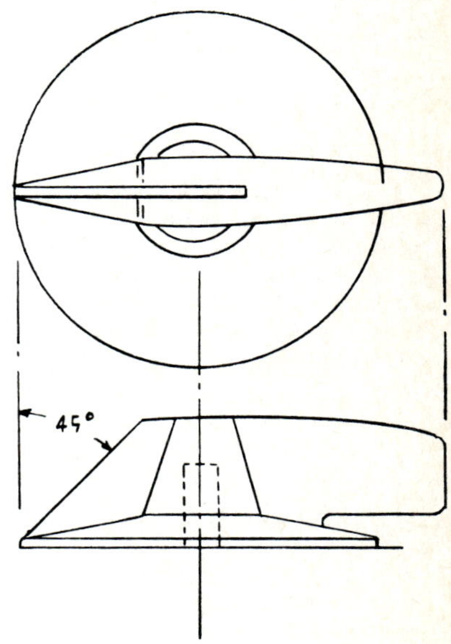

Example of Lighted Pointer Knob Design

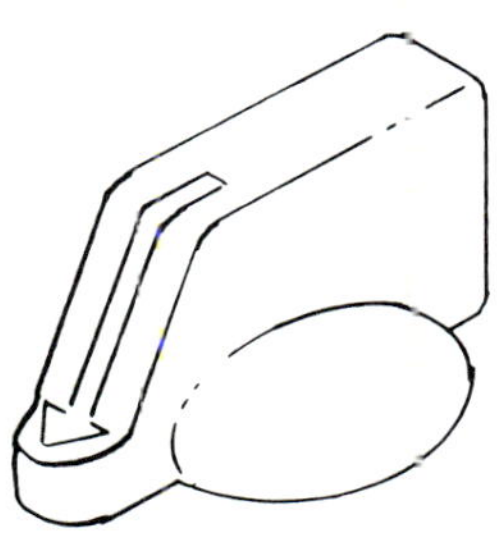

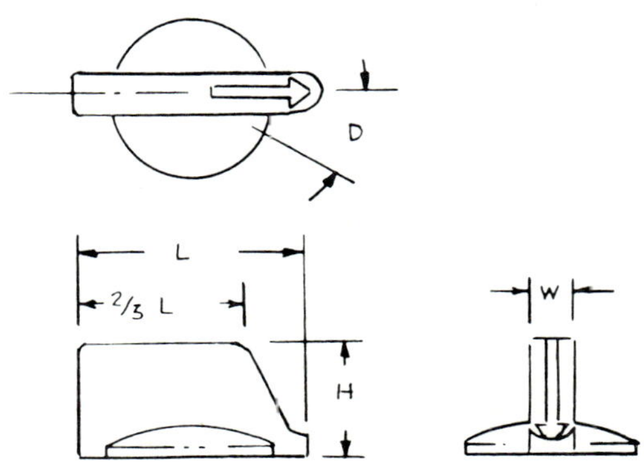

Operating Method	Pointer, in L (Length)	W (Width)	H (Depth)	D (Displacement in Degrees) Visual Positioning	Nonvisual Positioning	Resistance, oz*
Minimum	1.00	0.25	0.50	15	30	12
Maximum	3.00	1.00	3.00	40	40	48
				(90 when large separations are necessary)		

Notes: Military Standard MS 25166, *Knob, Pointer, Illuminated,* meets, to a certain degree, the above requirements. 1 in = 2.54 cm; 1 oz = 28 g.

*Use elastic resistance which builds up and then decreases as each detent is approached so that the control will snap into position and cannot easily stop between adjacent positions.

OTHER FINGER-OPERATED CONTROLS

Although a variety of control shapes are suitable for operation by the fingers, it is important to select controls that are large enough so that the operator can manipulate them conveniently and with a minimum of discomfort. The guidelines shown in the accompanying sketches provide useful information for selecting appropriate control devices.

Finger-operated levers should have resistances of 30 oz (850 g) or less. The action should include definite positional detents so that the operator knows the control is properly positioned.

The handle-length minimum for resistances of 30 oz (850 g) or less should be about 1.0 in (2.5 cm). When the resistance exceeds 30 oz (850 g), the handle length must be increased. The maximum resistance should not exceed about 3 lb (1.4 kg), in which case the handle length needs to be increased to at least 1.5 in (3.8 cm).

The slide switch resistance should not exceed about 10 oz (283 g).

A single "push-push" type of wall switch is preferred over the more typical two-button configuration because it removes the necessity for the user to decide which position the switch is in. The larger switch plate makes it possible to operate the switch with an arm or elbow when the hands are occupied.

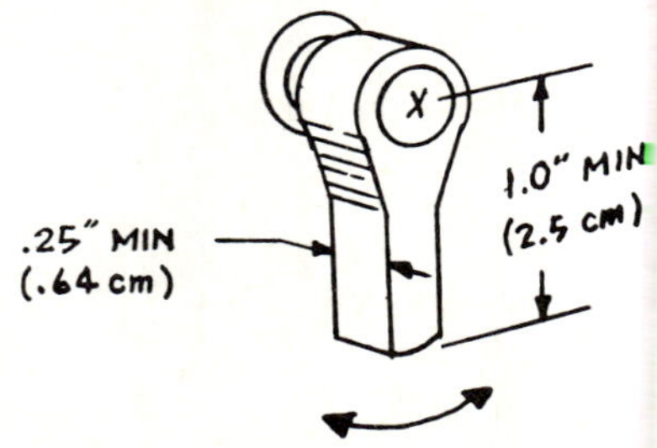

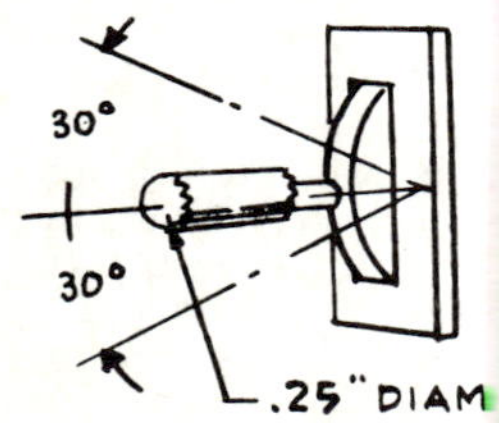

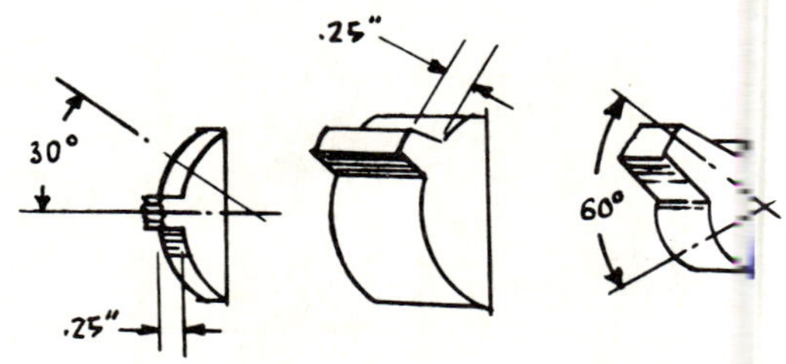

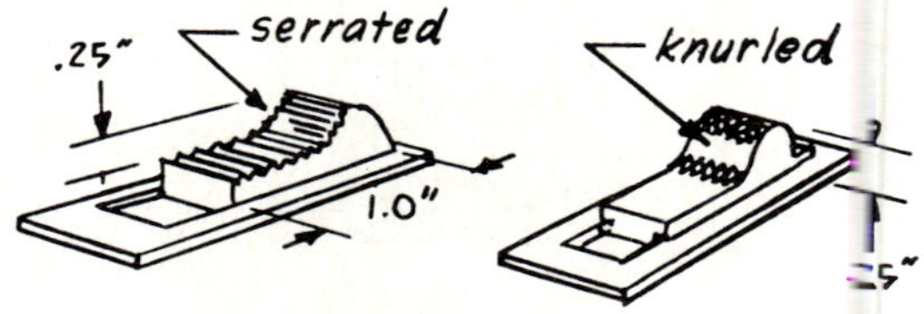

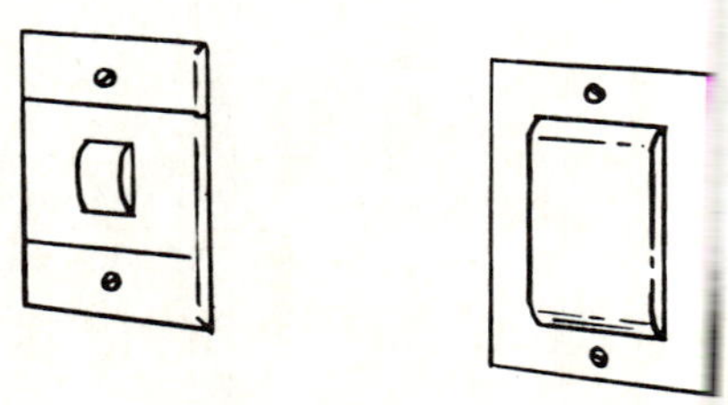

Fingertip-Operated Slide Controls

Although these controls can be used either for continuous-motion or for discrete, detented switching functions, they are not generally recommended for either one. However, they can be used for fairly coarse operating functions when operating error may be of little consequence and/or when the time taken by an operator to get the control positioned does not introduce other response delays. The basic problem with using these controls is that, in order to operate them precisely, the operator must steady his or her hand against the panel, using a long, total-arm movement from the shoulder, which is exceedingly difficult and very inaccurate.

Nevertheless, if such controls are selected, consider the following guidelines:

- The control knob or tab should be large enough to get hold of, e.g., minimums for *a* = 0.50 in (1.3 cm); *b* = 0.25 in (0.64 cm); and *c* = 0.50 in (1.3 cm).
- The handle should be offset on its shaft in the direction that will allow the operator to read whatever labeling is provided on the panel; i.e., the labeling will normally be placed above the horizontal slide or to the left or right of the vertical slide, depending on which hand is used to operate the slide.
- The slide resistance should not exceed about 2 lb (0.9 kg).
- The handle should be oriented as shown in the accompanying sketches, and there should normally be a pointer inscribed in the handle edges.
- When multiple slide arrangements are made, sufficient separation must be maintained so that labels can be seen and easily associated with the slide they apply to. (This is one of the most common errors observed in the design of slide assemblies. It occurs because of the desire to save space.)

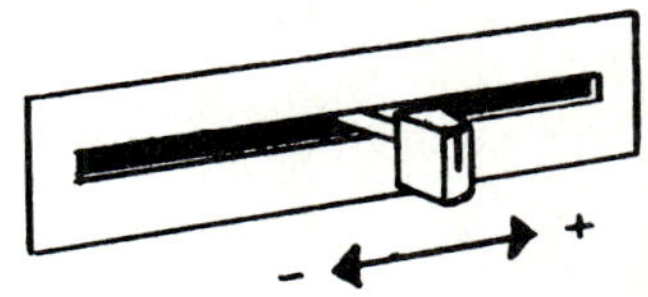

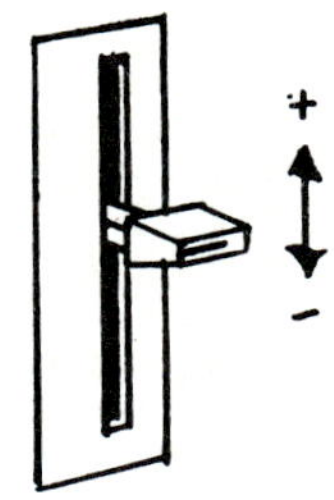

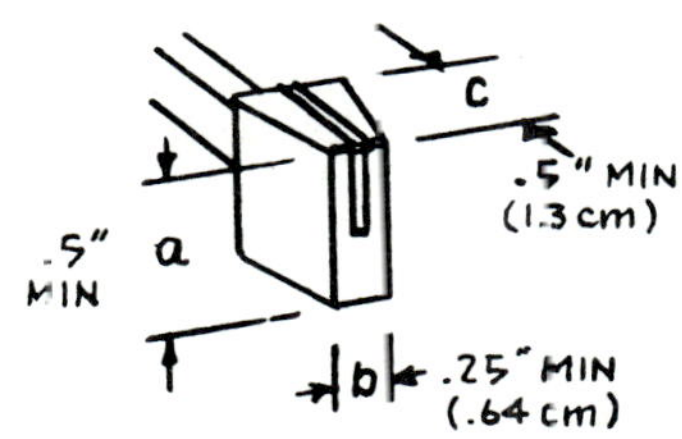

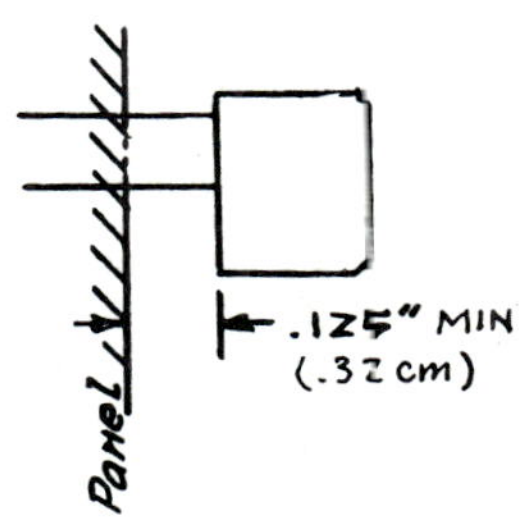

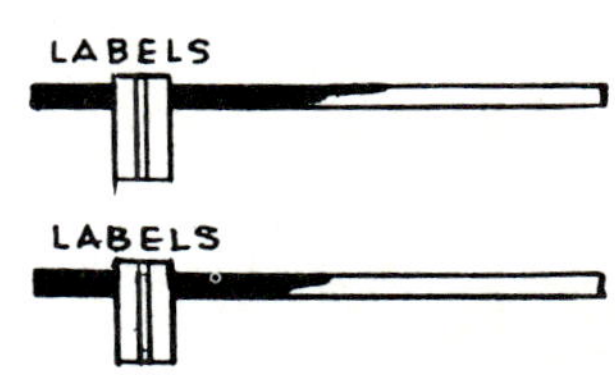

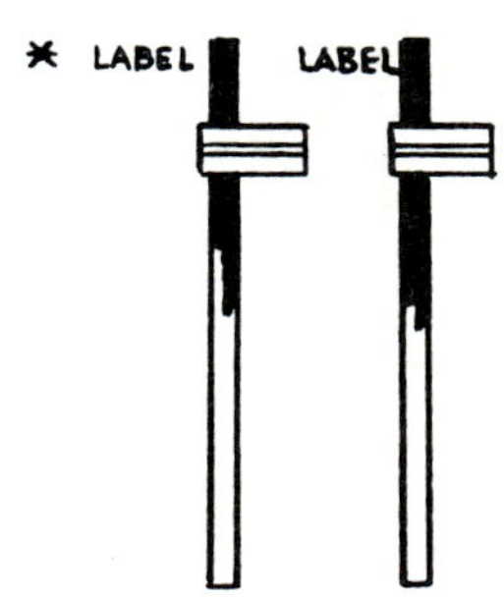

* Label on left when viewed from the left; right viewed from right.

CONTROL DYNAMICS

1. Feedback

All control systems should be designed to provide maximum information to the operator concerning the effects of the operator's input. This may consist of intrinsic feedback, extrinsic feedback, or both. Intrinsic feedback is that which is generated by the controller device, e.g., by "feel." Extrinsic feedback is that which emanates either from observation of an external visual condition or from a visual display especially provided to indicate the effect of controller inputs.

2. Control Input Mode

Continuous controlling or tracking is accomplished by either a compensatory or a pursuit mode. Compensatory tracking consists of adjusting a controller to maintain a vehicle, reticle, or pointer coincident with some visual reference. Pursuit tracking consists of "following" some randomly deviating reference. Compensatory tracking modes are more effective for acceleration control or when feedback inputs are infrequent. Otherwise, pursuit tracking modes are more efficient.

3. Controller Characteristics

Elastic resistance is desirable in a controller because it provides useful feedback from the controller movement or displacement; i.e., displacement can be quickly related to the changes in resistance. Design the controller system so that it provides distinct resistance gradients at critical control positions.

Viscous damping should be provided to aid in executing smooth control input at a prescribed rate.

Inertial resistance in the control system is desirable to enhance smooth control input and minimize the possibility of accidental or inadvertent activation. It should not be excessive, however, or it will become difficult to change adjustment direction. A certain amount of inertial resistance can, however, compensate for frictional drag that may be difficult to eliminate from the system.

Static friction should be minimized (e.g., resistance to control movement initiation). Not only does it provide no feedback, but also it often causes an excessive input force to overcome, which the operator may have difficulty correcting once he or she has observed the "overshoot."

In order for an operator to judge resistance, the input force required should be at least in the 5- to 10-lb (2.3- to 4.5-kg) range or above. Obviously, the resistance for manual controllers should be less than that for pedal controllers. Resistances above 30 to 40 lb (13.6 to 18 kg) should be avoided, however, in order to prevent fatigue.

Continuous tracking controllers should be designed to automatically return to a neutral reference position.

Sliding (coulomb) friction should be minimized in controller systems since it provides no meaningful input to the operator and may contribute to fatigue.

4. Control/Display Ratio

A low control/display (CD) ratio occurs when one unit of movement of the controller equals one unit of movement of the controlled element (e.g., the instrument pointer). Low CD ratios are desirable when it is important to obtain a gross system change with minimum controller movements. A high CD ratio is one in which the controller adjustment unit is greater than the unit motion of the system. High CD ratios are desirable for "fine" adjustment. Although specific control systems may require either a low or a high CD ratio because the nature of the control task is fairly constant, other systems may require a varying CD ratio program, such as for steering a vehicle or in using a joy-stick controller to manipulate an electronic "hook" to capture a target on a CRT display.

5. Detrimental Characteristics

Dead space, lag, backlash, and excessive resistance should be avoided in controller systems. Dead space occurs when movement of a controller does not immediately result in movement of the visual display element. Dead space should be eliminated as much as possible, especially for sensitive or high CD ratio systems. Lag is constant delay that may be inherent in the system design. With high CD ratios (1:6 to 1:3), lag is more serious for compensatory tracking. However, when CD ratios are low, lag may improve performance. Backlash is the tendency of a system to reverse when the control input ceases. If backlash cannot be minimized, the system display gain should be as low as possible.

6. System Aiding

Aided acceleration or rate tracking (i.e., allowing a single control movement to affect two variables, such as position and rate) should be considered in order to shift the burden of differentiation, integration, and algebraic addition from the operator. Predictor display concepts should be considered when the future state of the system is not directly predictable, i.e., when the control system computes and displays the predicted position at some selected future time. This permits operators to assess the predicted results of any current input they may desire to make, thus allowing them to readjust the input in the event the first input would have an undesired result. The predictor display provides a surrogate for operator prediction (e.g., when a driver looks ahead and predicts that his other steering input is too much or too little to properly negotiate a curve at the vehicle's current speed).[2]

[2]Because of the typically unique variations among more complex continuous-control systems, generalizations regarding specification of CD ratios and system dynamics are inadvisable. Therefore, the reader should consult someone who is experienced in complex control systems design and/or refer to the following excellent discussions of the general subject of control system dynamics: C. R. Kelley, *Manual and Automatic Control: A Theory of Manual Control and Its Application to Manual and to Automatic Systems,* John Wiley & Sons, Inc., New York, 1968 and E. C. Poulton, *Tracking Skill and Manual Control,* Academic Press, Inc., New York, 1974.

GENERAL CONSIDERATIONS

Human Factors Considerations in Materials-Handling Systems Conceptualization

Although a great deal of the handling of materials coming into, or being shipped from, the industrial complex, as well as the actual handling within the facility, may be accomplished by machines of one kind or another, humans interface with this process either directly or indirectly via machine controls and displays. The primary human factors considerations fall into two major categories, as described below.

1. Prevention of Materials Damage
Waste of materials due to damage is of considerable economic consequence to the manufacturer. Therefore, it is extremely important to examine the materials-handling concept in terms of the potential errors the human element could make with respect to movement of all materials in the receipt, manufacturing, and shipping phases of the proposed operation. This examination should include consideration not only of raw materials and substances but also of packaged items. The potential for human error often exists at the following points during materials handling:

a. Dock transfer to and from a transportation system (rail, truck, ship, etc.)
b. Shipping and storage-area transfer (i.e., to and from the loading dock and areas designated for temporary storage of either the preprocessed materials or the finished and packaged products)
c. Transfer within the plant.

2. Prevention of Personal Injuries
Personal injuries also have a considerable economic impact on an industrial operation in terms of lost time and possible legal liability. More important from a human factors point of view, however, is the direct consequence to the worker and his or her family. This latter consideration should guide the planner because it assumes that most workers do not want to be injured and that it is therefore incumbent on the systems designer to create as safe a place to work as is technologically possible. The typical areas in materials handling where personal injury often occurs include the following:

a. Direct interface with raw materials that are heavy, sharp, rough, toxic, breakable, or explosive
b. Storage and stacking, where materials fall on people
c. Automatic material movement, where workers may be struck by the material or machine or may fall into or be caught by machine elements
d. Toxic liquids or chemicals, which may spill onto workers during transfer

TRANSPORT AND LIFTING APPLIANCES
Narrow-gauge railway superseded by overhead trolleys or trackless electric trolleys with trailers. Long distance transport by lorries. Internal transport by traveling cranes of different types, fork lift trucks, etc.

CRANES Simplest lifting appliance for vertical elevation is the hoist (also electric) having a 0.5- to 5.0-ton capacity. Additional horizontal movement by wheel carriage (1) and (2). Wheel carriages controlled from shop floor or from driver's seat (3) or from remote control cabin (6).

Bracket jib cranes make load lifting above any part of an area possible. Exact positioning of load difficult, however, due to sloping extension device.

Combination of several wheel carriages possible for transport of heavy pieces (4).

Bracket cranes (8) are for transport along the length of the building. Jib cranes (7) are for transport of production parts to neighboring bay. Connecting platform (9) is for changeover of wheel carriage.

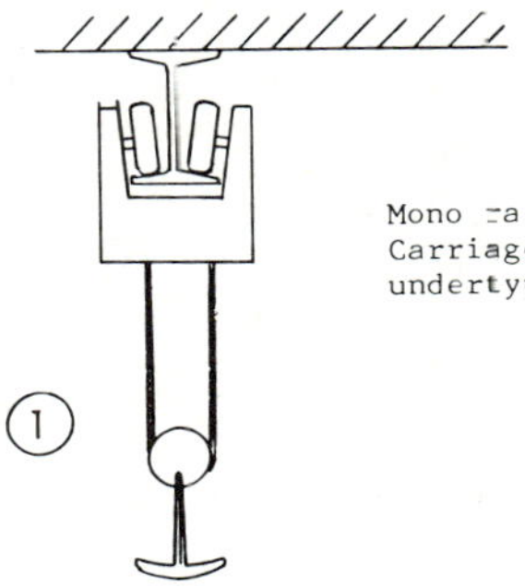
(1) Mono rail Carriage, undertype

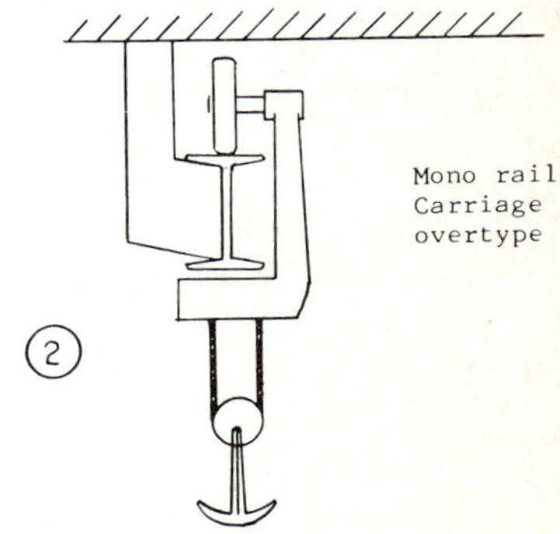
(2) Mono rail Carriage overtype

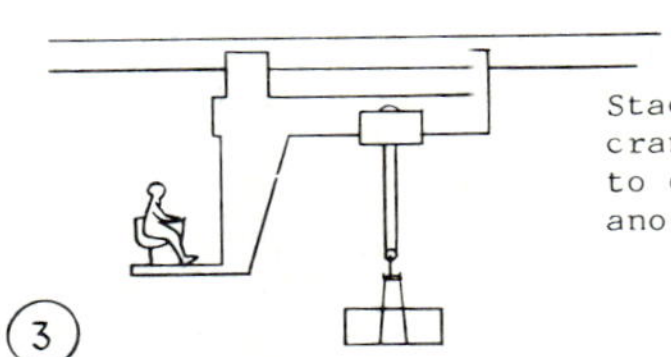
(3) Driving cab wheel carriage

Stacking crane as suspended crane lighter; also possible to cross from one crane to another.

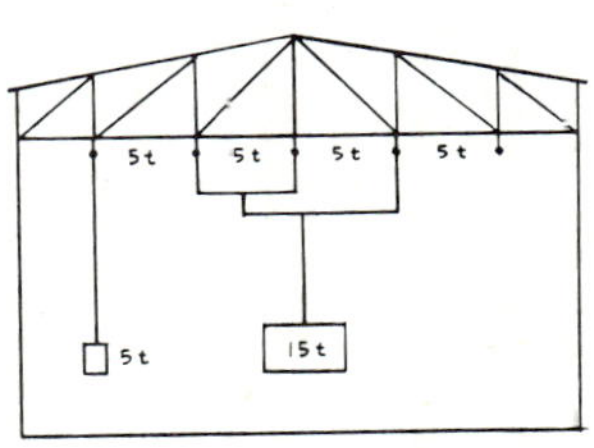

(4) Distribution of a large weight

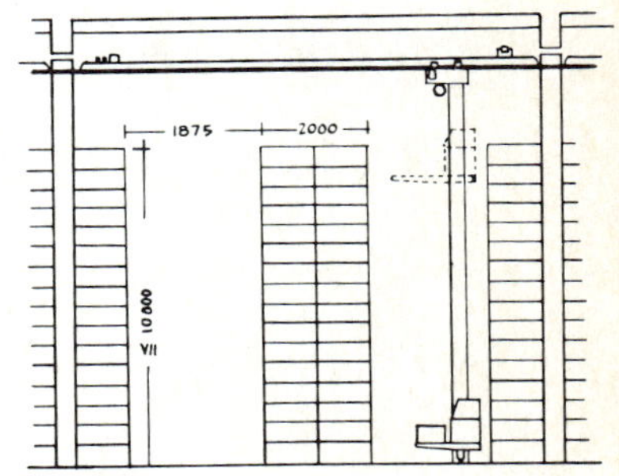

(5) Stacking crane as suspended crane

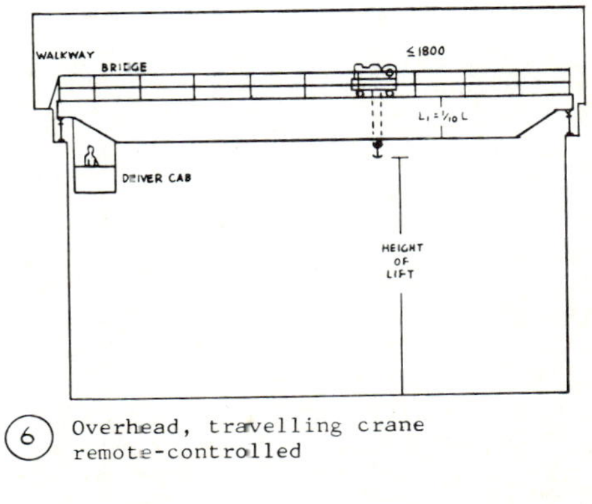

(6) Overhead, travelling crane remote-controlled

Stacking crane (5) is a combination fork-lift truck and crane. Utilization of unused roof space and thus better use of storage space. Height of stack up to 10 m (33 ft).

Outside cranes require enclosed or rain-protected motors.

Design of structure depends on height of lift and crane capacities.

For mechanical workshops a lifting capacity of 4.50 to 6.00 m (15 to 20 ft) is sufficient, thus room heights are from 6 to 10 m (20 to 33 ft). From top edge of crane bridge to bottom edge of roof girder > 1.80 m (6 ft).

The height of wheel carriage depends on size of products and not necessarily on height of lift.

Standard traveling speed of wheel carriage is approx 30 m/min (100 ft/min).

OTHER TYPES OF HANDLING PLANT

Other types include roller, curved roller, spiral roller, screw or worm, screw ribbon, wire mesh and belt, slat, portable belt conveyors, bucket conveyors, etc.

Fork-lift truck (11) loads from either side or is a lifting device on lorries which makes loading ramps unnecessary. (From E. Neufert, *Architects' Data,* Archon Books, 1970.)

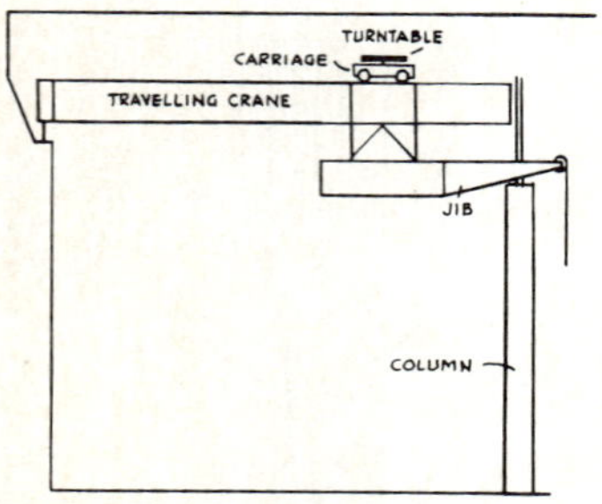

(7) Overhead, underhung jib crane will serve adjoining bay

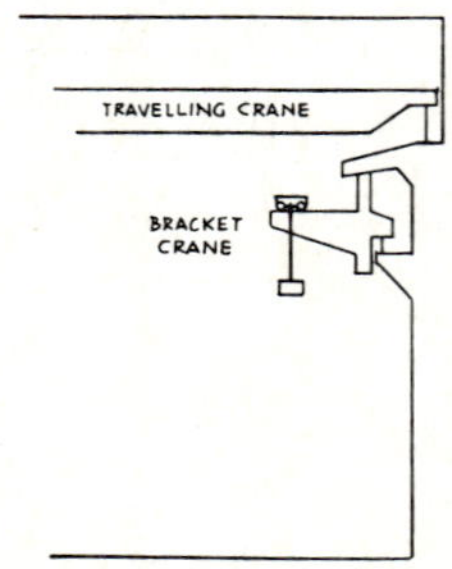

(8) Bracket crane below travelling crane

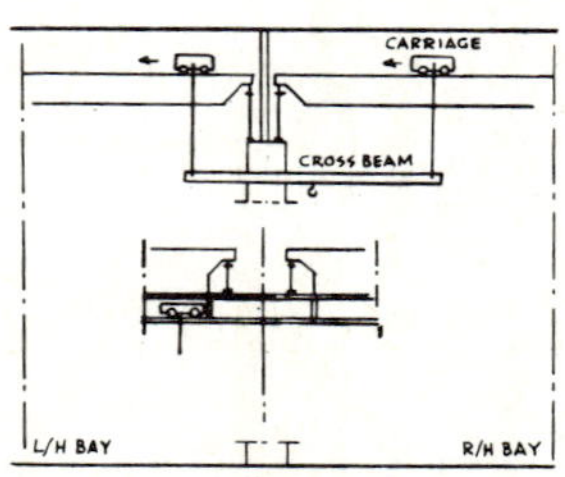

(9) Crane arrangement, multi-bay factories. Detail of connecting device through crane track girder.

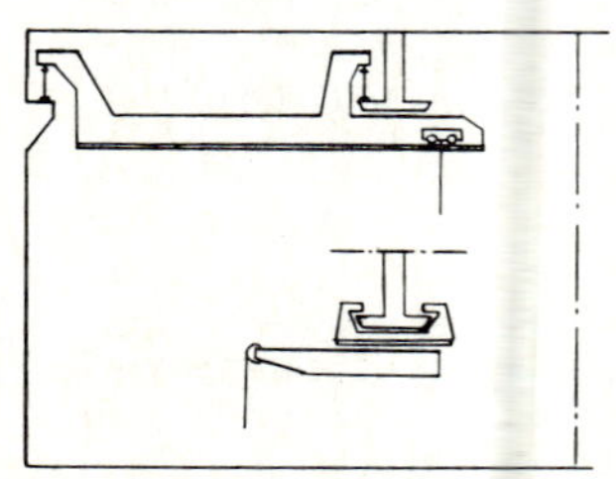

(10) Travelling crane with jib. Inset, fixed radius jib crane in center of building.

Capacity	Height	Length	Width
1000 kg	22[illegible]	1725	955
2000 kg	22[illegible]	2190	1400
3000 kg	22[illegible]	2885	1600

Length of fork lift: 0.80-1.4 m

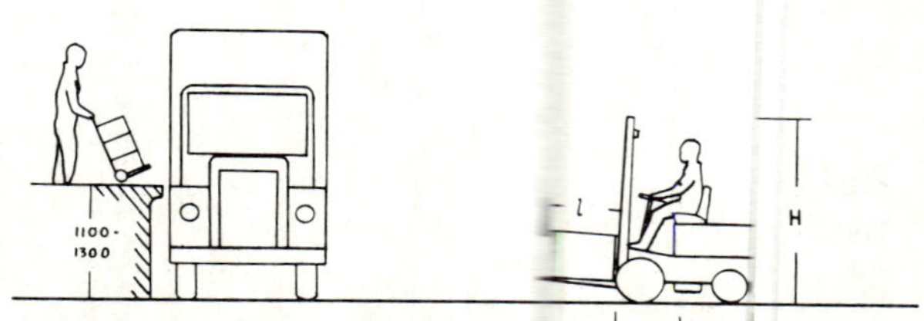

(11) Loading ramp and forklift truck

Semiautomatic Warehouse Systems

CONTINUOUS TROLLEY CONCEPTS Tripping and collision hazards must be considered. Design of the trolley track should be such that personnel cannot step into the track slot and turn an ankle. Forklifts should be painted with appropriate hazard markings, and audio alarms should make it possible for operators to warn other workers.

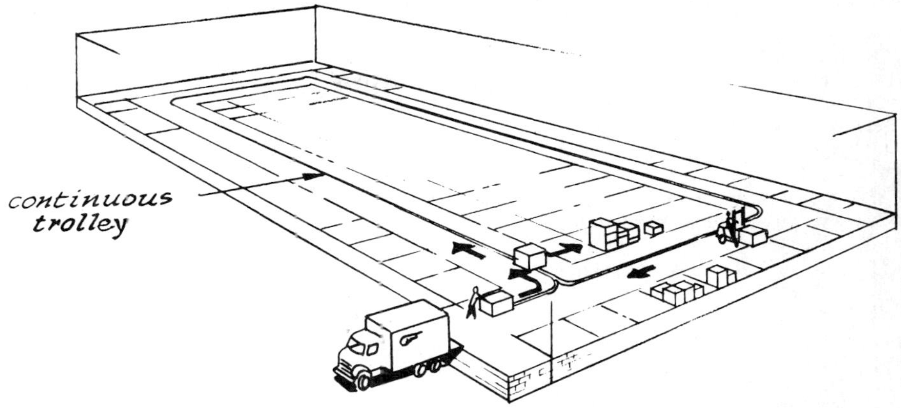

PARTIALLY COMPUTER-CONTROLLED TROLLEY, ELEVATOR, AND CONVEYOR CONCEPTS Inadvertent activation while performing maintenance tasks and limited visibility in high-density stack areas are common hazards. Manual operation during system down periods requires special consideration of access into areas and positions that may not have been planned for human occupation, and audio alarms should be provided to warn workers that conveyors are moving.

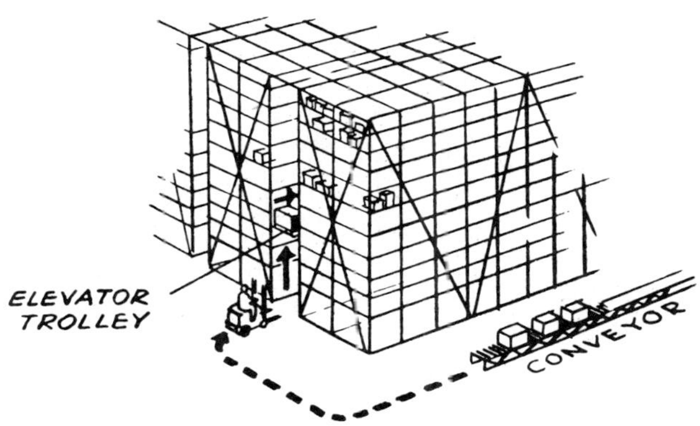

Human Factors Considerations in Conceptual Design of Materials-Handling Vehicles

Special vehicles designed for the express purpose of handling materials within the industrial plant should be approached from the operator's point of view, i.e., in terms of what the operator has to see and operate in order to manipulate the vehicle and its load in an efficient and safe manner. Some of the key issues are the following.

1. Visual Clearance of Vehicle and Load
The vehicle operator should be positioned so that he or she can see the perimeters of the vehicle and its load with reference to pathway obstructions.

2. Visual Observation of Load Transfer
The vehicle operator should be positioned so that he or she can maintain continuous observation of the load as it is raised and lowered and otherwise moved into position, i.e., onto a pallet, into a storage area, into a truck, etc.

3. Vehicle Controls and Displays
Controls and displays required to operate the vehicle should be located and positioned so that the operator can reach or see them, while at the same time visually monitoring vehicle and load clearance and load manipulation.

4. Operator Support and Restraint
Select the most efficient posture for the vehicle operator in terms of his or her ability to control the vehicle and load, considering alternate standing and seated modes. Consider the possible need to provide restraints to keep the operator from falling from the vehicle.

5. Operator Protection
Provide appropriate guards and/or screens to prevent materials or components being transported from toppling onto the operator. Consider weight and balance dynamics in terms of tipping over the vehicle. Provide a warning system to indicate when too heavy a load is being manipulated or when an impending load shift is approaching the imbalance point.

6. Fail-Safe Features
Consider the typical misuse behaviors of the average driver and provide fail-safe features to prevent the driver from leaving the vehicle insecured on an incline, leaving the vehicle in gear while he or she is away from the work position, driving too fast, etc.

GENERAL CONCEPTUAL PRINCIPLES

1. Avoid placing materials directly on the floor without a pallet or other support.
2. Assure adequate storage space at the workplace for the expected amount of materials, both before and after materials operations.
3. Use the same container throughout the system; i.e., avoid frequent changes of containers.
4. Provide adequate materials-handling clearances around each workplace.
5. Minimize the number of materials handlings; whenever practical, arrange for productive operations or inspections while materials are being moved.
6. Minimize the number of manual handlings, especially when materials are large or heavy.
7. Arrange for alternative handling methods in case of emergency.
8. Provide appropriate methods for anticipating bottlenecks or breakdowns of automatic materials-moving systems.
9. Plan movement in direct paths; i.e., avoid zigzagging and backtracking.
10. Minimize movements between floors and between buildings.
11. Minimize movement distance, especially for large, bulky materials.
12. Minimize rehandling.
13. Minimize the number of moves per process.
14. Minimize walking.
15. Utilize gravity whenever feasible.
16. Use stacking containers where practicable.
17. Use collapsible containers to reduce the amount of storage space required for empty containers.
18. Provide guards around automatic, moving materials-handling equipment.
19. Provide good lighting.
20. Provide "dead-man" switches for automatic equipment requiring operators.
21. Provide warning signs and signals around automatically operated hazardous equipment.
22. Deliver materials at a proper rate so that they do not pile up or make an operator wait.
23. Provide adequate communication between key operator positions.

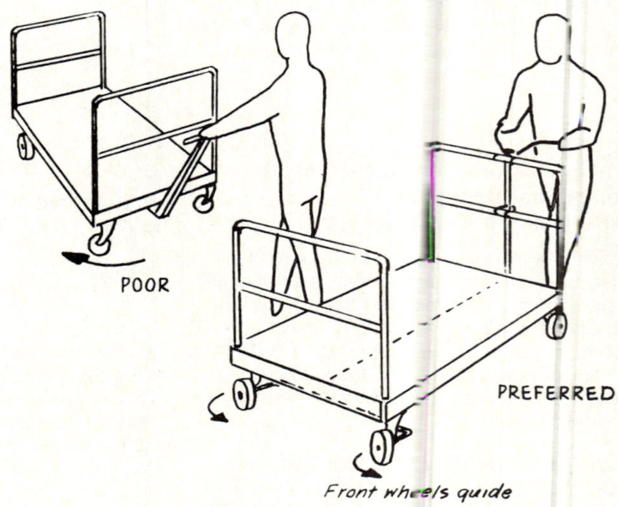

MATERIALS-HANDLING DEVICES

1. Place operators where they can see what they are doing.
2. Provide a control system that allows operators the best method for manipulating or guiding the device.
3. Provide positive control over the motion of the handler, i.e., putting it into motion, regulating its speed, and stopping it quickly.
4. In the case of a large, heavy device, provide the operator with an audio signaling device to warn bystanders.
5. Provide a parking brake.

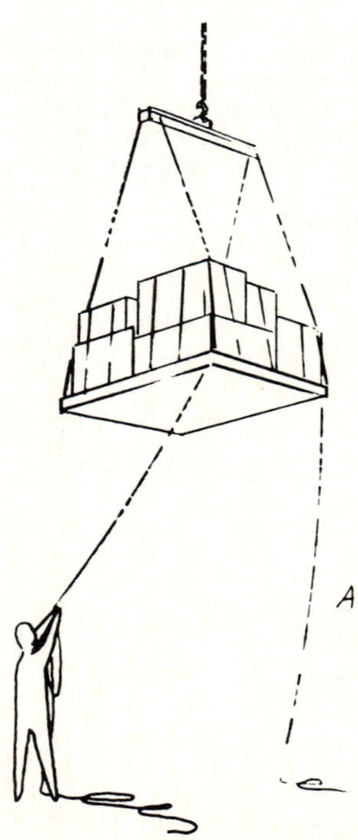

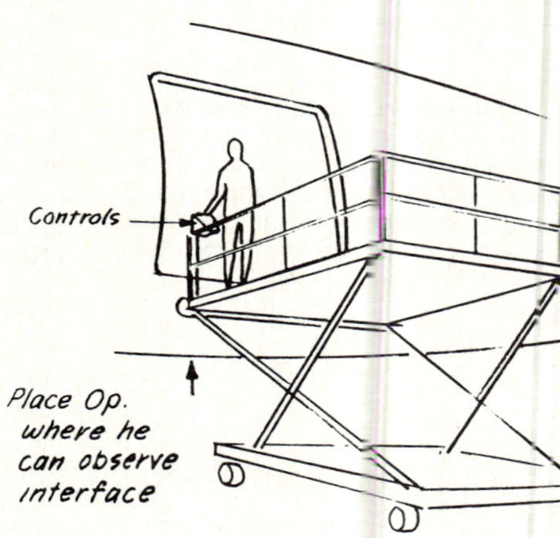

Example: Human Engineering Design Checklist

Although each human engineering checklist should be tailored to fit each particular program, the accompanying example may provide a useful point of departure for the individual who is attempting to construct a checklist for the first time. This example is provided courtesy of the Human Factors Group at the Electric Boat Division of General Dynamics.

HUMAN FACTORS CHECKLIST

FOR

DESIGN ENGINEERS

GENERAL DYNAMICS

Quincy Division

TABLE OF CONTENTS

USES

Use this checklist to perform a static evaluation to ensure that the constraints and recommendations of human engineering are in the design of components and total system development.

This checklist will assist engineering personnel in designing equipment in accordance with human capabilities and limitations. Used early in a design program, it will ensure that human engineering principles will be incorporated, at relatively low cost, in ultimate ship design. Used later in the program, the checklist serves as a final check that human engineering principles are in the design documentation.

The following procedure is recommended when evaluating components and systems with the checklist:

Review the table of contents to find the sections to be used for the check.

Compare the remaining checklist items with the design characteristics and check minimum acceptable, marginal acceptable, or reject, opposite each item.

Itemize the number of reject items and explain them in a memorandum to human factors personnel. Discrepancies not resolved in a conference between design engineers and human factors personnel will be submitted to the customer through Contracts for clarification.

HUMAN FACTORS CHECKLIST

PASS | MARGINAL | REJECT

1. *SAFETY*
   a. Conspicuous placards are mounted adjacent to high voltage, extremely cold, very hot, etc, equipment.
   b. A hazard-alerting device is provided to warn personnel of impending or existing hazards; e.g., fire, presence of combustible or asphyxiating gas, radiation, etc.
   c. A guard is provided on all moving parts of machinery and transmission equipment, including pulleys, belts, gears, and blades, etc., in which personnel may become injured or entangled.
   d. Self-locking or other foolproof devices are incorporated on elevating stands and work platforms to prevent accidental or inadvertent collapse.
   e. Some form of anchor or outrigger is employed on stands with high centers of gravity.
   f. Where applicable, the center of gravity of equipment is distinctly marked.
   g. Handrails are provided on platforms, stairs, and around floor openings, or wherever personnel may fall from an elevation.
   h. Safety bars or chains are attached across stair or step openings on a platform to prevent falling.
   i. Automatic shut-off devices are provided on fuel service equipment to prevent overflow and spillage.
   j. Portable hand-operated fire extinguishers are provided in areas where fire hazards exist or may be created.
   k. Emergency doors and exists are constructed so that they are readily accessible, unobstructed, and quick-opening; i.e., can be opened by a single motion of hand or foot.

   l. Eye baths, showers, and other first-aid equipment are readily available in areas where toxic materials are handled.
   m. Provision is made for neutralization or flushing of harmful materials spilled on equipment or personnel.
   n. Areas of operation or maintenance where special protective clothing, tools, or equipment are necessary (such as insulated shoes, nonsparking tools, gloves, or suits) are specifically identified, and appropriate action taken to ensure the availability of such items concurrent with complete development of the hardware.
   o. "No Step" markings are incorporated where applicable.
   p. Weight capacity is indicated on stands, hoists, lifts, jacks, and similar weight-bearing equipment to prevent overloading.
   q. Jacking and hoisting points are conspicuously and unambiguously identified.
   r. Wiring is routed through plugs and connectors so that removal of a plug or connector does not expose hot leads.
   s. Pipe lines carrying liquid, gas, steam, etc., are clearly and unambiguously labeled or coded as to contents, pressure, heat or cold, and any other hazardous properties.
   t. Skid-proof flooring and stair or step treads are provided.
   u. Clearance for fingers is provided in the design of telescopic steps or ladders.
   v. Provisions should be made to prevent personnel from coming into contact with voltages in excess of 70 volts rms. Do not locate adjustment screws or other commonly worked-on parts near unprotected high voltages or hot parts.
   w. Provide interlocks where potentials exceed 70 volts rms, with a means of bypassing for servicing with a proper warning indicator.
   x. Provide voltage dividers with test points for measurement of voltages in excess of 100 volts.

   y. Ground all external metal parts, control shafts, and bushings. Antenna or transmission line terminals should be at ground potential except with regard to the energy to be radiated.
   z. Provide rotating antenna assemblies with local power safety switch at the antenna.
   aa. Provide protection to personnel from imploding cathode ray tubes.
   bb. Doors or hinged covers should have rounded corners and should have positive-acting hold-open devices. The length of projecting and over-hanging edges should be held to a minimum, and all edges and corners should be rounded.
   cc. Design locking mechanisms for doors and drawers to prevent injury to the operator when the lock is released.

2. *AMBIENT ENVIRONMENT*

2.1 *Noise*
   a. No one is exposed to noise levels that exceed 150 db, no matter how much the noise level in the ear canal has been reduced.
   b. The amount of noise produced has been held at acceptable levels by careful acoustical design of new equipment, modification or redesign of present equipment, muffling of exhausts, etc.
   c. The amount of noise transmitted has been held at acceptable levels by increasing the distance between the work area and the source of noise, constructing barriers between the work area and the source of noise, sound-treating the work area to reduce vibration, placing equipment on vibration mounts, or other means.
   d. Where it is not possible to reduce noise levels through equipment design, personnel are provided with protective devices while working in high-intensity noise areas.

2.2 *Air-Conditioning*
   a. Air-conditioning is provided if the effective temperature exceeds 90 degrees Fahrenheit.

PASS | MARGINAL | REJECT

2.3 *Illumination*
a. Sharp gradients in illumination ratios (10 to 1 or greater) are avoided and gradients less than 8:1 surrounding displays should be avoided.
b. Local, direct lighting is provided for equipment which is not properly lighted by the general, diffuse lighting.
c. The intensity level and corresponding types of illumination for various visual tasks are in accord with Table IV, MIL-STD-803.

3. *WORK SPACE LAYOUT*

3.1 *Anthropometric Requirements*
a. The location, size, etc. of equipment is such that the equipment will be easily operated and maintained by at least the 5th to 95th percentile group of the Navy population.
b. Demands placed upon personnel in the performance of their duties lie within the following limits (it should be noted that these dimensions must be altered to allow for encumbrances):

Minimum overhead height for standing position: 73 inches
Maximum allowable overhead reach: 76 inches
Minimum height required for crawling: 31 inches
Maximum allowable depth of reach: 23 inches
Minimum dimension for passing body width: 20 inches
Maximum dimension for passing body thickness: 13 inches

c. Sufficient room to accommodate the hand is provided in the grasping of all handles.
d. Handles on cabinets and consoles are recessed, when practical, to eliminate projections on the cabinet surfaces.
e. All cabinets, consoles, and work surfaces requiring that an operator stand or sit in close proximity to their front surface contain a kick space four inches deep by four inches high at the base.

PASS | MARGINAL | REJECT

3.2 *Standing Operations*
a. Visual displays on vertical panels are mounted in an area no higher than 70 inches and no lower than 40 inches above the standing surface.
b. Precise reading indicators are placed in an area no higher than 64 inches and no lower than 48 inches above the standing surface.
c. Controls are mounted on vertical panels in an area no higher than 70 inches and no lower than 30 inches above the standing surface.
d. Precision controls, or those operated frequently, are located 40 to 55 inches from the standing surface.
e. Work benches and other work surfaces provided for standing operations are 32–38 inches above the floor.
f. Convenient work surfaces to support job instruction manuals, worksheets, etc., are provided where necessary for standing operators of control-display panels.

3.3 *Seated Operations*
a. Where continuous monitoring or control is required of a seated operator, controls and displays are mounted on a sloped console surface.
b. For normal seated operations, the slope of the control-display panel surface begins 30 inches from the floor, with the over-all console height not exceeding 48 inches, thereby allowing the operator's direct line of sight to extend beyond the console.
c. If the operator's direct line of sight is not required to extend beyond the console, the over-all console height does not exceed (but may extend) 65 inches from the floor.
d. If the over-all console height is between 48 and 65 inches, the upper panel surface is inclined from the vertical toward the operator.

e. Arm rests are provided at all consoles. These rests are a part of the console or a part of the operator's chair
f. Console arm supports provide at least 8 usable inches (preferably 12 inches) of resting surface projecting horizontally across the front of the console.
g. Arm rests integral with the back of the operator's chair are a minimum of two inches wide by 10 inches long.
h. If the operator must record data, a writing surface 12 inches in depth is provided.
i. Knee and foot room beneath the panel surfaces is provided, with minimum dimensions as follows: 26 inches high by 20 inches wide by 18 inches deep.
j. Desk tops, writing tables, and other work surfaces provided for seated operation are 30 inches above the floor.
k. Most important and frequently used displays near line of normal operator vision.

4. *CONSOLE AND PANEL DESIGN*

4.1 *Displays*

4.1.1 *PRINCIPLES*
a. The information displayed to each person who operates a piece of system equipment is limited to that information which is necessary to the specific actions or decisions requested of him.
b. Information is presented to the operator in directly usable form; that is, requirements for decoding, transpositioning, interpolation, etc., are minimized.
c. Displays are designed so that the failure of the display circuitry is immediately and readily apparent to the operator.

d. Failure of display circuitry does not cause a failure in the equipment associated with the display.
e. Crucial visual checks identified by attention-getting devices (e.g., visual or aural signals).
f. Information presentation is by means of most suitable type of display (e.g., pictorial, numerical, etc.) for task being performed.
g. Probability of confusion among instruments is minimal.
h. Instruments used only for maintenance made inconspicuous to operators not performing maintenance.
i. Precision of information is not greater or less than required to meet job task requirements.

4.1.2 *LABELING*
a. Trade names and other irrelevant information deleted.
b. Easy to read under expected conditions of illumination.
c. Labels do not obscure other needed information.
d. Labels in capital, extended copy in lower case letters (11 point type; line of 14–28 picas).
e. Each control and display is identified as to function.
f. Labels appear either on or immediately adjacent to (preferably above) the controls and displays to be identified.
g. Labels are located to preclude association of a label with the wrong control or display.
h. The location of labels in relation to controls and displays is consistent on all system equipment.
i. Labels are brief.
j. Labels clearly indicate the function being displayed or controlled.

PASS MARGINAL REJECT

k. Highly similar names are avoided.

l. Abbreviations, where required, are common or meaningful, and conform with MIL-STD-12 and ANA Bulletin 261.

m. Abstract symbols (squares, Greek alphabet, etc.) are not used as labels. Common, meaningful symbols such as the percent sign, plus sign, etc., are acceptable.

n. Lettering on panels is black, color No. 37038, Federal Standard 595.

4.1.3 *CODING*

a. Applicable coding techniques have been selected from the following methods.
   a. Color
   b. Size
   c. Location
   d. Shape

b. Optimum use is made of coding techniques for identification of functionally related displays.

c. Optimum use is made of coding techniques for indication of the relationship between displays (e.g. all emergency switches painted red). NOTE: Items d through g refer to color coding of indicator lights.

d. Red is used to alert an operator that the system or any portion of the system is inoperative. Also used for fire protection, danger, and stop warnings.

e. Amber is used to caution an operator that a condition exists which is marginal but that the system can still operate.

f. Green is used to indicate that a unit or component is within tolerance, or a condition is satisfactory or safe.

PASS MARGINAL REJECT

g. White is used to indicate those system conditions that are not intended to provide a right or wrong implication (such as indications of alternative functions), or are indicative of transitory conditions where such indication does not imply success of operations.

h. Lights used to denote emergency conditions (personnel or equipment disaster) and only those indicating such conditions, are coded in flashing red.

i. The flash rate for flashing warning lights is between three and five flashes per second, with "on" time being approximately equal of "off" time.

j. Colors for coding controls are chosen from the following list.

Red	11105	Amber	13538
Green	14187	White	17875
Black		Gray	

k. If it is considered imperative to relate a control to its corresponding display by means by color coding, the display and control are the same color.

l. If shape coding is selected as a coding scheme or combined with other codes for purposes of differentiating between controls, the application of the code is uniform throughout the system.

m. Control shapes, if used for coding purposes, are identifiable visually.

n. If blind operation of controls is a requirement, the shapes selected for coding are tactually discriminable.

o. Blue is used for covered electrical outlets and fuse boxes.

p. Flammable materials should be yellow.

q. Toxic and poisonous materials should be brown.

r. Glare shields, control columns, control wheels, control stick grips and plastic plate-edge lighted panels: black-colored.

s. Emergency exists and exit releases: orange yellow.

t. Instrument panels: medium gray.

4.1.4 *SCALES, DIALS, COUNTERS*

a. Numbers and letters are large enough for accurate reading at normal distance.

b. Reflected light does not create illusion warning is "ON" or obscure reading.

c. Where given, operating conditions always fall within a certain range on the scale; these areas are made readily identifiable by means of coding. Coding may be used to convey such information as desirable operating range, dangerous operating level, etc.

d. Reliance on color coding must be modified when other than white illumination is used.

e. Where scale color coding is used, the meaning of the colors conforms to paragraph 6.1.2 of MIL-STD-803.

f. Wherever possible, scales start at zero.

g. Scale graduations progress by one or five or 10 units, or decimal multiples thereof.

h. The increase in numerical progression reads clockwise, from left to right, or from the bottom up.

i. The number of minor or intermediate marks is not greater than nine.

j. Maximum contrast is used between scale face and markings.

k. On stationary scales, all numbers are upright.

l. On moving scales, all numbers are upright at the reading position.

m. The magnitude of the reading increases with a clockwise movement of the pointer.

n. Clockwise movement of a scale pointer results from clockwise movement of an associated rotary control, or movement forward, upward, or to the right of an associated lever or switch.

o. Where positive and negative values around a zero value are displayed, the zero is located at the 12 o'clock position (preferred) or the 9 o'clock position.

p. There is an obvious scale break between the two ends of a scale (not less than 1-½ divisions), except on multi-revolution instruments such as the clock.

q. For ease of monitoring a group of circular scaletype indicators, if a stable value of given operating conditions is present, the displays are arranged:
   1) in rows, with pointers normally aligned horizontally; i.e., the 9 o'clock position reflecting normal operating conditions; or
   2) in columns, with pointers normally aligned vertically; i.e., the 12 o'clock position reflecting normal operating conditions.

r. Where space is limited, numerals are placed inside of the graduation marks to avoid constriction of the scale.

s. Where space is not limited, the numbers are placed outside of the marks to avoid having the numbers covered by the pointer.

t. The pointer is located to the right of vertical scales, and at the bottom of horizontal scales.

u. For linear scales, numerals are placed on the side of the graduation marks away from the pointer to avoid having the numbers covered by the pointer.

v. The pointer or lubber line is at the 12 o'clock position for right-left direction information, or at the 9 o'clock position for up-down information.

w. If a scale is used for setting a given number, e.g., tuning a receiver to a desired wave-length, the unused portion of the dial face is covered, where advisable.

PASS | MARGINAL | REJECT

x. If the unused portion of a scale is covered, the open window is large enough to permit at least one numbered graduation to appear at each side of any setting.
y. If a scale is used in tracking, or as a heading indicator, the entire dial face is exposed.
z. Counters are used, where feasible, for presenting large ranges of quantitative data where continuous trend indication is not necessary and where quick, precise reading is required.
aa. Numbers change by snap action rather than by continuous movement.
bb. Counters are mounted as close to the panel surface as possible to maximize viewing angle and minimize parallax and shadows.
cc. The height-to-width ratio of numerals for counter displays is 1:1, rather than 5:3 as recommended for dials and scales.
dd. Numerals do not follow each other faster than about two per second if the observer is expected to read the numbers consecutively.
ee. Counters used to indicate sequencing of equipment are designed to reset automatically upon completion of the sequence; manual provision for resetting also is provided.
ff. Rotation of a counter reset knob is clockwise to increase the counter indication.
gg. Printers are used when a permanent record of quantitative data is desired.
hh. Plotters are used where a permanent record of continuous graphic data is desired.

4.1.5 *CATHODE-RAY TUBES*

a. Wherever possible, the scope face is perpendicular to the operator's normal line of sight.

PASS | MARGINAL | REJECT

b. The ambient illumination in the CRT area is sufficiently high for other visual functions (reading instruments), but does not interfere with the visibility of the signals on the CRT display.
c. Ambient room illumination will be minimized.
d. Scopes are adequately shielded from the light when room illumination is sufficiently high for other visual tasks.
e. The brightness of surfaces immediately adjacent to the scope falls in a range from screen brightness to 10 percent of screen brightness.
f. Surfaces immediately adjacent to the scope have a dull matte finish.
g. Shape conforms to configuration of presentation; PPI, round; A-scan, rectangular, etc.
h. Grid markers available for interpolation.
i. Adequate spacing used between range rings on polar displays (no less than ½U spacing).
j. Suggested viewing distance 16U for polar grid observation.

4.1.6 *INDICATOR AND LEGEND LIGHTS*

a. Legend lights are used primarily to display qualitative information that requires an immediate reaction by the operator or that his attention be called to an important system status.
b. Legend lights required to denote personnel or equipment disaster, caution or impending danger, and master summation go/no-go conditions are discriminately larger than all other legend-type displays.
c. Legend lights required to denote the conditions indicated in 3. above are brighter than other legend displays.
d. Legend light lettering is visible and legible whether or not the display is energized (applies to single-legend displays only).

e. Spacing between adjacent edges of round, simple indicator lights is sufficient for unambiguous labeling and convenient bulb removal.
f. Simple indicator lights used to denote emergency conditions are 1 inch in diameter and present a red flashing light.
g. Simple indicator lights used to denote cautionary conditions are 1 inch in diameter and present a steady amber light.
h. Simple indicator lights used to denote master summation conditions are 1 inch in diameter and present a steady red or green light.
i. Word warning lights are used where possible to specify an emergency condition.
j. All displays necessary to support an operator activity, or sequence of activities, are grouped together; e.g., subsystem grouping.
k. Displays are arranged in relation to one another to reflect the sequence of use or the functional relations of the components they represent, in that order of preference.
l. Distinct, functional areas set apart for purposes of ready identification are outlined by black lines approximately 1/16 inch wide.
m. Master caution, master warning, and summation lights used to indicate the condition of the entire subsystem are set apart from the lights which show the status of the subsystem components.
n. Indicators displaying critical functions are located within 30 degrees of the normal line of sight.
o. Transilluminated indicators are several times as bright as the background against which they appear, but not so bright as to dazzle the operator.
p. When transilluminated indicators are used under varied ambient illumination, a dimming control is provided.

q. If panels are to be used outdoors, provision is made to prevent reflected sunlight from making the indicators appear illuminated.
r. Where many lights are located on a control panel, a master light test control is incorporated.
s. Indicators which are not tested by a master light test control are designed for "press-to-test" bulb testing.
t. Provision is made for bulb removal from the front of the display panel without the use of tools, or other equally rapid and convenient means.
u. The filter caps and legend plates of indicator lights are physically coded or captive so as to preclude the possibility of a mismatch.
v. Pushbuttons are used where a control or an array of controls is needed for momentary contact or for activating a locking circuit in a high frequency-of-use situation.
w. Button surfaces are concave to fit the finger, or provide a high degree of frictional resistance to prevent slipping.
x. Buttons provide "snap feel" or an audible click to indicate that the control has been activated.
y. A channel or cover guard is provided when prevention of accidental activation is imperative.
z. The minimum diameter for fingertip operation is at least ½ inch.
aa. The minimum diameter for emergency controls which can be activated by thumb or heel of the hand is at least ¾ inch.
bb. Button displacement is between ⅛ inch and ¾ inches.
cc. Button resistance is between 10 ounces and 40 ounces.

PASS | MARGINAL | REJECT

dd. Foot pushbuttons are used only in those cases where an operator is likely to have both hands occupied at the time the pushbutton requires activation.
ee. Under normal conditions, foot pushbuttons are designed for the toe operation (by the ball of the foot) rather than heel operation.
ff. Foot pushbuttons provide "snap feel" or an audible click to indicate they have been actuated.
gg. The button diameter is at least ½ inch.
hh. Displacement for normal operation is at least ½ inch.
ii. Displacement for actuation by heavy boots is at least 1 inch.
jj. Displacement for controls operated by ankle flexion only does not exceed 2-½ inches.
kk. Displacement for controls operated by leg movement does not exceed 4 inches.
ll. Button resistance when foot will not rest on control is at least 4 pounds.
mm. Button resistance will range from 10 to 20 pounds.
nn. Toggle switches are used for those control functions which require two discrete positions when space limitations are severe.
oo. In general, toggle switches are vertically oriented, with the up position representing ON, and the down position representing OFF.
pp. Where inadvertent operation is considered serious, those toggle switches susceptible to accidental activation are guarded (usually with a channel guard).
qq. Where prevention of accidental activation is of primary importance, a cover guard is used.

16

PASS | MARGINAL | REJECT

rr. Control tip diameter is between ⅛ inch and 1 inch.
ss. Lever arm length is between ¾ inch and 2 inches.
tt. Minimum displacement between adjacent control positions is at least 30 degrees.
uu. Maximum displacement between adjacent control positions does not exceed 120 degrees.
vv. Switch resistance is between 10 ounces and 40 ounces.

4.1.7 *ROTARY SELECTOR SWITCH*

a. Rotary selector switches are used for discrete functions where three or more positions are required.
b. The number of positions incorporated into any one rotary switch does not exceed 24.
c. The pointer knob has a tapered tip, unless shape coding has been applied to the knobs.
d. Shape coding is utilized where a group of rotary switches, used for widely different functions, is placed on one panel and switch confusion is likely.
e. The position of the pointer knob relative to the scale minimizes parallax.
f. A moving pointer on a fixed scale is used, unless complete justification exists for a moving scale with a fixed index.
g. Maximum pointer width does not exceed 1 inch.
h. Displacement (between adjacent elements) for visual positioning is at least 15 degrees.
i. Displacement for non-visual positioning is at least 30 degrees.
j. Resistance is between 12 ounces and 48 ounces.

17

4.1.8 *LEVERS*

a. Levers are used where large amounts of force or displacement are involved.
b. Levers are used where multi-dimensional movement of the control is required.
c. Levers placed close to each other and not readily discriminable from each other are coded.
d. All levers are labeled as to function and direction of motion.
e. If fine or continuous adjustments must be made with the lever, support is provided for the elbow, forearm, or wrist.
f. The diameter of spherical handles grasped by the fingers is between ¾ and 3 inches.
g. The diameter of spherical handles grasped by the hand is between 1-½ inches and 3 inches.
h. Maximum displacement does not exceed 14 inches for fore-aft movement, or 38 inches for lateral movement.
i. Minimum resistance for aircraft joysticks is at least 5 pounds.
j. For one-hand operation along the fore-aft axis, maximum resistance of the lever handle does not exceed the following limits:
   1) Lever 10 inches forward from seat reference point: 30 pounds.
   2) Lever 16–24 inches forward from seat reference point: 40 pounds.
k. For operation along the lateral axis, maximum resistance of the lever handle does not exceed the following limits.
   1) One-hand operation (lever 10–19 inches forward of seat reference point): 20 pounds.

18

   2) Two-hand operation (lever 10–19 inches forward of seat reference point): 30 pounds.
l. Knobs are used where precise, accurate adjustments of a continuous variable are required, and where little force is necessary.
m. For knobs grasped by the fingertips:
   1) Minimum depth is at least ¾ inch.
   2) Minimum diameter is at least 1 inch.
   3) Maximum diameter does not exceed 4 inches.
n. For knobs grasped with thumb and finger encircled:
   1) Minimum diameter is at least 1 inch
   2) Maximum diameter does not exceed 3 inches.
o. For knobs grasped with the palm
   1) Minimum diameter is at least 1-½ inches.
   2) Maximum diameter does not exceed 3 inches.
p. Knob displacement is determined by the desired control/display ration.
q. Maximum resistance for fingertip operation of small (1-inch diameter) knobs does not exceed 4-½ ounces.
r. Maximum resistance for fingertip operation of larger knobs (greater than 1-inch diameter) does not exceed 6 ounces.

4.1.9 *CRANKS*

a. Cranks are used for tasks requiring many rotations of the control.
b. Minimum crank radius is at least 1 inch.
c. Maximum radius with minimum load and high rate (up to 275 rpm) does not exceed 4-½ inches.

19

PASS MARGINAL REJECT

- d. The grip handle is designed so that it turns freely around its shaft.

4.1.10 *HANDWHEEL*

- a Handwheels, designed for two-hand operation, are used where the breakout or rotational forces are too large to be overcome with a one-hand control.
- b. Minimum handwheel diameter is at least 7 inches.

4.1.11 *PEDALS*

- a. Pedals are used where a large amount of displacement and force is required and foot activation is desired.
- b. Pedals return to their null position when force is removed.
- c. Pedal size is at least 1-½ inch by 3-½ inches.
- d. Pedal displacement falls within the following limits.
  1) Minimum, for normal operation: 1 inch
  2) Minimum, with heavy boots: 1 inch
  3) Maximum, by ankle flexion only: 2-½ inches
  4) Maximum, by leg movement: 7 inches
- e. Pedal resistance falls within the following limits.
  1) Minimum, foot not resting on control: 4 pounds
  2) Minimum, where foot may rest on control: 10 pounds
  3) Maximum, with ankle flexion only: 20 pounds
  4) Maximum, with leg movement: 80 pounds

4.2 *Control/Display Relationships*

4.2.1 *ARRANGEMENTS*

- a. All controls having sequential relations, or having to do with a particular function or operation, or which are operated together, are grouped together, along with the associated displays.

PASS MARGINAL REJECT

- b. When a control is associated with a transilluminated indicator, the indicator is located so as to be immediately and unambiguously associated with the control.
- c. When a control is associated with a transilluminated indicator, the indicator is located above the control.
- d. If a control knob is adjacent to the instrument it controls, it is located so that the control or the hand normally used for setting does not obscure the indicator.
- e. Lights are used to display equipment response and not merely switch position, unless the switch position cannot be made apparent by proper design and labeling of the control.
- f. Warning lights are integral with, or adjacent to, the levers, switches, or other control devices by which the operator is to take action.

4.2.2 *PRECAUTIONS*

- a. The control is located or oriented so that the operator is not likely to hit it or move it accidentally in the normal sequence of control movements.
- b. Physical barriers are placed around the control, or it is recessed or shielded.
- c. The control is covered or guarded.
- d. Interlocks are provided so that extra movement or the prior operation of a related or locking control is required.
- e. Resistance is built into the control so that definite or sustained effort is required to actuate it.

5. *MAINTAINABILITY*

5.1 *Unitization*

- a. Unless structurally or functionally not feasible, all equipment is designed so that rapid and easy removal and replacement of malfunctioning units can be accomplished by one operator.

- b. Where possible, units serving the same function in different applications are designed to be interchangeable.
- c. The number of inputs and outputs from each unit is kept to a minimum by grouping of functions so that a minimum of crisscrossing of signals between units is required.
- d. Where consistent with system design requirements, functions are so utilized that it is possible to check and adjust each unit separately.

5.2 *Handling*

- a. Where possible, units are small and light enough for one man to handle and carry; i.e., weight of removable units is held below 45 pounds.
- b. Units in excess of 45 pounds have provision for two-man lift where the lifting height is not in excess of five feet and where the total weight is not in excess of 90 pounds.
- c. Units weighing over 90 pounds have provision for mechanical or power lift.
- d. All units weighing 45 pounds or more are prominently labeled with their weight.
- e. All units designed to be removed and replaced are provided with handles, or other suitable provision made for grasping, handling, and carrying.
- f. Wherever possible, handles or grasp areas are located over the center of gravity so that the unit does not swing or tilt when lifted.
- g. Handles which are to be used by the ungloved hand are at least 5-½ inches in length and 2 inches in depth (inside measurements), and ½-inch in diameter for units under 25 pounds, or ¾-inch in diameter for units over 25 lbs.
- h. Where feasible, rests or stands are provided and incorporate provisions for test equipment, tools, and manuals.

- i. Irregular, fragile, or awkward extensions, such as cables, wave guides, hoses, etc., are easily removable before a unit is handled.
- j. Where feasible, guides, tracks, and stops are provided to facilitate handling and to prevent damage to units and components.

5.3 *Covers and Cases*

- a. The proper orientation of a unit within its case is obvious, either through design of the case or by means of appropriate labels.
- b. Where possible, cases are designed to lift off units, rather than units lifted out of cases.
- c. The method of opening a cover is obvious, either from the construction of the cover itself or from an instruction plate permanently attached to the outside of the cover.
- d. It is obvious when a cover is in place but not secured.
- e. Sharp edges and corners on cases and covers are avoided.
- f. Where space permits, hinged covers are used to reduce the number of fasteners required.

PASS | MARGINAL | REJECT

5.4 *Mounting of Units*

a. Wherever possible, identical screw and bolt heads are used, thereby enabling various panels and components to be removed with one type of tool.

b. Units which are frequently pulled out of their installed position for checking are mounted on roll-out racks, slides, or hinges.

c. Guide pins or their equivalent are provided on units for alignment during mounting.

d. Limit stops which can be conveniently over-ridden are provided on roll-out racks and drawers.

e. Interchangeable units are coded (color, labels, etc.) to indicate the correct unit and its orientation for replacement.

5.5 *Location of Components*

a. Parts are mounted in an orderly array on a "two-dimensional" surface and not "stacked" on one another; e.g., the lower layer not supporting the upper layer of units.

b. Large parts which are difficult to remove are mounted so that they do not prevent access to other parts.

c. There is sufficient space to use test equipment and other required tools without difficulty or hazard.

d. Structural members of the units and chassis do not prevent access to components.

e. All throwaway assemblies or parts are accessible without removal of other components.

f. Delicate components are located or guarded so that they will not be damaged while the unit is being handled or worked on.

g. If screwdriver adjustments must be made blind, mechanical guides are provided, or the screws mounted so that the screwdriver will not fall out of line.

h. Sensitive adjustments are located or guarded so that they will not be disturbed.

PASS | MARGINAL | REJECT

i. Internal controls are not located close to dangerous voltages.

j. Units are laid out so that a minimum of place-to-place movement is required of the operator during checkout.

h. Bills of material or maintenance nomenclature are not placed on the front of the panel.

l. Where possible, bills of material or maintenance nomenclature are placed behind the control panel, in order to facilitate troubleshooting and maintenance.

5.6 *General Accessibility*

a. All access covers which are not completely removable are self-supporting in the open position.

b. If instructions applying to a covered unit are lettered on a hinged door, the lettering is properly oriented for reading when the door is open.

c. Sliding, rotating, or hinged units to which rear access is required are free to open or rotate their full distance and remain in the open position without being supported by hand.

d. Check points, adjustment points, cables and connectors, and labels are accessible, and where possible, face the maintenance man.

e. Bulkheads, brackets, other units, etc., do not interfere with removal or opening of covers of units within which work must be done.

f. Unless a unit is completely self-checking, provision is made for checking the operation of the unit in the operating condition without the use of special rigs and harnesses.

g. An opening with no cover is used unless this is likely to degrade system performance.

h. A sliding or hinged cap is used if dirt, moisture, or other foreign materials are a problem.

i. A quick-opening cover plate is used if a cap will not meet stress requirements.

j. Removal of any replaceable unit requires opening or removal of a minimum number of covers or panels (preferably one).

k. Wherever possible, units are located so that no other equipment must be removed to gain access.

l. Where it is necessary to place one unit behind another, the unit requiring most frequent access is most accessible to the user.

m. When equipment is of a highly critical nature and its maintenance requires highly specialized skills, access to units maintained by one operator does not require removal of equipment maintenance by a second operator.

n. Units are located and mounted so that access to them may be achieved without danger to personnel from electrical charge, heat, sharp edges and points, moving parts, chemical contamination, or other sources.

5.7 *Size of Apertures*

a. Openings and work spaces provided for adjusting and handling units are large enough to permit the required activity.

b. Where possible, openings and work spaces permit adequate view of the components being manipulated.

c. Inserting empty hand-held flat: 2-¼ by 4-¼ inches.

d. Smallest square hole through which empty hand can be inserted: 3-½ by 3-½ inches.

e. Inserting a miniature vacuum tube, held with the thumb and first two fingers, up to the center knuckle of the middle finger: 2-½ by 2-½ inches.

f. Inserting a large vacuum tube (CRC 1625) with base diameter 1-⅜ inches, height 4-½ inches, excluding pins and grid cap: 4 by 4 inches.

g. Inserting a box or electronic assembly, grasped by handles on front, into an aperture: ½-inch clearance on each side of the box.

h. Reaching through aperture with both hands to a depth of six to 25 inches: width—¾ depth of reach; height—5 inches, 6 with gloves.

i. Reaching in full arm's length (to shoulders) straight ahead with both arms: width—19-½ inches; height—5 inches, 6 with gloves.

5.8 Conductors

a. Cables are routed so that they cannot be walked on or used for hand holds.

b. Cables are routed so that they are accessible for inspection and repair.

c. Cables are routed so that they need not be bent or twisted sharply or repeatedly during repair.

d. Cables containing individually insulated conductors with a common sheath are coded as specified in ARDCM 80-5

5.9 Connectors

a. Plugs which require no more than one turn, or other quick-disconnect plugs, are used whenever feasible.

b. Connectors are located far enough apart so that they can be grasped firmly for connection and disconnection.

c. The rear of plug connectors is accessible for test and service, except where potting, sealing, or other considerations preclude this.

d. Plugs or receptacles are provided with alignment pins, or other alignment devices.

e. Alignment pins project beyond electrical pins in plugs or receptacle.

f. Where a reasonable possibility exists for unintentional interchange of connectors, plugs are designed so that it is impossible to insert the wrong plug in a receptacle.

g. Plugs or receptacles are arranged so that the alignment pins are oriented in the same direction throughout the system where possible.

h. Connecting plugs and receptacles are identified by color, shape, or other acceptable means.

	PASS	MARGINAL	REJECT
i. Plugs and receptacles have painted stripes, arrows, or other indications to show the position of alignment pins for proper insertion.			
j. The system is designed so that all "hot" contacts are socket contacts.			
5.10 *Fasteners*			
a. Maximum use is made of tongue-and-slot catches to minimize the number of fasteners required.			
b. The number and diversity of fasteners are the minimum commensurate with requirements for stress, bonding, etc.			
c. Where feasible, the same size and type of fasteners are used for all covers and cases.			
d. Captive fasteners are used where possible.			
e. Screws with different threads are of different sizes.			
f. If compatible with stress and load considerations, fasteners for mounting assemblies, subassemblies etc., fasten or unfasten with a maximum of one complete turn.			
g. Bolts requiring high torque are provided with external grip heads.			
h. Captive bolts and nuts are used in situations where the dropping of these small items into the equipment will cause damage or create a difficult removal problem.			
5.11 *Test Points*			
a. Where a unit is not completely self-checking in its operation condition, appropriate test points are provided which provide a measure of unit output.			
b. Only such primary test points as are necessary to determine that a unit is malfunctioning are provided.			
c. Primary test points are located and coded to be readily distinguishable from secondary test points.			
d. Where feasible, primary test points are grouped in a line or matrix reflecting the sequence of tests to be made.			

28

	PASS	MARGINAL	REJECT
e. Primary test points used in adjusting the unit are located close to the controls and displays used in the adjustment.			
f. Where feasible and not in conflict with other requirements, a secondary test point is supplied at the input and output of each part or throwaway component.			
g. Sufficient test points are provided so that it is not necessary to remove subassemblies from assemblies in troubleshooting.			
h. Each test point is marked to be readily identifiable.			
5.12 *Lubrication*			
a. Units containing mechanical components requiring lubrication is specified by a label at or near the lube port.			
b. Where lubrication is required, the type lubricant to be used and the frequency of lubrication is specified by a label at or near the lube port.			
5.13 *Unit Labeling*			
a. Wherever possible, the outside covering of manufactured parts such as resistors, condensers, and tubes, is stamped or coded with relevant information concerning electrical characteristics of the part.			
b. Where space permits, terminals are labeled with the same code symbol as the wire attached to it.			
c. If surface labels must be used, silk-screened or stamped labels are used in preference to stenciled labels or decals.			
d. Labels are not hidden by units or parts.			
5.14 *Equipment Color*			
a. The console interior of operational and maintenance equipment is Gray. Color 26622, Fed. Std. 595, where maintenance and troubleshooting are required within the console.			
b. Where maintenance and troubleshooting within the console will not be required, an economical internal protective finish is used.			

29

Sample Human Engineering Checklist (for Vehicles)

GENERAL ACCOMMODATIONS How appropriate are the general passenger, driver, and package carrying and storage accommodations?	Satisfactory	Unsatisfactory
Ingress and egress clearance for both normal and emergency use and for the handicapped and the elderly	______	______
Interior space dimensions (passenger width, head, and leg clearance)	______	______
Temperature and ventilation	______	______
Ride characteristics (sway, oscillation, and vibration)	______	______
Interior and exterior package storage space (amount, location, and accessibility)	______	______
Crash protection (restraint systems, noninjurious interior, handholds, and shatterproof glazing)	______	______
Noise minimization	______	______
Adequate seating (size, shape, seat-back angles, adjustability, properly secured)	______	______

DRIVER WORK STATION How appropriate are the layout of the driver's work station and the provisions for proper vehicle control?	Satisfactory	Unsatisfactory
Geometric relationship between seat, controls, and external viewing (to accommodate 5th-percentile female through 95th-percentile male drivers)	______	______
Control and display location and arrangement and detail interface designs	______	______
Exterior viewing capability (front, side, and rear, i.e., clear angle of view, including normal and adverse weather conditions)	______	______
Control system adequacy (steering force and rate, brake force, clutch force and excursion, accelerator force and excursion, and transmission lever pattern and force characteristics)	______	______

GENERAL SAFETY	Satisfactory	Unsatisfactory
Cab integrity (front, rear, and side crash; roll-over)	______	______
Nonflammable interior materials	______	______
Flame barriers between passenger compartment engine and rear fuel compartment or tank	______	______
Fail-safe systems (power windows, seats, steering, and brakes)	______	______
Non-brake-lockup and skid control	______	______
Blowout-proof tires	______	______
Energy-absorption systems (front, rear, and side impact)	______	______
Noninjurious pedestrian-vehicle interfaces (sheet metal, handles, and trim)	______	______
Rupture-proof fuel tank	______	______
Steering-wheel lock	______	______
Rearview mirrors	______	______

Index